SATELLITE GRAVIMETRY AND THE SOLID EARTH

SATELLITE GRAVIMETRY AND THE SOLID EARTH

Mathematical Foundations

MEHDI ESHAGH

ELSEVIER

Elsevier
Radarweg 29, PO Box 211, 1000 AE Amsterdam, Netherlands
The Boulevard, Langford Lane, Kidlington, Oxford OX5 1GB, United Kingdom
50 Hampshire Street, 5th Floor, Cambridge, MA 02139, United States

Library of Congress Cataloging-in-Publication Data
A catalog record for this book is available from the Library of Congress

British Library Cataloguing-in-Publication Data
A catalogue record for this book is available from the British Library

ISBN: 978-0-12-816936-0

Publisher: Candice Janco
Acquisitions Editor: Amy Shapiro
Editorial Project Manager: Liz Heijkoop
Production Project Manager: Omer Mukthar
Cover Designer: Miles Hitchen

Typeset by TNQ Technologies

Dedicated to my wife
Elsa
and our beautiful daughters
Vendela and Helen

Contents

Preface

Satellite gravimetry and its geophysical applications have been my main subjects of research, teaching and supervising students for over 15 years. I felt that a mathematically oriented book covering all aspects of such a subject is needed for higher education students and young researchers in geodesy and geophysics. I had in mind to write such a book and explain the mathematical foundations behind various satellite gravimetry observables in relation to the Earth's gravity field and the solid Earth's geophysics until I talked to one of my close friends, Professor Robert Tenzer, about it. He highly encouraged me to write such a book; also, Dr Martin Pitonak, who had been with me for a short postdoctoral research sabbatical, motivated me towards this goal. However, due to my huge amount of academic tasks and duties, taking such an extra and hard job was difficult, until 2 years ago, when Mrs Marisa LaFleur from Elsevier proposed that I write a book about the Earth's gravity field for publication. Elsevier was one of my favourite publishers, and I had published the majority of my scientific articles in its journals. This was a serious step towards my idea, and after writing a book proposal and its approval after the reviewing process and revisions, my writing was started in June 2018.

My main goal was that the book should be simple, complete, didactic and pedagogic and suitable for students and researchers in the related fields. However, I have also developed new ideas in the book for the readers' benefit. In my opinion, showing how satellite gravimetry data, which are collected from platforms far from the Earth's surface, are related to the gravity field and geophysical phenomena inside the Earth is an interesting subject. This is what I have tried to present and explain in this book step by step.

I have done all I could during these 2 years to present this book in the best form, and checked and rederived formulae several times. In addition, I restructured the book a couple of times for its better presentation, which led to extra work on the part of the Elsevier Production Division. I would be indeed appreciative if all colleagues, students and readers assist me in improving the book through their constructive and useful comments, so that the next edition of the book will be clearer and free of any probable errors.

Mehdi Eshagh
Summer 2020

Acknowledgements

I would like to express my gratitude to those without whose support and help my idea could never come true. I am grateful to Professor Robert Tenzer and Dr Martin Pitonak for inspiring me with their positive words and encouragement about the idea of this book. I appreciate all support from my closest friend and colleague, Dr Majid Abrehdary, especially all the kind and scientific discussions about satellite altimetry and the Moho and density contrast determination. My special thanks go to Mr Andenet Ashagrie Gedamu and Mr Farzam Fatolazadeh for help in writing Chapters 8 and 10. Professor Bernhard Steinberger is acknowledged for his help by reading the draft version of Chapter 8 and commenting on it. Dr S.M. Ali Noori Rahimabadi is appreciated for the scientific discussion and advice regarding mantle flow theory.

I am indeed appreciative to Dr Michal Sprlak, who acted as the technical editor and reviewer of this book, for his careful reading and detailed and constructive comments on the whole book. I thank Mrs Marisa LaFleur from Elsevier, who contacted me about proposing the book idea, handling administrative processes with Mr Michael Lutz for up to half of the book. Mrs Amy Shapiro and Mrs Liz Heijkoop are cordially acknowledged for their support and help in handling all the administrative processes of the book and for help in the design of the cover page of the book. Mr Omer Mukthar, the production project manager and the Production Division of Elsevier are sincerely thanked for their excellent cooperation in producing and revising the whole book.

A heartfelt word of thanks goes to my wife, Elsa, for providing a suitable atmosphere at home and taking care of the family during my writing from the start to the end, especially during the pandemic times.

Mehdi Eshagh
Summer 2020

CHAPTER 1

Spherical harmonics and potential theory

1.1 General solution of Laplace equation in spherical coordinates

It is known in physics, geophysics and physical geodesy that the earth's gravitational potential fulfils the Poisson partial differential equation. Since satellite gravimetry data are collected outside the earth's surface, the gravitational potential becomes harmonic there, meaning that it satisfies the Laplace equation instead. Such a harmonic potential, which we show by V in this book, has the following mathematical form (Heiskanen and Moritz, 1967, p. 12):

$$\Delta V = 0 \quad \text{for } r >= R, \tag{1.1}$$

where, as mentioned, V is the potential, R stands for the radius of the spherical earth and r the geocentric distance, the distance from the earth's centre of mass and any point outside it. Δ stands for the Laplace operator with the following form in the spherical coordinate system (Heiskanen and Moritz, 1967, p. 19):

$$\Delta = \frac{\partial^2}{\partial r^2} + \frac{2}{r}\frac{\partial}{\partial r} + \frac{1}{r^2}\frac{\partial^2}{\partial \theta^2} + \frac{\cot\theta}{r^2}\frac{\partial}{\partial \theta} + \frac{1}{r^2 \sin^2\theta}\frac{\partial^2}{\partial \lambda^2}, \tag{1.2}$$

where θ and λ are, respectively, the co-latitude and longitude of any point with a geocentric distance of r outside the earth. By assuming that the earth is spherical, the solution of Eq. (1.1) will be (e.g., cf. Heiskanen and Moritz, 1967, p. 21):

$$V(r,\theta,\lambda) = \sum_{n=0}^{\infty}\sum_{m=-n}^{n}\left(\frac{R}{r}\right)^{n+1} v_{nm} Y_{nm}(\theta,\lambda) = \sum_{n=0}^{\infty}\left(\frac{R}{r}\right)^{n+1} v_n(\theta,\lambda) \quad \text{for } r > R, \tag{1.3}$$

Satellite Gravimetry and the Solid Earth
ISBN 978-0-12-816936-0
https://doi.org/10.1016/B978-0-12-816936-0.00001-3

where $Y_{nm}(\theta, \lambda)$ is the fully-normalised spherical harmonic function of degree n and order m with the following expressions:

$$Y_{nm}(\theta,\lambda)=\begin{cases}\sin m\lambda\overline{P}_{nm}(\cos\theta) & \text{for } m>0\\ \cos m\lambda\overline{P}_{n|m|}(\cos\theta) & \text{for } m\leq 0\end{cases}, \tag{1.4}$$

and $\overline{P}_{nm}(\cos\theta)$ is the fully-normalised associated Legendre functions of degree n and order m and argument of $\cos\theta$, and v_{nm} the fully-normalised spherical harmonic coefficients (SHCs) of the gravitational potential. The term v_n is called the Laplace coefficient with the following relation to v_{nm}:

$$v_n(\theta,\lambda)=\sum_{m=-n}^{n} v_{nm}Y_{nm}(\theta,\lambda). \tag{1.5}$$

As observed, unlike v_{nm}, the Laplace coefficient v_n is position dependent.

The coefficient $(R/r)^{n+1}$ in Eq. (1.3) is called the upward continuation factor. The ratio R/r is always smaller than 1 for $r>R$; this means that it becomes even smaller at the power of $n+1$. When n increases, this factor reduces the values of v_{nm}, and finally when n goes to infinity, the ratio goes to zero and significantly reduces the value of v_{nm}. Obviously, when r is much larger than R, or in other words, the points are farther away from the spherical earth, this factor diminishes faster and reduces more of v_{nm}. This means that by increasing the geocentric distance r, the potential contains fewer frequencies and it will be smoother.

The spherical harmonic functions presented in Eq. (1.4) have the following orthogonality property:

$$\iint_{\sigma} Y_{nm}(\theta,\lambda)Y_{n'm'}(\theta,\lambda)d\sigma=4\pi\delta_{nn'}\delta_{mm'} \quad \text{and } \delta_{ij}=\begin{cases}1 & i=j\\ 0 & i\neq j\end{cases}, \tag{1.6}$$

where σ stands for the sphere on which the integration is performed, $d\sigma$ is the surface integration element at the sphere and δ is the Kronecker delta.

Now, consider $r=R$ in Eq. (1.3), multiply both sides by $Y_{n'm'}(\theta',\lambda')$ and take the surface integration over the sphere. According to Eq. (1.6), the result will be:

$$\iint_{\sigma} V(R,\theta',\lambda')Y_{n'm'}(\theta',\lambda')d\sigma=\sum_{n=0}^{\infty}\sum_{m=-n}^{n} v_{nm}\iint_{\sigma} Y_{nm}(\theta,\lambda)Y_{n'm'}(\theta',\lambda')d\sigma = 4\pi v_{nm}. \tag{1.7}$$

Solution of Eq. (1.7) for v_{nm} yields:

$$v_{nm} = \frac{1}{4\pi} \iint_{\sigma} V(R, \theta', \lambda') Y_{nm}(\theta', \lambda') d\sigma. \tag{1.8}$$

Now, let us introduce another important property of the spherical harmonics, which is known as the addition theorem. This theorem relates the product of two spherical harmonic functions of the same degree and order, but at two different points, to the spherical geocentric distance between them as an argument for the Legendre function. This addition theorem is (Heiskanen and Moritz, 1967, p. 33):

$$\sum_{m=-n}^{n} Y_{nm}(\theta', \lambda') Y_{nm}(\theta, \lambda) = (2n+1) P_n(\cos \psi), \tag{1.9}$$

where $P_n(\cos \psi)$ is the Legendre polynomial with argument $\cos \psi$, where ψ is the spherical geocentric angle between the computation and integration points (θ, λ) and (θ', λ'), which can be computed from the spherical coordinates of both points by (see Fig. 1.5):

$$\cos \psi = \cos \theta \cos \theta' + \sin \theta \sin \theta' \cos(\lambda' - \lambda). \tag{1.10}$$

By inserting Eq. (1.8) into Eq. (1.5) and after applying the addition theorem of spherical harmonics (Eq. 1.9), we obtain another formula for the Laplace coefficient of the potential:

$$v_n(\theta, \lambda) = \frac{2n+1}{4\pi} \iint_{\sigma} V(R, \theta', \lambda') P_n(\cos \psi) d\sigma. \tag{1.11}$$

Eq. (1.11) is very useful for transforming the integral formula to spherical harmonic expansions and vice versa.

Eq. (1.3) is the solution of the gravitational potential outside the spherical earth. When v_{nm} are available, the gravitational potential can be computed by summing up the series. This process is known forward computation or spherical harmonic synthesis. Different methods have been presented by researchers for performing this synthesis effectively and fast. However, in gravimetry the goal is to determine v_{nm}, carrying physical properties of the gravitational potential. This process is known as spherical harmonic analysis; see Eq. (1.8). The rest of the parameters in Eq. (1.3), i.e., the spherical harmonics and the upward continuation factor, are mathematical functions depending on the position of the computation points.

1.2 Solving potential from potential outside the earth

Here, the mathematical foundation of recovering potential from potentials at higher levels than the earth's surface is presented. It is explained how the spherical harmonics and their properties can be used for determining gravitational potential. A boundary-value problem can be organised and solved for recovering the potential outside the sphere from some boundary values of potential at its surface, Dirichlet's problem. However, in satellite gravimetry the goal is to recover the potential at the surface from the measured potential outside the sphere. This is the reverse case of Dirichlet's problem and is called the inverse problem. In this section, and the rest of this chapter, we discuss such subjects in spectral and spatial forms as well as integral equations.

1.2.1 Spectral solution

Suppose that $V(r,\theta,\lambda)$ is known in Eq. (1.3) outside a sphere with the radius R. If both sides of Eq. (1.3) are multiplied by $Y_{n'm'}(\theta',\lambda')$ and a surface integration is performed all over this sphere, we obtain:

$$\iint_\sigma V(r,\theta',\lambda')Y_{n'm'}(\theta',\lambda')d\sigma = \sum_{n=0}^{\infty}\sum_{m=-n}^{n}\left(\frac{R}{r}\right)^{n+1} v_{nm}\iint_\sigma Y_{nm}(\theta,\lambda)Y_{n'm'}(\theta',\lambda')\mathrm{d}\sigma. \tag{1.12}$$

According to Eq. (1.6), Eq. (1.12), will change to:

$$\begin{aligned}\iint_\sigma V(r,\theta',\lambda')Y_{n'm'}(\theta',\lambda')\mathrm{d}\sigma &= 4\pi\sum_{n=0}^{\infty}\sum_{m=-n}^{n}\left(\frac{R}{r}\right)^{n+1} v_{nm}\delta_{nn'}\delta_{mm'}\\ &= 4\pi\left(\frac{R}{r}\right)^{n+1} v_{n'm'}.\end{aligned} \tag{1.13}$$

Solving Eq. (1.13) for v_{nm} reads:

$$v_{nm} = \frac{1}{4\pi}\left(\frac{r}{R}\right)^{n+1}\iint_\sigma V(r,\theta',\lambda')Y_{nm}(\theta',\lambda')\mathrm{d}\sigma. \tag{1.14}$$

This integral (Eq. 1.14) is a direct mathematical formula connecting $V(r,\theta,\lambda)$, over a sphere with radius r and v_{nm} at the sphere. The term

$(r/R)^{n+1}$ is called the downward continuation factor, meaning that it will bring the gravitational potential from the sphere with radius r down to the potential at a sphere with radius R. Since $r > R$, this ratio is always larger than 1, and when it rises to the power of $n + 1$, it will increase. This means that the gravitational potential is amplified degree by degree during the downward continuation process and this amplification is considerable for high degrees. In this case, any small error in the measured potential causes large variations in the solution of v_{nm}. Limiting the maximum degree of computation is a way to control this signal amplification and obtain a smooth solution. This is quite normal, because the gravitational potential gets weaker by the distance from the spherical earth and it will be smooth at higher elevations. This is due to the presence of the upward continuation factor $(R/r)^{n+1}$ in Eq. (1.3), which reduces high frequencies of the signal. Therefore, these frequencies will be at the level of measurement noise and hard to recognise. However, for using Eq. (1.14) a dense set of a grid of potential data with a global coverage is required to minimise at least the discretisation error of the integral formula. Such a solution is not optimal from the statistical perspective.

To determine v_{nm} optimally, Eq. (1.3) can be used directly as an observation equation. The maximum degree of the solution should be specified and the values of potential with a global coverage are required. For each measured potential, one equation is constructed connecting the potential to v_{nm}. When more measurements than the number of unknown coefficients v_{nm} are included in the system, the least-squares method can be applied to optimally estimate the best values for v_{nm}. In addition, errors of the coefficients can also be estimated in such a case, which is not possible to do using Eq. (1.14). One problem of the optimal solution is the aliasing effect of the truncated higher frequencies in the solution. This means that the effect of truncated frequencies will be seen as errors in the low degrees and orders, which are recovered. However, such a problem does not occur in the integral solution (Eq. 1.14). Another advantage of applying the least-squares approach is that the values of potential can have different distributions and they are not required to be in a grid form.

1.2.2 Solving Dirichlet's problem

The integral formula (1.14) is considered for the global gravity field recovery requiring the potential data with a global coverage. If the data are available in a spatially restricted area, the mathematical models need to be transformed to integrals connecting the measured gravimetric quantities to the earth's gravitational potential and not the SHCs v_{nm}. Such type of integrals are also called integral transformations.

Now, consider that the goal is to determine the potential outside the surface of the spherical earth $V(r,\theta,\lambda)$ from some values measured at a spherical earth, i.e., $V(R,\theta,\lambda)$. By inserting Eq. (1.8), which is the solution of v_{nm} into Eq. (1.3) at the surface of the sphere, we obtain:

$$V(r,\theta,\lambda)=\frac{1}{4\pi}\sum_{n=0}^{\infty}\sum_{m=-n}^{n}\left(\frac{R}{r}\right)^{n+1}\iint_{\sigma}V(R,\theta',\lambda')Y_{nm}(\theta',\lambda')Y_{nm}(\theta,\lambda)d\sigma. \tag{1.15}$$

Interchanging the integral and summations yields:

$$V(r,\theta,\lambda)=\frac{1}{4\pi}\sum_{n=0}^{\infty}\left(\frac{R}{r}\right)^{n+1}\iint_{\sigma}V(R,\theta',\lambda')\sum_{m=-n}^{n}Y_{nm}(\theta',\lambda')Y_{nm}(\theta,\lambda)d\sigma. \tag{1.16}$$

Therefore, according to Eq. (1.9), Eq. (1.16) will change to:

$$V(r,\theta,\lambda)=\frac{1}{4\pi}\iint_{\sigma}V(R,\theta',\lambda')P(r,\psi)d\sigma, \tag{1.17}$$

where $P(r,\psi)$ is called the kernel of the integral, or the Green function, and it will have the following spectral form, or Legendre expansion:

$$P(r,\psi)=\sum_{n=0}^{\infty}(2n+1)\left(\frac{R}{r}\right)^{n+1}P_n(\cos\psi). \tag{1.18}$$

Eq. (1.17) is known as the Poisson integral and Eq. (1.18) as the Poisson kernel, which is convergent, and a closed-form formula can be derived for that. Consider the following expression for the Legendre polynomials (Hobson, 1965):

$$\frac{1}{D}=\sum_{n=0}^{\infty}t^nP_n(\tilde{x})\quad \text{where } D=\sqrt{1+t^2-2t\tilde{x}},\ \tilde{x}=\cos\psi \text{ and } t=\frac{R}{r}. \tag{1.19}$$

Now, we should see by which type of operation the Poisson kernel on the right-hand side (rhs) of Eq. (1.18) is constructed from Eq. (1.19). It will not be difficult to show that:

$$\sum_{n=0}^{\infty}(2n+1)\left(\frac{R}{r}\right)^{n+1}P_n(\cos\psi)=\frac{t-t^3}{D^3}\quad r>R; \tag{1.20}$$

when $r=R$, then $t=1$, and the kernel will change to the spherical Diract delta function, which gives ∞ when $\psi=0$, and for $\psi\neq 0$, the kernel has zero value:

$$\delta(\psi)=\sum_{n=0}^{\infty}(2n+1)P_n(\cos\psi). \tag{1.21}$$

Eq. (1.17) can be considered as an integral equation having the unknown potentials $V(R,\theta',\lambda')$ inside the integral. In this case, the integral should be discretised according to the desired resolution of $V(R,\theta',\lambda')$ in a grid form. This discretised integral is in fact an observation equation connecting each value of $V(r,\theta,\lambda)$ to $V(R,\theta',\lambda')$ over a regular grid. Therefore, a system of equations is constructed and solved to recover $V(R,\theta',\lambda')$. This system will be ill-conditioned, meaning that the solution will be highly sensitive to the errors in $V(r,\theta,\lambda)$. There are numerical methods for stabilising this system, which are called regularisation. In Chapter 4, this subject will be discussed in detail.

Fig. 1.1 illustrates the behaviour of the Poisson kernel, presented in Eq. (1.20) for the case in which $R<r$ and the difference between R and r is respectively 10 and 50 km. Both kernels are plotted to $\psi=5^\circ$. For the 10 km difference the kernel has its largest value at $\psi=0^\circ$ and decays fast and gets closer to zero. This means that the contribution of $V(R,\theta',\lambda')$ in the Poisson integral (Eq. 1.17) to the result of integration is very large compared with those which are at the locations with larger ψ. When the difference is 50 km, the kernel still has its large value at $\psi=0^\circ$ but it is not as large as the case in which the difference is 10 km. This plot can be interpreted in another way. If the potential is measured at the spherical boundary with radius r and it is convolved with the Poisson integral (Eq. 1.17), a larger portion of the signal will be visible for 10 km than for 50 km. This means that, for example, the potential at a level of 10 km contains more frequencies, or the signal is stronger, than at 50 km. Therefore, by looking at the kernel of such integral, we can see how significant the contribution of the data is in the integration domain and how many frequencies can be recovered from this integral.

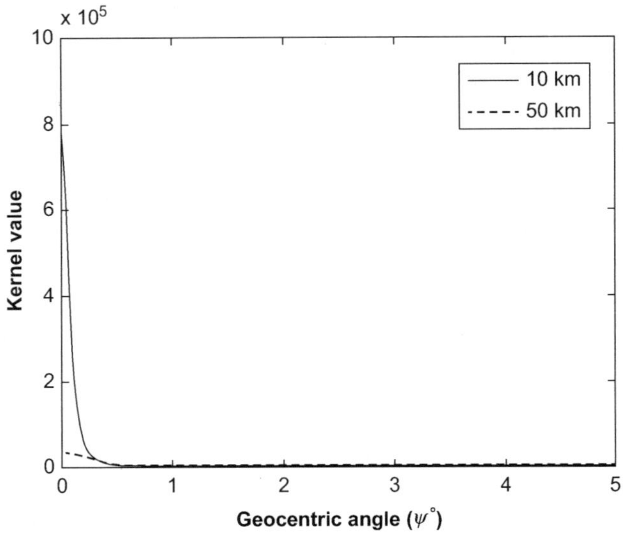

Figure 1.1 The behaviour of the Poisson kernel with respect to different geocentric angles ψ.

1.3 Solving potential from its first-order derivatives

In practice, the potential is not observed directly; instead, gravity is measured with gravimeters, and is in fact the first-order derivative of the potential. Gravity can be measured in two different ways: scalar or vector form. In this section, the methods of determining the potential from these gravimetric data are theoretically discussed.

1.3.1 Solving potential from its radial derivative

Gravity can be regarded as the radial derivative of potential along the plumb line, or the vector of gravity, and it is considered as a vector quantity with one dimension. In the following, it will be discussed how the potential can be determined from this measurement.

1.3.1.1 Spectral solution

Here, it is assumed that the radial derivative of the potential $V_r(r,\theta,\lambda)$ is given outside the spherical earth and the problem is to determine v_{nm} at the sphere with radius R. The first-order radial derivative is:

$$\frac{\partial V(r,\theta,\lambda)}{\partial r} = V_r(r,\theta,\lambda). \tag{1.22}$$

Therefore, taking the derivative of Eq. (1.3) with respect to r gives us the spherical harmonic expression of gravity:

$$V_r(r,\theta,\lambda) = -\sum_{n=0}^{\infty}\sum_{m=-n}^{n}\frac{n+1}{r}\left(\frac{R}{r}\right)^{n+1} v_{nm}Y_{nm}(\theta,\lambda). \tag{1.23}$$

If we multiply both sides of Eq. (1.23) by $Y_{n'm'}(\theta',\lambda')$ and take the integral of the results over a unit sphere according to the addition theorem (Eq. 1.9), we can write

$$v_{nm} = -\frac{r}{n+1}\left(\frac{r}{R}\right)^{n+1}\frac{1}{4\pi}\iint_{\sigma} V_r(r,\theta',\lambda')Y_{nm}(\theta',\lambda')\mathrm{d}\sigma. \tag{1.24}$$

Eq. (1.24) is an integral for solving v_{nm} from $V_r(r,\theta,\lambda)$ above the spherical earth, i.e., $r > R$. The factor $(r/R)^{n+1}$ continues $V_r(r,\theta,\lambda)$ downward to sea level and $r/(n+1)$ coverts the spherical harmonics of $V_r(r,\theta,\lambda)$ to those of the potential. Since $r > R$, the ratio (r/R) is larger than 1, and this means that by the power $n+1$, it will be even larger, and when n goes to infinity so does this downward continuation factor. However, $r/(n+1)$ decreases by increasing the power and when it is multiplied to $(r/R)^{n+1}$, it will reduce the power of this factor. This means that more frequencies, or v_{nm} with higher degrees and orders, are recoverable from $V_r(r,\theta,\lambda)$ than $V(r,\theta,\lambda)$; see Eq. (1.14).

1.3.1.2 Spatial solution

For solving Eq. (1.24) $V_r(r,\theta,\lambda)$ with a global coverage is required, but a simple integral formula can be derived with a limited coverage above the sphere. To do so, $Y_{nm}(\theta,\lambda)$ is multiplied to both sides of Eq. (1.24) and a summation over m from $-n$ to n is taken from the results. By applying the addition theorem of spherical harmonics, Eq. (1.9), we obtain:

$$v_n(\theta,\lambda) = -\frac{r}{4\pi}\iint_{\sigma}\frac{2n+1}{n+1}\left(\frac{r}{R}\right)^{n+1}P_n(\cos\psi)V_r(r,\theta',\lambda')\mathrm{d}\sigma. \tag{1.25}$$

By taking summation over n of both sides, we come to:

$$V(R,\theta,\lambda) = -\frac{r}{4\pi}\iint_{\sigma} V_r(r,\theta',\lambda')H^*(r,\psi)\mathrm{d}\sigma, \tag{1.26}$$

with the kernel:

$$H^{*}(r,\psi)=\sum_{n=0}^{\infty}\frac{2n+1}{n+1}\left(\frac{r}{R}\right)^{n+1}P_{n}(\cos\psi), \tag{1.27}$$

which is a divergent function. However, if the maximum degree of this kernel is limited, a smooth solution can be obtained.

Now, let us look at the problem in another way, where $V_r(R,\theta,\lambda)$ is given at the boundary and the potential $V(r,\theta,\lambda)$ outside the sphere is sought. This is known as Neumann's boundary-value problem. If $r=R$ in Eq. (1.24), it means that v_{nm} is recovered from $V_r(R,\theta,\lambda)$, and when the result is inserted into Eq. (1.3) and the addition theorem (Eq. 1.9) is applied, the following integral formula is achieved:

$$V(r,\theta,\lambda)=\frac{R}{4\pi}\iint_{\sigma}V_r(R,\theta',\lambda')H(r,\psi)\mathrm{d}\sigma, \tag{1.28}$$

with

$$H(r,\psi)=\sum_{n=0}^{\infty}\frac{2n+1}{n+1}\left(\frac{R}{r}\right)^{n+1}P_{n}(\cos\psi), \tag{1.29}$$

which are, respectively, known as the extended Hotine integral and function. Unlike Eq. (1.27), the extended Hotine function is convergent with the following closed-form formula (Pick et al., 1973):

$$H(r,\psi)=\frac{2}{D}-\ln\left(\frac{t-\widetilde{x}+D}{1-\widetilde{x}}\right). \tag{1.30}$$

For the case of $t=1$, or $R=r$, Hotine (1969) presented:

$$H(\psi)=\frac{1}{\sin\frac{\psi}{2}}-\ln\left(1+\frac{1}{\sin\frac{\psi}{2}}\right). \tag{1.31}$$

Fig. 1.2 represents the behaviour of the extended Hotine kernel, Eq. (1.30), to $\psi=10^{\circ}$. The kernel shows that when $V_r(R,\theta',\lambda')$ is located closer to the computation point, or $\psi=0^{\circ}$, it has more significant contributions than those $V_r(R,\theta',\lambda')$ which are far from it. In addition, the figure shows that the kernel does not decrease to zero even up to $\psi=10^{\circ}$. When the difference between the measurement and the computation levels is 10 km more frequencies are recoverable than when it is 50 km.

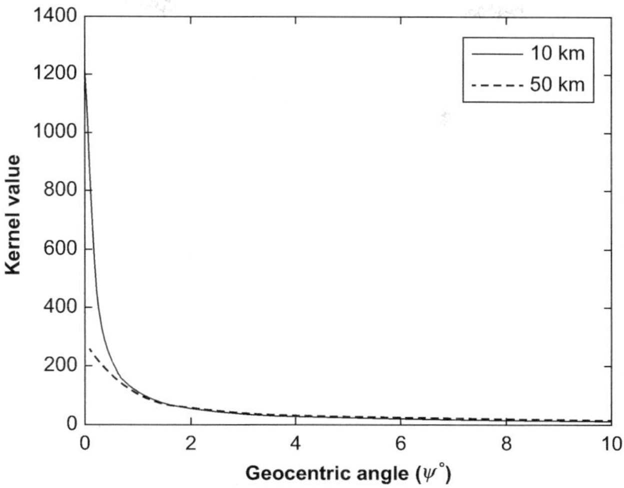

Figure 1.2

1.3.1.3 Integral equation

In Eq. (1.26), $V_r(r,\theta',\lambda')$ should be given in a regular grid at a sphere with the radius r so that the integration can be performed. However, finding another integral formula connecting $V_r(r,\theta',\lambda')$ to the potential $V(R,\theta',\lambda')$ is also possible. Such an integral can be derived simply by taking the radial derivative of the Poisson formula, Eq. (1.17):

$$V_r(r,\theta,\lambda) = \frac{1}{4\pi}\iint\limits_{\sigma} V(R,\theta',\lambda')P_r(r,\psi)\mathrm{d}\sigma, \tag{1.32}$$

where

$$P_r(r,\psi) = -\frac{1}{r}\sum_{n=0}^{\infty}(2n+1)(n+1)\left(\frac{R}{r}\right)^{n+1}P_n(\cos\psi). \tag{1.33}$$

The closed-form formula for such a kernel can be derived directly by taking the derivative of the kernel (Eq. 1.20) with respect to r:

$$P_r(r,\psi) = -\frac{t}{r}\left[\frac{1-3t^2}{D^3} - 3\frac{(t-t^3)(t-\tilde{x})}{D^5}\right]. \tag{1.34}$$

Fig. 1.3 illustrates the behaviour of the kernel (Eq. 1.34) to $\psi = 3^{\circ}$ for $R < r$ by 10 and 30 km. This kernel is used in the integral, which is in fact an integral for obtaining $V_r(r,\theta,\lambda)$ from $V(R,\theta',\lambda')$. However, in

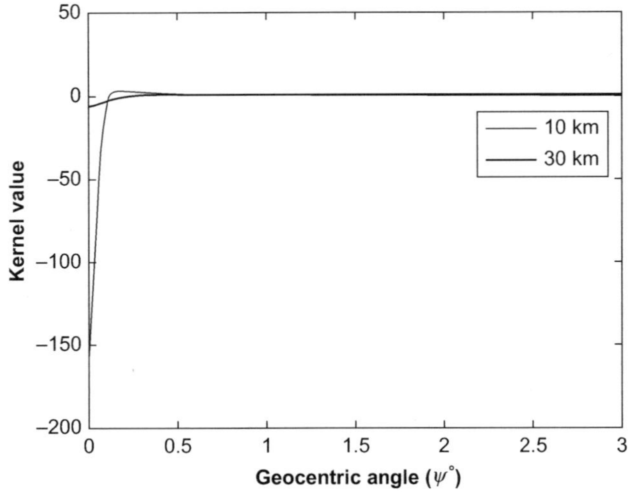

Figure 1.3 Behaviour of the first-order derivative of the Poisson kernel, $P_r(r, \psi)$.

gravimetry the problem is to find $V(R, \theta', \lambda')$ from $V_r(r, \theta, \lambda)$, and Eq. (1.32) is an integral equation having the unknown inside the integral formula. The behaviour of the kernel of this integral is important to check how much the potential can be recovered. Similar to those kernels we discussed before, this kernel has its largest value at $\psi = 0^\circ$, but for the case of 10 km the contribution of near-zone potentials is more significant than that of those which are far. For the case of 30 km the largest value is closer to zero than that of 10 km.

1.3.1.4 Solving the potential from a linear combination of the potential and its radial derivative

Now, suppose that we have a linear combination of the potential $V(r, \theta, \lambda)$ and its derivative $V_r(r, \theta, \lambda)$ at the surface of a sphere with radius $r > R$. The problem is to determine the potential at the surface of the sphere with the radius R. Let us write this linear combination as:

$$V^{\text{comb}}(r, \theta, \lambda) = \left(a + b\frac{\partial}{\partial r}\right) V(r, \theta, \lambda) = aV(r, \theta, \lambda) + bV_r(r, \theta, \lambda). \quad (1.35)$$

The superscript "comb" means combined, and a and b are two constants. The term $V^{\text{comb}}(r, \theta, \lambda)$ is given at the whole surface of the spherical boundary.

1.3.1.5 Spectral solution

By inserting Eqs. (1.3) and (1.23) into Eq. (1.35) and further simplification, it will not be difficult to show that the spherical harmonic expansion of this linear combination is:

$$V^{\mathrm{comb}}(r,\theta,\lambda)=\sum_{n=0}^{\infty}\sum_{m=-n}^{n}\left(\frac{R}{r}\right)^{n+1}\left(a-b\frac{n+1}{r}\right)v_{nm}Y_{nm}(\theta,\lambda). \qquad (1.36)$$

By integrating the product of both sides of Eq. (1.36) with $Y_{nm}(\theta',\lambda')$ and simplifying the results based on the orthogonality property of spherical harmonics given in Eq. (1.6), we have:

$$\iint_{\sigma} V^{\mathrm{comb}}(r,\theta',\lambda')Y_{nm}(\theta',\lambda')\mathrm{d}\sigma=4\pi\left(\frac{R}{r}\right)^{n+1}\left(a-b\frac{n+1}{r}\right)v_{nm}. \qquad (1.37)$$

Solving Eq. (1.37) for v_{nm} reads:

$$v_{nm}=\frac{1}{4\pi}\left(\frac{r}{R}\right)^{n+1}\left(a-b\frac{n+1}{r}\right)^{-1}\iint_{\sigma} V^{\mathrm{comb}}(r,\theta',\lambda')Y_{nm}(\theta',\lambda')\mathrm{d}\sigma. \qquad (1.38)$$

1.3.1.6 Spatial solutions

Multiplying both sides of Eq. (1.38) by $Y_{nm}(\theta,\lambda)$, applying the addition theorem (Eq. 1.9) on the rhs of the result, and finally taking summation from 0 to infinity over n leads to:

$$V^{\mathrm{comb}}(R,\theta,\lambda)=\frac{1}{4\pi}\iint_{\sigma} K(r,\psi)V(r,\theta',\lambda')\mathrm{d}\sigma, \qquad (1.39)$$

where

$$K(r,\psi)=\sum_{n=0}^{\infty}(2n+1)\left(\frac{r}{R}\right)^{n+1}\left(a-b\frac{n+1}{R}\right)P_n(\cos\psi), \qquad (1.40)$$

which is a divergent kernel without any closed-form formula.

Now, assume that $V^{\mathrm{comb}}(R,\theta',\lambda')$ is given at the reference sphere and the problem is to find the potential $V(r,\theta,\lambda)$ outside the sphere; in this case, we obtain:

$$V(r,\theta,\lambda)=\frac{1}{4\pi}\iint_{\sigma} S^{*}(r,\psi)V^{\mathrm{comb}}(R,\theta',\lambda')\mathrm{d}\sigma, \qquad (1.41)$$

with

$$S^*(r,\psi)=\sum_{n=0}^{\infty}(2n+1)\left(\frac{R}{r}\right)^{n+1}\left(a-b\frac{n+1}{R}\right)^{-1}P_n(\cos\psi). \tag{1.42}$$

This kernel is convergent, and a closed-form formula can be found for that if a and b are given.

1.3.1.7 Integral equation

If $V^{\text{comb}}(r,\theta,\lambda)$ is given outside the sphere and the goal is to determine $V(R,\theta,\lambda)$ at the sphere, we can simply apply the operator given in Eq. (1.35) to the Poisson integral (Eq. 1.17):

$$\begin{aligned}V^{\text{comb}}(r,\theta,\lambda)&=\left(a+b\frac{\partial}{\partial r}\right)V(r,\theta,\lambda)\\&=\frac{1}{4\pi}\iint_{\sigma}V(R,\theta',\lambda')\left(a+b\frac{\partial}{\partial r}\right)P(r,\psi)\mathrm{d}\sigma,\end{aligned} \tag{1.43}$$

where according to Eqs. (1.20) and (1.34), the kernel of this integral will be:

$$\left(a+b\frac{\partial}{\partial r}\right)P(r,\psi)=a\frac{t-t^3}{D^3}-b\frac{t}{r}\left[\frac{1-3t^2}{D^3}-3\frac{(t-t^3)(t-\tilde{x})}{D^5}\right]. \tag{1.44}$$

Eq. (1.43) is an integral equation with $V(R,\theta,\lambda)$ as unknown and $V^{\text{comb}}(r,\theta,\lambda)$ as known.

1.3.2 Solving the potential from its gradients

The derivative of a function is dependent on the type of coordinate systems in which the derivative is taken. Here, a local frame is used for defining the gradient operator. The frame has its z-axis radially upward from the centre of the spherical earth. The x-axis is pointing to the north and the system is left handed, meaning that the y-axis is towards the east. The gradient operator in such a coordinate system is defined by:

$$\nabla=\frac{\partial}{\partial r}\mathbf{e}_z+\frac{\partial}{r\partial\theta}\mathbf{e}_x+\frac{\partial}{r\sin\theta\partial\lambda}\mathbf{e}_y, \tag{1.45}$$

where $\mathbf{e}_i$, $i=x,y,z$, are the unit vectors along the axes of the frame at the point for which the gradient is calculated. The gradient of a potential field will then be:

$$\begin{aligned}v(r,\theta,\lambda)&=\nabla V(r,\theta,\lambda)=V_z(r,\theta,\lambda)\mathbf{e}_z+V_x(r,\theta,\lambda)\mathbf{e}_x+V_y(r,\theta,\lambda)e_y\\&=V_r(r,\theta,\lambda)\mathbf{e}_z+\frac{1}{r}V_\theta(r,\theta,\lambda)\mathbf{e}_x+\frac{1}{r\sin\theta}V_\lambda(r,\theta,\lambda)\mathbf{e}_y.\end{aligned} \tag{1.46}$$

In this section, the discussion is about the transformations between the potential and its derivatives. To do so, it is necessary to take advantage of the vector spherical harmonics.

1.3.2.1 Vector spherical harmonics

Generally, any vector field $\mathbf{v}_f$ can be represented by the series of vector spherical harmonics (see e.g., Eshagh, 2016):

$$\mathbf{v}_f(r,\theta,\lambda)=\sum_{n=0}^{\infty}\sum_{m=-n}^{n}v_{f,nm}\left\{\frac{\partial}{\partial r}\left(\frac{R}{r}\right)^{n+1}\mathbf{X}_{nm}^{(1)}(\theta,\lambda)+\frac{1}{r}\left(\frac{R}{r}\right)^{n+1}\mathbf{X}_{nm}^{(2)}(\theta,\lambda)+\frac{1}{r}\left(\frac{R}{r}\right)^{n+1}\mathbf{X}_{nm}^{(3)}(\theta,\lambda)\right\}, \tag{1.47}$$

where $v_{nm}^{(i)}$, $i=1, 2, 3$, are the SHCs of the vector field derived from each element of the gradient of potential and $\mathbf{X}_{nm}^{(i)}(\theta,\lambda)$ the vector spherical harmonics with the following expression in terms of scalar spherical harmonics (cf. Eshagh, 2014):

$$\mathbf{X}_{nm}^{(1)}(\theta,\lambda)=Y_{nm}(\theta,\lambda)\mathbf{e}_z, \tag{1.48}$$

$$\mathbf{X}_{nm}^{(2)}(\theta,\lambda)=\frac{\partial Y_{nm}(\theta,\lambda)}{\partial\theta}\mathbf{e}_x+\frac{\partial Y_{nm}(\theta,\lambda)}{\sin\theta\partial\lambda}\mathbf{e}_y, \tag{1.49}$$

$$\mathbf{X}_{nm}^{(3)}(\theta,\lambda)=-\frac{\partial Y_{nm}(\theta,\lambda)}{\sin\theta\partial\lambda}\mathbf{e}_x+\frac{\partial Y_{nm}(\theta,\lambda)}{\partial\theta}\mathbf{e}_y. \tag{1.50}$$

The vector spherical harmonics have the following orthogonality property:

$$\iint_{\sigma}\mathbf{X}_{nm}^{(i)}(\theta',\lambda')\cdot\mathbf{X}_{n'm'}^{(i')}(\theta',\lambda')d\sigma=4\pi\left(N_n^{(i)}\right)_{\mathbf{X}}^2\delta_{nn'}\delta_{mm'}\delta_{ii'}, \tag{1.51}$$

where the dot $(\cdot)$ means the inner product operator, δ stands for the Kronecker delta and $\left(N_n^{(i)}\right)_{\mathbf{X}}^2$ the squared of the norm of vector spherical harmonics:

$$\left(N_n^{(i)}\right)_{\mathbf{X}}^2=\begin{cases}1 & i=1\\ n(n+1) & i=2\text{ and }3\end{cases}. \tag{1.52}$$

The term $v_{f,nm}^{(i)}$ is derived by multiplying $\mathbf{X}_{nm}^{(i)}(\theta',\lambda')$ to both sides of Eq. (1.47) and taking the integral of the result over the sphere with radius r.

If the orthogonality property (Eq. 1.51) is applied and the result is solved for $v_{f,nm}^{(i)}$, Eq. (1.53) will be obtained:

$$v_{f,nm}^{(i)} = \frac{1}{4\pi\left(N_n^{(i)}\right)_{\mathbf{X}}^2} \iint_\sigma \mathbf{v}_f(r,\theta',\lambda')\cdot\mathbf{X}_{nm}^{(i)}(\theta',\lambda')\mathrm{d}\sigma, \quad i = 1, 2 \text{ and } 3. \tag{1.53}$$

This means that there are three solutions for the SHCs of the vector field. For the gradient of a scalar field $\mathbf{v}_f = \mathbf{v}$, see Eq. (1.46) and compare it with Eqs. (1.48)—(1.50); only the vector spherical harmonics $\mathbf{X}_{nm}^{(1)}(\theta,\lambda)$ and $\mathbf{X}_{nm}^{(2)}(\theta,\lambda)$ are required. For other types of fields, $\mathbf{X}_{nm}^{(2)}(\theta,\lambda)$ gives the poloidal components of the field and $\mathbf{X}_{nm}^{(3)}(\theta,\lambda)$ the toroidal ones. This will be discussed in Chapter 8.

1.3.2.2 Spectral solutions

When the gradient operator, presented in Eq. (1.45), is applied to a scalar potential field, the gradient of the potential will be obtained, and therefore, two sets of $v_{nm}^{(i)}$ can be obtained from this gradient. This means that only the cases $i = 1$ and 2 are considered for deriving the spectral solution of $v_{nm}^{(i)}$. In the case in which the vector field $\mathbf{v}(r,\theta',\lambda')$ is given outside the sphere and the goal is to determine $v_{nm}^{(i)}$, $i = 1$ and 2, at the spherical boundary, we can write:

$$v_{nm}^{(1)} = \frac{1}{4\pi} \iint_\sigma \mathbf{v}(r,\theta',\lambda')\cdot\mathbf{X}_{nm}^{(1)}(\theta',\lambda')\mathrm{d}\sigma, \tag{1.54}$$

$$v_{nm}^{(2)} = \frac{1}{4\pi n(n+1)} \iint_\sigma \mathbf{v}(r,\theta',\lambda')\cdot\mathbf{X}_{nm}^{(2)}(\theta',\lambda')\mathrm{d}\sigma. \tag{1.55}$$

The solution of Eq. (1.54), which is related to recovery of $v_{nm}^{(1)}$ from $V_r(r,\theta,\lambda)$, is the same as was discussed in Section 1.3.1.1, and the spectral, spatial, and integral equations have been presented. Therefore, we continue our discussion only with the horizontal derivatives. By inserting Eq. (1.46) into Eq. (1.55) and using the orthogonality property (Eq. 1.51), $v_{nm}^{(2)}$ is derived:

$$v_{nm}^{(2)} = \frac{R}{n(n+1)}\left(\frac{r}{R}\right)^{n+2}\frac{1}{4\pi}\iint_\sigma\left[\frac{\partial Y_{nm}(\theta',\lambda')}{\partial\theta'}V_x(r,\theta',\lambda') + \frac{\partial Y_{nm}(\theta',\lambda')}{\sin\theta'\partial\lambda'}V_y(r,\theta',\lambda')\right]\mathrm{d}\sigma. \tag{1.56}$$

Since n is in the denominator of Eq. (1.56), when it becomes zero the result will be singular. This means that recovering the zero-degree harmonic coefficient of the potential is not possible from its horizontal derivatives.

Theoretically, the solutions, Eqs. (1.54) and (1.55), are the same in the presence of no error; i.e., $v_{nm}^{(1)} = v_{nm}^{(2)}$. If Eqs. (1.54) and (1.55) are equated we obtain:

$$V_{r,nm} = \frac{1}{n} V_{H,nm}, \quad \text{for } n > 0, \tag{1.57}$$

with

$$V_{r,nm} = \frac{1}{4\pi} \iint_{\sigma} V_r(r, \theta', \lambda') Y_{nm}(\theta', \lambda') \mathrm{d}\sigma, \tag{1.58}$$

$$V_{H,nm} = \frac{1}{4\pi} \iint_{\sigma} \left[\frac{\partial Y_{nm}(\theta', \lambda')}{\partial \theta'} V_x(r, \theta', \lambda') + \frac{\partial Y_{nm}(\theta', \lambda')}{\sin \theta' \partial \lambda'} V_y(r, \theta', \lambda') \right] \mathrm{d}\sigma. \tag{1.59}$$

However, practically, it is not so, as:

(1) two measurements are used in Eq. (1.55), with different noise characteristics; and

(2) discretisation error of the integrals in Eqs. (1.24) or (1.55) is different as well.

1.3.2.3 Spatial solutions

Similar to the process used for finding integral relations between the radial derivative and the potential, an integral formula relating horizontal derivatives of $V_x(r, \theta', \lambda')$ and $V_y(r, \theta', \lambda')$ to $V(R, \theta, \lambda)$ can be obtained. By multiplying $Y_{nm}(\theta, \lambda)$ to Eq. (1.56), taking summation over m from $-n$ to n, and over n from 1 to infinity we obtain:

$$\begin{aligned} V(R, \theta, \lambda) = \frac{r}{4\pi} \iint_{\sigma} \sum_{n=1}^{\infty} \frac{1}{n(n+1)} \left(\frac{r}{R}\right)^{n+1} \Bigg[V_x(r, \theta', \lambda') \\ \sum_{m=-n}^{n} \frac{\partial Y_{nm}(\theta', \lambda')}{\partial \theta'} Y_{nm}(\theta, \lambda) + V_y(r, \theta', \lambda') \\ \sum_{m=-n}^{n} \frac{\partial Y_{nm}(\theta', \lambda')}{\sin \theta' \partial \lambda'} Y_{nm}(\theta, \lambda) \Bigg] \mathrm{d}\sigma. \end{aligned} \tag{1.60}$$

Taking derivatives of the addition theorem (Eq. 1.9) with respect to θ' and λ' reads:

$$\sum_{m=-n}^{n} \frac{\partial Y_{nm}(\theta', \lambda')}{\partial \theta'} Y_{nm}(\theta, \lambda) = (2n+1) \frac{\partial P_n(\cos \psi)}{\partial \theta'}, \tag{1.61}$$

$$\sum_{m=-n}^{n} \frac{\partial Y_{nm}(\theta', \lambda')}{\sin \theta' \partial \lambda'} Y_{nm}(\theta, \lambda) = (2n+1) \frac{\partial P_n(\cos \psi)}{\sin \theta' \partial \lambda'}. \tag{1.62}$$

Substituting Eqs. (1.61) and (1.62) into Eq. (1.60) and simplification yields:

$$V(R, \theta, \lambda) = \frac{r}{4\pi} \iint_{\sigma} \left[V_\theta(r, \theta', \lambda') G_\theta^*(r, \psi) + V_\lambda(r, \theta', \lambda') G_\lambda^*(r, \psi) \right] \mathrm{d}\sigma, \tag{1.63}$$

where $G_\theta(r, \psi)$ and $G_\lambda(r, \psi)$ are the kernels of the integral (Eq. 1.63):

$$G_\theta^*(r, \psi) = \sum_{n=1}^{\infty} \frac{2n+1}{n(n+1)} \left(\frac{r}{R} \right)^{n+1} \frac{\partial P_n(\cos \psi)}{\partial \theta}, \tag{1.64}$$

$$G_\lambda^*(r, \psi) = \sum_{n=1}^{\infty} \frac{2n+1}{n(n+1)} \left(\frac{r}{R} \right)^{n+1} \frac{\partial P_n(\cos \psi)}{\sin \theta \partial \lambda}. \tag{1.65}$$

From the relation between the associated Legendre functions and the Legendre polynomials (Abramowitz and Stegun, 1964, p. 334):

$$P_{nm}(\cos \psi) = (-1)^m \sin^m \psi \frac{\mathrm{d}^m P_n(\cos \psi)}{\mathrm{d}(\cos \psi)^m}, \tag{1.66}$$

and for $m = 1$, we can write:

$$\begin{aligned} \frac{\partial P_n(\cos \psi)}{\partial \theta} = \frac{\mathrm{d}P_n(\cos \psi)}{\mathrm{d}\cos \psi} \frac{\mathrm{d}\cos \psi}{d\psi} \frac{\partial \psi}{\partial \theta'} &= -\sin \psi \frac{\mathrm{d}P_n(\cos \psi)}{\mathrm{d}\cos \psi} \cos \alpha' \\ &= P_{n1}(\cos \psi) \cos \alpha', \end{aligned} \tag{1.67}$$

where α' is called the backward azimuth, meaning the azimuth from the integration point to the computation point (see Fig. 1.5). Similarly:

$$\begin{aligned} \frac{\partial P_n(\cos \psi)}{\sin \theta' d\lambda'} &= \frac{\mathrm{d}P_n(\cos \psi)}{\mathrm{d}\cos \psi} \frac{\mathrm{d}\cos \psi}{\mathrm{d}\psi} \frac{\partial \psi}{\sin \theta' \partial \lambda'} \\ &= -\sin \psi \frac{\mathrm{d}P_n(\cos \psi)}{\mathrm{d}\cos \psi} (-\sin \alpha') \\ &= -P_{n1}(\cos \psi) \sin \alpha'. \end{aligned} \tag{1.68}$$

Finally, Eq. (1.60) will be simplified to:

$$V(R,\theta,\lambda)=\frac{r}{4\pi}\iint_{\sigma}[V_x(r,\theta',\lambda')\cos\alpha'-V_y(r,\theta',\lambda')\sin\alpha']G_{\mathrm{H}}^{*}(r,\psi)d\sigma, \tag{1.69}$$

where

$$G_{\mathrm{H}}^{*}(r,\psi)=\sum_{n=1}^{\infty}\frac{2n+1}{n(n+1)}\left(\frac{r}{R}\right)^{n+1}P_{n1}(\cos\psi). \tag{1.70}$$

If the gradients are given at the surface of the spherical boundary, we can determine the potential outside sphere. To do so, we consider $r=R$ in Eq. (1.56) and insert the result back into Eq. (1.3). According to Eqs. (1.64) and (1.65) we obtain:

$$\begin{aligned}V(r,\theta,\lambda)=&\frac{R}{4\pi}\iint_{\sigma}\sum_{n=0}^{\infty}\frac{2n+1}{n(n+1)}\left(\frac{R}{r}\right)^{n+1}\\&\left[V_x(R,\theta',\lambda')\frac{\partial P_n(\cos\psi)}{\partial\theta'}+V_y(R,\theta',\lambda')\frac{\partial P_n(\cos\psi)}{\sin\theta' d\lambda'}\right]d\sigma.\end{aligned} \tag{1.71}$$

After repeating the whole process to get Eq. (1.63), for Eq. (1.68) we reach:

$$V(r,\theta,\lambda)=\frac{R}{4\pi}\iint_{\sigma}[V_x(R,\theta',\lambda')\cos\alpha'-V_y(R,\theta',\lambda')\sin\alpha']G_{\mathrm{H}}(r,\psi)d\sigma, \tag{1.72}$$

where

$$G_{\mathrm{H}}(r,\psi)=\sum_{n=1}^{\infty}\frac{2n+1}{n(n+1)}\left(\frac{R}{r}\right)^{n+1}P_{n1}(\cos\psi), \tag{1.73}$$

which is a convergence with the following closed-form formula:

$$G_{\mathrm{H}}(r,\psi)=\left(\frac{t^2(D+1)}{D(D-t\widetilde{x}+1)}-\frac{t+D}{(D-\widetilde{x}+t)D}-\frac{1}{\widetilde{x}-1}\right)\sqrt{1-\widetilde{x}^2}, \tag{1.74}$$

Fig. 1.4 shows the behaviour of the kernel (Eq. 1.74) for the cases in which r is larger than R by 10 and 50 km.

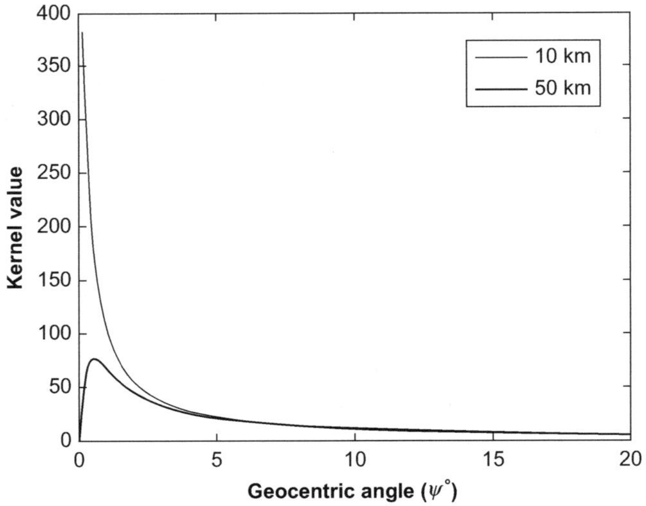

Figure 1.4 Behaviour of $G_H(r, \psi)$.

For $r = R$, this kernel will have the following spectral and closed-form expression (see Hwang, 1998):

$$G_H(\psi) = \sum_{n=1}^{\infty} \frac{2n+1}{n(n+1)} \frac{dP_n(\cos\psi)}{d\psi} = -\cot\frac{\psi}{2} - 1. \tag{1.75}$$

So far, we have presented integral formulae, which are dependent on the backward azimuth. Now, our goal is to explain how to derive this azimuth from the coordinates of the computation and integration points and how the forward and backward azimuths are related to each other. Fig. 1.5 shows these two azimuths clearly. A spherical triangle is seen in the figure, with all its angles and sides.

According to the cosine rule of spherical trigonometry, we can write two equations for connecting the forward and backward azimuths to the coordinates of the points:

$$\cos\theta' = \cos\theta \cos\psi + \sin\theta \sin\psi \cos\alpha, \tag{1.76a}$$

$$\cos\theta = \cos\theta' \cos\psi + \sin\theta' \sin\psi \cos\left(360^\circ - \alpha'\right). \tag{1.76b}$$

The solutions of Eqs. (1.76a) and (1.76b), respectively, for α and $360^\circ - \alpha'$ are:

$$\cos\alpha = \frac{\cos\theta' - \cos\theta\cos\psi}{\sin\theta\sin\psi}, \tag{1.77}$$

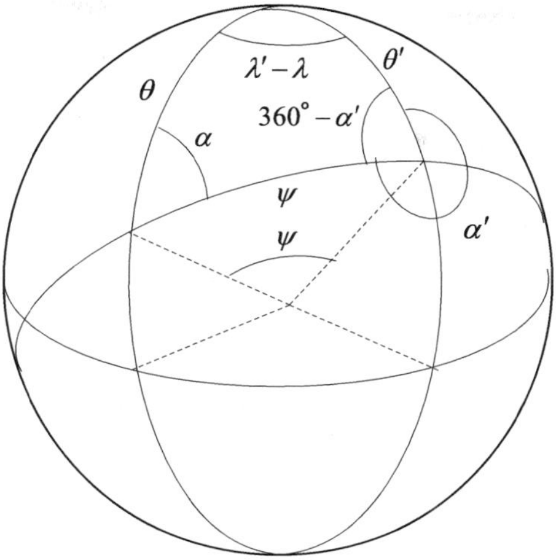

Figure 1.5 Forward and backward azimuths.

$$\cos\left(360^\circ - \alpha'\right) = \cos\alpha' = \frac{\cos\theta - \cos\theta' \cos\psi}{\sin\theta' \sin\psi}. \tag{1.78}$$

Therefore, these azimuths can be determined easily from Eqs. (1.77) and (1.78). The term $\cos\psi$ has already been presented in Eq. (1.10), and $\sin\psi = \sqrt{1 - \cos^2\psi}$. Now, according to the sine rule, we can write:

$$\frac{\sin\alpha}{\sin\theta'} = \frac{\sin(360^\circ - \alpha')}{\sin\theta} = \frac{\sin(\lambda' - \lambda)}{\sin\psi}. \tag{1.79}$$

After simplification, we obtain:

$$\sin\alpha \sin\theta + \sin\alpha' \sin\theta' = 0. \tag{1.80}$$

Eq. (1.80) is the mathematical relation between the forward and the backward azimuths. If one of them is given, the other can be simply derived from this equation.

1.3.2.4 Integral equations

Now, imagine that the derivatives $V_x(r, \theta, \lambda)$ and $V_y(r, \theta, \lambda)$ are given and the potential at the sphere $V(R, \theta', \lambda')$ is to be determined. To find integral equations connecting them to $V(R, \theta', \lambda')$ it is enough to apply the gradient operator (Eq. 1.45) to the Poisson integral Eq. (1.17). In this case, three integral equations will be obtained, and the first one, which is related

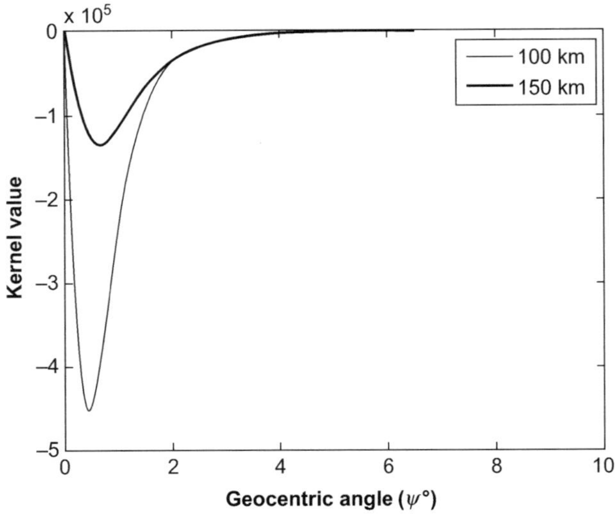

Figure 1.6 Behaviour of $P_\psi(r,\psi)$.

to the first-order radial derivative, is the same as Eq. (1.32), but the other two are:

$$\begin{pmatrix} V_x(r,\theta,\lambda) \\ V_y(r,\theta,\lambda) \end{pmatrix} = \frac{1}{4\pi r}\iint_\sigma V(R,\theta',\lambda')P_\psi(r,\psi)\begin{pmatrix} \cos\alpha \\ -\sin\alpha \end{pmatrix} d\sigma, \tag{1.81}$$

where α is the forward azimuth, the azimuth from the computation to the integration point. The kernel of the integral is derived by taking the derivative of Eq. (1.20) with respect to ψ:

$$P_\psi(r,\psi) = -\frac{3t(t-t^3)\sqrt{1-\tilde{x}^2}}{D^5}. \tag{1.82}$$

Fig. 1.6 shows the behaviour of this kernel to $\psi = 10^\circ$ for two levels of 100 and 150 km above the spherical boundary.

1.4 Solving the potential from its second-order derivatives

If the second-order derivatives of the potential can be summarised as a second-order tensor:

$$\boldsymbol{\tau}(r,\theta,\lambda) = \nabla\nabla V(r,\theta,\lambda) = \begin{bmatrix} V_{xx} & V_{xy} & V_{xz} \\ V_{yx} & V_{yy} & V_{yz} \\ V_{zx} & V_{zy} & V_{zz} \end{bmatrix}. \tag{1.83}$$

However, it should be mentioned that the choice of coordinate system is of vital importance for taking the derivatives. Here, the local north-oriented frame, which was already defined for the gradient of the potential, is used for the definition of the second-order derivatives. In the case of applying the operator defined in Eqs. (1.83) to (1.3), we obtain (Martinec, 2003):

$$\boldsymbol{\tau}(r,\theta,\lambda)=\sum_{n=0}^{\infty}\sum_{m=-n}^{n}v_{nm}\left\{\frac{\partial^2}{\partial r^2}\left(\frac{R}{r}\right)^{n+1}\mathbf{Z}^{(1)}_{nm}(\theta,\lambda)+2\frac{\partial}{\partial r}\left(\frac{R}{r}\right)^{n+1}\mathbf{Z}^{(2)}_{nm}(\theta,\lambda)\right.$$
$$\left.+\frac{1}{2r^2}\left(\frac{R}{r}\right)^{n+1}\mathbf{Z}^{(3)}_{nm}(\theta,\lambda)-\frac{1}{r}\left[\frac{1}{n(n+1)}\frac{\partial}{\partial r}-\frac{1}{2r}\right]\left(\frac{R}{r}\right)^{n+1}\mathbf{Z}^{(4)}_{nm}(\theta,\lambda)\right\},\tag{1.84}$$

where $\mathbf{Z}^{(i)}_{nm}(\theta,\lambda)$, $i=1$, 2, 3 and 4, stand for tensor spherical harmonics. Therefore, it is necessary to define these functions and explain how to use them for determining the potential.

1.4.1 Tensor spherical harmonics

Any second-order tensor $\boldsymbol{\tau}_f(R,\theta,\lambda)$, whose components are square-integrable functions, can be expanded in a series of the tensor spherical harmonics $\mathbf{Z}^{(i)}_{nm}(\theta,\lambda)$ as (Martinec, 2003):

$$\boldsymbol{\tau}_f(R,\theta,\lambda)=\sum_{n=0}^{\infty}\sum_{m=-n}^{n}\sum_{i=1}^{6}v^{(i)}_{nm}\mathbf{Z}^{(i)}_{nm}(\theta,\lambda),\tag{1.85}$$

where $v^{(i)}_{nm}$ are the expansion coefficients, which can be obtained using the application of the orthogonality of $\mathbf{Z}^{(i)}_{nm}(\theta,\lambda)$ (Martinec, 2003):

$$\iint_{\sigma}\mathbf{Z}^{(i)}_{nm}(\theta',\lambda'):\left[\mathbf{Z}^{(i')}_{n'm'}(\theta',\lambda')\right]\mathrm{d}\sigma=4\pi\left[N^{(i)}_{n}\right]^2_{\mathbf{Z}}\delta_{nn'}\delta_{mm'}\delta_{ii'},\tag{1.86}$$

where the colon denotes the double-dot product of the tensors, σ is the unit sphere of integration, $\mathrm{d}\sigma$ is the integration elements, δ stands for Kronecker's delta and $\left[N^{(i)}_{n}\right]^2_{\mathbf{Z}}$ is the squared norm of base functions $\mathbf{Z}^{(i)}_{nm}(\theta,\lambda)$:

$$
\left[N_n^{(i)}\right]_{\mathbf{Z}}^2 = \begin{cases} 1 & i=1 \\ \frac{1}{2}n(n+1) & i=2 \\ 2(n-1)n(n+1)(n+2) & i=3 \\ 2n^2(n+1)^2 & i=4 \\ \frac{1}{2}n(n+1) & i=5 \\ 2(n-1)n(n+1)(n+2) & i=6 \end{cases}. \tag{1.87}
$$

According to Zerilli (1970), the tensor spherical harmonics have the following forms:

$$
\mathbf{Z}_{nm}^{(1)}(\theta,\lambda) = Y_{nm}(\theta,\lambda)\mathbf{e}_{rr}, \tag{1.88}
$$

$$
\mathbf{Z}_{nm}^{(2)}(\theta,\lambda) = \frac{\partial Y_{nm}(\theta,\lambda)}{\partial\theta}\mathbf{e}_{zx} + \frac{\partial Y_{nm}(\theta,\lambda)}{\sin\theta\partial\lambda}\mathbf{e}_{zy}, \tag{1.89}
$$

$$
\begin{aligned} \mathbf{Z}_{nm}^{(3)}(\theta,\lambda) = {} & \left(\frac{\partial^2}{\partial\theta^2} - \cot\theta\frac{\partial}{\partial\theta} - \frac{1}{\sin^2\theta}\frac{\partial^2}{\partial\lambda^2}\right) Y_{nm}(\theta,\lambda)(\mathbf{e}_{xx} - \mathbf{e}_{yy}) \\ & + 2\left[2\frac{\partial}{\partial\theta}\left(\frac{1}{\sin\theta}\frac{\partial Y_{nm}(\theta,\lambda)}{\partial\lambda}\right)\right]\mathbf{e}_{xy}, \end{aligned} \tag{1.90}
$$

$$
\mathbf{Z}_{nm}^{(4)}(\theta,\lambda) = -n(n+1)Y_{nm}(\theta,\lambda)(\mathbf{e}_{xx} + \mathbf{e}_{yy}), \tag{1.91}
$$

$$
\mathbf{Z}_{nm}^{(5)}(\theta,\lambda) = -\frac{\partial Y_{nm}(\theta,\lambda)}{\sin\theta\partial\lambda}\mathbf{e}_{zx} + \frac{\partial Y_{nm}(\theta,\lambda)}{\partial\theta}\mathbf{e}_{zy}, \tag{1.92}
$$

$$
\begin{aligned} \mathbf{Z}_{nm}^{(6)}(\theta,\lambda) = {} & \left(\frac{\partial^2}{\partial\theta^2} - \cot\theta\frac{\partial}{\partial\theta} - \frac{1}{\sin^2\theta}\frac{\partial^2}{\partial\lambda^2}\right) Y_{nm}(\theta,\lambda)\mathbf{e}_{xy} \\ & - 2\frac{\partial}{\partial\theta}\left(\frac{1}{\sin\theta}\frac{\partial Y_{nm}(\theta,\lambda)}{\partial\lambda}\right)(\mathbf{e}_{xx} - \mathbf{e}_{yy}), \end{aligned} \tag{1.93}
$$

where $\mathbf{e}_{ij} = (\mathbf{e}_i \otimes \mathbf{e}_j)$, $i,j = (x,y,z)$ are the symmetric spherical dyadics.

By multiplying $\mathbf{Z}_{n'm'}^{(i')}(\theta',\lambda')$ to both sides of Eq. (1.85), taking the integral over the sphere and implying the orthogonality property of the tensor spherical harmonics. (Eq. 1.86), we obtain:

$$
v_{nm}^{(i)} = \frac{1}{4\pi\left[N_n^{(i)}\right]_{\mathbf{Z}}^2}\iint_\sigma \boldsymbol{\tau}_f(R,\theta',\lambda') : \mathbf{Z}_{nm}^{(i)}(\theta',\lambda')d\sigma \quad i = 1,2,\ldots,6. \tag{1.94}
$$

This means that there are six solutions for SHCs of the tensor field.

1.4.2 Spectral solutions

If we apply the gradient operator (Eq. 1.45) twice to Eq. (1.3), the second-order derivatives of the potential will be (Reed, 1973):

$$V_{zz} = \frac{\partial^2}{\partial r^2} V, \tag{1.95}$$

$$V_{xx} = \left(\frac{1}{r}\frac{\partial}{\partial r} + \frac{1}{r^2}\frac{\partial^2}{\partial \theta^2}\right) V = \frac{1}{r} V_r + \frac{1}{r^2} V_{\theta\theta}, \tag{1.96}$$

$$\begin{aligned} V_{yy} &= \left(\frac{1}{r}\frac{\partial}{\partial r} + \frac{1}{r^2 \tan\theta}\frac{\partial}{\partial \theta} + \frac{1}{r^2 \sin^2\theta}\frac{\partial^2}{\partial \lambda^2}\right) V \\ &= \frac{1}{r} V_r + \frac{1}{r^2 \tan\theta} V_\theta + \frac{1}{r^2 \sin^2\theta} V_{\lambda\lambda}, \end{aligned} \tag{1.97}$$

$$V_{xy} = \left(\frac{1}{r^2 \sin\theta}\frac{\partial^2}{\partial\theta\partial\lambda} - \frac{\cos\theta}{r^2 \sin^2\theta}\frac{\partial}{\partial\lambda}\right) V = \frac{1}{r^2 \sin\theta} V_{\theta\lambda} - \frac{\cos\theta}{r^2 \sin^2\theta} V_\lambda, \tag{1.98}$$

$$V_{xz} = \left(\frac{1}{r^2}\frac{\partial}{\partial\theta} - \frac{1}{r}\frac{\partial^2}{\partial r\partial\theta}\right) V = \frac{1}{r^2} V_\theta - \frac{1}{r} V_{r\theta}, \tag{1.99}$$

$$V_{yz} = \left(\frac{1}{r^2 \sin\theta}\frac{\partial}{\partial\lambda} - \frac{1}{r\sin\theta}\frac{\partial^2}{\partial r\partial\lambda}\right) V = \frac{1}{r^2 \sin\theta} V_\lambda - \frac{1}{r \sin\theta} V_{r\lambda}. \tag{1.100}$$

Here, we consider $V = V(r, \theta, \lambda)$ to keep the formulae short.

A second-order tensor can be written in terms of a symmetric and antisymmetric tensor. For the purpose of determining the gravity field of the earth the symmetric part is used. The gravitational tensor can be written in terms of symmetrical dyadics as:

$$\begin{aligned} \boldsymbol{\tau} = &V_{zz}\mathbf{e}_{zz} + 2V_{xz}\mathbf{e}_{xz} + 2V_{yz}\mathbf{e}_{yz} + \frac{1}{2}(V_{xx} - V_{yy})(\mathbf{e}_{xx} - \mathbf{e}_{yy}) + 2V_{xy}\mathbf{e}_{xy} \\ &+ \frac{1}{2}(V_{xx} + V_{yy})(\mathbf{e}_{xx} + \mathbf{e}_{yy}). \end{aligned} \tag{1.101}$$

According to the orthogonality property (Eq. 1.86) and comparing Eq. (1.101) with it, we have:

$$v_{nm}^{(1)} = \frac{r^2}{(n+1)(n+2)}\left(\frac{r}{R}\right)^{n+1}\frac{1}{4\pi}\iint_\sigma V_{zz}(r, \theta', \lambda') Y_{nm}(\theta', \lambda') d\sigma, \tag{1.102}$$

$$v_{nm}^{(2)} = \frac{-r^2}{n(n+1)(n+2)} \left(\frac{r}{R}\right)^{n+1} \frac{1}{4\pi} \iint_{\sigma} \left[V'_{xz} \frac{\partial Y_{nm}(\theta', \lambda')}{\partial \theta'} + V'_{yz} \frac{\partial Y_{nm}(\theta', \lambda')}{\sin\theta' \partial \lambda'} \right] d\sigma, \tag{1.103}$$

where $V'_{xz} = V_{xz}(r', \theta', \lambda')$ and $V'_{yz} = V_{yz}(r', \theta', \lambda')$ to keep the formula shorter, but the prime over the derivatives means that they are at the integration point. Similarly, we can obtain another solution for the SHCs:

$$\begin{aligned} v_{nm}^{(3)} = & \frac{r^2}{(n-1)n(n+1)(n+2)} \left(\frac{r}{R}\right)^{n+1} \\ & \frac{1}{4\pi} \iint_{\sigma} \left[\left(V'_{xx} - V'_{yy} \right) \times \left(\frac{\partial^2}{\partial \theta'^2} - \cot\theta' \frac{\partial}{\partial \theta'} - \frac{1}{\sin^2\theta'} \frac{\partial^2}{\partial \lambda'^2} \right) \right. \\ & \left. Y_{nm}(\theta', \lambda') - 2V'_{xy} \left[2 \frac{\partial}{\partial \theta'} \left(\frac{1}{\sin\theta'} \frac{\partial Y_{nm}(\theta', \lambda')}{\partial \lambda'} \right) \right] \right] d\sigma, \end{aligned} \tag{1.104}$$

where $V'_{xx} = V_{xx}(r, \theta', \lambda')$, $V'_{yy} = V_{yy}(r, \theta', \lambda')$ and $V'_{xy} = V_{xy}(r, \theta', \lambda')$. Three independent solutions for the SHCs of the earth's gravitational potential can be derived. If we look at Eq. (1.103), we will observe that for $n = 0$ it is singular. This means that the zero-degree harmonic of the potential cannot be solved from the derivatives V'_{zx} and V'_{zy}. In a very similar manner we can conclude that the zero- and first-degree harmonics cannot be recovered from V'_{xx}, V'_{yy} and V'_{xy}.

Again, theoretically, these spectral solutions are equivalent, meaning that $v_{nm}^{(1)} = v_{nm}^{(2)} = v_{nm}^{(3)}$. Therefore, we can write:

$$v_{zz,nm} = -\frac{1}{n} v_{\mathrm{VH},nm} = \frac{1}{n(n-1)} v_{\mathrm{VH},nm} \quad \text{for } n > 1, \tag{1.105}$$

with

$$v_{zz,nm} = \frac{1}{4\pi} \iint_{\sigma} V_{zz}(r, \theta', \lambda') Y_{nm}(\theta', \lambda') d\sigma, \tag{1.106}$$

$$v_{\mathrm{VH},nm} = \frac{1}{4\pi} \iint_{\sigma} \left[V'_{xz} \frac{\partial Y_{nm}(\theta', \lambda')}{\partial \theta'} + V'_{yz} \frac{\partial Y_{nm}(\theta', \lambda')}{\sin\theta' \partial \lambda'} \right] d\sigma, \tag{1.107}$$

$$v_{\mathrm{HH,nm}} = \frac{1}{4\pi}\iint\limits_{\sigma}\left[\left(V'_{xx} - V'_{yy}\right)\left(\frac{\partial^2}{\partial\theta'^2} - \cot\theta'\frac{\partial}{\partial\theta'} - \frac{1}{\sin^2\theta'}\frac{\partial^2}{\partial\lambda'^2}\right)Y_{nm}(\theta',\lambda')\right.$$
$$\left. - 2V'_{xy}\left[2\frac{\partial}{\partial\theta'}\left(\frac{1}{\sin\theta'}\frac{\partial Y_{nm}(\theta',\lambda')}{\partial\lambda'}\right)\right]\right]d\sigma. \tag{1.108}$$

1.4.3 Spatial solutions

The following integral formula can be derived based on the same principle, presented in the former sections, whereby $V_{rr}(r,\theta,\lambda)$ is given outside the sphere and recovery of the potential at the sphere $V(R,\theta,\lambda)$ is going to be solved:

$$V(R,\theta,\lambda) = \frac{r^2}{4\pi}\iint\limits_{\sigma} V_{rr}(r,\theta',\lambda')G^*_{rr}(r,\psi)d\sigma, \tag{1.109}$$

where

$$G^*_{rr}(r,\psi) = \sum_{n=0}^{\infty}\frac{2n+1}{(n+1)(n+2)}\left(\frac{r}{R}\right)^{n+1}P_n(\cos\psi) \tag{1.110}$$

is the kernel of the integral.

Similarly, by inserting Eq. (1.103) into Eq. (1.3) and taking advantage of Eqs. (1.61) and (1.62) we obtain:

$$V(R,\theta,\lambda) = \frac{r^2}{4\pi}\iint\limits_{\sigma}\sum_{n=1}^{\infty}\frac{2n+1}{n(n+1)(n+2)}\left(\frac{r'}{R}\right)^{n+1}$$
$$\left[V'_{xz}\frac{\partial P_n(\cos\psi)}{\partial\theta} + V'_{yz}\frac{\partial P_n(\cos\psi)}{\sin\theta\partial\lambda}\right]d\sigma. \tag{1.111}$$

According to Eqs. (1.67, 1.68), Eq. (1.111) will change to:

$$V(R,\theta,\lambda) = \frac{r^2}{4\pi}\iint\limits_{\sigma}\left[V_{xz}(r,\theta',\lambda')\cos\alpha' - V_{yz}(r,\theta',\lambda')\sin\alpha'\right]G^*_{\mathrm{VH}}(r,\psi)d\sigma, \tag{1.112}$$

where the divergent kernel is:

$$G^*_{\mathrm{VH}}(r,\psi) = \sum_{n=1}^{\infty}\frac{2n+1}{n(n+1)(n+2)}\left(\frac{r}{R}\right)^{n+1}P_{n1}(\cos\psi). \tag{1.113}$$

Finally, by inserting Eq. (1.104) into Eq. (1.3) and performing similar manipulations we obtain:

$$
\begin{aligned}
V(R,\theta,\lambda) = \frac{r^2}{4\pi}\iint_\sigma \Bigg[& \sum_{n=2}^{\infty} \frac{1}{(n-1)n(n+1)(n+2)} \left(\frac{r}{R}\right)^{n+1} \\
& \Bigg(V_{xx}(r,\theta',\lambda') - V_{yy}(r,\theta',\lambda') - \\
& \times \sum_{m=-n}^{n} \left(\frac{\partial^2}{\partial \theta'^2} - \cot\theta' \frac{\partial}{\partial \theta'} - \frac{1}{\sin^2\theta'} \frac{\partial^2}{\partial \lambda'^2} \right) Y_{nm}(\theta',\lambda') Y_{nm}(\theta,\lambda) \\
& - 2V_{xy}(r,\theta',\lambda') \sum_{m=-n}^{n} \left[2\frac{\partial}{\partial \theta'} \left(\frac{1}{\sin\theta'} \frac{\partial}{\partial \lambda'} \right) \right] Y_{nm}(\theta',\lambda') Y_{nm}(\theta,\lambda) \Bigg) \Bigg] \mathrm{d}\sigma.
\end{aligned}
\tag{1.114}
$$

Martinec (2003) presented the following relations for the addition theorem of the spherical harmonics (see also Eq. 1.66):

$$
\begin{aligned}
& \sum_{m=-n}^{n} \left(\frac{\partial^2}{\partial \theta'^2} - \cot\theta' \frac{\partial}{\partial \theta'} - \frac{1}{\sin^2\theta'} \frac{\partial^2}{\partial \lambda'^2} \right) Y_{nm}(\theta',\lambda') Y_{nm}(\theta,\lambda) \\
& \quad = (2n+1)\cos 2\alpha' \sin^2\psi \frac{\mathrm{d}^2 P_n(\cos\psi)}{\mathrm{d}(\cos\psi)^2},
\end{aligned}
\tag{1.115}
$$

$$
\begin{aligned}
& \sum_{m=-n}^{n} \left[2\frac{\partial}{\partial \theta'} \left(\frac{1}{\sin\theta'} \frac{\partial Y_{nm}(\theta',\lambda')}{\partial \lambda'} \right) \right] Y_{nm}(\theta,\lambda) = \\
& \quad - (2n+1)\sin 2\alpha' \sin^2\psi \frac{\mathrm{d}^2 P_n(\cos\psi)}{\mathrm{d}(\cos\psi)^2}.
\end{aligned}
\tag{1.116}
$$

Considering Eq. (1.114), Eq. (1.116) will be simplified to:

$$
V(R,\theta,\lambda) = \frac{r^2}{4\pi}\iint_\sigma \left[\left(V'_{xx} - V'_{yy} \right) \cos 2\alpha' - V'_{xy} \sin 2\alpha' \right] G^*_{\mathrm{HH}}(r,\psi)\mathrm{d}\sigma,
\tag{1.117}
$$

where

$$
G^*_{HH}(r,\psi) = \sum_{n=2}^{\infty} \frac{2n+1}{(n-1)n(n+1)(n+2)} \left(\frac{r}{R}\right)^{n+3} P_{n2}(\cos\psi).
\tag{1.118}
$$

None of the kernels presented in Eqs. (1.111), (1.113) and (1.118) is convergent. Therefore, finding closed-form formulae for them is not possible.

Martinec (2003) considered $r = R$ in Eqs. (1.111), (1.113) and (1.118) and inserted them back into Eq. (1.3) to obtain the potential outside the sphere from the boundary values, which are the gradients, at the spherical boundary. This type of problem is called a gradiometric boundary-value problem. The summary of his formulae is:

$$V(r,\theta,\lambda) = \frac{r^2}{4\pi} \iint_\sigma V_{rr}(R,\theta',\lambda') G_{rr}(r,\psi) d\sigma, \tag{1.119}$$

$$V(r,\theta,\lambda) = \frac{r^2}{4\pi} \iint_\sigma [V_{xz}(R,\theta',\lambda')\cos\alpha' - V_{yz}(R,\theta',\lambda')\sin\alpha'] G_{\mathrm{VH}}(r,\psi) d\sigma, \tag{1.120}$$

$$\begin{aligned} V(r,\theta,\lambda) = \frac{r^2}{4\pi} \iint_\sigma \Big[\Big(V'_{xx}(R,\theta,\lambda) - V'_{yy}(R,\theta,\lambda) \Big) \cos 2\alpha' \\ - V'_{xy}(R,\theta,\lambda) \sin 2\alpha' \Big] G_{\mathrm{HH}}(r,\psi) d\sigma, \end{aligned} \tag{1.121}$$

with the kernels

$$G_{rr}(r,\psi) = \sum_{n=0}^{\infty} \frac{2n+1}{(n+1)(n+2)} \left(\frac{R}{r}\right)^{n+1} P_n(\cos\psi), \tag{1.122}$$

$$G_{\mathrm{VH}}(r,\psi) = -\sum_{n=1}^{\infty} \frac{2n+1}{n(n+1)(n+2)} \left(\frac{R}{r}\right)^{n+1} P_{n1}(\cos\psi). \tag{1.123}$$

$$G_{HH}(r,\psi) = \sum_{n=2}^{\infty} \frac{2n+1}{(n-1)n(n+1)(n+2)} \left(\frac{R}{r}\right)^{n+1} P_{n2}(\cos\psi), \tag{1.124}$$

having the following closed-form expressions:

$$G_{rr}(r,\psi) = \frac{3}{t}(D-1) + \left(\frac{3\tilde{x}}{t} - 1\right) \ln\left(\frac{D+t-\tilde{x}}{1-\tilde{x}}\right), \tag{1.125}$$

$$G_{VH}(r,\psi)=\left[\frac{3}{2D}+\frac{t^2(D+1)}{2D(D+1-t\widetilde{x})}+\left(1-\frac{3\widetilde{x}}{2t}\right)\right.$$
$$\left.\times\left(\frac{1}{1-\widetilde{x}}-\frac{D+t}{D(D+t-\widetilde{x})}\right)-\frac{3}{2t}\ln\left(\frac{D+t-\widetilde{x}}{1-\widetilde{x}}\right)\right]\sqrt{1-\widetilde{x}^2}, \quad (1.126)$$

$$G_{HH}(r,\psi)=-\frac{t}{2}+\frac{3}{2}\widetilde{x}t^2+Dt+\frac{1}{t}(1-D)+\frac{\widetilde{x}^2t^3}{D+1-t\widetilde{x}}+\frac{\widetilde{x}(\widetilde{x}-t)}{t(1-\widetilde{x})}$$
$$-\frac{\widetilde{x}^2}{t(D+t-\widetilde{x})}. \quad (1.127)$$

Fig. 1.7A–C, respectively, shows the behaviour of the kernels $G_{rr}(r,\psi)$, $G_{VH}(r,\psi)$ and $G_{HH}(r,\psi)$. As observed, none of these kernels approaches the zero value even up to $\psi=20^\circ$. This means that the contributions of the far-zone data are very significant. Martinec (2003) showed their

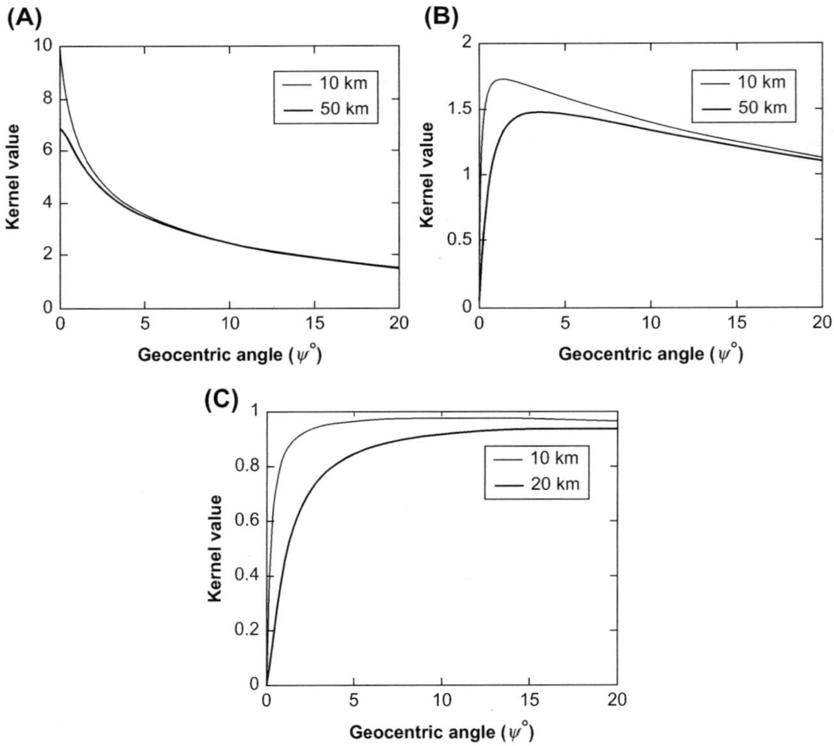

Figure 1.7 Behaviour of the kernels (A) $G_{rr}(r,\psi)$, (B) $G_{VH}(r,\psi)$ and (C) $G_{HH}(r,\psi)$.

significance even up to $\psi = 180^\circ$. Therefore, solving a potential outside the reference sphere from the second-order derivatives of the potential at the spherical boundary is successful if the values are given with global coverages.

1.4.4 Integral equations

By applying the operators presented in Eqs. (1.95)—(1.100) to the Poisson integral (Eq. 1.17), integral equations for connecting the derivatives of the potential can be constructed having the following general form (see e.g. Reed, 1973):

$$V_i(r,\theta,\lambda) = \frac{1}{4\pi}\iint_\sigma V(R,\theta',\lambda')P_i(r,\psi)d\sigma, \quad i = zz, xx, yy, xz, yz, xy \tag{1.128}$$

where the kernels are:

$$P_{zz}(r,\psi) = -\frac{t(-15t^4\tilde{x}^2 + 5t^4 + 28t^3\tilde{x} - t^2\tilde{x}^2 - 23t^2 + 4t^3\tilde{x} + 2)}{r^2D^7}, \tag{1.129}$$

$$\begin{bmatrix} P_{xx}(r,\psi) \\ P_{yy}(r,\psi) \end{bmatrix} = \frac{P_r(r,\psi)}{r} + \frac{1}{2r^2}\{[P_{\psi\psi}(r,\psi) + \cot\psi P_\psi(r,\psi)] \pm [P_{\psi\psi}(r,\psi) - \cot\psi P_\psi(r,\psi)]\cos 2\alpha\}, \tag{1.130}$$

$$P_{xy}(r,\psi) = -\frac{1}{2r^2}[P_{\psi\psi}(r,\psi) + \cot\psi P_\psi(r,\psi)]\sin 2\alpha, \tag{1.131}$$

$$\begin{bmatrix} P_{xz}(r,\psi) \\ P_{yz}(r,\psi) \end{bmatrix} = \frac{1}{r}\left[\frac{1}{r}P_\psi(r,\psi) - P_{r\psi}(r,\psi)\right]\begin{pmatrix} \cos\alpha \\ -\sin\alpha \end{pmatrix}, \tag{1.132}$$

where

$$P_{r\psi}(r,\psi) = -\frac{t^2}{r}\left[\frac{6(2t^2-1)}{D^5} + \frac{15(t-t^3)(t-\tilde{x})}{D^7}\right]\sqrt{1-\tilde{x}^2}, \tag{1.133}$$

$$P_{\psi\psi}(r,\psi) = 3t^2(t^2-1)\left[\frac{5t(1-\tilde{x}^2)}{D^7} - \frac{\tilde{x}}{D^5}\right]. \tag{1.134}$$

As seen, all kernel functions are functions of the azimuth between the computation point and the integration points except $P_{zz}(r,\psi)$. Fig. 1.8

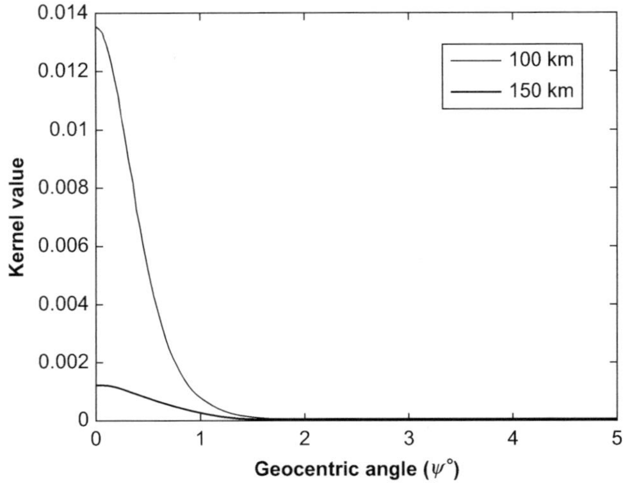

Figure 1.8 Behaviour of $P_{zz}(r,\psi)$.

represents the behaviour of this kernel for the computation point, which is located at 100 and 150 km above the reference sphere to $\psi = 5^{\circ}$. The kernel has large values at the computation point, where $\psi = 0^{\circ}$ and decrease fast to zero.

However, the important issue in integration is the contribution of the far-zone data, which can be seen on the plot of the isotropic parts of the kernel, or those parts which are not azimuth dependent. The kernel functions presented in Eq. (1.130) contain two isotropic parts:

$$P_{\psi\psi}(r,\psi) + \cot\psi P_{\psi}(r,\psi) = 3t^2(t^2-1)\frac{5t(1-\tilde{x}^2)}{D^7}, \tag{1.135}$$

$$P_{\psi\psi}(r,\psi) - \cot\psi P_{\psi}(r,\psi) = 3t^2(t^2-1)\left[\frac{5t(1-\tilde{x}^2)}{D^7} - \frac{2\tilde{x}}{D^5}\right]. \tag{1.136}$$

Fig. 1.9A and B shows, respectively, the behaviour of $P_{\psi\psi}(r,\psi) + \cot\psi P_{\psi}(r,\psi)$ and $P_{\psi\psi}(r,\psi) - \cot\psi P_{\psi}(r,\psi)$, the kernels (Eqs. 1.135 and 1.136), up to $\psi = 10^{\circ}$. The former represents that the kernel has a zero value at the computation point and increases towards its maximum value, which is seen around a geocentric angle of about 1.5° and thereafter decreases down to zero. This kernel has a bell form and Eshagh (2011) called such kernels bell-shaped kernels. The latter is the plot of the kernel

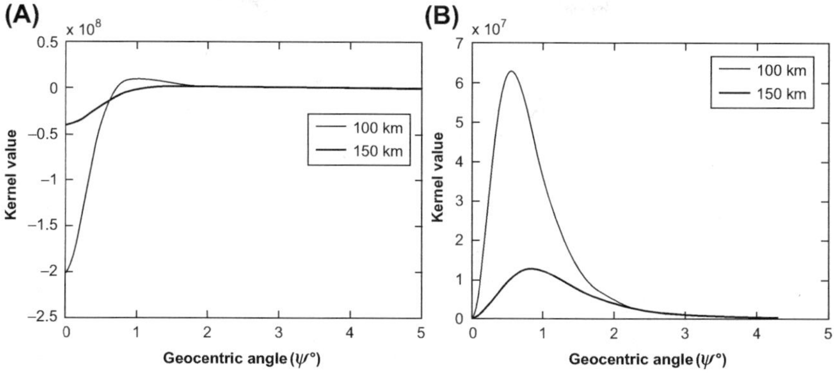

Figure 1.9 (A) $P_{\psi\psi}(r,\psi) + \cot\psi P_{\psi}(r,\psi)$ and (B) $P_{\psi\psi}(r,\psi) - \cot\psi P_{\psi}(r,\psi)$.

Eq. (1.136) and, as we observed, the kernel has its largest minimum at the computation point and increases up zero.

The kernel Eq. (1.132) is also azimuth dependent, but its isotropic part has the following formula:

$$\frac{1}{r}P_{\psi}(r,\psi) - P_{r\psi}(r,\psi) = \left\{\frac{15t^3 - 9t}{D^5} + t\left[\frac{15(t-t^3)(t-\tilde{x})}{D^7}\right]\right\}\frac{t\sqrt{1-\tilde{x}^2}}{r}, \tag{1.137}$$

with $P_{\psi}(r,\psi)$ presented in Eq. (1.82). Eq. (1.136) can also be derived from Eqs. (1.137) and (1.82).

Fig. 1.10 represents the behaviour of this kernel for heights of 100 and 150 km above the surface of the sphere. The kernel is bell shaped and approaches zero rather fast.

1.4.5 Product of a spherical harmonic series

Consider a point at the surface of the sphere for two quantities with the following spherical harmonic expansions:

$$V^{(1)}(R,\theta,\lambda) = \sum_{i=0}^{\infty}\sum_{j=-i}^{i} v_{ij}^{(1)} Y_{ij}(\theta,\lambda), \tag{1.138}$$

$$V^{(2)}(R,\theta,\lambda) = \sum_{k=0}^{\infty}\sum_{l=-k}^{k} v_{kl}^{(2)} Y_{kl}(\theta,\lambda). \tag{1.139}$$

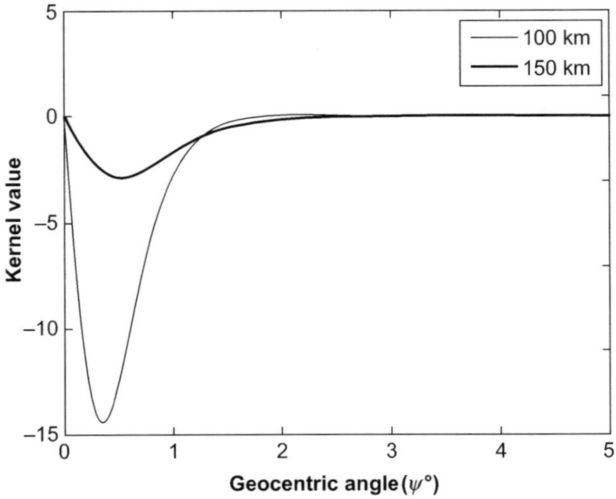

Figure 1.10 $\frac{1}{r}P_{\psi}(r,\psi) - P_{r\psi}(r,\psi)$

The product of these two expansions will be:

$$V^{(1)}(R,\theta,\lambda)V^{(2)}(R,\theta,\lambda) = \sum_{i=0}^{\infty}\sum_{j=-i}^{i}\sum_{k=0}^{\infty}\sum_{l=-k}^{k} v_{ij}^{(1)}v_{kl}^{(2)}Y_{ij}(\theta,\lambda)Y_{kl}(\theta,\lambda)$$
$$= \sum_{n'=0}^{\infty}\sum_{m'=-n'}^{n'} v_{n'm'}^{(1,2)}Y_{n'm'}(\theta,\lambda). \tag{1.140}$$

As seen, the product of two spherical harmonic expansions can be written as a new spherical harmonic series with $v_{n'm'}^{(1,2)}$ as the SHCs, which can be derived from the SHCs of both series by the Clebsch–Gordan series:

$$v_{n'm'}^{(1,2)} = \sum_{i=0}^{\infty}\sum_{j=-i}^{i}\sum_{k=0}^{\infty}\sum_{l=-k}^{k} v_{ij}^{(1)}v_{kl}^{(2)}Q_{ijkl}^{n'm'}, \tag{1.141}$$

where $Q_{ijkl}^{n'm'}$ are the Gaunt coefficients, which appear in the product of two associated Legendre's functions (Xu, 1996):

$$P_{ij}(\cos\theta)P_{kl}(\cos\theta) = \sum_{n'=|i-k|}^{i+k} Q_{ijkl}^{n'm'}P_{n'm'}(\cos\theta). \tag{1.142}$$

or, based on the spherical harmonics (Pail et al., 2001):

$$Y_{ij}(\theta,\lambda)Y_{kl}(\theta,\lambda)=\sum_{n'=|i-k|}^{i+k}\sum_{m'=-n'}^{n'}Q_{ijkl}^{n'm'}Y_{n'm'}(\theta,\lambda). \tag{1.143}$$

There is a challenge among geoscientists to find a relevant algorithm to speed up the Gaunt coefficient generation; see, e.g., Xu (1996) and Sebilleau (1998). These coefficients have the following relation to the Clebsch–Gordan coefficients (Xu, 1996):

$$Q_{ijkl}^{n'm'}=\sqrt{\frac{(i+j)!(k+l)!(n-m)!}{(i-j)!(k-l)!(n+m)!}}v_{ijkl}^{n'm'}v_{i0k0}^{n'0}, \tag{1.144}$$

and the Clebsch–Gordan coefficients have the following relation to the Wigner 3*j* coefficients (Varshalovich et al., 1989) (shown by parentheses):

$$v_{ijkl}^{n'm'}=(-1)^{i-k+m'}\sqrt{2m'+1}\begin{pmatrix} i & k & n' \\ j & l & -m' \end{pmatrix}. \tag{1.145}$$

If we require the product of two fully-normalised associated Legendre functions, it is enough to replace the non-normalised functions with the fully-normalised ones and consider their normalisation factor. In this case, the Gaunt coefficients will have the following relation to the Clebsch–Gordan coefficients (e.g., Eshagh, 2009):

$$Q_{ijkl}^{n'm'}=\sqrt{\frac{(2i+1)(2-\delta_{j0})(2k+1)(2-\delta_{l0})}{(2n'+1)(2-\delta_{m'0})}}v_{ijkl}^{n'm'}v_{i0k0}^{n'0}. \tag{1.146}$$

1.5 Spectra of the potential field

Spectra show the average power of the potential all over the sphere per degree and it can be derived from the spherical harmonic expansions of the potential or its derivatives. Spectra are good measures to study the downward and upward continuation properties of different gravimetric quantities. Here, we present how to derive such spectra in both global and local perspectives.

1.5.1 Global spectra of the potential

Generally, the Laplace coefficients of the potential can be written in terms of spherical harmonics Eq. (1.3):

$$\left(\frac{R}{r}\right)^{n+1} v_n(\theta,\lambda) = \left(\frac{R}{r}\right)^{n+1} \sum_{m=-n}^{n} v_{nm} Y_{nm}(\theta,\lambda). \tag{1.147}$$

The signal spectra are defined as the mean value of the square of each Laplace coefficient all over the sphere, i.e.:

$$\overline{v}_n(r) = \frac{1}{4\pi} \iint_{\sigma} \left(\frac{R}{r}\right)^{n+1} v_n(\theta',\lambda') \left(\frac{R}{r}\right)^{n'+1} v_{n'}(\theta',\lambda') d\sigma. \tag{1.148}$$

To find a simpler mathematical model Eq. (1.147) is inserted into Eq. (1.148), and we obtain:

$$\overline{v}_n(r) = \frac{1}{4\pi} \iint_{\sigma} \left(\frac{R}{r}\right)^{n+1} \sum_{m=-n}^{n} v_{nm} Y_{nm}(\theta',\lambda') \left(\frac{R}{r}\right)^{n'+1} \sum_{m'=-n'}^{n'} v_{n'm'} Y_{n'm'}(\theta',\lambda') d\sigma. \tag{1.149}$$

By interchanging the summations and the integral in Eq. (1.149) we arrive at:

$$\overline{v}_n(r) = \frac{1}{4\pi} \left(\frac{R}{r}\right)^{n+1} \left(\frac{R}{r}\right)^{n'+1} \sum_{m=-n}^{n} v_{nm} \sum_{m'=-n'}^{n'} v_{n'm'} \iint_{\sigma} Y_{nm}(\theta',\lambda') Y_{n'm'}(\theta',\lambda') d\sigma. \tag{1.150}$$

Since the spherical harmonics are orthogonal, based on Eq. (1.6), Eq. (1.150) will be simplified to:

$$\overline{v}_n(r) = \left(\frac{R}{r}\right)^{2n+2} \sum_{m=-n}^{n} v_{nm}^2, \tag{1.151}$$

which is the signal spectra of the potential at a sphere with the radius r.

1.5.1.1 Global first- and second-order radial derivatives of the potential

The Laplace coefficients of the first- and second-order radial derivatives of the potential are, respectively:

$$\left(\frac{R}{r}\right)^{n+1} v_{r,n}(\theta,\lambda) = -\frac{n+1}{r}\left(\frac{R}{r}\right)^{n+1}\sum_{m=-n}^{n} v_{nm} Y_{nm}(\theta,\lambda), \tag{1.152}$$

$$\left(\frac{R}{r}\right)^{n+1} v_{rr,n}(\theta,\lambda) = \frac{(n+1)(n+2)}{r^2}\left(\frac{R}{r}\right)^{n+1}\sum_{m=-n}^{n} v_{nm} Y_{nm}(\theta,\lambda). \tag{1.153}$$

Inserting each of them into Eq. (1.148), and after performing simplifications, we obtain:

$$\overline{v}_{r,n}(r) = \left(\frac{n+1}{r}\right)^2\left(\frac{R}{r}\right)^{2n+2}\sum_{m=-n}^{n} v_{nm}^2, \tag{1.154}$$

$$\overline{v}_{rr,n}(r) = \left(\frac{(n+1)(n+2)}{r^2}\right)^2\left(\frac{R}{r}\right)^{2n+2}\sum_{m=-n}^{n} v_{nm}^2. \tag{1.155}$$

If we compare these spectra with the one derived from the potential, Eq. (1.151), we see that there exist some degree-dependent coefficients behind the upward continuations factor $(R/r)^{2n+2}$ for the spectra of derivatives of the potential. This means that these coefficients, which are increasing by degree, reduce the power of this factor and do not allow the factor to smooth the signal. These degree-dependent coefficients of the spectra of the second-order radial derivative are stronger and decrease the power of the upward continuation factor more than the first derivative does.

1.5.1.2 Global horizontal derivatives and vertical–horizontal derivatives of the potential

Determining the spectra from the horizontal derivatives and the vertical–horizontal derivatives is slightly more complicated than the radial derivatives, as the vector and tensor spherical harmonics and their orthogonality properties should be used, which is not as simple as using the scalar spherical harmonics. However, the principle is the same, and the Laplace coefficients should be derived and inserted into Eq. (1.148). From Eqs. (1.47) and (1.84), it will not be difficult to show that these coefficients have the following relations to the vector and tensor spherical harmonics:

$$\left(\frac{R}{r}\right)^{n+1}\mathbf{v}_{\mathrm{H},n}(\theta,\lambda)=\frac{1}{r}\left(\frac{R}{r}\right)^{n+1}\sum_{m=-n}^{n}v_{nm}\mathbf{X}_{nm}^{(2)}(\theta,\lambda), \tag{1.156}$$

$$\left(\frac{R}{r}\right)^{n+1}\mathbf{v}_{\mathrm{VH},n}(\theta,\lambda)=-\frac{n+2}{r^2}\left(\frac{R}{r}\right)^{n+1}\sum_{m=-n}^{n}v_{nm}\mathbf{Z}_{nm}^{(2)}(\theta,\lambda). \tag{1.157}$$

Let us start with Eq. (1.156) and insert into Eq. (1.148). The result will be:

$$\begin{aligned}\overline{v}_{\mathrm{H},n}(r)&=\frac{1}{4\pi}\frac{1}{r^2}\left(\frac{R}{r}\right)^{n+1}\left(\frac{R}{r}\right)^{n'+1}\iint_{\sigma}\sum_{m=-n}^{n}v_{nm}\mathbf{X}_{nm}^{(2)}(\theta,\lambda)\cdot\sum_{m'=-n'}^{n'}v_{n'm'}\mathbf{X}_{n'm'}^{(2)}(\theta,\lambda)\mathrm{d}\sigma\\&=\frac{1}{r^2}\left(\frac{R}{r}\right)^{n+1}\left(\frac{R}{r}\right)^{n'+1}\sum_{m=-n}^{n}v_{nm}\sum_{m'=-n'}^{n'}v_{n'm'}\frac{1}{4\pi}\iint_{\sigma}\mathbf{X}_{nm}^{(2)}(\theta,\lambda).\mathbf{X}_{n'm'}^{(2)}(\theta,\lambda)\mathrm{d}\sigma.\end{aligned} \tag{1.158}$$

Now, the orthogonality of the vector spherical harmonics (Eq. 1.51) will be applied, and after simplification, the spectra will be:

$$\overline{v}_{\mathrm{H},n}(r)=\frac{n(n+1)}{r^2}\left(\frac{R}{r}\right)^{2n+2}\sum_{m=-n}^{n}v_{nm}^2. \tag{1.159}$$

Similarly, we can obtain the following spectra from the vertical—horizontal derivatives, according to the orthogonality of the tensor spherical harmonics Eq. (1.86):

$$\overline{v}_{\mathrm{VH},n}(r)=\frac{n(n+1)(n+2)^2}{r^4}\left(\frac{R}{r}\right)^{2n+2}\sum_{m=-n}^{n}v_{nm}^2. \tag{1.160}$$

As we observe, the degree-dependent coefficients behind the upward continuation factor reduce the power of the factor. The coefficient for the spectra obtained from vertical—horizontal derivatives is stronger than the one from only the horizontal derivatives.

1.5.1.3 Global spectra of horizontal–horizontal derivatives of the potential

Here, the Laplace coefficients of the horizontal–horizontal derivative is derived from Eq. (1.84):

$$\left(\frac{R}{r}\right)^{n+1}\tau_{\mathrm{HH},n}(\theta,\lambda)=\frac{1}{2r^2}\left(\frac{R}{r}\right)^{n+1}\sum_{m=-n}^{n}v_{nm}\mathbf{Z}^{(3)}_{nm}(\theta,\lambda). \tag{1.161}$$

By inserting Eq. (1.161) into Eq. (1.148), we arrive at:

$$\begin{aligned}\bar{v}_{\mathrm{HH},n}(r)&=\frac{1}{4\pi}\frac{1}{4r^4}\iint_{\sigma}\sum_{m=-n}^{n}v_{nm}\mathbf{Z}^{(3)}_{nm}(\theta',\lambda'):\sum_{m'=-n'}^{n'}v_{n'm'}\mathbf{Z}^{(3)}_{n'm'}(\theta',\lambda')\mathrm{d}\sigma\\&=\sum_{m=-n}^{n}v_{nm}\sum_{m'=-n'}^{n'}v_{n'm'}\frac{1}{4\pi}\iint_{\sigma}\mathbf{Z}^{(3)}_{nm}(\theta',\lambda'):\mathbf{Z}^{(3)}_{n'm'}(\theta',\lambda')\mathrm{d}\sigma,\end{aligned} \tag{1.162}$$

and finally by applying the orthogonality of the tensor spherical harmonics (Eq. 1.86), the spectra will be:

$$\bar{v}_{\mathrm{HH},n}(r)=\frac{(n-1)n(n+1)(n+2)}{r^4}\left(\frac{R}{r}\right)^{2n+2}\sum_{m=-n}^{n}v^2_{nm}. \tag{1.163}$$

1.5.2 Local spectra of the potential

In the case in which the potential or its derivatives are available over limited areas, the spectra can be determined locally. Here, we present two approaches for this purpose, one using the products of the spherical harmonic and one presented by Sjöberg (1982).

1.5.2.1 Local spectra of the potential based on the product of spherical harmonics and the Gaunt coefficient

The limited area does not allow us to use the orthogonality property of the spherical harmonics; in this case, Eq. (1.150) will change to:

$$\bar{v}_n(r)=\frac{1}{4\pi}\left(\frac{R}{r}\right)^{n+1}\left(\frac{R}{r}\right)^{n'+1}\sum_{m=-n}^{n}v_{nm}\sum_{m'=-n'}^{n'}v_{n'm'}\iint_{\sigma_0}Y_{nm}(\theta',\lambda')Y_{n'm'}(\theta',\lambda')\mathrm{d}\sigma, \tag{1.164}$$

where σ_0 is a part of the surface of sphere σ.

According to Eq. (1.143), Eq. (1.164) will change to:

$$\bar{v}_n(r) = \frac{1}{4\pi}\left(\frac{R}{r}\right)^{n+1}\left(\frac{R}{r}\right)^{n'+1}\sum_{m=-n}^{n} v_{nm}\sum_{m'=-n'}^{n'} v_{n'm'}\sum_{n''=|n-n'|}^{n+n'} \times \sum_{m''=-n''}^{n''} Q_{nmn'm'}^{n''m''}\iint_{\sigma_0} Y_{n''m''}(\theta',\lambda')\mathrm{d}\sigma. \quad (1.165)$$

The main issue is that the limited area σ_0 may not have a regular and rectangular form. In this case, the area should be divided into small rectangular cells and the integral should be discretised. Let us present the discretised form of this integral by:

$$\iint_{\sigma_0} Y_{n''m''}(\theta',\lambda')\mathrm{d}\sigma = \sum_{i=1}^{P}\iint_{\mathrm{Cell}_i} Y_{n''m''}(\theta',\lambda')\mathrm{dCell}_i. \quad (1.166)$$

Here P is the total number of the cells over the area.

Therefore, for a rectangular cell limited between the latitudes θ_1 and θ_2 and the longitudes λ_1 and λ_2, we can write:

$$\iint_{\mathrm{Cell}_i} Y_{n''m''}(\theta',\lambda')\mathrm{dCell}_i = \iint_{\mathrm{Cell}_i}\left\{\begin{matrix}\cos m''\lambda' \\ \sin m''\lambda'\end{matrix}\right\} P_{n''m''}(\cos\theta')\,\mathrm{dCell}_i = \int_{\lambda_1}^{\lambda_2}\left\{\begin{matrix}\cos m''\lambda' \\ \sin m''\lambda'\end{matrix}\right\}\mathrm{d}\lambda'\int_{\theta_1}^{\theta_2} P_{n''m''}(\cos\theta')\sin\theta'\mathrm{d}\theta'. \quad (1.167)$$

Further simplification leads to:

$$\iint_{\mathrm{Cell}_i} Y_{n''m''}(\theta',\lambda')\mathrm{dCell}_i = \int_{\theta_1}^{\theta_2} P_{n''m''}(\cos\theta')\sin\theta'\mathrm{d}\theta' \; \frac{1}{m''}\left\{\begin{matrix}\sin m''\lambda_2 - \sin m''\lambda_1 & m'' \geq 0 \\ \cos m''\lambda_1 - \cos m''\lambda_2 & m'' < 0\end{matrix}\right. . \quad (1.168)$$

Note that for $m'' = 0$, Eq. (1.168) is singular, meaning that the zonal SHCs cannot be resolved.

The solution of the integral of the associated Legendre function is rather well known, and different iterative formulae have been developed for its computations; see e.g., Paul (1978), Mainville (1986) or Hwang (1995). The recursion formulae for the integral of fully-normalised associated Legendre functions, presented by Paul (1978), are:

$$I_{n-1,m-1}(\tilde{x}_1,\tilde{x}_2) = -\frac{1}{n}a_{nm}\tilde{y}^2(P_{n-2,m-1}(\tilde{x}_2) - P_{n-2,m-1}(\tilde{x}_1)) + \frac{n-3}{n}b_{nm}I_{n-3,m-1}(\tilde{x}_1,\tilde{x}_2) \quad \text{for} \quad m \neq n, \tag{1.169}$$

$$I_{n-1,n-1}(\tilde{x}_1,\tilde{x}_2) = \frac{1}{2n}a_n\tilde{y}^2(P_{n-2,n-3}(\tilde{x}_2) - P_{n-2,n-3}(\tilde{x}_1)) + \frac{1}{2n}b_nI_{n-3,n-3}(\tilde{x}_1,\tilde{x}_2) \quad \text{for} \quad m = n, \tag{1.170}$$

where I_{ij} stands for the integral of the associated Legendre functions with degree i and order j and argument $\tilde{x} = \cos\theta$.

The recursion formulae for generating the fully-normalised associated Legendre functions are:

$$P_{n-1,m-1}(\tilde{x}) = a_{nm}\tilde{x}P_{n-2,m-1}(\tilde{x}) - b_{nm}P_{n-3,m-1}(\tilde{x}) \quad \text{for} \quad m \neq n, \tag{1.171}$$

$$P_{n-1,n-1}(\tilde{x}) = \sqrt{\frac{2n-1}{2n-2}}\tilde{y}P_{n-2,n-2}(\tilde{x}) \quad \text{for} \quad m = n, \tag{1.172}$$

where $\tilde{x} = \cos\theta, \tilde{x}_i = \cos\theta_i$ for $i = 1$ or 2, and $\tilde{y} = \sqrt{1-\tilde{x}^2}, \tilde{y}_i = \sqrt{1-\tilde{x}_i^2}$:

$$a_{nm} = \sqrt{\frac{(2n-1)(2n-3)}{(n-m)(n+m-2)}} \quad \text{and} \quad b_{nm} = \sqrt{\frac{(2n-1)(n+m-3)(n-m-1)}{(2n-5)(n+m-2)(n-m)}}, \tag{1.173}$$

$$a_n = \sqrt{\frac{2n-1}{(n-1)(n-2)}} \quad \text{and} \quad b_n = \sqrt{\frac{(n-1)(2n-1)(2n-3)}{n-2}}. \tag{1.174}$$

The following initial values are required for performing the recursions:

$$P_{00}(\tilde{x}) = 1, \qquad P_{10}(\tilde{x}) = \sqrt{3}\tilde{x}, \qquad P_{11}(\tilde{x}) = \sqrt{3}\tilde{y},$$

$$P_{20}(\tilde{x}) = \frac{\sqrt{5}}{2}(3\tilde{x}^2 - 1), \quad P_{21}(\tilde{x}) = \sqrt{15}\tilde{x}\tilde{y}, \quad P_{22}(\tilde{x}) = \frac{\sqrt{15}}{2}\tilde{y}^2,$$

$$I_{00}(\tilde{x}_1,\tilde{x}_2) = \tilde{x}_2 - \tilde{x}_1, \qquad I_{10}(\tilde{x}_1,\tilde{x}_2) = \frac{\sqrt{3}}{2}(\tilde{x}_2^2 - \tilde{x}_1^2),$$

$$I_{11}(\widetilde{x}_1,\widetilde{x}_2)=\frac{\sqrt{3}}{2}\left(\widetilde{x}_2\widetilde{y}_2+\sin^{-1}\widetilde{x}_2-\widetilde{x}_1\widetilde{y}_1+\sin^{-1}\widetilde{x}_1\right),$$

$$I_{20}(\widetilde{x}_1,\widetilde{x}_2)=\frac{\sqrt{5}}{2}\left(\widetilde{x}_1\widetilde{y}_1^2-\widetilde{x}_2\widetilde{y}_2^2\right),$$

$$I_{21}(\widetilde{x}_1,\widetilde{x}_2)=\sqrt{\frac{5}{3}}\left(\widetilde{y}_1^3-\widetilde{y}_2^3\right),$$

$$I_{22}(\widetilde{x}_1,\widetilde{x}_2)=\sqrt{\frac{5}{12}}\left(3\widetilde{x}_2-\widetilde{x}_2^3-3\widetilde{x}_1+\widetilde{x}_1^3\right).$$

1.5.2.2 Local spectra of the potential based on Sjöberg's approach

Sjöberg's (1982) approach is applied directly to the spectra result after using the orthogonality property of the spherical harmonics. In this method, the integral formula for determining the SHCs, Eq. (1.8), is used, but with a limited integration domain, and inserted into Eq. (1.151):

$$\overline{v}_n(r)=\left(\frac{R}{r}\right)^{2n+2}\sum_{m=-n}^{n}\frac{1}{4\pi}\iint_{\sigma_0}V(R,\theta,\lambda)Y_{nm}(\theta,\lambda)\mathrm{d}\sigma \\ \frac{1}{4\pi}\iint_{\sigma_0}V(R,\theta',\lambda')Y_{n'm'}(\theta',\lambda')\mathrm{d}\sigma. \tag{1.175}$$

By reordering the integral and the summation, we obtain:

$$\overline{v}_n(r)=\frac{1}{16\pi^2}\left(\frac{R}{r}\right)^{2n+2}\iint_{\sigma_0}V(R,\theta,\lambda) \\ \iint_{\sigma_0}V(R,\theta',\lambda')\mathrm{d}\sigma\sum_{m=-n}^{n}Y_{nm}(\theta,\lambda)Y_{n'm'}(\theta',\lambda')\mathrm{d}\sigma. \tag{1.176}$$

According to the addition theorem of the spherical harmonics Eq. (1.9), Eq. (1.176) will change to:

$$\overline{v}_n(r)=\frac{2n+1}{16\pi^2}\left(\frac{R}{r}\right)^{2n+2}\iint_{\sigma_0}V(R,\theta,\lambda)\iint_{\sigma_0}V(R,\theta',\lambda')\mathrm{d}\sigma P_n(\cos\psi)\mathrm{d}\sigma. \tag{1.177}$$

Now, the integral should be discretised and the integration domain should be partitioned into small cells. Assuming that V is constant in each small cell, we can write the discretised form of Eq. (1.177) as:

$$\overline{v}_n(r) = \frac{2n+1}{16\pi^2}\left(\frac{R}{r}\right)^{2n+2}\sum_{i=1}^{I_{\max}} V_i \sum_{j=1}^{J_{\max}} V_j \iint_{\sigma_i}\iint_{\sigma_j} P_n(\cos\psi_{ij})\,d\sigma_i d\sigma_j. \quad (1.178)$$

Here $I_{\max}$ and $J_{\max}$ are the numbers of cells in areas σ_i and σ_j, which are assumed to be equal. The term ψ_{ij} is the spherical geocentric angle between σ_i and σ_j.

As we observe, two double integrals appear in Eq. (1.178), but Sjöberg (1982) proposed the following approximation to their solution:

$$\iint_{\sigma_i}\iint_{\sigma_j} P_n(\cos\psi_{ij})\,d\sigma_i d\sigma_j \approx \beta'^2_n P_n(\cos\psi_{ij})\Delta\sigma_i\Delta\sigma_j, \quad (1.179)$$

where

$$\beta'_n = \frac{1}{1-\cos\psi_0}\frac{1}{2n+1}\left[P_{n-1}(\cos\psi_0) - P_{n+1}(\cos\psi_0)\right]. \quad (1.180)$$

The term ψ_0 is the radius of the small cap covering each cell. By considering that the areas of the cap and the cell are equal, the area of the cell and the radius of the spherical cap ψ_0 will be:

$$\Delta\sigma = 2\pi(1-\cos\psi_0) \text{ or } \psi_0 = \cos^{-1}\left(1-\frac{\Delta\sigma}{2\pi}\right). \quad (1.181)$$

Sjöberg (1982) also proposed the following recursion formula for computing β'_n:

$$\beta'_n = \frac{2n-1}{n+1}\cos\psi_0\beta'_{n-1} - \frac{n-2}{n+1}\beta'_{n-2} \quad \text{for} \quad n \geq 2, \quad (1.182)$$

with the initial values

$$\beta'_0 = 1 \text{ and } \beta'_1 = (1+\cos\psi_0)/2. \quad (1.183)$$

Finally, the spectra can be determined locally by:

$$\overline{v}_n(r) = \frac{2n+1}{16\pi^2}\left(\frac{R}{r}\right)^{2n+2}\beta'^2_n\sum_{i=1}^{I} V_i \sum_{j=1}^{J} V_j P_n(\cos\psi_{ij})\Delta\sigma_i\Delta\sigma_j. \quad (1.184)$$

References

Abramowitz, M., Stegun, I.A., 1964. Handbook of Mathematical Functions with Formulas, Graphs, and Mathematical Tables, National Bureau of Standards Applied Mathematics Series 55, United States Department of Commerce.

Eshagh, M., 2009. Comparison of two approaches for considering laterally varying density in topographic effect on satellite gravity gradiometry data. Acta Geophys. 58 (4), 661–686.

Eshagh, M., 2011. The effect of spatial truncation error on integral inversion of satellite gravity gradiometry data. Adv. Space Res. 47, 1238–1247.

Eshagh, M., 2014. From satellite gradiometry data to the sub-crustal stress due to the mantle convection. Pure Appl. Geophys. 171, 2391–2406.

Eshagh, M., 2016. Integral approaches to determine sub-crustal stress from terrestrial gravimetric data. Pure Appl. Geophys. 173, 805–825.

Hobson, E.W., 1965. The Theory of Spherical and Ellipsoidal Harmonics, second ed. Chelsea, New York.

Hotine, M., 1969. Mathematical Geodesy. US Environmental Science Services Administration, Rockville.

Hwang, C., 1995. A method for computing the coefficients in the product-sum formula of associated Legendre functions. J. Geod. 70, 110–116.

Hwang, C., 1998. Inverse Vening Meinesz formula and deflection-geoid formula: applications to the predictions of gravity and geoid over the South China Sea. J. Geod. 72, 304–312.

Heiskanen, W., Moritz, H., 1967. Physical Geodesy. W. H. Freeman and Company, San Francisco.

Mainville, A., 1986. The Altimetry-Gravimetry Problem Using Orthonormal Base Functions [pdf]. Ph.D. dissertation 203.

Martinec, Z., 2003. Green's function solution to spherical gradiometric boundary-value problems. J. Geodyn. 77, 41–49.

Pail, R., Plank, G., Schuh, W.D., 2001. Spatially restricted data distributions on the sphere: the method of orthonormalised funcations and applications. J. Geodyn. 75, 44–56.

Paul, M.K., 1978. Recurrence relations for integrals of associated Legendre functions. Bull. Géodésique 52, 177–190.

Pick, M., Picha, J., Vyskocil, V., 1973. Theory of the Earth Gravity Field. Elsevier, Amsterdam.

Reed, G.B., 1973. Application of Kinematical Geodesy for Determining the Shorts Wavelength Component of the Gravity Field by Satellite Gradiometry. Ohio state University, Department of Geodetic Science, Report No. 201, Columbus, Ohio.

Sjöberg, L.E., 1982. Studies on land uplift and its implication on the geoid in Fennoscandia. Report No. 14, Department of Geodesy, Uppsala University.

Sebilleau, D., 1998. On the computation of the integrated products of three spherical harmonics. J. Phys. A Math. Gen. 31, 7157–7168.

Varshalovich, D.A., Moskalev, A.N., Khersonskii, V.K., 1989. Quantum Theory of Angular Momentum. World Scientific Publishing, Singapore.

Xu, Y.L., 1996. Fast evaluation of the Gaunt coefficients. Math. Comput. 65, 1601–1612.

Zerilli, F.J., 1970. Tensor harmonics in canonical form for gravitational radiation and other applications. J. Math. Phys. 11, 2203–2208.

CHAPTER 2

Satellite gravimetry observables

2.1 Satellite orbit and the Earth's gravitational potential

According to Kepler's law a satellite's orbit is an ellipse in which the Earth is located at one of its focuses; see e.g. Seeber (2003) or Montenbruck and Gill (2000). This law is valid for the case where the Earth is assumed as a point mass with a constant gravity and no other force than the gravitational force of the Earth acts on the satellite. Nevertheless, in reality, the Earth has almost a spherical form; it has flattening and topographic masses and its interior mass is not homogeneous, meaning that some parts of the Earth are heavier than other parts. This causes the Earth's gravity field to change from one point to another all over the globe. On the other hand, there are other forces acting on a satellite, which can have gravitational or non-gravitational sources. The gravitational ones consist of the gravity forces of the Earth, other planets, solid and ocean tides and rotational deformation; and the best known non-gravitational ones are the solar and Earth radiation pressures, atmospheric drag and thermal effects. Consequently, the satellite's orbit deviates from that ideal ellipse mentioned by Kepler. Here, we call this ideal orbit the Keplerian orbit. If the departure of the orbit from its Keplerian is solely due to the Earth's gravity field, then by measuring it, the gravity field of the Earth can be determined. Removal of the effects of the other forces is possible, as the mathematical models of them are known, so they can be computed and removed from the real orbit or satellite gravimetry observation to make them purely gravitational from the Earth system. Some non-gravitational forces can be eliminated practically. Some non-gravitational forces can be eliminated practically.

Low Earth-orbiting (LEO) satellites can be positioned via measurements to other high orbiting satellites and/or the stations at the Earth's surface. Fig. 2.1 shows schematically the principle of these methods. For example, today there are different satellite missions for positioning, the global positioning system (GPS), global navigation satellite system (GNSS), Galileo and BeiDou. Tracking these LEO satellites by these systems is called

Satellite Gravimetry and the Solid Earth
ISBN 978-0-12-816936-0
https://doi.org/10.1016/B978-0-12-816936-0.00002-5

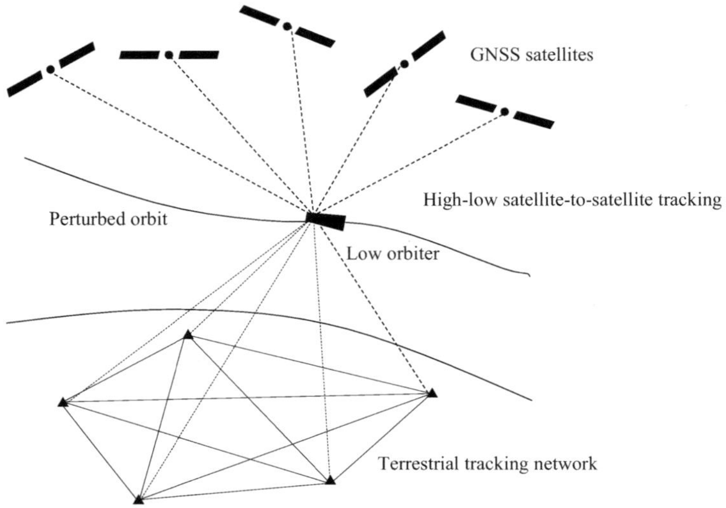

Figure 2.1 High–low satellite-to-satellite tracking and terrestrial tracking network. *GNSS*, global navigation satellite system.

high–low satellite-to-satellite tracking. In addition, the positions of these LEO satellites can be estimated from large-extent precise terrestrial geodetic networks, having precise coordinates already determined from other geodetic methods. When a satellite passes over the network some measurements between each point and the satellite are made, for example, satellite laser ranging, whereby the distance between the station and the reflectors mounted on the satellite is measured precisely, or Doppler data. In both cases, the least-squares approach is applied for estimating the coordinate of the LEO satellites and its uncertainties.

The equation of motion of a satellite is expressed by a second-order vector differential equation in a celestial reference frame (CRF). The CRF is an inertial system; it is heliocentric, meaning that the frame origin coincides with the centre of the sun, but we consider this frame geocentric. This is just a shift from heliocentre to geocentre. Its z-axis coincides with the rotation axis of the Earth, corrected for the irregularities due to precession, nutation and polar motion. This axis points towards the north celestial pole. The x-axis is defined on the celestial equator and points towards the vernal equinox, and the y-axis is considered so that a right-handed frame is completed. This differential equation of motion of a satellite in the CRF is:

$$\ddot{\mathbf{r}}_{\mathrm{CRF}} = \begin{bmatrix} \ddot{x} \\ \ddot{y} \\ \ddot{z} \end{bmatrix}_{\mathrm{CRF}} = -\frac{GM}{\left(r_{\mathrm{CRF}}\right)^3}\mathbf{r}_{\mathrm{CRF}} + \ddot{\mathbf{r}}_{\mathrm{CRF}}^{\mathrm{Gravitational}} + \ddot{\mathbf{r}}_{\mathrm{CRF}}^{\mathrm{Non-gravitational}}, \quad (2.1)$$

where $\mathbf{r}_{CRF}$ is the position vector of the satellite in the absence of perturbing accelerations and $\ddot{\mathbf{r}}_{CRF}$ is the acceleration vector of the satellite, with the elements $\ddot{x}_{CRF}$, $\ddot{y}_{CRF}$ and $\ddot{z}_{CRF}$; the double dot means second-order derivative with respect to time. x_{CRF}, y_{CRF} and z_{CRF} are the coordinates of the satellite in the CRF. r_{CRF} stands for the geocentric distance of the satellite, G is the Newtonian gravitational constant and M is the mass of the Earth. $\ddot{\mathbf{r}}_{CRF}^{Gravitational}$ and $\ddot{\mathbf{r}}_{CRF}^{Non-gravitational}$ are, respectively, interactions of all gravitational and non-gravitational forces.

The first term on the right-hand side (rhs) of Eq. (2.1) is the gravity of a spherical solid Earth with a constant density. Therefore, by ignoring the second and third terms on the rhs of this equation, and solving this differential equation by integrating it twice with respect to time, the position and velocity vector of the satellite can be estimated in a Keplerian orbit. For discussions and methods for integrating satellite orbits and the perturbing forces acting on satellites the reader is referred to Montenbruck and Gill (2000) and Eshagh and Najafi-Alamdary (2007).

The terrestrial reference frame (TRF) is geocentric and its z-axis points towards the north; it has its x-axis as the intersection line between the equator and the meridian passing through Greenwich, and the system is right—handed. Because of the Earth's rotation the coordinates of points in frame change with respect to the CRF. When the precession and nutation as well as the Earth's free nutation are considered for the z-axis of the TRF, then the z-axes of the CRF and TRF coincide. Therefore, only one rotation around this common axis is required to transform the x-axis of the CRF, which is towards the vernal equinoxes, to that of the TRF, or the Greenwich meridian. This angle is estimated by time and called Greenwich apparent sidereal time (GAST); see e.g. Seeber (2003). Fig. 2.2 shows these frames in addition to the local north-oriented frame (LNOF).

Therefore, it is of vital importance to be aware of which reference frame is used for gravimetry and know the transformation from one frame to another. In orbit integration, the goal is to compute satellite orbits from the forces acting on the satellite. In this case, the rhs of Eq. (2.1) is assumed as known and the satellite acceleration vector considers all gravitational and non-gravitational forces in the CRF and is integrated twice with respect to time for estimating the position and velocity vectors of a satellite. In the problem of satellite gravimetry, all forces are known except the one due to the Earth's gravitation. Now, we present how Eq. (2.1), after removal of the other forces, is related to the Earth's gravitational potential. According to Newton's law, the gravitational acceleration is nothing other than the

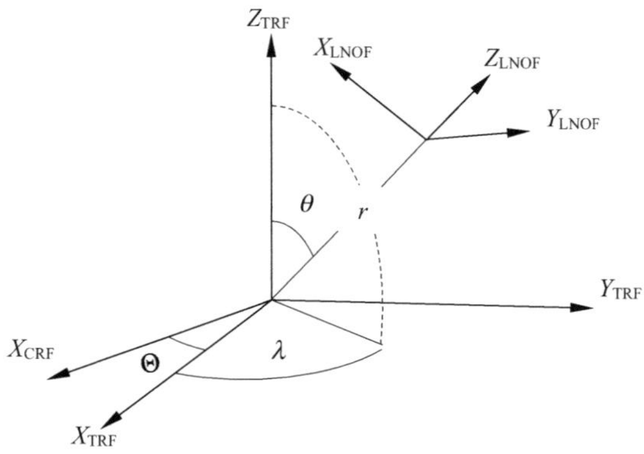

Figure 2.2 Celestial reference frame (*CRF*), terrestrial reference frame (*TRF*) and local north-oriented frame (*LNOF*), Θ is Greenwich apparent sidereal time.

gradient of the potential. Therefore, by taking the gradient operator presented in Eq. (1.45) to the spherical harmonic expansion of potential (Eq. 1.3), the satellite accelerations in the LNOF are derived:

$$V_z(r,\theta,\lambda) = -\frac{1}{r}\sum_{n=0}^{\infty}(n+1)\left(\frac{R}{r}\right)^{n+1}\sum_{m=-n}^{n} v_{nm}Y_{nm}(\theta,\lambda), \tag{2.2}$$

$$V_x(r,\theta,\lambda) = \frac{1}{r}\sum_{n=1}^{\infty}\left(\frac{R}{r}\right)^{n+1}\sum_{m=-n}^{n} v_{nm}\frac{\partial Y_{nm}(\theta,\lambda)}{\partial\theta}, \tag{2.3}$$

$$V_y(r,\theta,\lambda) = \frac{1}{r}\sum_{n=1}^{\infty}\left(\frac{R}{r}\right)^{n+1}\sum_{m=-n}^{n} v_{nm}\frac{\partial Y_{nm}(\theta,\lambda)}{\sin\theta\ \partial\lambda}. \tag{2.4}$$

Now, these accelerations are transformed from the LNOF to the CRF:

$$\ddot{\mathbf{r}}_{CRF} = \begin{pmatrix} \ddot{z}(r,\theta,\lambda^*) \\ \ddot{x}(r,\theta,\lambda^*) \\ \ddot{y}(r,\theta,\lambda^*) \end{pmatrix}_{CRF} = \begin{pmatrix} \sin\theta\cos\lambda^* & \cos\theta\cos\lambda^* & -\sin\lambda^* \\ \sin\theta\sin\lambda^* & \sin\theta\sin\lambda^* & \cos\lambda^* \\ \cos\theta & -\sin\theta & 0 \end{pmatrix}\begin{pmatrix} V_z(r,\theta,\lambda) \\ V_x(r,\theta,\lambda) \\ V_y(r,\theta,\lambda) \end{pmatrix}_{LNOF}, \tag{2.5}$$

where $\ddot{x}$, $\ddot{y}$ and $\ddot{z}$ are the satellite accelerations in the CRF and

$$\lambda^* = \lambda + \Theta \quad \text{with } \Theta \text{ as GAST} \tag{2.6}$$

and

$$\lambda^* = \tan^{-1}\left(\frac{y_{\mathrm{CRF}}}{x_{\mathrm{CRF}}}\right), \tag{2.7}$$

where x_{CRF} and y_{CRF} are the x and y coordinates of the satellite in these CRF. λ^* is the right ascension of the satellite in the CRF.

Eq. (2.5) is the differential equation of motion of a satellite in the Earth's gravitational field. Note that Eqs. (2.3) and (2.4) do not contain zero-degree spherical harmonic coefficient (SHC). If only the zero-degree term of Eq. (2.2) is considered, the Keplerian orbit will be obtained by solving this differential equation (Eq. 2.1). Therefore, for computing the perturbations due solely to the Earth's gravity field, it is straightforward to compute the Keplerian orbit and subtract it from the perturbed one. When this geopotential perturbation is determined from a real satellite gravimetry mission, it can be used for recovering the Earth's gravitational field.

2.2 Geometry of orbit and geopotential perturbation

The orbital ellipse of a satellite has the semi-major axis a and the semi-minor axis b, from which the orbital eccentricity $e^2 = (a^2 - b^2)/a^2$ is defined. An orbital reference frame (ORF) is defined from this ellipse (Fig. 2.3). This frame is geocentric, and its x-axis passes through the nearest and farthest points of the ellipse, which are, respectively, the so-called perigee and apogee. The line passing through these points is called the line of apsides, and in fact the x-axis of the ORF coincides with it. The z-axis of the system is perpendicular to the orbital ellipse, and from the

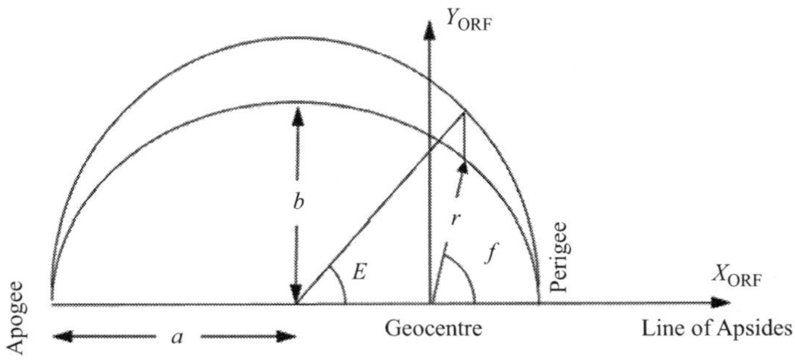

Figure 2.3 Orbital reference frame (*ORF*).

geocentre, finally, the y-axis is selected in such a way that the system becomes right handed. The position of a satellite in such a system is computed from its geocentric distance r and an angle from the x-axis of the ORF, which is well known as the true anomaly, f:

$$r = \frac{a(1 - e^2)}{1 + e \cos f} = a(1 - e \cos E), \tag{2.8}$$

whereas the angle E is called the eccentric anomaly. To define this angle, the satellite is projected onto a circle with the same centre as that of the orbital ellipse with the radius a, parallel to the y-axis; the angle between the line passing through the centre of this orbit and this point and the X_{ORF} is the eccentric anomaly (E).

According to the third law of Kepler, which says that the ratio of the square of a satellite's revolution period (P) and the cube of the semi-major axis of the orbital ellipse (a) is constant, the mean motion of satellite is defined:

$$\overline{n} = \frac{2\pi}{P} = \sqrt{\frac{GM}{a^3}}, \tag{2.9}$$

where G stands for the Newtonian gravitational constant and M the mass of the Earth. In fact, GM is the constant mentioned in Kepler's third law.

From this mean motion, the mean anomaly $\overline{M}$ is defined as:

$$\overline{M} = \overline{n}(t - t_0), \tag{2.10}$$

with t as the time and t_0 the time at which the satellite passes through the perigee. $\overline{M}$ is used for estimating the eccentric anomaly E by the Kepler equation; e.g. Seeber (2003, p.73):

$$E = \overline{M} + e \sin E. \tag{2.11}$$

The problem in solving this equation is that the unknown parameter E is on both sides of the equation. Therefore, an iterative method should be applied for its solution:

$$E_{k+1} = \overline{M} + e \sin E_k, \tag{2.12}$$

where k is iteration number. An approximate value for E_0 is required for starting the process. At the first step, it is assumed that $E_0 = \overline{M}$, and thereafter E_1 is computed; later on E_1 is used on the rhs and E_2 is computed. This process is repeated until the solution converges, meaning that by further iterations E does not change. This method works simply for the circular or near-circular orbits, otherwise, the Newton–Raphson method can be applied; see Montenbruck and Gil (2000). The final or converged value of E is used for computing the true anomaly (f) (Hoffmann-Wellenhof et al., 2001, p. 29):

$$f = 2\arctan\left[\sqrt{\frac{1+e}{1-e}}\tan\frac{E}{2}\right]. \tag{2.13}$$

The geocentric distance r is obtained from Eq. (2.8). Therefore, the position and velocity vectors can simply be computed by (Hoffmann-Wellenhof et al., 2011, p. 30):

$$\mathbf{r}_{\mathrm{ORF}} = \begin{bmatrix} x \\ y \end{bmatrix}_{\mathrm{ORF}} = \begin{bmatrix} r\cos f \\ r\sin f \end{bmatrix}, \tag{2.14}$$

$$\dot{\mathbf{r}}_{\mathrm{ORF}} = \begin{bmatrix} \dot{x} \\ \dot{y} \end{bmatrix}_{\mathrm{ORF}} = \sqrt{\frac{GM}{a(1-e^2)}}\begin{bmatrix} -\sin f \\ \cos f + e \end{bmatrix} \tag{2.15}$$

As we observe, these vectors are two dimensional and without any z-component, as the orbital ellipse is a two-dimensional geometric form. For three-dimensional computations, the z-component is considered to be zero. The vectors $\mathbf{r}_{\mathrm{ORF}}$ and $\dot{\mathbf{r}}_{\mathrm{ORF}}$ need to be transformed to the CRF by (see e.g. Hofmann-Wellenhof et al., 2011, p. 31):

$$\mathbf{r}_{\mathrm{CRF}} = \mathbf{R}_3(-\Omega)\mathbf{R}_1(-i)\mathbf{R}_3(-\omega)\mathbf{r}_{\mathrm{ORF}}, \tag{2.16}$$

$$\dot{\mathbf{r}}_{\mathrm{CRF}} = \mathbf{R}_3(-\Omega)\mathbf{R}_1(-i)\mathbf{R}_3(-\omega)\dot{\mathbf{r}}_{\mathrm{ORF}}, \tag{2.17}$$

where Ω is the right ascension of the ascending node of the satellite; this is an angle between the nodal line of the orbital plane and the celestial equator towards the vernal equinox, or the X_{CRF}. i is called the orbital inclination, the angle between the orbital plane and the celestial equator, and finally ω is the perigee argument as the angle between the line of apsides, or X_{ORF}, and the nodal line of the orbital and equatorial planes. $\mathbf{R}_1$ and $\mathbf{R}_3$ are, respectively, the rotation matrices around X_{CRF} and Z_{CRF}. Fig. 2.4 shows the TRF and CRF and ORF and these angles, presenting the orbit orientation in space.

An orbit can be represented by either the satellite position and velocity vectors (state vector) or the parameters $\overline{M}$, a, e and Ω, i and ω, which are known as orbital elements. The first three represent the satellite position in

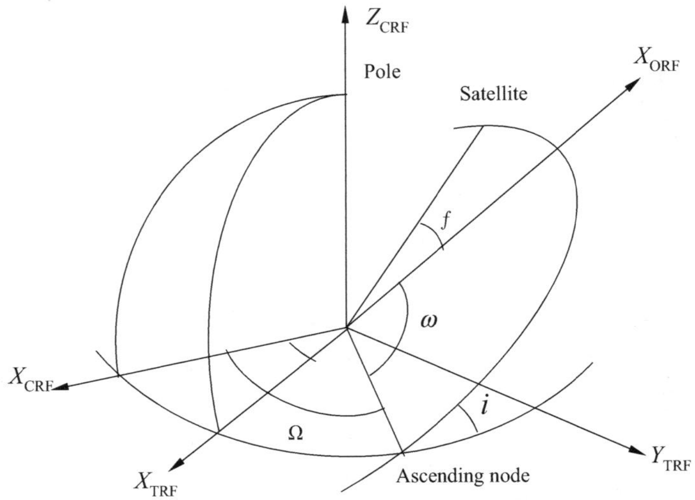

Figure 2.4 Celestial reference frame (*CRF*), terrestrial reference frame (*TRF*) and orbital reference frame (*ORF*) and orbital orientation angles.

the ORF and the rest the orbit orientation angles in space. Therefore, according to the transformations (Eqs. 2.16 and 2.17) the satellite state vectors can be computed from these orbital elements. In other words, for computing the satellite state vector, the size and the shape of the orbit, which are represented by a and e of the orbit, are required. In addition, the mean motion and the time are required to compute the mean anomaly $\overline{M}$, from which the true anomaly f is computed, and finally the transformations (Eqs. 2.16 and 2.17) with the orientation angles $\boldsymbol{\Omega}$, i and ω.

Also, from $\mathbf{r}_{\mathrm{CRF}}$ and $\dot{\mathbf{r}}_{\mathrm{CRF}}$ the orbital elements can be computed by (e.g. Seeber, 2003, p. 80):

$$a = -\left(\frac{\|\dot{\mathbf{r}}_{\mathrm{CRF}}\|^2}{GM} - \frac{2}{|\mathbf{r}|}\right)^{-1}, \tag{2.18}$$

$$e = \sqrt{\left(1 - \frac{\|\mathbf{r}_{\mathrm{CRF}}\|}{a}\right)^2 + \frac{\|\mathbf{r}_{\mathrm{CRF}} \cdot \dot{\mathbf{r}}_{\mathrm{CRF}}\|^2}{GMa}}, \tag{2.19}$$

$$E = \tan^{-1}\left[\mathbf{r}_{\mathrm{CRF}} \cdot \dot{\mathbf{r}}_{\mathrm{CRF}} / \left((1 - \|\mathbf{r}_{\mathrm{CRF}}\| / a)\sqrt{GMa}\right)\right], \tag{2.20}$$

$$\overline{M} = E - e \sin E, \tag{2.21}$$

$$i = \cos^{-1}\left(\frac{(\mathbf{r}_{\text{CRF}} \times \dot{\mathbf{r}}_{\text{CRF}})\cdot[0 \quad 0 \quad 1]}{\|\mathbf{r}_{\text{CRF}} \times \dot{\mathbf{r}}_{\text{CRF}}\|}\right), \tag{2.22}$$

$$\Omega = \cos^{-1}\left(\frac{([0 \quad 0 \quad 1]\cdot(\mathbf{r}_{\text{CRF}} \times \dot{\mathbf{r}}_{\text{CRF}}))\cdot[1 \quad 0 \quad 0]}{\|[0 \quad 0 \quad 1]\cdot(\mathbf{r}_{\text{CRF}} \times \dot{\mathbf{r}}_{\text{CRF}})\|}\right), \tag{2.23}$$

$$\omega = \tan^{-1}\left(\frac{z_{\text{CRF}}}{(x_{\text{CRF}}\cos\Omega + y_{\text{CRF}}\sin\Omega)\sin i}\right) - f, \tag{2.24}$$

where $\|\cdot\|$ is the L_2 norm of the vector, a dot (.) stands for the inner product and $\times$ stands for the cross-product operator.

Fig. 2.5 shows the behaviour of the orbital elements of a Gravity Field and Steady-State Ocean Circulation Explorer (GOCE)-type satellite, the orbit of which has been generated by the Runge-Kutta method of fourth-order integrator and the gravity model EGM96 (Lemoine et al., 1998) with the necessary initial values. The generated accelerations have been integrated twice numerically for estimating the velocity and position vectors by Eqs. (2.18)–(2.24) for determining the orbital elements and their variation under one day.

To obtain the perturbations in orbital elements, it suffices to compute a Keplerian orbit by solving Eq. (2.5) considering only the zero-degree SHC to determine $\mathbf{r}_{\text{CRF}}$ and $\dot{\mathbf{r}}_{\text{CRF}}$. Thereafter, these vectors are converted to the orbital elements using Eqs. (2.18)–(2.24). In this case, unperturbed orbital elements, or the Keplerian elements, will be obtained. In the second step, $\mathbf{r}_{\text{CRF}}$ and $\dot{\mathbf{r}}_{\text{CRF}}$ are computed considering all harmonics in solving Eq. (2.5) and they are converted to orbital elements, which are now perturbed. Consequently, the geopotential perturbation in the orbital elements can be simply derived by subtracting the Keplerian elements from the perturbed ones. This was the strategy used by Eshagh and Najafi-Alamdary (2007) for investigating the orbital perturbation due to different perturbing forces.

The satellite perturbations can be presented in another way. Here, we define a reference frame, having its centre at the satellite centre. Its u-axis is along the geocentric distance of satellite and positive outwards, the v-axis of the system is along the velocity vector of the satellite and tangent to the orbit, and finally the w-axis is perpendicular to the orbital ellipse. Here and after, we name this frame the track-oriented frame (TOF). Fig. 2.6 shows the TOF and LNOF systems and their relations.

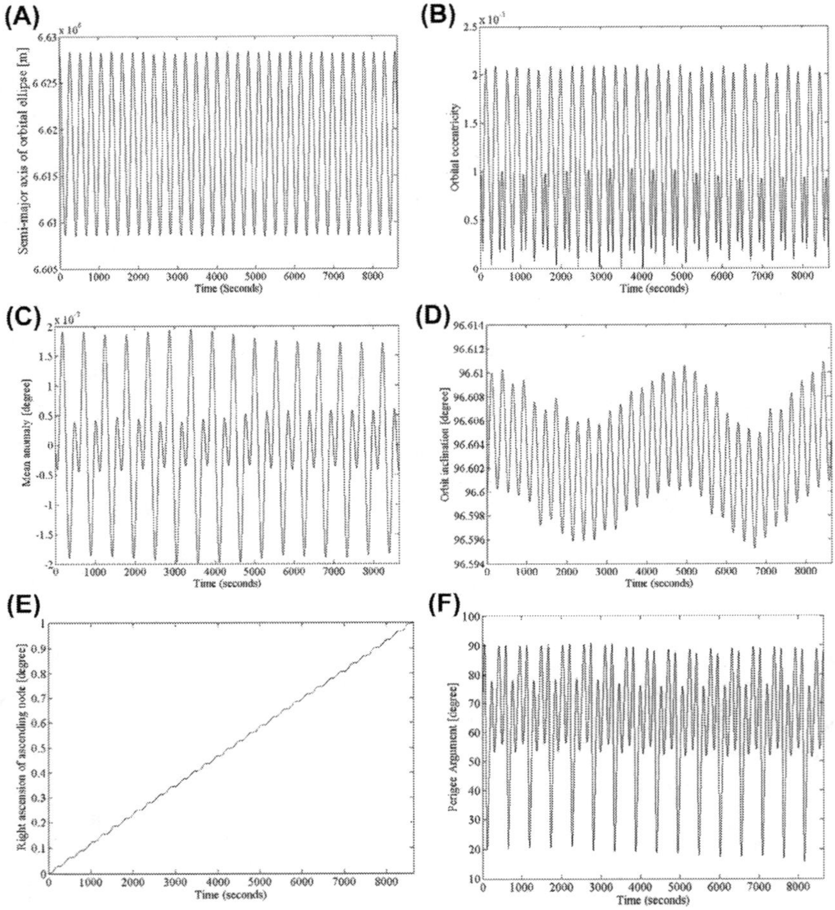

Figure 2.5 Variations of the orbital element, (A) a, (B) e, (C) $\overline{M}$, (D) i and (E) Ω, during one day revolution of a GOCE-type satellite.

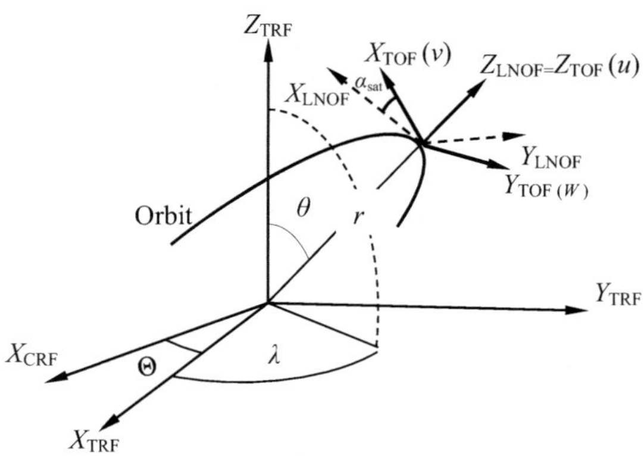

Figure 2.6 Terrestrial reference frame (*TRF*), celestial reference frame (*CRF*) and track-oriented frame (*TOF*). *LNOF*, local north-oriented frame.

The u-axis of the TOF coincides with the z-axis of the LNOF; therefore, by a rotation around this common axis, which is the satellite track azimuth, α_{Sat}, these two frames can be transformed to each other:

$$\mathbf{r}_{TOF} = \mathbf{R}_3(\alpha_{Sat})\mathbf{r}_{LNOF} = \begin{bmatrix} x_{TOF} \\ y_{TOF} \\ z_{TOF} \end{bmatrix} = \begin{bmatrix} \cos\alpha_{Sat} & \sin\alpha_{Sat} & 0 \\ -\sin\alpha_{Sat} & \cos\alpha_{Sat} & 0 \\ 0 & 0 & 1 \end{bmatrix} \begin{bmatrix} x_{LNOF} \\ y_{LNOF} \\ z_{LNOF} \end{bmatrix}, \tag{2.25}$$

where $\mathbf{R}_3$ is the rotation matrix around the z-axis (Fig. 2.7).

According to Fig. 2.7 and by writing the sine rule of the spherical trigonometry, we get:

$$\frac{\sin(90^\circ - i)}{\sin\theta} = \frac{\sin(180^\circ - \alpha_{Sat})}{\sin 90^\circ}. \tag{2.26}$$

Simplification of Eq. (2.26) yields:

$$\sin\alpha_{Sat} = \frac{\cos i}{\sin\theta}. \tag{2.27}$$

Also, the cosine rule can be applied:

$$\cos 90^\circ = \cos(\omega + f)\cos\theta + \sin(\omega + f)\sin\theta\cos(180^\circ - \alpha_{Sat}), \tag{2.28}$$

after simplifications:

$$\sin(\omega + f)\sin\theta\cos\alpha_{Sat} = \cos(\omega + f)\cos\theta. \tag{2.29}$$

On the other hand, according to the sine rule of the spherical triangle limited between the equator, the orbit and the satellite meridian, we obtain:

$$\frac{\cos\theta}{\sin i} = \sin(\omega + f). \tag{2.30}$$

By inserting Eq. (2.30) into Eq. (2.29) and further simplifications, we obtain:

$$\cos\alpha_{Sat} = \frac{\cos(\omega + f)\sin i}{\sin\theta}. \tag{2.31}$$

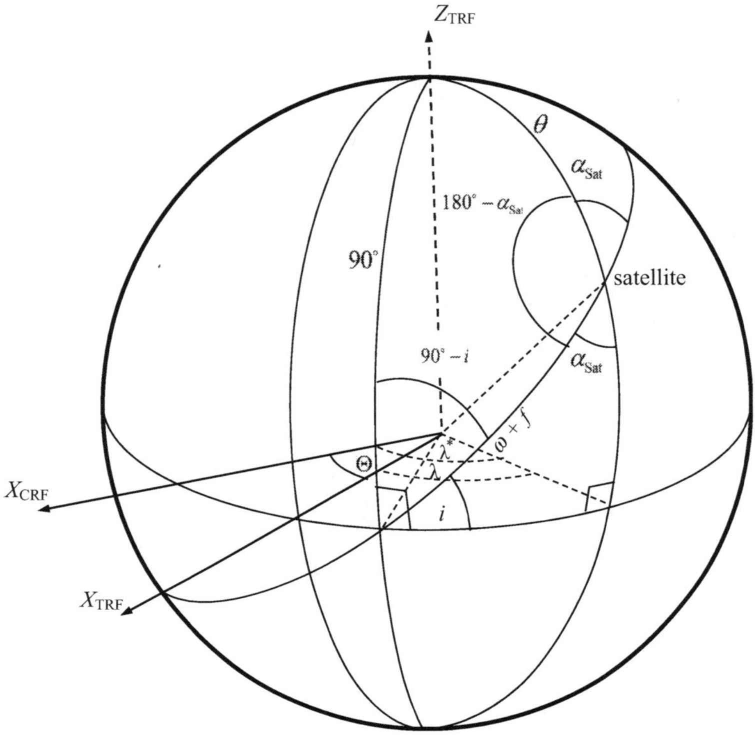

Figure 2.7 Satellite track azimuth on the celestial sphere. *CRF*, celestial reference frame; *TRF*, terrestrial reference frame.

By using either Eq. (2.27) or Eq. (2.31) the satellite track azimuth is computed.

The departure vector of a perturbation from a reference orbit (e.g., the Keplerian orbit), radially upward ($\overline{u}$), along the satellite track ($\overline{v}$) and perpendicular to the orbital plane ($\overline{w}$) can be defined as (see e.g. Santos, 1995, p. 23):

$$\overline{u} = \frac{\mathbf{r} \cdot (\mathbf{r} - \mathbf{r}_{\text{Ref}})}{\|\mathbf{r}\|}, \tag{2.32}$$

$$\overline{v} = \frac{\dot{\mathbf{r}} \cdot (\mathbf{r} - \mathbf{r}_{\text{Ref}})}{\|\dot{\mathbf{r}}\|}, \tag{2.33}$$

$$\overline{w} = \frac{(\mathbf{r} \times \dot{\mathbf{r}}) \cdot (\mathbf{r} - \mathbf{r}_{\mathrm{Ref}})}{\|\mathbf{r} \times \dot{\mathbf{r}}\|}, \tag{2.34}$$

where $\mathbf{r}_{\mathrm{Ref}}$ stands for the reference satellite position vector, which can be considered the satellite position on the Keplerian orbit. Santos (1995) has defined the radial track towards the Earth's centre and he considered a minus sign behind Eq. (2.32). However, since we have defined the Z_{TOF} in the opposite direction, we did not consider that minus sign. In addition, he considered u as along track and v as radial, but we did it the opposite in this book to have all formulae consistent. However, the principle of both cases is the same.

The TOF is dependent on the frame in which the satellite coordinates are computed. The satellite position can be given in the CRF or TRF, which are different because of the direction of their x-axes. By assuming their z-axes coincide, therefore, their x-axes will coincide by a single rotation around the common z-axis with $\boldsymbol{\Theta}$.

To present the geopotential perturbation, the state vector, $\mathbf{r}_{\mathrm{CRF}}$ and $\dot{\mathbf{r}}_{\mathrm{CRF}}$ are estimated twice. In the first case, Eq. (2.5) is solved considering only the zero-degree SHC, and in the second round it is solved considering all SHCs. The position vectors, determined in the first case, present the reference orbit, which should be subtracted from the perturbed one for presenting the perturbation in the TOF.

Now, suppose that two satellites start revolving around the Earth from a common point, but one without perturbing potential and the other one with, computed by the gravity model EGM96 to degree and order 250. The satellite accelerations are computed and integrated using the Runge–Kutta method of fourth order during one day. The first case, where no Earth perturbing potential is used, a Keplerian orbit is obtained and considered as a reference orbit. The departures of these orbits in the TOF are presented in Fig. 2.8. Fig. 2.8A shows that the radial-track perturbation $\overline{u}$ has some short periodic variation for each revolution around the Earth and a trend. The along-track $\overline{v}$ is shown in Fig. 2.8B with its short and secular trend; it shows that the satellite moving based on the perturbing potential is in front of the other one moving on the Keplerian orbit. Fig. 2.8C illustrates that the cross-track perturbation $\overline{w}$ is periodic with an amplitude increasing by time.

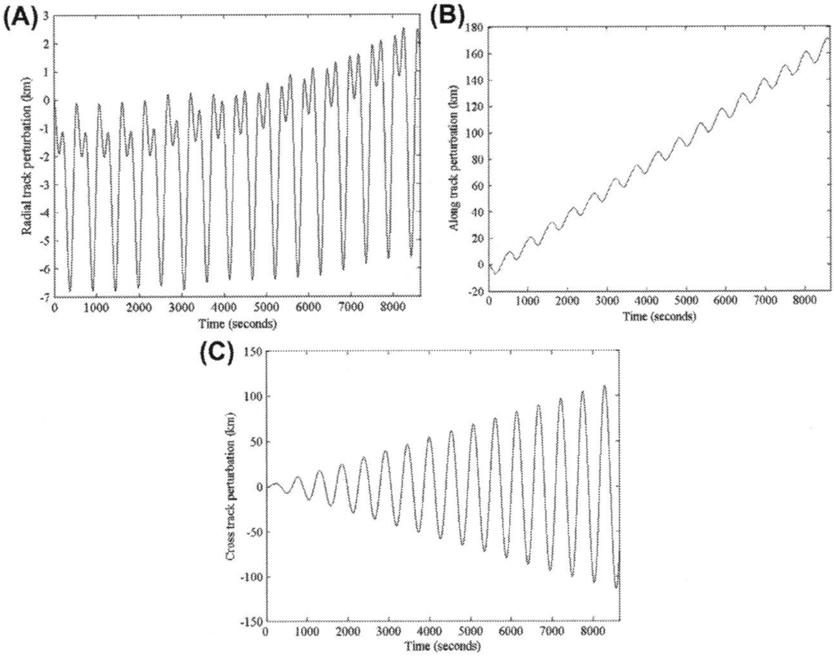

Figure 2.8 (A) Radial-track $\overline{u}$, (B) along-track $\overline{v}$ and (C) cross-track $\overline{w}$ perturbation during one day of a GOCE-type orbit.

2.3 Orbital elements

The temporal variations in orbital elements can be considered as satellite gravimetry observations. Generating the geopotential perturbations in the TOF and orbital elements from the Earth's gravity field has been presented before. Now, the goal is to determine the gravitational potential from the orbital elements, which is in fact an inverse problem in satellite gravimetry. In this section, such a discussion is followed and the relation between the temporal variations in orbital elements and the gravitational potential is formulated mathematically.

Let us follow our discussion using Eq. (1.3), which is the spherical harmonic expansion of the gravitational potential, but excluding the zero- and first-degree harmonics, and call it perturbing potential:

$$V^{\mathrm{P}}(r,\theta,\lambda)=\sum_{n=2}^{\infty}\left(\frac{R}{r}\right)^{n+1}\sum_{m=-n}^{n}v_{nm}Y_{nm}(\theta,\lambda). \tag{2.35}$$

This potential can also be expressed in terms of the orbital elements (see Kaula, 1966):

$$V^{\mathrm{P}}\left(r,i,e,\overline{\Psi}\right)=\frac{1}{R}\sum_{n=2}^{\infty}\sum_{m=-n}^{n}\sum_{p=0}^{n}\sum_{q=-\infty}^{\infty}\left(\frac{R}{r}\right)^{n+1}F_{nmp}(i)G_{npq}(e)s'_{nmpq}\left(\overline{\Psi}\right), \tag{2.36}$$

where $F_{nmp}(i)$ is called the inclination and $G_{npq}(e)$ the eccentricity function; and they have well-known formulae in many textbooks in celestial mechanics and satellite geodesy (see e.g. Kaula, 1966, and Seeber, 2003, pp. 90–92). For computing $F_{nmp}(i)$ only the orbital inclination and for $G_{npq}(e)$ only the orbital eccentricity is required. Therefore, they are known, and

$$s'_{nmpq}\left(\overline{\Psi}\right)=\frac{\mathrm{d}s_{nmpq}\left(\overline{\Psi}\right)}{\mathrm{d}\overline{\Psi}} \tag{2.37}$$

with

$$s_{nmpq}\left(\overline{\Psi}\right)=\begin{cases} v_{nn}\begin{cases}\sin\overline{\Psi} & m>0\\ -\cos\overline{\Psi} & m\leq 0\end{cases} & n-m \text{ even}\\ v_{n,-m}\begin{cases}-\sin\overline{\Psi} & m>0\\ -\cos\overline{\Psi} & m\leq 0\end{cases} & n-m \text{ odd}\end{cases}, \tag{2.38}$$

$$\overline{\Psi}=(n-2p)\left(\omega+\overline{M}\right)+m(\Omega-\Theta)+q\overline{M}. \tag{2.39}$$

As observed, the SHCs of the gravitational potential (v_{nm}) are involved in s_{nmpq}, which is also a function of the other orbital elements and the GAST (Θ). Now, the question is, how can these orbital geopotential perturbations be used to determine v_{nm}? To discuss this issue, we need to use Lagrange, Gaussian or Hill equations, which are presented next.

2.3.1 Lagrange equations

The Lagrange planetary equations represent the relations between the temporal variations in orbital elements and the derivatives of the perturbing potential V^{P} with respect to the orbital elements. Kaula (1966) has presented them in the following form:

$$\dot{a}=\frac{2}{\overline{n}a}\frac{\partial V^{\mathrm{P}}}{\partial\overline{M}}, \tag{2.40}$$

$$\dot{e} = \frac{1-e^2}{\overline{n}a^2 e}\frac{\partial V^{\mathrm{P}}}{\partial \overline{M}} - \frac{\sqrt{1-e^2}}{\overline{n}a^2 e}\frac{\partial V^{\mathrm{P}}}{\partial \omega}, \tag{2.41}$$

$$\dot{\omega} = -\frac{\cos i}{\overline{n}a^2\sqrt{1-e^2}\sin i}\frac{\partial V^{\mathrm{P}}}{\partial i} + \frac{\sqrt{1-e^2}}{\overline{n}a^2 e}\frac{\partial V^{\mathrm{P}}}{\partial e}, \tag{2.42}$$

$$\dot{i} = \frac{\cos i}{\overline{n}a^2\sqrt{1-e^2}\sin i}\frac{\partial V^{\mathrm{P}}}{\partial \omega} - \frac{1}{\overline{n}a^2\sqrt{1-e^2}\sin i}\frac{\partial V^{\mathrm{P}}}{\partial \Omega}, \tag{2.43}$$

$$\dot{\Omega} = \frac{1}{\overline{n}a^2\sqrt{1-e^2}\sin i}\frac{\partial V^{\mathrm{P}}}{\partial i}, \tag{2.44}$$

$$\dot{\overline{M}} = \frac{1-e^2}{\overline{n}a^2 e}\frac{\partial V^{\mathrm{P}}}{\partial e} - \frac{2}{\overline{n}a}\frac{\partial V^{\mathrm{P}}}{\partial a}. \tag{2.45}$$

By inserting Eq. (2.36) into the Lagrange Eqs. (2.40)–(2.45), integrating both sides of the results with respect to time, and linearising, the orbital elements will be obtained (Visser, 1992, pp. 1, 14 and 15):

$$\Delta a_{nmpq} = \frac{2\overline{n}}{R\dot{\overline{\psi}}}\left(\frac{R}{a}\right)^{n+1} F_{nmp}G_{npq}(n-2p+q)s'_{nmpq}(\overline{\psi}), \tag{2.46}$$

$$\begin{aligned}\Delta e_{nmpq} = {} & \frac{\overline{n}}{\dot{\overline{\psi}}}\frac{\sqrt{1-e^2}}{e}\left(\frac{R}{a}\right)^{n} F_{nmp}G_{npq}\times \\ & \left[(n-2p+q)\sqrt{1-e^2}-(n-2p)\right]s'_{nmpq}(\overline{\psi}),\end{aligned} \tag{2.47}$$

$$\Delta\omega_{nmpq} = \frac{\overline{n}}{\dot{\overline{\psi}}}\left(\frac{R}{a}\right)^{n}\left[\frac{\sqrt{1-e^2}}{e}F_{nmp}G'_{npq} - \frac{\cot iF'_{nmp}G_{npq}}{\sqrt{1-e^2}}\right]s_{nmpq}(\overline{\psi}), \tag{2.48}$$

$$\Delta i_{nmpq} = \frac{\overline{n}}{\dot{\overline{\psi}}\sqrt{1-e^2}\sin i}\left(\frac{R}{a}\right)^{n} F_{nmp}G_{npq}[(n-2p)\cos i - m]s'_{nmpq}(\overline{\psi}), \tag{2.49}$$

$$\Delta\Omega_{nmpq} = \frac{\overline{n}}{\dot{\overline{\psi}}\sqrt{1-e^2}\sin i}\left(\frac{R}{a}\right)^{n} F'_{nmp}G_{npq}s_{nmpq}(\overline{\psi}), \tag{2.50}$$

$$\begin{aligned}\Delta\overline{M}_{nmpq} = {} & \frac{\overline{n}}{\dot{\overline{\psi}}}\left(\frac{R}{a}\right)^{n} F_{nmp}\left[2(n+1)G_{npq} - \frac{\sqrt{1-e^2}}{e}G'_{npq} - 3G_{npq}(n-2p+q)\frac{\overline{n}}{\dot{\overline{\psi}}}\right]\times \\ & s_{nmpq}(\overline{\psi}),\end{aligned} \tag{2.51}$$

where $G'_{npq} = \frac{\mathrm{d}G_{npq}(e)}{\mathrm{d}e}$, $G_{npq} = G_{npq}(e)$, $F'_{nmp} = \frac{\mathrm{d}F_{nmp}}{\mathrm{d}i}$ and $F_{nmp} = F_{nmp}(i)$.

These solutions hold under the assumption that the effect of a particular perturbation has no significant influence on the formulation of another perturbation (Seeber, 2003, p. 93). By taking summations on n, m, p and q the perturbation of each orbital element will be obtained. Therefore, the left-hand side (lhs) of Eqs. (2.46)–(2.51) can be considered as observations, or measured geopotential perturbations, and v_{nm} as the unknown parameter on their rhs. Finally, a system of equations can be established for solving v_{nm}. Note that $(R/a)^n$ in Eqs. (2.46)–(2.51) is in fact an upwards continuation factor of the perturbing potential from the surface of the spherical Earth, with radius R, to the satellite flying in an orbit with semi-major axis a about the Earth.

Visser (1992) has presented simplified forms of the linearised Lagrange equations for near-circular orbits, where $e \approx 0$; in such a case, it is enough to calculate the first-order perturbations for such an orbit, only $q = -1$, 0 and 1 need be considered for calculating $G_{npq}(e)$, and therefore (Schrama, 1986):

$$G_{np,0}(e) \approx 1, \tag{2.52}$$

$$G_{np,-1}(e) \approx (-n + 4p + 1)\frac{e}{2}, \tag{2.53}$$

$$G_{np,+1}(e) \approx (3n - 4p + 1)\frac{e}{2}. \tag{2.54}$$

Rosborough (1986) presented the following relations connecting the perturbations of the orbital element to the departures $\overline{u}$, $\overline{v}$ and $\overline{w}$ from the Keplerian orbit in the CRF; see Eqs. (2.32)–(2.34) for near-circular orbits:

$$\overline{u} = \Delta a - a\Delta e \cos\overline{M} + ae\Delta\overline{M}\sin\overline{M}, \tag{2.55}$$

$$\overline{v} = a\left(\Delta\omega + \Delta\Omega \cos i + \Delta\overline{M} + 2\Delta e \sin\overline{M} + 2e\Delta\overline{M}\cos\overline{M}\right), \tag{2.56}$$

$$\overline{w} = a\left(\Delta i \sin\left(\omega + \overline{M}\right) - \Delta\Omega \sin i \cos\left(\omega + \overline{M}\right)\right). \tag{2.57}$$

Therefore, if the perturbations of the orbital elements are known, these perturbations are achievable from Eqs. (2.55)–(2.57). Solving Δa, Δe, $\Delta\omega$, $\Delta\Omega$, $\Delta\overline{M}$ and Δi from $\overline{u}$, $\overline{v}$ and $\overline{w}$ is not possible, because we have three equations with six unknowns. However, what these equations show is which orbital element contributes to which components of perturbation. For example, Δa contributes only to the radial-track perturbation $\overline{u}$, $\Delta\Omega$ has no effect on $\overline{u}$, and Δi is seen only in the cross-track perturbation $\overline{w}$.

Nevertheless, these equations are derived based on linearisation and the assumption of having circular or near-circular orbits.

One problem in using the Lagrange equations is singularity for the circular orbits. For example, $\dot{\overline{\psi}}$ is in the denominators of these equations, and when $n = 2p$, $m = 0$ and $q = 0$, then $\dot{\overline{\psi}}$ will be zero. In addition, if these equations are solved for $s_{nmpq}(\overline{\psi})$ or $s'_{nmpq}(\overline{\psi})$, G_{npq} and/or G'_{npq} will be seen in the denominators of the results, and when $e = 0$ these terms will be zero, and again, therefore, singularity occurs in the equations. Also, the small value of $\dot{\overline{\psi}}$ in the denominator amplifies drastically the temporal variations in orbital elements, in which case it is said that the orbit is in resonance. Such resonance between the mean motion of a satellite and the Earth's rotation originally contributed to the evaluation and assessment of v_{nm} (Klokočník et al., 2013).

2.3.2 Gaussian equations

The Gaussian equations of satellite motion relate the temporal changes in orbital elements to the geopotential perturbation defined in Eq. (2.2)–(2.4). These equations have the following mathematical expressions (Moulton, 1914, p. 404):

$$\dot{a} = 2a\mu\left(V_u^P e \sin f + \frac{p}{r} V_v^P\right), \tag{2.58}$$

$$\dot{e} = 2a\overline{m}\left(V_u^P \sin f + V_v^P \overline{\Theta}\right), \tag{2.59}$$

$$\dot{i} = \frac{r}{a}\mu \cos(\omega + f) V_w^P, \tag{2.60}$$

$$\dot{\Omega} = \frac{r}{a}\mu \frac{\sin(\omega + f)}{\sin i} V_w^P, \tag{2.61}$$

$$\dot{\omega} = \overline{m}\left[-\frac{\cos f}{e} V_u^P + V_v^P \frac{r + p}{ep} \sin f - \frac{r}{p} V_w^P \sin(\omega + f) \cot i\right], \tag{2.62}$$

$$\dot{\overline{M}} = \overline{n} - \Gamma V_u^P - \Pi V_v^P, \tag{2.63}$$

where

$$\Gamma = \overline{m}\left(\frac{2r}{a} - \frac{1 - e^2}{e} \cos f\right), \tag{2.64}$$

$$\Pi = \frac{1 - e^2}{e} \overline{m}\left(1 + \frac{r}{p}\right) \sin f, \tag{2.65}$$

$$\overline{\Theta} = \frac{r+p}{p}\cos f + \frac{er}{p}, \tag{2.66}$$

$$\mu = \sqrt{\frac{a}{GM(1-e^2)}}, \tag{2.67}$$

$$\overline{m} = \sqrt{\frac{p}{GM}} \quad \text{and} \quad p = a(1-e^2). \tag{2.68}$$

V_u^P, V_v^P and V_w^P are the derivatives of the perturbing potential in the directions radial-, along- and cross-track to the orbit, which are obtained from the position and velocity vector of the perturbed and unperturbed orbits in the CRF. From Eqs. (2.2)−(2.4) and the transformation matrix in Eq. (2.25) we obtain:

$$V_u^P\left(r,\theta,\lambda^*\right) = -\frac{1}{r}\sum_{n=2}^{\infty}(n+1)\left(\frac{R}{r}\right)^{n+1}\sum_{m=-n}^{n} v_{nm}Y_{nm}(\theta,\lambda), \tag{2.69}$$

$$V_v^P\left(r,\theta,\lambda^*\right) = \frac{1}{r}\sum_{n=2}^{\infty}\left(\frac{R}{r}\right)^{n+1}\sum_{m=-n}^{n} v_{nm}A_{nm}(\alpha_{\mathrm{Sat}},\theta,\lambda), \tag{2.70}$$

$$V_w^P\left(r,\theta,\lambda^*\right) = \frac{1}{r}\sum_{n=2}^{\infty}\left(\frac{R}{r}\right)^{n+1}\sum_{m=-n}^{n} v_{nm}A'_{nm}(\alpha_{\mathrm{Sat}},\theta,\lambda), \tag{2.71}$$

where

$$\begin{aligned} A_{nm}(\alpha_{\mathrm{Sat}},\theta,\lambda) &= \cos\alpha_{\mathrm{Sat}}\frac{\partial Y_{nm}(\theta,\lambda)}{\partial\theta} \\ &\quad + \sin\alpha_{\mathrm{Sat}}\frac{\partial Y_{nm}(\theta,\lambda)}{\sin\theta\,\partial\lambda} \quad \text{and} \\ A_{00}(\alpha_{\mathrm{Sat}},\theta,\lambda) &= 0, \end{aligned} \tag{2.72}$$

$$\begin{aligned} A'_{nm}(\alpha_{\mathrm{Sat}},\theta,\lambda) &= -\sin\alpha_{\mathrm{Sat}}\frac{\partial Y_{nm}(\theta,\lambda)}{\partial\theta} \\ &\quad + \cos\alpha_{\mathrm{Sat}}\frac{\partial Y_{nm}(\theta,\lambda)}{\sin\theta\partial\lambda} \quad \text{and} \\ A'_{00}(\alpha_{\mathrm{Sat}},\theta,\lambda) &= 0. \end{aligned} \tag{2.73}$$

Note that V_u^P, V_v^P and V_w^P are in the CRF but v_{nm} will be computed in the TRF, and $\lambda^* = \lambda + \Theta$ with Θ as GAST.

After inserting Eqs. (2.69)–(2.71) into the Gaussian equations, we obtain:

$$\dot{a}=\frac{2a\mu}{r}\sum_{n=2}^{\infty}\left(\frac{R}{r}\right)^{n+1}\sum_{m=-n}^{n}v_{nm}\left[-e\sin f(n+1)Y_{nm}(\theta,\lambda)+\frac{p}{r}A_{nm}(\alpha_{\mathrm{Sat}},\theta,\lambda)\right], \tag{2.74}$$

$$\dot{e}=\frac{2a\overline{m}}{r}\sum_{n=2}^{\infty}\left(\frac{R}{r}\right)^{n+1}\sum_{m=-n}^{n}v_{nm}\left[-\sin f(n+1)Y_{nm}(\theta,\lambda)+\overline{\Theta}A_{nm}(\alpha_{\mathrm{Sat}},\theta,\lambda)\right], \tag{2.75}$$

$$\dot{i}=\frac{\mu\cos(\omega+f)}{a}\sum_{n=2}^{\infty}\left(\frac{R}{r}\right)^{n+1}\sum_{m=-n}^{n}v_{nm}A'_{nm}(\alpha_{\mathrm{Sat}},\theta,\lambda), \tag{2.76}$$

$$\dot{\Omega}=\frac{\mu}{a}\frac{\sin(\omega+f)}{\sin i}\sum_{n=2}^{\infty}\left(\frac{R}{r}\right)^{n+1}\sum_{m=-n}^{n}v_{nm}A'_{nm}(\alpha_{\mathrm{Sat}},\theta,\lambda), \tag{2.77}$$

$$\dot{\overline{M}}=\overline{n}+\frac{1}{r}\sum_{n=2}^{\infty}\left(\frac{R}{r}\right)^{n+1}\sum_{m=-n}^{n}v_{nm}[\Gamma(n+1)Y_{nm}(\theta,\lambda)-\Pi A_{nm}(\alpha_{\mathrm{Sat}},\theta,\lambda)], \tag{2.78}$$

$$\dot{\omega}=\frac{\overline{m}}{r}\sum_{n=2}^{\infty}\left(\frac{R}{r}\right)^{n+1}\sum_{m=-n}^{n}v_{nm}\left[\frac{\cos f}{e}(n+1)Y_{nm}(\theta,\lambda)+\frac{r+p}{ep}A_{nm}(\alpha_{\mathrm{Sat}},\theta,\lambda)\times\right.$$
$$\left.\sin f-\frac{r}{p}\sin(\omega+f)A'_{nm}(\alpha_{\mathrm{Sat}},\theta,\lambda)\cot i\right]. \tag{2.79}$$

We emphasise that the temporal changes in the orbital elements are calculated in the CRF, but by using their rectangular coordinates and Θ (GAST), the corresponding position in the TRF is derived.

Eqs. (2.74)–(2.79), the Gaussian equations, connect the temporal changes of orbital elements directly to v_{nm} with simpler mathematical models than the Lagrangian equations. With the singularities for $e=0$ ($e\approx 0$), the problems about the near-circular orbits and resonances do not influence the recovery of v_{nm}. Note that recovering the zero-degree SHC of the gravitational potential from $\dot{i}$ and $\dot{\Omega}$ is not possible, as they involve only V_w^p, which does not contain this term.

For applying the Gaussian Eqs. (2.74)–(2.79), temporal variations in the orbital elements with a global coverage are required. These equations play the role of observation equations and they form a large system of equations having v_{nm} as their unknowns. All parameters on the rhs of these equations are computable from the state vector of satellites at the perturbed and Keplerian orbits. These temporal changes are estimable by fitting a polynomial to the time series of the elements and mathematically modelling the equation of motion of each element as a function of time. The first time derivative of each equation of each orbital element leads to an equation of velocity for the element. The value of temporal changes at any desired geocentric position can be obtained according to the time at that position. Švehla and Földvary (2006) showed that the Newton–Gregory interpolation proved to be more accurate than the smoothing spline for such a purpose.

In the case in which the potential is desired directly, but not v_{nm}, these Gaussian equations can be transformed to integral formulae. If Eq. (1.8) is inserted into Eqs. (2.74)–(2.79) and the addition theorem (Eq. 1.9) is applied to the results, we obtain:

$$\dot{a}=\frac{2a\mu}{4\pi r}\iint_{\sigma} V(R,\theta',\lambda')\left(-e\sin f P_r(r,\psi)+\frac{p}{r^2}P_{\psi}(r,\psi)\cos(\alpha_{\mathrm{Sat}}+\alpha)\right)\mathrm{d}\sigma, \tag{2.80}$$

$$\dot{e}=\frac{2a\overline{m}}{4\pi r}\iint_{\sigma} V(R,\theta',\lambda')\left(-\sin f P_r(r,\psi)+\frac{\overline{\Theta}}{r}P_{\psi}(r,\psi)\cos(\alpha_{\mathrm{Sat}}+\alpha)\right)\mathrm{d}\sigma, \tag{2.81}$$

$$\begin{pmatrix}\dot{i}\\ \dot{\Omega}\end{pmatrix}=\frac{-\mu}{4\pi a}\begin{pmatrix}\cos(\omega+f)\\ \dfrac{\sin(\omega+f)}{\sin i}\end{pmatrix}\iint_{\sigma} V(R,\theta',\lambda')P_{\psi}(r,\psi)\sin(\alpha_{\mathrm{Sat}}+\alpha)\mathrm{d}\sigma, \tag{2.82}$$

$$\dot{\overline{M}}-\overline{n}=\frac{1}{4\pi r}\iint_{\sigma} V(R,\theta',\lambda')\left(\Gamma P_r(r,\psi)-\frac{\pi}{r}P_{\psi}(r,\psi)\cos(\alpha_{\mathrm{Sat}}+\alpha)\right)\mathrm{d}\sigma, \tag{2.83}$$

$$\dot{\omega} = \frac{\overline{m}}{4\pi r} \iint_{\sigma} V(R, \theta', \lambda')$$

$$\times \left\{ \frac{\cos f}{e} P_r(r, \psi) + \frac{P_\psi(r, \psi)}{r} \left(\frac{r+p}{ep} \sin f \cos(\alpha_{\text{Sat}} + \alpha) \right) \right.$$

$$\left. - \frac{r}{p} \sin(\omega + f) \cot i \sin(\alpha_{\text{Sat}} + \alpha) \right\} d\sigma, \tag{2.84}$$

which are integral equations having the potential in their integration domains and the temporal changes in the orbital elements as measurements. The kernels of these integral formulae are not only a function of position, but also the orbital elements. Therefore, their behaviour will be different from one satellite gravimetry mission to another.

Fig. 2.9 shows the temporal variations in the orbital elements of a satellite for a period of 6 h. These velocities have rather periodic variations.

2.3.3 Velocity and acceleration of the perturbations

An alternative way to avoid the singularity problem in the Lagrangian equations is to express the geopotential perturbations as the Hill equations, simplified by Schrama (1986) to the following forms:

$$\ddot{\overline{u}} + 2\overline{n}\,\dot{\overline{v}} = V_u^P, \tag{2.85}$$

$$\ddot{\overline{v}} - 2\overline{n}\dot{\overline{u}} - 3\overline{n}^2\overline{v} = V_v^P, \tag{2.86}$$

$$\ddot{\overline{w}} + \overline{n}^2\overline{w} = V_w^P. \tag{2.87}$$

If V_u^P, V_v^P and V_w^P are given, these three differential equations can be solved to obtain $\overline{u}, \overline{v}$ and $\overline{w}$. Note that Eq. (2.87) is solvable independent of the two others. Schrama (1986) solved these equations and presented different solutions by considering Eq. (2.36) as the Earth's perturbing potential. However, in gravimetry the goal is to determine v_{nm} from these perturbations. If a reference gravity model or Keplerian orbit is available, the perturbing accelerations V_u^P, V_v^P and V_w^P can be computed from it and these differential equations can be solved to obtain $\overline{u}, \overline{v}$ and $\overline{w}$ caused by that reference gravity model. The differences between the measured perturbations and those obtained from this gravity model can be analysed to determine the changes in the reference gravity model for improvement or recovering higher degrees and orders of v_{nm}.

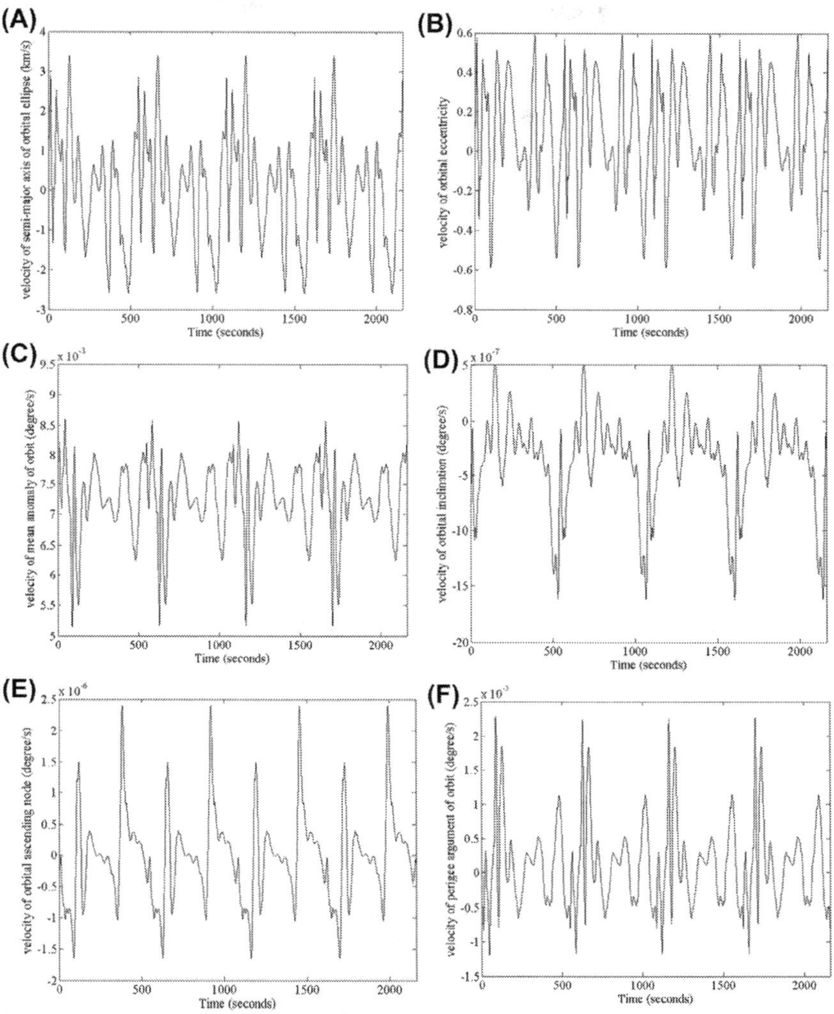

Figure 2.9 (A) $\dot{a}$, (B) $\dot{e}$, (C) $\dot{\bar{M}}$, (D) $\dot{i}$, (E) $\dot{\Omega}$ and (F) $\dot{\omega}$ of orbit during 6 h.

In this case, the lhs of these equations can be considered as knowns, and v_{nm}, which are inherently inside V_u^P, V_v^P and V_w^P, as unknowns.

According to Eqs. (2.69)–(2.71) the Hill equations will be simplified to:

$$\ddot{u} + 2\bar{n}\,\dot{v} = -\frac{1}{r}\sum_{n=2}^{\infty}(n+1)\left(\frac{R}{r}\right)^{n+1}\sum_{m=-n}^{n} v_{nm} Y_{nm}(\theta,\lambda), \tag{2.88}$$

$$\ddot{\overline{v}} - 2\overline{n}\,\dot{\overline{u}} - 3\overline{n}^2\overline{v} = \frac{1}{r}\sum_{n=2}^{\infty}\left(\frac{R}{r}\right)^{n+1}\sum_{m=-n}^{n} v_{nm}A_{nm}(\alpha_{\mathrm{Sat}}, \theta, \lambda), \tag{2.89}$$

$$\ddot{\overline{w}} + \overline{n}^2\overline{w} = \frac{1}{r}\sum_{n=2}^{\infty}\left(\frac{R}{r}\right)^{n+1}\sum_{m=-n}^{n} v_{nm}A'_{nm}(\alpha_{\mathrm{Sat}}, \theta, \lambda). \tag{2.90}$$

See Eqs. (2.6) and (2.7) for the definition of λ^* and its relation with λ.

For solving v_{nm} from these equations, the velocity and acceleration of the perturbations should be determined with global coverages, and each equation will be considered as an observation equation connecting the perturbations in the Hill equations to v_{nm}. A system of equations can be constructed for each equation and for estimating v_{nm}; when the number on the lhs is larger than the number v_{nm} the least-squares approach can be applied. Three integral equations can also be presented according to Eq. (1.8) and the addition theorem of spherical harmonics (Eq. 1.9),

$$\ddot{\overline{u}} + 2\overline{n}\,\dot{\overline{v}} = \frac{1}{4\pi}\iint_{\sigma} V(R, \theta', \lambda')P_r(r, \psi)\mathrm{d}\sigma, \tag{2.91}$$

$$\ddot{\overline{v}} + 2\overline{n}\,\dot{\overline{u}} - 3\overline{n}^2\overline{v} = \frac{1}{4\pi}\iint_{\sigma} V(R, \theta', \lambda')P_\psi(r, \psi)\cos(\alpha_{\mathrm{Sat}} + \alpha)\mathrm{d}\sigma, \tag{2.92}$$

$$\ddot{\overline{w}} + \overline{n}^2\overline{w} = -\frac{1}{4\pi}\iint_{\sigma} V(R, \theta', \lambda')P_\psi(r, \psi)\sin(\alpha_{\mathrm{Sat}} + \alpha)\mathrm{d}\sigma. \tag{2.93}$$

The isotropic parts of the kernels of these integrals have been already presented in Eq. (1.82).

Fig. 2.10A, B and C shows, respectively, V_u^P, V_v^P and V_w^P along an orbit during 6 h. These perturbing accelerations have been generated from an EGM96 gravity model and the orbit is a simulated GOCE-type orbit from the same model.

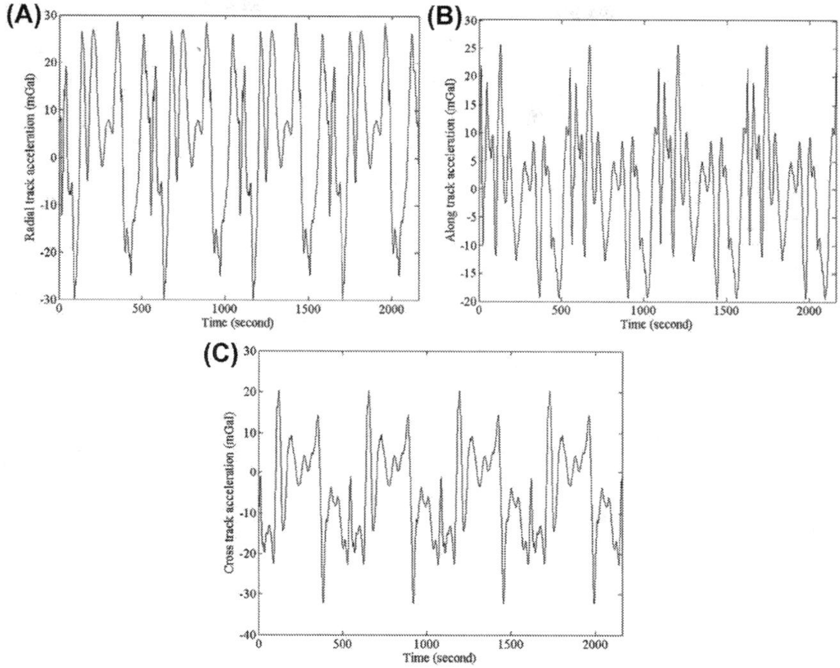

Figure 2.10 (A) radial-track, (B) along-track and (C) cross-track perturbing accelerations.

2.4 Satellite acceleration

If the pure geopotential acceleration vectors of a satellite are measured, they can be used for gravity field recovery purposes directly. There are two ways to recover the Earth's gravitational potential:

(a) The accelerations can be transformed from the TRF to the LNOF based on the position of the satellite, and then the theory presented in Section 1.3.2 is applied.

(b) Mathematical formulae developed for relating the acceleration vector in the TRF are used for recovery purposes.

An acceleration approach was proposed by Ditmar and Eck der Sluijus (2004) and developed further by Guo et al. (2017) from orbits measured by GPS phase measurements. If the position vector of satellite $\mathbf{r}(t)$ is measured at epoch t with intervals Δt along the orbit, the velocity vector of satellite can be computed by:

$$\dot{\mathbf{r}} = \frac{\mathbf{r}(t + \Delta t) - \mathbf{r}(t)}{\Delta t}, \tag{2.94}$$

where $\dot{\mathbf{r}}$ stands for the velocity vector in the interval Δt. However, for estimating the acceleration vector $\ddot{\mathbf{r}}$, three position vectors should be used, and from each two successive position vectors:

$$\ddot{\mathbf{r}}(t) = \frac{\dot{\mathbf{r}}(t+\Delta t) - \dot{\mathbf{r}}(t-\Delta t)}{\Delta t}. \tag{2.95}$$

According to Eqs. (2.94) and (2.95), the acceleration can be written in terms of position vectors at three epochs, $t+\Delta t$, t and $t-\Delta t$; see Guo et al. (2017):

$$\ddot{\mathbf{r}}(t) = \frac{\mathbf{r}(t+\Delta t) - 2\mathbf{r}(t) + \mathbf{r}(t-\Delta t)}{(\Delta t)^2}. \tag{2.96}$$

Eq. (2.96) is a simple three-point numerical differentiation scheme for computing acceleration from a satellite's orbit. Noise in the orbit-derived accelerations is strongly dependent on frequency. Therefore, the key element of the proposed technique is frequency-dependent data weighting (Ditmar and Eck der Sluijus, 2004). In this approach, the satellite positions must be differentiated in an inertial frame like the CRF to yield the accelerations without centrifugal and Coriolis terms. It is important to remember that the total satellite accelerations derived from a satellite orbit have to be converted into residual accelerations by subtracting the contribution of the reference gravity field. Another way to determine the acceleration is to fit a polynomial to the elements of the position vector and find analytical expressions for them. By taking double differentiation of these, analytical forms can lead to an average value for the acceleration in the interval in which the polynomial fitting is performed.

The satellite acceleration in the TRF is given by:

$$\ddot{\mathbf{r}}_{\mathrm{TRF}} = \begin{pmatrix} -\cos\theta\cos\lambda & -\sin\lambda & \sin\theta\cos\lambda \\ -\cos\theta\sin\lambda & \cos\lambda & \sin\theta\sin\lambda \\ \sin\theta & 0 & \cos\theta \end{pmatrix} \begin{pmatrix} V_x(r,\theta,\lambda) \\ V_y(r,\theta,\lambda) \\ V_z(r,\theta,\lambda) \end{pmatrix}_{\mathrm{LNOF}}. \tag{2.97}$$

After substitution of Eqs. (2.2)–(2.4), and considering Eqs. (2.2)–(2.4) with rather long mathematical simplifications, we arrive at:

$$\ddot{\mathbf{r}}_{\mathrm{TRF}} = \frac{1}{r}\begin{bmatrix} \sum_{n=0}^{\infty}\left(\frac{R}{r}\right)^{n+1}\sum_{m=-n}^{n} v_{nm}\left[-\cos\theta\cos\lambda\frac{\partial}{\partial\theta} - \sin\lambda\frac{\partial}{\sin\theta\,\partial\lambda} - (n+1)\cos\lambda\sin\theta\right]Y_{nm}(\theta,\lambda) \\ \sum_{n=0}^{\infty}\left(\frac{R}{r}\right)^{n+1}\sum_{m=-n}^{n} v_{nm}\left[-\cos\theta\sin\lambda\frac{\partial}{\partial\theta} + \cos\lambda\frac{\partial}{\sin\theta\,\partial\lambda} - (n+1)\sin\lambda\sin\theta\right]Y_{nm}(\theta,\lambda) \\ \sum_{n=0}^{\infty}\left(\frac{R}{r}\right)^{n+1}\sum_{m=-n}^{n} v_{nm}\left[\sin\theta\frac{\partial}{\partial\theta} - (n+1)\cos\theta\right]Y_{nm}(\theta,\lambda) \end{bmatrix}. \tag{2.98}$$

The geocentric coordinates of the satellite are known on the rhs of Eq. (2.98), which means that all parameters are known except v_{nm}. If the acceleration vector of the satellite in the TRF, $\ddot{\mathbf{r}}_{\mathrm{TRF}}$, is measured, a system of equations can be constructed connecting v_{nm} to each measured element of the acceleration vector during satellite revolution around the Earth.

By inserting the integral definition of v_{nm}, Eq. (1.8), into Eq. (2.98) and applying the addition theorems of spherical harmonics, Eqs. (1.9), (1.61) and (1.62), the following integral equations are obtained for determining the Earth's gravitational potential $V(R,\theta',\lambda')$ at the surface of the geoid:

$$\ddot{\mathbf{r}}_{\mathrm{TRF}} = \frac{1}{4\pi}\iint_{\sigma} V(R,\theta',\lambda') \times \begin{bmatrix} (-\cos\theta\cos\lambda\cos\alpha + \sin\lambda\sin\alpha)\frac{1}{r}P_{\psi}(r,\psi) + \cos\lambda\sin\theta P_r(r,\psi) \\ (-\cos\theta\sin\lambda\cos\alpha - \cos\lambda\sin\alpha)\frac{1}{r}P_{\psi}(r,\psi) + \sin\lambda\sin\theta P_r(r,\psi) \\ \sin\theta\cos\alpha\frac{1}{r}P_{\psi}(r,\psi) + \cos\theta P_r(r,\psi) \end{bmatrix} d\sigma. \tag{2.99}$$

The satellite accelerations in the TRF on the orbit are now connected to the potential at the surface of the spherical Earth. The terms $P_r(r,\psi)$ and $P_{\psi}(r,\psi)$ have been presented already in Eqs. (1.34) and (1.82) and α is the azimuth between the computation point, or the position of the satellite in the TRF, and the integration point at which the potential is being recovered.

Fig. 2.11 shows the computed acceleration of a satellite with the properties of the GOCE orbit during one day. Obviously, one day of coverage does not give a dense distribution of the satellite orbit. Therefore,

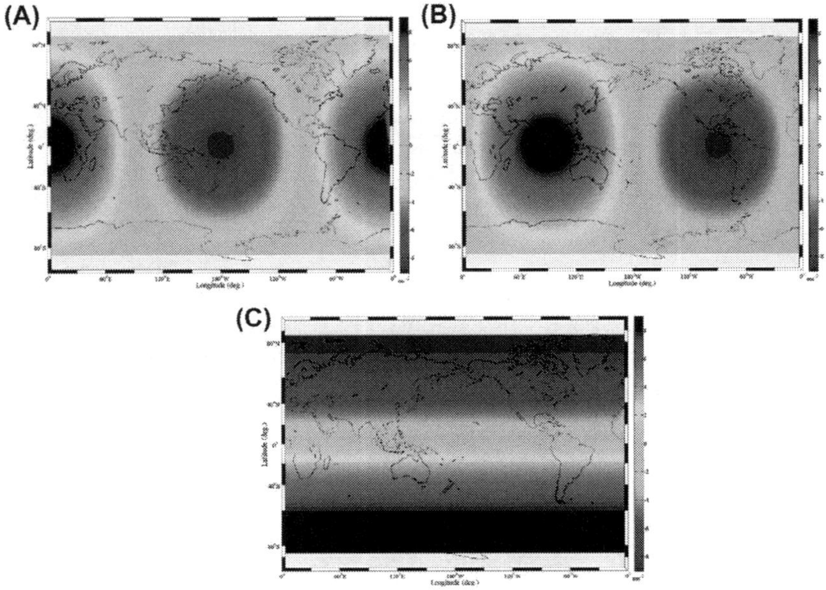

Figure 2.11 (A) $\ddot{x}_{TRF}$, (B) $\ddot{y}_{TRF}$ and (C) $\ddot{z}_{TRF}$.

we had to interpolate the generated accelerations to obtain a grid of them with a resolution of 0.5 × 0.5 degrees all over the globe. Since the GOCE orbit does not cover the polar regions, these areas are empty. Fig. 2.11A, B and C shows, respectively, the accelerations $\ddot{x}_{TRF}$, $\ddot{y}_{TRF}$ and $\ddot{z}_{TRF}$ in units of ms^{-2}. As observed, the map of $\ddot{z}_{TRF}$ differs significantly from those of $\ddot{x}_{TRF}$ and $\ddot{y}_{TRF}$ according to their mathematical definitions (Eq. 2.98).

2.5 Satellite velocity

Bjerhammar (1968) presented the energy integral method for recovering the potential from the kinetic energy of a satellite. In this approach, the satellite velocity is used to determine the kinetic energy and later is applied to determine the potential according to Lagrange:

$$L = \frac{1}{2}\dot{\mathbf{r}}_{CRF}\cdot\dot{\mathbf{r}}_{CRF} + V, \tag{2.100}$$

or Hamiltonian (see Visser et al., 2003):

$$H = \frac{1}{2}\dot{\mathbf{r}}_{TRF}\cdot\dot{\mathbf{r}}_{TRF} - \frac{1}{2}(\boldsymbol{\omega}\times\mathbf{r}_{TRF})\cdot(\boldsymbol{\omega}\times\mathbf{r}_{TRF}) - V, \tag{2.101}$$

where

$\boldsymbol{\omega} = [0 \quad 0 \quad \omega_e]^{\mathrm{T}}$ and $\mathbf{r}_{\mathrm{TRF}} = [x_{\mathrm{TRF}}\ y_{\mathrm{TRF}}\ z_{\mathrm{TRF}}]^{\mathrm{T}}$, where ω_e stands for the Earth's rotation rate. Therefore:

$$\boldsymbol{\omega} \times \mathbf{r}_{\mathrm{TRF}} = [-\omega_e y_{\mathrm{TRF}}\ \omega_e x_{\mathrm{TRF}}\ 0]^{\mathrm{T}}, \tag{2.102}$$

$$(\boldsymbol{\omega} \times \mathbf{r}_{\mathrm{TRF}}) \cdot (\boldsymbol{\omega} \times \mathbf{r}_{\mathrm{TRF}}) = \omega_e^2 \left(x_{\mathrm{TRF}}^2 + y_{\mathrm{TRF}}^2\right). \tag{2.103}$$

Finally:

$$\frac{1}{2}\left(x_{\mathrm{TRF}}^2 + y_{\mathrm{TRF}}^2 + z_{\mathrm{TRF}}^2\right) - \frac{1}{2}\omega_e^2\left(x_{\mathrm{TRF}}^2 + y_{\mathrm{TRF}}^2\right) = -V + H, \tag{2.104}$$

where the Hamiltonian H is constant in the case of no energy dissipation (not only unavailable but irrecoverable); therefore, it can be considered as an extra unknown in addition to v_{nm} in the system of equations formed from the observation Eq. (2.104). If Eq. (1.3) is inserted into Eq. (2.104) and we remove energy dissipation from Eq. (2.104), we obtain (see also Jekeli, 1999):

$$\frac{1}{2}\left(\dot{x}_{\mathrm{TRF}}^2 + \dot{y}_{\mathrm{TRF}}^2 + \dot{z}_{\mathrm{TRF}}^2\right) - \frac{1}{2}\omega_e^2\left(x_{\mathrm{TRF}}^2 + y_{\mathrm{TRF}}^2\right) - \int_0^t \mathbf{a}_{\mathrm{TRF}}.\dot{\mathbf{r}}_{\mathrm{TRF}}\mathrm{d}t$$

$$= -\sum_{n=0}^{\infty}\left(\frac{R}{r}\right)^{n+1}\sum_{m=-n}^{n} v_{nm} Y_{nm}(\theta, \lambda) + H, \tag{2.105}$$

where the integral on the lhs is the energy dissipation of the system in the time interval of 0 to t and $\mathbf{a}_{\mathrm{TRF}}$ is the vector of dissipative acceleration (Jekeli, 1999).

It should be stated that from the position and velocity vectors of the satellite only one equation can be made for recovering v_{nm}. Comparing with the acceleration and other methods presented so far, we can see that the energy integral approach has a direct relation to the potential and not to its derivatives. Therefore, a smoother gravity field can be obtained by this approach than by those working with gravitational accelerations, or the derivatives of the potential. In other words, higher degrees and orders of v_{nm} than those from the accelerations cannot be estimated. The integral form of this equation is obtained by replacing the first term on the rhs with the Poisson integral, or the rhs of Eq. (1.17).

Fig. 2.12 shows the map of the generated lhs of Eq. (2.105), except the integral term, from the positions and velocities of a GOCE-type satellite

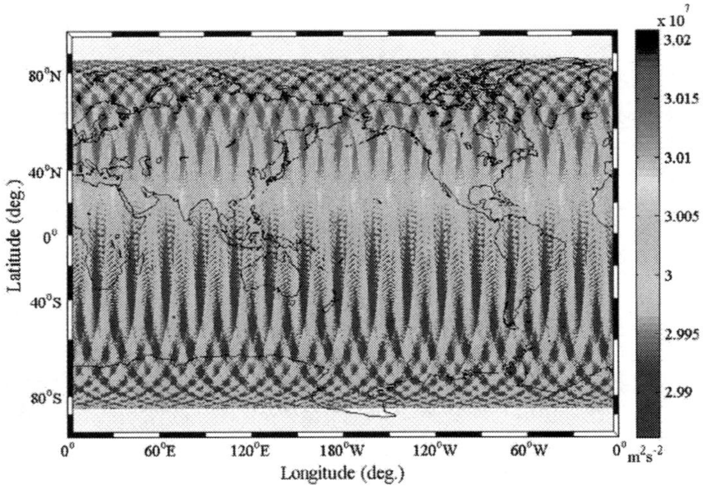

Figure 2.12 $\frac{1}{2}\left(\dot{x}_{\mathrm{TRF}}^{2}+\dot{y}_{\mathrm{TRF}}^{2}+\dot{z}_{\mathrm{TRF}}^{2}\right)-\frac{1}{2}\omega_{e}^{2}\left(x_{\mathrm{TRF}}^{2}+y_{\mathrm{TRF}}^{2}\right)$ of a GOCE-type orbit during 3 months.

after transforming them to the TRF. The orbit has been generated for 3 months with a sampling integration of 10 s. The Earth's gravitational accelerations have been computed from the EGM96 to degree and order 360 and integrated using the Runge—Kutta method of fourth integrator. Larger values are seen in the northern part of the globe, but the values are all about $3 \times 10^7\ \mathrm{m^2s^{-2}}$.

2.6 Inter-satellite range rate

The inter-satellite range, or the distance between satellites, is measured in two different scenarios. In high—low mode the higher orbiting satellites track the LEO ones. For such a purpose, a precise GNSS receiver is mounted on the LEO satellites to receive the GNSS satellite signals. The same principle is used for the other methods presented earlier. The low—low mode of ranging means two low orbiters moving at the same orbit track each other. Due to the mass heterogeneity of the Earth, the distance between the twin satellites will not be constant during their manoeuvres and the changes in this distance are the direct consequence of the gravitational attraction of the Earth's mass and its irregularities. Therefore, the change in the distance or the range rate is an important quantity for gravity field determination. Fig. 2.13 shows this principle. Now, we explain how this range or range rate is related to the gravitational field of the Earth.

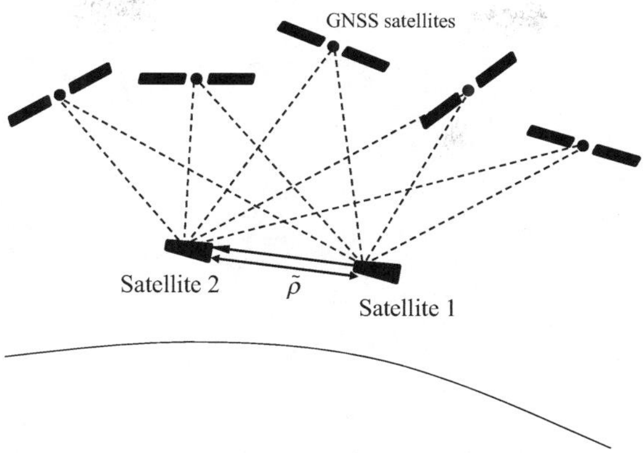

Figure 2.13 High–low and low–low tracking principle. *GNSS*, global navigation satellite system.

2.6.1 Approach 1

Wolff (1969) and Fischell and Pisacane (1978) considered the velocity vector of each satellite in the TOF for finding the mathematical relation between the range rate and the Earth's gravitational potential. For each satellite, they write the kinetic energy as:

$$\frac{1}{2}\left[\left(v_0+\dot{\overline{v}}\right)^2+\dot{\overline{u}}^2+\dot{\overline{w}}^2\right]=\frac{1}{2}v_0^2+v_0\dot{\overline{v}}+\frac{1}{2}\left(\dot{\overline{v}}^2+\dot{\overline{u}}^2+\dot{\overline{w}}^2\right)\approx\frac{1}{2}v_0^2+v_0\dot{\overline{v}}, \tag{2.106}$$

where v_0 is the mean velocity of the satellite and $\overline{u}$, $\overline{v}$ and $\overline{w}$ are, respectively, the radial-, along- and cross-track geopotential perturbations, and $\dot{\overline{u}}$, $\dot{\overline{v}}$ and $\dot{\overline{w}}$ their temporal variations, respectively.

Fischell and Pisacane (1978) mentioned that $\frac{1}{2}\left(\dot{\overline{v}}^2+\dot{\overline{u}}^2+\dot{\overline{w}}^2\right)$ is very small and negligible as $\dot{\overline{v}}$, $\dot{\overline{u}}$ and $\dot{\overline{w}}$ are each significantly smaller than the mean velocity of the satellites.

The interaction of the kinetic and potential energy $V(r,\theta,\lambda)$ should be constant. This means that:

$$\frac{1}{2}v_0^2+v_0\dot{\overline{v}}+V(r,\theta,\lambda)=c, \tag{2.107}$$

where c is the constant, and for a close circular orbit it will be equal to $GM/2a$, where a is the semi-major axis of the orbital ellipse (Fischell and Pisacane, 1978).

Eq. (2.107) expresses the energy compensation for a single satellite. If two satellites are following each other in the same orbit, this equation can be written for each one of them. When they are subtracted from each other, and after some simplifications, we obtain:

$$V(r_2, \theta_2, \lambda_2) - V(r_1, \theta_1, \lambda_1) = -v_0\left(\dot{\bar{v}}_2 - \dot{\bar{v}}_1\right), \tag{2.108}$$

meaning that the difference between the along-track velocities of the satellite pair is proportional to the potential difference between them.

Now, according to Eq. (1.3), we obtain:

$$-v_0\left(\dot{\bar{v}}_2 - \dot{\bar{v}}_1\right) = \sum_{n=0}^{\infty} \sum_{m=-n}^{n} v_{nm} \left(\left(\frac{R}{r_2}\right)^{n+1} Y_{nm}(\theta_2, \lambda_2) - \left(\frac{R}{r_1}\right)^{n+1} Y_{nm}(\theta_1, \lambda_1)\right). \tag{2.109}$$

Again, by inserting the integral definition of v_{nm}, Eq. (1.8), into Eq. (2.109) and applying the addition theorem of spherical harmonics, Eq. (1.9), the following integral equations are obtained for determining $V(R, \theta', \lambda')$:

$$-v_0\left(\dot{\bar{v}}_2 - \dot{\bar{v}}_1\right) = \frac{1}{4\pi} \iint_{\sigma} V(R, \theta', \lambda')\left(P(r_2, \psi_2) - P(r_1, \psi_1)\right) \mathrm{d}\sigma. \tag{2.110}$$

Eq. (2.107) relates the potential to the velocities of the along-track perturbations of both satellites. Now, we should find a mathematical model for the connections between these along-track velocity differences to the range rate between satellites. To do so, consider Fig. 2.14, which

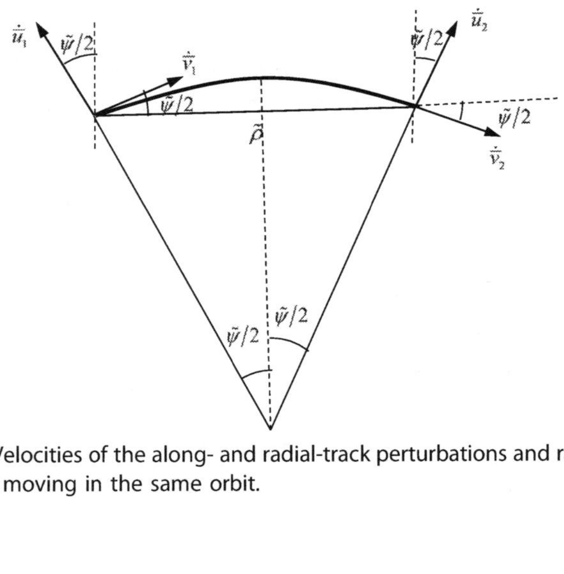

Figure 2.14 Velocities of the along- and radial-track perturbations and range between two satellites moving in the same orbit.

shows two satellites at positions 1 and 2 along the same orbit. The term $\widetilde{\rho}$ is the distance or range between the satellites and $\widetilde{\psi}$ is the geocentric angle between them. According to Fig. 2.14 it is not difficult to show that the range rate due to perturbations between the satellites can be expressed by the rates of the along- and radial-track components of perturbations. Fischell and Pisacane (1978) showed that:

$$\dot{\widetilde{\rho}} = \left(\dot{\bar{u}}_1 + \dot{\bar{u}}_2\right)\sin\frac{\widetilde{\psi}}{2} + \left(\dot{\bar{v}}_1 - \dot{\bar{v}}_2\right)\cos\frac{\widetilde{\psi}}{2}. \tag{2.111}$$

The term $\widetilde{\psi}$ will be very small, as the distance between satellites is considerably smaller than the Earth's radius and the satellite altitude. In this case, Eq. (2.111) is approximated further to:

$$\dot{\widetilde{\rho}} \approx \dot{\bar{v}}_1 - \dot{\bar{v}}_2. \tag{2.112'}$$

Therefore, the range rate between the satellites is almost the difference between the velocities of their along-track perturbations. In this case, Eqs. (2.109) and (2.110) will change to:

$$v_0\dot{\widetilde{\rho}} = \sum_{n=0}^{\infty}\sum_{m=-n}^{n} v_{nm}\left(\left(\frac{R}{r_2}\right)^{n+1} Y_{nm}(\theta_2, \lambda_2) - \left(\frac{R}{r_1}\right)^{n+1} Y_{nm}(\theta_1, \lambda_1)\right) \tag{2.113'}$$

and

$$v_0\dot{\widetilde{\rho}} = \frac{1}{4\pi}\iint_{\sigma} V(R, \theta', \lambda')(P(r_2, \psi_2) - P(r_1, \psi_1))d\sigma. \tag{2.114'}$$

Fig. 2.15 shows the map of the potential difference between the twin satellites of the GRACE mission generated from the EGM2008 (Pavlis et al., 2012) up to degree and order 180, in the period from 1 January 2010 to 30 June 2010, over an area bounded by the spherical coordinates latitude between 5 and 57 degrees and longitude between 55 and 115 degrees. The restricted domain covers the area of India, Himalayas and Southeast Asia and contains 30,036 points in total.

2.6.2 Approach 2

To see how the range rates are related to the gravity field of the Earth, we present the approach used at the French Space Agency (CNES), which was

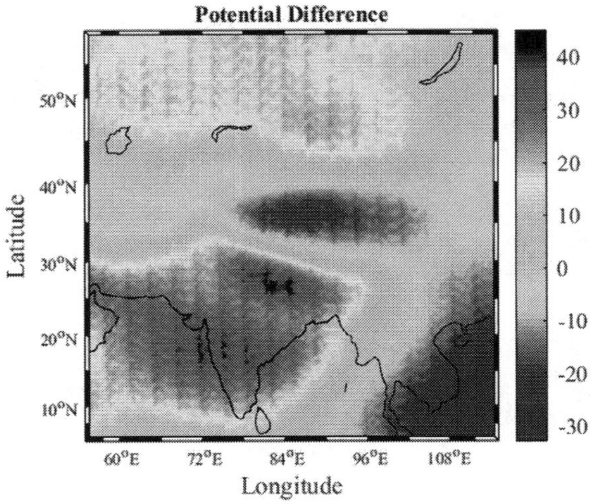

Figure 2.15 Map of the potential difference between the twin satellites of the GRACE mission from 1 January to 1 June 2010 (m^2/s^2) *(From Eshagh M., Sprlak M., 2016. On the integral inversion of satellite-to-satellite velocity differences for local gravity field recovery: a theoretical study. Celestial Mech. Dyn. Astron. 124, 27–144.)*

applied by Eshagh et al. (2013) to investigate the quality of smoothing methods for the GRACE range-rate data inversion:

$$\dot{\tilde{\rho}} = \|\dot{\mathbf{r}}_2 - \dot{\mathbf{r}}_1\|, \tag{2.112'}$$

where $\dot{\mathbf{r}}_1$ and $\dot{\mathbf{r}}_2$ are the velocity or position rate of the vectors $\mathbf{r}_1$ and $\mathbf{r}_2$, and $\|\cdot\|$ stands for L_2-norm. Some a priori values of the estimated parameter are needed, which can be estimated using numerical integration of the twin satellites' orbits. Eq. (2.112') should be linearised around the approximate values of $\dot{\mathbf{r}}_1$ and $\dot{\mathbf{r}}_2$:

$$\dot{\tilde{\rho}} - \dot{\tilde{\rho}}_0 = \sum_{i=1}^{2N+1} \frac{\partial \dot{\tilde{\rho}}}{\partial \Upsilon_i} \delta \Upsilon_i, \tag{2.113'}$$

where Υ_i is a vector containing the unknown parameters v_{nm} to be estimated. The term $\dot{\tilde{\rho}}_0$ is the approximate value of the range rate computed from the initial values of Υ_i^0, and $\delta\Upsilon_i = \Upsilon_i - \Upsilon_i^0$ are the corrections to the initial values. The main problem is to find $\partial\dot{\tilde{\rho}}/\partial\Upsilon_i$, as the range rate does not have a direct mathematical relation with the unknowns. According to the chain rule of partial derivative, one can write:

$$\frac{\partial \dot{\tilde{\rho}}}{\partial \Upsilon_i} = \frac{\partial \dot{\tilde{\rho}}}{\partial \mathbf{r}} \frac{\partial \mathbf{r}}{\partial \Upsilon_i} + \frac{\partial \dot{\tilde{\rho}}}{\partial \dot{\mathbf{r}}} \frac{\partial \dot{\mathbf{r}}}{\partial \Upsilon_i}, \tag{2.114'}$$

where $\boldsymbol{\gamma}$ is the vector of unknown parameters. $\partial\dot{\widetilde{\rho}}/\partial\mathbf{r}$ and $\partial\dot{\widetilde{\rho}}/\partial\dot{\mathbf{r}}$ can be derived analytically from Eq. (2.112'). To do that, let the satellite acceleration be a function of $\mathbf{r}$, $\dot{\mathbf{r}}$, $\boldsymbol{\gamma}$ and time t:

$$\ddot{\mathbf{r}}=f(\mathbf{r},\dot{\mathbf{r}},\boldsymbol{\gamma},t), \tag{2.115}$$

where $f(\mathbf{r},\dot{\mathbf{r}},\boldsymbol{\gamma},t)$ is a function relating the position vector, time and $\boldsymbol{\gamma}$ (which is assumed independent of time); see, e.g., Eq. (2.5), which is the mathematical model of $f(\mathbf{r},\dot{\mathbf{r}},\boldsymbol{\gamma},t)$. The partial derivatives of this function with respect to $\boldsymbol{\gamma}$ will be:

$$\frac{\partial\ddot{\mathbf{r}}}{\partial\boldsymbol{\gamma}}=\frac{\partial f(\mathbf{r},\dot{\mathbf{r}},\boldsymbol{\gamma},t)}{\partial\mathbf{r}}\frac{\partial\mathbf{r}}{\partial\boldsymbol{\gamma}}+\frac{\partial f(\mathbf{r},\dot{\mathbf{r}},\boldsymbol{\gamma},t)}{\partial\dot{\mathbf{r}}}\frac{\partial\dot{\mathbf{r}}}{\partial\boldsymbol{\gamma}}+\frac{\partial f(\mathbf{r},\dot{\mathbf{r}},\boldsymbol{\gamma},t)}{\partial\boldsymbol{\gamma}}. \tag{2.116}$$

Using the following relationships:

$$\begin{cases}\dfrac{\partial\ddot{\mathbf{r}}}{\partial\boldsymbol{\gamma}}=\dfrac{\mathrm{d}^2}{\mathrm{d}t^2}\left(\dfrac{\partial\mathbf{r}}{\partial\boldsymbol{\gamma}}\right)\\[2ex] \dfrac{\partial\dot{\mathbf{r}}}{\partial\boldsymbol{\gamma}}=\dfrac{\mathrm{d}}{\mathrm{d}t}\left(\dfrac{\partial\mathbf{r}}{\partial\boldsymbol{\gamma}}\right)\end{cases}, \tag{2.117}$$

Eq. (2.116) becomes:

$$\frac{\mathrm{d}^2}{\mathrm{d}t^2}\left(\frac{\partial\mathbf{r}}{\partial\boldsymbol{\gamma}}\right)=\frac{\partial f(\mathbf{r},\dot{\mathbf{r}},\boldsymbol{\gamma},t)}{\partial\mathbf{r}}\frac{\partial\mathbf{r}}{\partial\boldsymbol{\gamma}}+\frac{\partial f(\mathbf{r},\dot{\mathbf{r}},\boldsymbol{\gamma},t)}{\partial\dot{\mathbf{r}}}\frac{\mathrm{d}}{\mathrm{d}t}\left(\frac{\partial\mathbf{r}}{\partial\boldsymbol{\gamma}}\right)+\frac{\partial f(\mathbf{r},\dot{\mathbf{r}},\boldsymbol{\gamma},t)}{\partial\boldsymbol{\gamma}}. \tag{2.118}$$

The derivatives of $f(\mathbf{r},\dot{\mathbf{r}},\boldsymbol{\gamma},t)$ with respect to $\mathbf{r}$, $\dot{\mathbf{r}}$ and $\boldsymbol{\gamma}$ are obtained using the mathematical models of $f(\mathbf{r},\dot{\mathbf{r}},\boldsymbol{\gamma},t)$. Numerical integration of the vector differential equation (see Eq. 2.118) with respect to the time t will yield $\partial\mathbf{r}/\partial\boldsymbol{\gamma}$ and $\partial\dot{\mathbf{r}}/\partial\boldsymbol{\gamma}$. Different numerical integrators can be used in this respect, e.g. Runge–Kutta, Runge–Kutta–Nyström, predictor–corrector methods of Adams–Bashforth and Adams–Moulton (cf. Somodi and Földvary, 2011) or Cowell (cf. Parrot, 1989 and Santos, 1995). By substituting numerically solved $\partial\mathbf{r}/\partial\boldsymbol{\gamma}$ and $\partial\dot{\mathbf{r}}/\partial\boldsymbol{\gamma}$ into Eq. (2.114), the coefficient matrix of the system is derived. Here, the initial values of the partial derivatives are set at zero for the parameters controlling the dynamics of satellites. The identity matrix for the Jacobian between Cartesian and Keplerian coordinates at the initial state of Keplerian coordinates are used instead. A similar discussion about numerically finding the partial derivatives was provided by Rummel (1975) for downward continuation of satellite-to-satellite tracking data.

The system of Eq. (2.113') can contains millions of observations when the data have a global and dense coverage, and working with such a system is not always possible due to the extremely large dimensions of its coefficient matrix. Such a large over-determined system should be solved using least-squares.

2.7 Line-of-sight measurements

As Fig. 2.14 illustrates, the distance between satellites 1 and 2 is $\widetilde{\rho}$. This distance can be simply presented based on the coordinates of both satellites (see e.g. Keller and Sharifi, 2005; Novak, 2007):

$$\widetilde{\rho} = \|\delta\mathbf{r}_{12}\| = \|\mathbf{r}_2 - \mathbf{r}_1\|, \tag{2.119}$$

where $\mathbf{r}_1$ and $\mathbf{r}_2$ are respectively the position vectors of satellites 1 and 2 in the TRF. If we take the derivative of Eq. (2.119) with respect to time, the range rate will be:

$$\dot{\widetilde{\rho}} = \frac{\mathrm{d}\|\delta\mathbf{r}_{12}\|}{\mathrm{d}t} = \delta\dot{\mathbf{r}}_{12}\frac{\delta\mathbf{r}_{12}}{\widetilde{\rho}} = \delta\dot{\mathbf{r}}_{12}\cdot\mathbf{e}_{\mathrm{LOS}} \quad \text{and} \quad \delta\dot{\mathbf{r}}_{12} = \dot{\mathbf{r}}_2 - \dot{\mathbf{r}}_1, \tag{2.120}$$

where $\mathbf{e}_{\mathrm{LOS}} = \delta\mathbf{r}_{12}/\widetilde{\rho}$ is the line-of-sight (LOS) unit vector pointing from satellite 1 to satellite 2. Let us take the derivative of Eq. (2.120) one more time with respect to time (see also Rummel, 1980):

$$\begin{aligned}\ddot{\widetilde{\rho}} &= \frac{\mathrm{d}}{\mathrm{d}t}(\delta\dot{\mathbf{r}}_{12}.\mathbf{e}_{\mathrm{LOS}}) = \delta\ddot{\mathbf{r}}_{12}\mathbf{e}_{\mathrm{LOS}} + \delta\dot{\mathbf{r}}_{12}.\dot{\mathbf{e}}_{\mathrm{LOS}} \\ &= \delta\ddot{\mathbf{r}}_{12}\mathbf{e}_{\mathrm{LOS}} + \frac{(\delta\dot{\mathbf{r}}_{12})^2 - \dot{\widetilde{\rho}}^2}{\widetilde{\rho}} \quad \text{and} \quad \delta\ddot{\mathbf{r}}_{12} = \ddot{\mathbf{r}}_2 - \ddot{\mathbf{r}}_1.\end{aligned} \tag{2.121}$$

Therefore, a mathematical model for the range acceleration between the satellite pair is derived. This acceleration is a function of relative position and acceleration and velocity vectors between the satellites in addition to the range and range rate.

Generally, the acceleration is the gradient of gravitational potential according to Newton's law. Therefore, the relative acceleration between the satellite is:

$$\delta\ddot{\mathbf{r}}_{12} = \nabla\delta V_{12} \quad \text{where} \quad \delta V_{12} = V(r_2, \theta_2, \lambda_2) - V(r_1, \theta_1, \lambda_1), \tag{2.122}$$

and ∇ is the gradient operator (see Eq. 1.45) and δV_{12} is the potential difference between the satellites.

By inserting Eq. (2.122) into Eq. (2.121) and solving the results for $\nabla\delta V_{12}$ we obtain:

$$\ddot{\widetilde{\rho}} - \frac{(\delta\dot{\mathbf{r}}_{12})^2 - \dot{\widetilde{\rho}}^2}{\widetilde{\rho}} = \nabla\delta V_{12}.\mathbf{e}_{\mathrm{LOS}}. \tag{2.123}$$

To find v_{nm}, the spherical harmonic expansion of δV_{12} is used:

$$\delta V_{12} = \sum_{n=0}^{\infty}\sum_{m=-n}^{n} v_{nm}\left(\left(\frac{R}{r_2}\right)^{n+1} Y_{nm}(\theta_2, \lambda_2) - \left(\frac{R}{r_1}\right)^{n+1} Y_{nm}(\theta_1, \lambda_1)\right). \tag{2.124}$$

By applying the gradient operator (see Eq. (1.45)) to Eq. (2.124) we arrive at:

$$\nabla\delta V_{12} = \sum_{n=0}^{\infty}\sum_{m=-n}^{n} v_{nm}\sum_{i=1}^{2}\left(-(-1)^i\frac{n+1}{r_i}\left(\frac{R}{r_i}\right)^{n+1}\mathbf{X}_{nm}^{(1)}(\theta_i, \lambda_i) + \frac{(-1)^i}{r_i}\left(\frac{R}{r_i}\right)^{n+1}\left(\mathbf{X}_{nm}^{(2)}(\theta_i, \lambda_i)\right)\right), \tag{2.125}$$

where $\mathbf{X}_{nm}^{(1)}(\theta_i, \lambda_i)$ and $\mathbf{X}_{nm}^{(2)}(\theta_i, \lambda_i)$ are vector spherical hamronics, see Section 1.3.2, at the positions of the satellites 1 and 2.

Inserting Eq. (2.125) into Eq. (2.123) gives the mathematical relation between the LOS measurements and v_{nm}:

$$\ddot{\widetilde{\rho}} - \frac{(\delta\dot{\mathbf{r}}_{12})^2 - \dot{\widetilde{\rho}}^2}{\widetilde{\rho}} = \sum_{n=0}^{\infty}\sum_{m=-n}^{n} v_{nm}\sum_{i=1}^{2}\left(-(-1)^i\frac{n+1}{r_i}\left(\frac{R}{r_i}\right)^{n+1}\mathbf{X}_{nm}^{(1)}(\theta_i, \lambda_i) + \frac{(-1)^i}{r_i}\left(\frac{R}{r_i}\right)^{n+1}\mathbf{X}_{nm}^{(2)}(\theta_i, \lambda_i)\right).\mathbf{e}_{\mathrm{LOS}}. \tag{2.126}$$

By taking the gradient of the potential the acceleration is obtained in the LNOF, but the origin of this frame depends on the position of satellite. The inner product of these gradients and $\mathbf{e}_{\mathrm{LOS}}$ will project them to the LOS direction from satellites 1 and 2, or in the direction in which the range and range rate are measured.

Eq. (2.126) is considered an observation equation in which the observations are on the lhs and the unknowns are v_{nm} on the rhs. Note that the vector spherical harmonics and $\mathbf{e}_{\text{LOS}}$ can be simply computed from the satellites' positions. A system of equations can be organised by this equation and along the twin satellite orbits, and when the measurements cover the whole globe, the system becomes solvable by the least-squares method. An integral equation can also be derived for inverting the range:

$$\ddot{\widetilde{\rho}} - \frac{(\delta\dot{\mathbf{r}}_{12})^2 - \dot{\widetilde{\rho}}^2}{\widetilde{\rho}} = \frac{1}{4\pi}\iint_{\sigma} V(R,\theta',\lambda')K(r_1,\psi_1,\alpha_1,r_2,\psi_2,\alpha_2)\mathrm{d}\sigma, \tag{2.127}$$

where

$$\begin{aligned} K(r_1,\psi_1,\alpha_1,r_2,\psi_2,\alpha_2) &= \left(P_{r_2}(r_2,\psi_2)\mathbf{e}_{r_2} + \frac{1}{r_2}P_{\psi_2}(r_2,\psi_2)\right. \\ &\left.\times\left(\cos\alpha_2\mathbf{e}_{x_2} - \sin\alpha_2\mathbf{e}_{y_2}\right)\right)\cdot\mathbf{e}_{\text{LOS}} - \left(P_{r_1}(r_1,\psi_1)\mathbf{e}_{r_1} + \frac{1}{r_1}P_{\psi_1}(r_1,\psi_1)\right. \\ &\left.\times\left(\cos\alpha_1\mathbf{e}_{x_1} - \sin\alpha_1\mathbf{e}_{y_1}\right)\right)\cdot\mathbf{e}_{\text{LOS}} \end{aligned} \tag{2.128}$$

and α_1 and α_2 are respectively the azimuth between satellites 1 and 2 and the integration point at which the potential is being recovered.

We name this kernel function bipolar as there are two points from which the kernel should be computed, which are the positions of satellites 1 and 2. The closed-form formulae for the derivative of the Poisson kernel have been presented in Eq. (1.34).

Fig. 2.16 shows the map of the LOS disturbing potential between the twin satellites of the GRACE mission, produced by Eshagh and Sprlak (2016) for recovering the sub-lithospheric stress over the area; see the explanation for Fig. 2.15.

2.8 Satellite gravity gradiometry

Gravity gradiometry means measuring the components of the gravitation tensor, or simply the gradients, which are the second-order derivatives of the potential. Determination of v_{nm} up to higher degrees and orders is possible by gravity gradiometry as the gradients are more sensitive to local

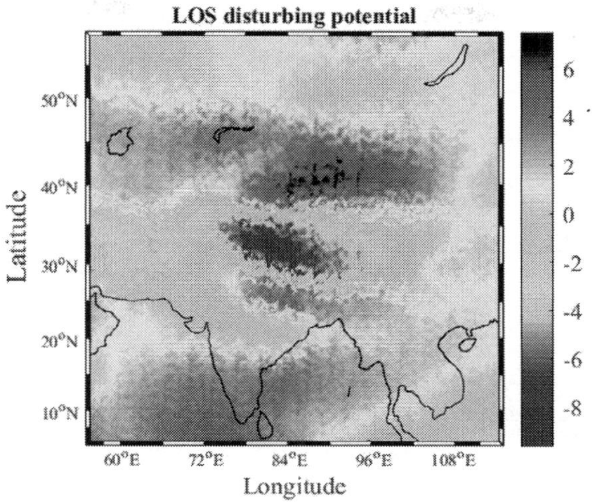

Figure 2.16 The map of the line-of-sight (*LOS*) disturbing potential between the twin satellites of the GRACE mission from 1 January to 1 June 2010 (mGal).

features of the gravity field; the frame in which the gradients are measured is called the gradiometer frame (GF). In Fig. 2.17 a gradiometer is presented schematically. As observed, two accelerometers are mounted along each axis of the GF at a specified distance. When a satellite is carrying such a gradiometer, the accelerations measured at each two accelerations will be different. We already know from Newton's law that the acceleration is the gradient of the potential, but in gradiometry the gradient of the gradient of the potential is measured. A gradiometer is, in other words, a differential accelerometer.

The tensor of gravitation has been presented in Eq. (1.83), and the operators, which are used for deriving these gradients, have been presented

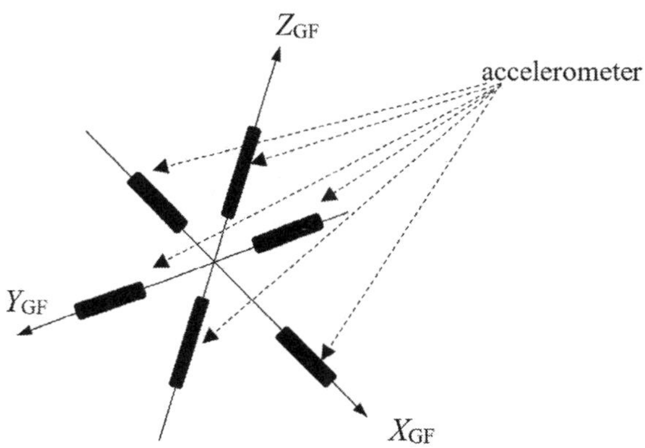

Figure 2.17 Gradiometer frame

in Eqs. (1.95)–(1.100). By inserting the spherical harmonic expression of the gravitational potential, Eq. (1.3), into these equations, the spherical harmonic expansions of the gradients in the LNOF are derived:

$$V_{zz} = \frac{1}{r^2}\sum_{n=0}^{\infty}(n+1)(n+2)\left(\frac{R}{r}\right)^{n+1}\sum_{m=-n}^{n} v_{nm}Y_{nm}(\theta,\lambda), \tag{2.129}$$

$$V_{xx} = \frac{1}{r^2}\sum_{n=0}^{\infty}\left(\frac{R}{r}\right)^{n+1}\sum_{m=-n}^{n} v_{nm}\left(-(n+1)Y_{nm}(\theta,\lambda)+\frac{\partial^2 Y_{nm}(\theta,\lambda)}{\partial\theta^2}\right), \tag{2.130}$$

$$V_{yy} = \frac{1}{r^2}\sum_{n=0}^{\infty}\left(\frac{R}{r}\right)^{n+1}\sum_{m=-n}^{n}\left(-(n+1)Y_{nm}(\theta,\lambda)+\frac{1}{\tan\theta}\frac{\partial Y_{nm}(\theta,\lambda)}{\partial\theta}\right.$$
$$\left.+\frac{1}{\sin^2\theta}\frac{\partial^2 Y_{nm}(\theta,\lambda)}{\partial\lambda^2}\right), \tag{2.131}$$

$$V_{xy} = \frac{1}{r^2}\sum_{n=2}^{\infty}\left(\frac{R}{r}\right)^{n+1}\sum_{m=-n}^{n} v_{nm}\left(\frac{1}{\sin\theta}\frac{\partial^2 Y_{nm}(\theta,\lambda)}{\partial\theta\partial\lambda}-\frac{\cos\theta}{\sin^2\theta}\frac{\partial Y_{nm}(\theta,\lambda)}{\partial\lambda}\right), \tag{2.132}$$

$$V_{xz} = \frac{1}{r^2}\sum_{n=1}^{\infty}(n+2)\left(\frac{R}{r}\right)^{n+1}\sum_{m=-n}^{n} v_{nm}\frac{\partial Y_{nm}(\theta,\lambda)}{\partial\theta}, \tag{2.133}$$

$$V_{yz} = \frac{1}{r^2}\sum_{n=1}^{\infty}(n+2)\left(\frac{R}{r}\right)^{n+1}\sum_{m=-n}^{n} v_{nm}\frac{\partial Y_{nm}(\theta,\lambda)}{\sin\theta\partial\lambda}. \tag{2.134}$$

Spectral and integral solutions for v_{nm} from the gradients in the LNOF have been presented in Section 1.4, therefore, we do not repeat them here.

Fig. 2.18 presents the maps of the gravity gradients reduced for the normal gravity field GRS80 (Moritz, 1980), which were generated by Eshagh (2010) in the LNOF at 250 km above sea level. The mathematical models applied for generating these maps are Eqs. (2.129)–(2.134). The gravitational potential reduced for the normal potential is called the disturbing potential (T), and this is the reason for seeing the derivatives of T in the LNOF rather than V. However, the principle is the same in both cases. Note that the unit of the gravity gradients is Eötvös (E), which is equivalent to s^{-2}.

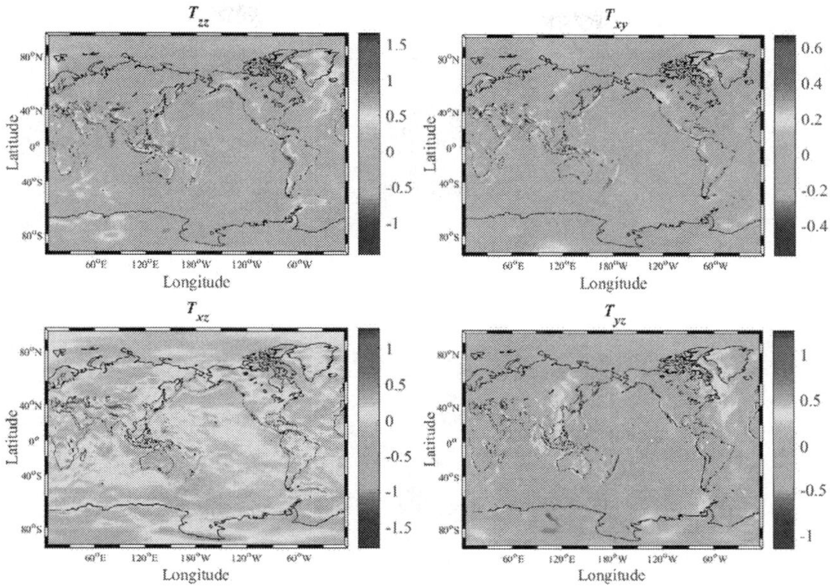

Figure 2.18 T_{xx}, T_{yy}, T_{zz}, T_{xy}, T_{zz} and T_{yz} (E).

Petrovskaya and Vershkov (2006) and Eshagh (2010) presented non-singular formulae for Eqs. (2.129)–(2.134), and Eshagh (2010) presented similar formulae for the tensor spherical harmonics. Petrovskaya and Vershkov (2006) also defined the gravity gradients in a frame, which is equivalent to the TOF but in the TRF. It is an LNOF rotated around its z-axis by the satellite track azimuth α_{Sat}.

The gravitational tensor can be presented in the TOF as well by:

$$\begin{bmatrix} V_{vv} & V_{vw} & V_{vu} \\ V_{vw} & V_{ww} & V_{wu} \\ V_{vu} & V_{wu} & V_{uu} \end{bmatrix} = \begin{bmatrix} \cos\alpha_{\mathrm{Sat}} & \sin\alpha_{\mathrm{Sat}} & 0 \\ -\sin\alpha_{\mathrm{Sat}} & \cos\alpha_{\mathrm{Sat}} & 0 \\ 0 & 0 & 1 \end{bmatrix}$$
$$\times \begin{bmatrix} V_{xx} & V_{xy} & V_{xz} \\ V_{xy} & V_{yy} & V_{yz} \\ V_{xz} & V_{yz} & V_{zz} \end{bmatrix} \begin{bmatrix} \cos\alpha_{\mathrm{Sat}} & \sin\alpha_{\mathrm{Sat}} & 0 \\ -\sin\alpha_{\mathrm{Sat}} & \cos\alpha_{\mathrm{Sat}} & 0 \\ 0 & 0 & 1 \end{bmatrix}^{\mathrm{T}}, \qquad (2.135)$$

where α_{Sat} is the satellite track azimuth; see Eq. (2.27). After the matrix multiplications are performed, the elements of the gravitational tensor in the TOF will be:

$$V_{uu} = V_{zz}, \tag{2.136}$$

$$V_{vv} = \cos^2\alpha_{Sat} V_{xx} + 2\sin\alpha_{Sat}\cos\alpha_{Sat} V_{xy} + \sin^2\alpha_{Sat} V_{yy}, \tag{2.137}$$

$$V_{ww} = \sin^2\alpha_{Sat} V_{xx} - 2\sin\alpha_{Sat}\cos\alpha_{Sat} V_{xy} + \cos^2\alpha_{Sat} V_{yy}, \tag{2.138}$$

$$V_{vw} = V_{xy}(\cos^2\alpha_{Sat} - \sin^2\alpha_{Sat}) + \cos\alpha_{Sat}\sin\alpha_{Sat}(V_{yy} - V_{xx}), \tag{2.139}$$

$$V_{vu} = \cos\alpha_{Sat} V_{xz} + \sin\alpha_{Sat} V_{yz}, \tag{2.140}$$

$$V_{wu} = -\sin\alpha_{Sat} V_{xz} + \sin\alpha_{Sat} V_{yz}. \tag{2.141}$$

It will not be difficult to show that:

$$V_{vv} + V_{ww} = V_{xx} + V_{yy}, \tag{2.142}$$

and since the gravitational potential is harmonic outside the Earth, we can write:

$$V_{zz} = -V_{vv} - V_{ww}. \tag{2.143}$$

2.9 Satellite altimetry data

Satellite altimetry data can also be used for determining the Earth's gravity field over oceans. The altimetry satellites transfer radar signals down to the surface of the water over oceans, seas and lakes and receive their reflections. Since the velocity of the signal is known, the distance between the satellite and the water surface can be estimated by measuring the signal travel time. Therefore, the sea surface topography (SST), showing the water level variations over the oceans, can be measured by this technique; see Fig. 2.19.

In the gravimetry problem using satellite altimetry, there is no downward continuation of data. So, the problem is to convert the altimetry data to gravity information. However, what makes satellite gravimetry complicated is the downward continuation process, which is different from one satellite gravimetry data point to another.

Suppose an SST model is available; then it can be used to determine geoid height, which is approximately the mean sea level (see Chapter 3).

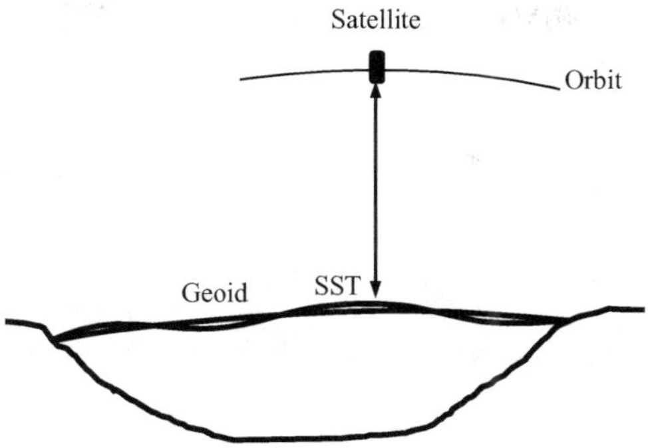

Figure 2.19 Satellite altimetry. *SST*, sea surface topography.

The geoid has a direct relation with the Earth's gravitational potential, as it is an equipotential surface, which has the best fit to mean sea level. The measured geoid height and its slopes along the north and east, which are known as the deflections of vertical, Here, our goal is to explain how the gravity field can be determined from the satellite altimetry data over oceans in a general sense.

Consider that the geoid height is the representative of the gravity field and it has the following spherical harmonic expansion:

$$N(R,\theta,\lambda)=\sum_{n=2}^{\infty}\sum_{m=-n}^{n}N_{nm}Y_{nm}(\theta,\lambda), \tag{2.144}$$

where N is the geoid height and N_{nm} its SHCs. The geoid height is defined directly at the surface of the spherical Earth, and therefore, the upward continuation factor will be 1. Since these data have limited coverage, according to Eq. (1.8), we can write:

$$N^{\mathrm{O}}_{nm}=\frac{1}{4\pi}\iint_{\sigma_{\mathrm{Oceans}}}N(R,\theta',\lambda')Y_{nm}(\theta',\lambda')\mathrm{d}\sigma, \tag{2.145}$$

where N^{O}_{nm} means the SHCs of the altimetry geoid over oceans. This is the reason that the integration domain is not σ but σ_{Oceans}. The lack of satellite altimetry data over the continents causes the determined N^{O}_{nm} to contain a bias from its true values. To estimate this bias, let us insert Eq. (2.145) into Eq. (2.144), and then:

$$N_{nm}^{O} = \frac{1}{4\pi} \iint_{\sigma_{\text{Oceans}}} \sum_{n'=2}^{\infty} \sum_{m'=-n'}^{n'} N_{n'm'} Y_{n'm'}(\theta', \lambda') Y_{nm}(\theta', \lambda') d\sigma. \tag{2.146}$$

In fact, by Eq. (2.146) our goal is to estimate the signal over the ocean (see the integration domain). By interchanging the integral and summations, we obtain:

$$N_{nm}^{O} = \sum_{n'=2}^{\infty} \sum_{m'=-n'}^{n'} N_{n'm'} \frac{1}{4\pi} \iint_{\sigma_{\text{Oceans}}} Y_{n'm'}(\theta', \lambda') Y_{nm}(\theta', \lambda') d\sigma. \tag{2.147}$$

However, according to Eq. (1.143) we get:

$$Y_{n'm'}(\theta', \lambda') Y_{nm}(\theta', \lambda') = \sum_{n''=|n'-n|}^{n'+n} \sum_{m''=-n''}^{n''} Q_{n'm'nm}^{n''m''} Y_{n''m''}(\theta', \lambda'), \tag{2.148}$$

where $Q_{n'm'nm}^{n''m''}$ are the well-known Gaunt coefficients.

By inserting Eq. (2.148) into Eq. (2.147) we obtain:

$$N_{nm}^{O} = \sum_{n'=2}^{\infty} \sum_{m'=-n'}^{n'} N_{n'm'} \sum_{n''=|n'-n|}^{n'+n} \sum_{m''=-n''}^{n''} Q_{n'm'nm}^{n''m''} J_{n''m''}. \tag{2.149}$$

It should be stated that the oceans do not have regular borders with the continents. Therefore, it is not possible to express the oceanic areas only by two parallels and two meridians. Consequently the altimetry data should be gridded according to the resolution of the gravity field recovery, and this integral is solved for each cell of the grid separately.

Now, we simplify the integral of Eq. (2.149) and write it in a discrete form:

$$J_{n''m''} = \frac{1}{4\pi} \iint_{\sigma_{\text{Oceans}}} Y_{n''m''}(\theta', \lambda') d\sigma = \frac{1}{4\pi} \sum_{i=1}^{P} \iint_{\text{Cell}_i} Y_{n''m''}(\theta', \lambda') d\text{Cell}_i, \tag{2.150}$$

where P is the number of cells over oceans. The solution of the integral on the rhs of Eq. (2.150) has been presented in Eq. (1.168). Fig. 2.20 represents a gridding over the oceanic geoid computed from EGM96.

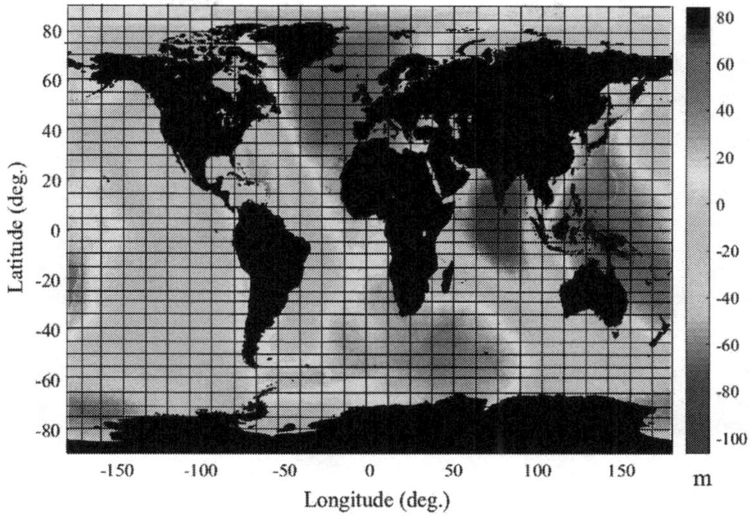

Figure 2.20 Gridding oceanic geoid.

References

Bjerhammar, A., 1968. On the Energy Integral for Satellites. The Royal Institute of Technology (KTH). Division of Geodesy, Stockholm, Sweden.

Ditmar, P., Eck, van der S.A., 2004. A technique for modeling the Earth's gravity field on the basis of satellite accelerations. J. Geodyn. 78, 12–33.

Eshagh, M., Najafi-Alamdari, M., 2007. Perturbations in orbital elements of a low Earth orbiting (LEO) satellite. J. Earth Space Phys. 33 (1), 1–12.

Eshagh, M., 2010. Alternative expressions for gravity gradients in local north-oriented frame and tensor spherical harmonics. Acta Geophys. 58, 215–243. https://doi.org/10.2478/s11600-009-0048-z.

Eshagh, M., Sprlak, M., 2016. On the integral inversion of satellite-to-satellite velocity differences for local gravity field recovery: a theoretical study. Celestial Mech. Dyn. Astron. 124, 127–144.

Eshagh, M., Lemoine, J.M., Gegout, P., Biancale, R., 2013. On regularized time varying gravity field models based on GRACE data and their comparisons with hydrological models. Acta Geophys. 61 (1), 1–17.

Fischell, R.E., Pisacane, V.L., 1978. A drag-free lo- lo satellite system for improved gravity field measurements. In: Mueller, I. (Ed.), Applications of Geodesy to Geodynamics, an International Symposium, Proceedings of the Ninth Geodesy/Solid Earth and Ocean Physics (GEOP), Research Conference, October 2–5, 1978, Columbus Ohio, Report No. 280. Department of Geodetic Science, Ohio State University, Columbus, USA, pp. 213–218.

Guo, X., Ditmar, P., Zhao, Q., Klees, R., Farahani, H., 2017. Earth's gravity field modelling based on satellite accelerations derived from onboard GPS phase measurements. J. Geodyn. 91 (9), 1049–1068.

Hofmann-Wellenhof, B., Lichtenegger, H., Wasle, E., 2011. GNSS – Global Navigation Satellite Systems. Springer.

Jekeli, C., 1999. The determination of gravitational potential differences from satellite-to-satellite tracking. Celestial Mech. Dyn. Astron. 75 (2), 85–101.

Kaula, W., 1966. Theory of Satellite Geodesy. Blaisdell Publishing Company, Waltham, Massachusetts, USA.
Keller, W., Sharifi, M.A., 2005. Satellite gradiometry using a satellite pair. J. Geodyn. 78 (9), 544–557.
Klokocnık, J., Gooding, R.H., Wagner, C.A., Kostelecky, J., Bezdek J, A., 2013. The use of resonant orbits in satellite Geodesy: a review. Surv. Geophys. 34 (1), 43–72.
Lemoine, F.G., Kenyon, S.C., Factor, J.K., Trimmer, R.G., Pavlis, N.K., Chinn, D.S., Cox, C.M., Klosko, S.M., Luthcke, S.B., Torrence, M.H., Wang, Y.M., Williamson, R.G., Pavlis, E.C., Rapp, R.H., Olson, T.R., 1998. The Development of the Joint NASA GSFC and the National Imagery and Mapping Agency (NIMA) Geopotential Model EGM96. NASA/TP—1998–206861.
Montenbruck, O., Gill, E., 2000. Satellite Orbits, Models, Methods and Applications. Springer, p. 369.
Moritz, H., 1980. Geodetic reference system. Bulletin Géodésique 54, 395–405.
Moulton, F.R., 1914. An Introduction to Celestial Mechanics. Dover Publications Inc., New York.
Novák, P., 2007. Integral inversion of GRACE-type data. Studia Geophys. Geod. 51, 351–367.
Parrot, D., 1989. Short Arc Orbit Improvement for GPS Satellites. MSc thesis, Department of Surveying Engineering, University of New Brunswick, Canada.
Pavlis, N.K., Holmes, S.A., Kenyon, S.C., Factor, J.K., 2012. The development and evaluation of the Earth Gravitational Model 2008 (EGM2008). J. Geophys. Res. 117, B04406. https://doi.org/10.1029/2011JB008916.
Petrovskaya, M.S., Vershkov, A.N., 2006. Non-Singular Expressions for the Gravity Gradients in the Local North-Oriented and Orbital Reference Frames. J Geodesy 80, 117–127. https://doi.org/10.1007/s00190-006-0031-2.
Rosborough, G.W., 1986. Satellite Orbit Perturbations Due to the Geopotential. CRS. Technical Report CRS-86-1.
Rummel, R., 1975. Downward Continuation of Gravity Field Information from Satellite to Satellite Tracking or Satellite Gradiometry in Local Areas. Report No. 221. Department of Geodetic Science, Ohio State University, Columbus, USA, 1975.
Rummel, R., 1980. Geoid Heights, Geoid Height Differences, and Mean Gravity Anomalies from Low-Low Satellite-To-Satellite Tracking – an Error Analysis. Report No. 306. Department of Geodetic Science, Ohio State University, Columbus, USA.
Santos, M.C., 1995. Real-Time Orbit Improvements for GPS Satellites. Ph.D. dissertation, Department of Geodesy and Geomatics Engineering, Technical Report No. 178. University of New Brunswick, Fredericton, New Brunswick, Canada, p. 125.
Schraman, E.J.O., 1986. A Study of a Satellite-To-Satellite Tracking Configuration by Application of Linear Perturbation Theory. Delft University of Technology.
Seeber, G., 2003. Satellite Geodesy, second ed. Walter de Gruyter, Berlin, p. 589.
Somodi, B., Földvary, L., 2011. Application of numerical integration techniques for orbit determination of state-of-the-art LEO satellites. Per. Pol. Civil Eng. 55 (2), 99–106.
Sveha, D., Földvary, L., 2006. From Kinematic orbit determination to derivation of satellite velocity and gravity field. In: Flury, J., Rummel, R., Reigber, C., Rotacher, M., Boedecker, G., Schreiber, U. (Eds.), Observation of the Earth System from Space. Springer.
Visser, P.N.A.M., 1992. The Use of Satellites in Gravity Field Determination and Adjustment. PhD dissertation. Delft University Press, Delft, Netherlands.
Visser, P.N.A.M., Sneeuw, N., Gerlach, C., 2003. Energy integral method for gravity field determination from satellite orbit coordinates. J. Geodyn. 77 (3–4), 207–216.
Wolff, M., 1969. Direct measurements of the Earth's gravitational potential using a satellite pair. J. Geophys. Res. 74, 5295–5300.

CHAPTER 3

Integral equations for inversion of satellite gravimetry data

3.1 Anomalous parameters of the Earth's gravity field

To compute the spherical harmonic coefficients (SHCs) of the gravitational potential (v_{nm}), satellite gravimetry data with global coverage are required. However, because of variations of the gravitational potential from one point to another, a mean or normal gravitational potential is defined and departures of the real field from it are used to study the gravity field of the Earth. This means that the known part of the gravitational signal is removed to highlight the signal and increase the precision of computations. There are textbooks explaining these anomalous parameters. such as those by Heiskanen and Moritz (1967), Pick et al. (1973), Vanicek and Krakiwski (1982) and Torge (2001). In this section, we present a brief review of these parameters, which are needed later to construct integral relations amongst the satellite gravimetry data and them.

3.1.1 Normal gravity field and disturbing potential

Generally, three shapes can be defined for the Earth: the natural shape, where we live on it; the physical shape, which is called the geoid and is an equipotential surface, and which is the best approximation to mean sea level; and finally, a mathematical shape, defined from the geoid. At the first approximation, a best-fitting sphere was considered to approximate the geoid surface, but a discrepancy of up to 11 km was found in some places. Later, a biaxial ellipsoid was chosen to approximate it, and a maximum departure of 100 m was found and finally, a triaxial ellipsoid with a maximum departure of about 80 m was found, but with a complicated mathematical description. Therefore, it is wise to select the biaxial ellipsoid as the best mathematical figure for the Earth. This ellipsoid is geocentric and has the same mass and rotation rate as that of the Earth, but with a constant density. Such a smooth body can also attract the other masses towards itself

Satellite Gravimetry and the Solid Earth
ISBN 978-0-12-816936-0
https://doi.org/10.1016/B978-0-12-816936-0.00003-7

by its own smooth gravitational potential, known as the normal gravitational potential. Fig. 3.1 shows the physical figure of the Earth, the geoid, and the biaxial ellipsoid generating the normal gravitational potential. Therefore, for any point with spherical coordinates (r, θ, λ) outside these surfaces, two types of potential are defined: the Earth's gravitational potential $V(r, \theta, \lambda)$ and the normal gravitational potential $U(r, \theta, \lambda)$. The difference between them is well-known as the disturbing potential $T(r, \theta, \lambda)$.

Consider the following spherical harmonic expansion for the normal potential:

$$U(r, \theta, \lambda) = \sum_{n=2}^{\infty} \left(\frac{R}{r}\right)^{n+1} \sum_{m=-n}^{n} u_{nm} Y_{nm}(\theta, \lambda), \tag{3.1}$$

in which u_{nm} is the SHCs of the normal potential. However, because the ellipsoid is geocentric, the first-degree harmonics of the field vanish. The normal gravitational potential is defined by the parameters of the best-fitting ellipsoid to the geoid. In practice, we compute the normal field, whilst we observe the actual one. Because the ellipsoid is biaxial, the normal gravity field will be constant along parallels; this means the potential field is zonal, $m = 0$. On the other hand, because of the symmetry of the field with respect to the equator, the harmonics will be of even degrees. In this case, Eq. (3.1) will change to:

$$U(r, \theta) = \sum_{n=2,4,6,\ldots} \left(\frac{R}{r}\right)^{n+1} u_{n0} P_n(\cos \theta). \tag{3.2}$$

In addition, the normal potential series will converge quickly; this means that its first few SHCs suffice to compute the potential.

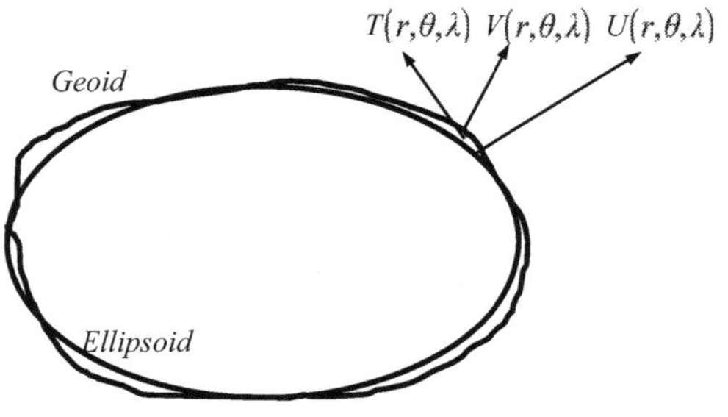

Figure 3.1 Geoid, ellipsoid, normal, real and disturbing gravitational potentials.

The disturbing potential can be expressed mathematically by:

$$T(r,\theta,\lambda) = V(r,\theta,\lambda) - U(r,\theta,\lambda). \tag{3.3}$$

To define the normal gravitational potential, it is assumed that the ellipsoid has the same mass as the Earth; therefore, this constant term vanishes when the normal potential is subtracted from the true one. In this case, the spherical harmonic expansion of the disturbing potential starts from $n = 2$:

$$T(r,\theta,\lambda) = \sum_{n=2}^{\infty}\left(\frac{R}{r}\right)^{n+1}\sum_{m=-n}^{n} t_{nm}Y_{nm}(\theta,\lambda) \quad \text{where } t_{nm} = v_{nm} - u_{n0}. \tag{3.4}$$

The disturbing potential (T) is an important quantity because it connects the physical shape of the Earth to its mathematical form and helps us define anomalous parameters for gravimetry.

3.1.2 Geoid height, gravity disturbance and anomaly as anomalous quantities

The geoid height is the distance along the vector of gravitation between the geoid and the ellipsoid. According to Bruns' formula, in spherical approximation, it is (Heiskanen and Mortiz, 1967, p. 85):

$$N(R,\theta,\lambda) = \frac{T(R,\theta,\lambda)}{\gamma}, \tag{3.5}$$

where γ stands for the normal gravity at the surface of the ellipsoid.

Gravity disturbance (δg) is defined as the radial derivative of T with a negative sign. In spherical approximation, it can be expressed by (e.g., Hesiskanen and Moritz, 1967, p. 88):

$$\delta g(r,\theta,\lambda) = -\frac{\partial T(r,\theta,\lambda)}{\partial r}, \tag{3.6}$$

and finally, the fundamental equation of physical geodesy, which gives the relation between T and gravity anomaly Δg (Heiskanen and Moritz, 1967, p. 92), in spherical approximation, is:

$$\Delta g(r,\theta,\lambda) = -\frac{\partial T(r,\theta,\lambda)}{\partial r} - 2\frac{T(r,\theta,\lambda)}{r}. \tag{3.7}$$

Gravity is a measurable quantity; therefore, δg and Δg are assumed to be known parameters. According to Eq. (3.4) and the definitions of a geoid, gravity disturbance and anomaly, for Eqs. (3.5)—(3.7) the spherical harmonic expansion of these anomalous quantities will be:

$$N(r,\theta,\lambda)=\frac{1}{\gamma}\sum_{n=2}^{\infty}\sum_{m=-n}^{n}\left(\frac{R}{r}\right)^{n+1}t_{nm}Y_{nm}(\theta,\lambda), \tag{3.8}$$

$$\delta g(r,\theta,\lambda)=\frac{1}{R}\sum_{n=2}^{\infty}(n+1)\left(\frac{R}{r}\right)^{n+2}\sum_{m=-n}^{n}t_{nm}Y_{nm}(\theta,\lambda), \tag{3.9}$$

$$\Delta g(r,\theta,\lambda)=\frac{1}{R}\sum_{n=2}^{\infty}(n-1)\left(\frac{R}{r}\right)^{n+2}\sum_{m=-n}^{n}t_{nm}Y_{nm}(\theta,\lambda). \tag{3.10}$$

From Eqs. (3.9) and (3.10), we can write the following relation between SHCs of T and those of gravity disturbance and anomaly (δg_{nm} and Δg_{nm}):

$$\delta g_{nm}=\frac{n+1}{R}t_{nm} \text{ and } \Delta g_{nm}=\frac{n-1}{R}t_{nm}. \tag{3.11}$$

3.1.3 Deflections of the vertical

The geoid and ellipsoid are not parallel; this means that any point at the geoid or outside the normal to the ellipsoid does not coincide with the tangent to the plumb line, which is a curve with curvature and twist perpendicular to the geoid. Assume a point at the surface of the geoid; the tangent to the plumb line and normal to the ellipsoid at this point do not coincide because of the slope of the geoid with respect to the ellipsoid (Fig. 3.2A). The angle between this tangent line and the normal to the ellipsoid is called the deflection of the vertical, ϑ.

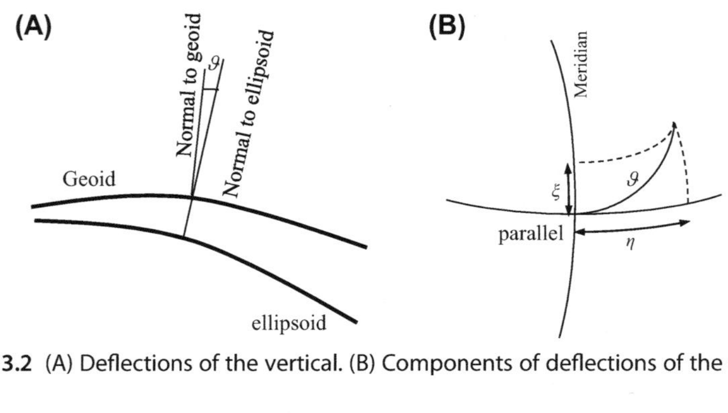

Figure 3.2 (A) Deflections of the vertical. (B) Components of deflections of the vertical.

A vertical deflection has many different applications; for example, it can be used to study geostrophic flows over oceans and in airborne gravimetry. This angle can also represent lateral variations in the gravity field. It changes from one point to another and in different directions; therefore, to use this angle more easily, it is projected onto the local meridian and parallel. The projection on the meridian is called the north–south component of ϑ and is shown by ξ; the one on the parallel is the east–west component by η; and according to the Pythagorean rule, $\vartheta = \sqrt{\xi^2 + \eta^2}$ (Fig. 3.2B). In fact, these components represent the slopes of the geoid along the local meridian and parallel, which can be mathematically presented in spherical approximation by:

$$\xi(R, \theta, \lambda) = \frac{\partial N(R, \theta, \lambda)}{R \partial \theta}, \tag{3.12}$$

$$\eta(R, \theta, \lambda) = -\frac{\partial N(R, \theta, \lambda)}{R \sin \theta \partial \lambda}. \tag{3.13}$$

According to Eq. (3.8), which is the spherical harmonic expansion of the geoid height, it may be shown that

$$\xi(R, \theta, \lambda)\mathbf{e}_x - \eta(R, \theta, \lambda)\mathbf{e}_y = \frac{1}{R\gamma} \sum_{n=2}^{\infty} \sum_{m=-n}^{n} t_{nm} \mathbf{X}_{nm}^{(2)}(\theta, \lambda), \tag{3.14}$$

where $\mathbf{X}_{nm}^{(2)}(\theta, \lambda)$ stands for the vector spherical harmonics (see Eq. 1.49). Deflections of the vertical are observable by astronomical methods. The geodetic latitude and longitude, which are defined based on the ellipsoid mathematical shape of the Earth, of any point at the surface of the Earth are determined precisely by classical geodetic measurements and ellipsoid geometry, and the astronomical latitude and longitude of the same point are determined by observing stars. The difference between the latitudes and the longitudes has a direct relation to the components of the deflection of the vertical; the mathematical relations between the geodetic and astronomical coordinates are well-known in geodesy and are available in textbooks.

3.1.4 From geoid height to other anomalous parameters

Now, assume that the geoid height is given at the surface of the spherical Earth and the goal is to determine T outside the Earth. This process is similar to determining the potential from the potential outside the Earth, which was discussed in Section 1.2. Therefore, the spectral solution will be (see Eq. 1.8):

$$t_{nm} = \frac{\gamma}{4\pi} \iint_{\sigma} N(R, \theta', \lambda') Y_{nm}(\theta', \lambda') \mathrm{d}\sigma = \gamma N_{nm}. \tag{3.15}$$

This is the spectral solution of the disturbing potential leading to t_{nm}; obtaining the spatial solution is easier according to Bruns' formulae (Eq. 3.5). If the goal is to determine δg from geoid height N, the spectral relations between δg and N are needed, which are derived by comparing Eq. (3.15) and Eq. (3.11):

$$\delta g_{nm} = \gamma \frac{n+1}{R} N_{nm}. \tag{3.16}$$

Derivation of the integral formula for determining δg from N is similar to that presented in Section 1.3.1.3, and the result will be:

$$\delta g = \frac{\gamma}{4\pi R} \iint_{\sigma} N(R, \theta', \lambda') P_r'(\psi) \mathrm{d}\sigma, \tag{3.17}$$

where

$$P_r'(\psi) = \sum_{n=2}^{\infty} (2n+1)(n+1) P_n(\cos\psi). \tag{3.18}$$

Now, compare the kernel function of this integral with that presented in Eq. (1.33) for $r = R$. Note that the zero- and first-degree terms do not exist in this kernel; the closed-form formula of the kernel is:

$$P_r'(\psi) = \frac{-1}{4\sin^3\frac{\psi}{2}} - 1 - 6\cos\psi. \tag{3.19}$$

Δg can be obtained similarly by:

$$\Delta g = \frac{\gamma}{4\pi R} \iint_{\sigma} N(R, \theta', \lambda') K'(\psi) \mathrm{d}\sigma, \tag{3.20}$$

where

$$K'(\psi) = \sum_{n=2}^{\infty} (2n+1)(n-1) P_n(\cos\psi). \tag{3.21}$$

To obtain Δg at a point outside the spherical Earth (geoid), the operator presented in Eq. (3.7), which is the fundamental equation of physical geodesy, is applied to the Poisson integral (Eq. 1.17), replacing potential V with the disturbing potential (T) and considering Bruns' formula, Eq. (3.5). In this case, we obtain:

$$\Delta g(r,\theta,\lambda) = \frac{\gamma}{4\pi R}\iint_{\sigma} N(R,\theta',\lambda')K''(r,\psi)\mathrm{d}\sigma, \tag{3.22}$$

where

$$K''(r,\psi) = -P_r(r,\psi) - \frac{2}{r}P(r,\psi) \tag{3.23}$$

and when $r = R$,

$$K''(\psi) = \frac{-1}{4\sin^3\frac{\psi}{2}} + 1. \tag{3.24}$$

Similarly, by applying the operator presented in Eq. (3.6), the radial derivative of the potential with a minus sign, to the Poisson integral, Eq. (1.17), the corresponding integral formula to determine δg outside the geoid will be obtained.

3.1.5 From gravity disturbance/anomaly and deflections of the vertical to geoid height

To obtain the disturbing potential (T) outside the geoid from gravity disturbance δg at the geoid, t_{nm} should be derived for δg_{nm} by solving Eq. (3.11) for t_{nm}, and the result should be inserted back into Eq. (3.4). This will give us the spherical harmonic expansion of T in terms of δg_{nm}. By writing δg_{nm} in integral form (see Eq. 1.8), applying the addition theorem of spherical harmonics (Eq. 1.9) and considering the orthogonality property of the spherical harmonics (Eq. 1.6), the following integral formula will be derived:

$$T(r,\theta,\lambda) = \frac{R}{4\pi}\iint_{\sigma} \delta g(R,\theta',\lambda')H(r,\psi)\mathrm{d}\sigma, \tag{3.25}$$

which is almost the same with Eq. (1.28). Eq. (3.25) continues δg upward from level R to r while transforming it to T. $H(r,\psi)$ is the well-known extended Hotine function; Eqs. (1.29) and (1.30) show its spectral and closed-form formulas. To determine the geoid height, the Bruns formula (Eq. 3.5) is used, and by considering $r = R$ in the kernel function, Eq. (3.25) will change to the well-known Hotine integral:

$$N(R,\theta,\lambda) = \frac{R}{4\pi\gamma} \iint_{\sigma} \delta g(R,\theta',\lambda') H(\psi) d\sigma. \tag{3.26}$$

In this equation, there is no upward continuation process and δg is converted directly to N. The Hotine kernel $H(\psi)$ is presented in Eq. (1.31).

Similarly, T can be derived from Δg by:

$$T(r,\theta,\lambda) = \frac{R}{4\pi} \iint_{\sigma} \Delta g(R,\theta',\lambda') S(r,\psi) d\sigma. \tag{3.27}$$

This is the well-known extended Stokes integral, and $S(r,\psi)$ is the extended Stokes function:

$$S(r,\psi) = 2\frac{t}{D} + t - 3tD - t^2\tilde{x}\left(5 + 3\ln\frac{1 - t\tilde{x} + D}{2}\right). \tag{3.28}$$

To obtain the well-known Stokes integral for geoid determination, the Bruns formula (Eq. 3.5) is applied, and we assume that $r = R$ in the kernel function (Heiskanen and Moritz, 1967, p. 94):

$$N(R,\theta,\lambda) = \frac{R}{4\pi\gamma} \iint_{\sigma} \Delta g(R,\theta',\lambda') S(\psi) d\sigma, \tag{3.29}$$

The kernel of this integral, known as the Stokes kernel, is:

$$S(\psi) = \frac{1}{\sin\frac{\psi}{2}} + 1 - 6\sin\frac{\psi}{2} - \cos\psi\left(5 + 3\ln\left(\sin^2\frac{\psi}{2} + \sin\frac{\psi}{2}\right)\right). \tag{3.30}$$

The principle for determining the potential from its horizontal derivatives was presented in Section 1.3.2. Analogous to Eq. (1.69) we obtain:

$$N(R,\theta,\lambda) = \frac{R}{4\pi\gamma} \iint_{\sigma} [\xi(R,\theta',\lambda')\cos\alpha' - \eta(R,\theta',\lambda')\sin\alpha'] G_{\mathrm{H}}(\psi) d\sigma, \tag{3.31}$$

in which $G_H(\psi)$ is the kernel function and was presented in Eq. (1.75) and α' is the backward azimuth from the integration point to the computation point (Eq. 1.80).

3.1.6 Integral relations amongst gravity disturbance/anomaly and deflections of the vertical

The integral formulas for the horizontal derivatives of the disturbing potential (T) (i.e., T_x and T_y) and relations from gravity disturbance/anomaly (δg or Δg) can be derived simply by taking the horizontal derivatives of the Hotine and Stokes formulae (Eqs. 3.25 and 3.27):

$$\left.\begin{pmatrix} T_x(r,\theta,\lambda) \\ T_y(r,\theta,\lambda) \end{pmatrix}\right\} = \begin{cases} \dfrac{R}{4\pi}\iint\limits_{\sigma} \delta g(R,\theta',\lambda')\begin{pmatrix} H_\theta(r,\psi) \\ H_\lambda(r,\psi) \end{pmatrix} d\sigma \\ \dfrac{R}{4\pi}\iint\limits_{\sigma} \Delta g(R,\theta',\lambda')\begin{pmatrix} S_\theta(r,\psi) \\ S_\lambda(r,\psi) \end{pmatrix} d\sigma \end{cases}, \tag{3.32}$$

in which $H_\theta(r,\psi)$ and $H_\lambda(r,\psi)$ are the derivatives of the extended Hotine function with respect to θ and λ and $S_\theta(r,\psi)$ and $S_\lambda(r,\psi)$ are the corresponding ones of the extended Stokes function. Eq. (3.32) can also be written as:

$$\begin{pmatrix} T_x(r,\theta,\lambda) \\ T_y(r,\theta,\lambda) \end{pmatrix} = \begin{cases} \dfrac{R}{4\pi}\iint\limits_{\sigma} \delta g(R,\theta',\lambda')H_\psi(r,\psi)\begin{pmatrix} \cos\alpha \\ -\sin\alpha \end{pmatrix} d\sigma \\ \dfrac{R}{4\pi}\iint\limits_{\sigma} \Delta g(R,\theta',\lambda')S_\psi(r,\psi)\begin{pmatrix} \cos\alpha \\ -\sin\alpha \end{pmatrix} d\sigma \end{cases}, \tag{3.33}$$

in which α stands for the forward azimuth from computation to the integration point (Eq. 1.80). The kernel of this integral is (Kern and Haagmans, 2005; Heiskanen and Moritz, 1967, p. 235):

$$H_\psi(r,\psi) = \left(-\frac{2t^2}{D^3} - \frac{t+D}{D(t-\widetilde{x}+g)} + \frac{1}{1-\widetilde{x}}\right)\sqrt{1-\widetilde{x}^2}, \tag{3.34}$$

$$S_\psi(r,\psi) = -t^2\sqrt{1-\tilde{x}^2}\left(\frac{2}{D^3}+\frac{6}{D}-8-3\frac{1-t\tilde{x}-D}{D(1-\tilde{x}^2)}-3\ln\frac{1-t\tilde{x}+D}{2}\right). \tag{3.35}$$

If the idea is to obtain ξ and η at the surface of the sphere, the Hotine and Stokes integrals will change to the Vening Meinesz formula by considering $r = R$:

$$\begin{pmatrix}\xi(R,\theta,\lambda)\\ \eta(R,\theta,\lambda)\end{pmatrix} = \begin{cases}\dfrac{R}{4\pi\gamma}\iint\limits_\sigma \delta g(R,\theta',\lambda')H_\psi(\psi)\begin{pmatrix}\cos\alpha\\ -\sin\alpha\end{pmatrix}d\sigma\\ \dfrac{R}{4\pi\gamma}\iint\limits_\sigma \Delta g(R,\theta',\lambda')S_\psi(\psi)\begin{pmatrix}\cos\alpha\\ -\sin\alpha\end{pmatrix}d\sigma\end{cases}, \tag{3.36}$$

in which (Heiskanen and Moritz, 1967, p. 114):

$$S_\psi(\psi) = -\frac{\cos\dfrac{\psi}{2}}{2\sin^2\dfrac{\psi}{2}} + 8\sin\psi - 6\cos\frac{\psi}{2} - 3\frac{1-\sin\dfrac{\psi}{2}}{\sin\psi} + 3\sin\ln\left[\left(\sin\frac{\psi}{2}+\sin^2\frac{\psi}{2}\right)\right], \tag{3.37}$$

$$H_\psi(\psi) = -\frac{\cos\dfrac{\psi}{2}}{2\sin^2\dfrac{\psi}{2}} + \frac{\cos\dfrac{\psi}{2}}{2\left(\sin^2\dfrac{\psi}{2}+\sin\dfrac{\psi}{2}\right)}. \tag{3.38}$$

According to Eq. (1.56), which is the spectral solution of the SHCs from the horizontal derivatives of the potential, we can write another, similar expression for t_{nm}:

$$t_{nm} = \frac{R}{n(n+1)}\frac{1}{4\pi}\iint\limits_\sigma\left[\frac{\partial Y_{nm}(\theta',\lambda')}{\partial\theta'}\xi(R,\theta',\lambda') + \frac{\partial Y_{nm}(\theta',\lambda')}{\sin\theta'\partial\lambda'}\eta(R,\theta',\lambda')\right]d\sigma. \tag{3.39}$$

By solving Eq. (3.11) for t_{nm} and inserting the result into Eq. (3.39), we get:

$$\left\{ \begin{array}{c} \delta g_{nm} \\ \Delta g_{nm} \end{array} \right\} = \frac{1}{n} \left\{ \begin{array}{c} 1 \\ (n-1) \end{array} \right\} \frac{1}{4\pi} \iint_{\sigma} \left[\frac{\partial Y_{nm}(\theta', \lambda')}{\partial \theta'} \xi(R, \theta', \lambda') + \frac{\partial Y_{nm}(\theta', \lambda')}{\sin \theta' \partial \lambda'} \eta(R, \theta', \lambda') \right] d\sigma. \tag{3.40}$$

Now, let us insert Eq. (3.40) into Eq. (3.9), and using the addition theorem Eqs. (1.61) and (1.62), we obtain:

$$\left\{ \begin{array}{c} \delta g \\ \Delta g \end{array} \right\} = \frac{1}{4\pi\gamma} \iint_{\sigma} \sum_{n=2}^{\infty} \left\{ \begin{array}{c} 1 \\ \frac{n-1}{n+1} \end{array} \right\} \frac{2n+1}{n} \left[\xi(R, \theta', \lambda') \frac{\partial P_n(\cos \psi)}{\partial \theta'} + \eta(R, \theta', \lambda') \frac{\partial P_n(\cos \psi)}{\sin \theta' \partial \lambda} \right] d\sigma, \tag{3.41}$$

and finally:

$$\left\{ \begin{array}{c} \delta g(r, \theta, \lambda) \\ \Delta g(r, \theta, \lambda) \end{array} \right\} = \frac{R}{4\pi\gamma} \iint_{\sigma} [\xi(R, \theta', \lambda') \cos \alpha' - \eta(R, \theta', \lambda') \sin \alpha'] \left\{ \begin{array}{c} G_{\delta g}(r, \psi) \\ G_{\Delta g}(r, \psi) \end{array} \right\} d\sigma, \tag{3.42}$$

with

$$G_{\delta g}(r, \psi) = \sum_{n=2}^{\infty} \frac{2n+1}{n} \left(\frac{R}{r} \right)^{n+2} \frac{\partial P_n(\cos \psi)}{\partial \psi} = -\frac{2t^3\sqrt{1-\tilde{x}^2}}{D^{3/2}} - \frac{t^2(D+t)\sqrt{1-\tilde{x}^2}}{D(D+1-t\tilde{x})}, \tag{3.43}$$

$$G_{\Delta g}(r, \psi) = \sum_{n=2}^{\infty} \frac{(2n+1)(n-1)}{n(n+1)} \left(\frac{R}{r} \right)^{n+2} \frac{\partial P_n(\cos \psi)}{\partial \psi} = -\sqrt{1-\tilde{x}^2} \left[\frac{2t^3}{D^{3/2}} + \frac{t^3(D+1)}{D(D+1-t\tilde{x})} - \frac{2t}{1-\tilde{x}} - \frac{2t(t+D)}{D(D+t-\tilde{x})} \right]. \tag{3.44}$$

At the surface of the sphere where $r = R$, the kernels will change to (Pick et al., 1973, pp. 476, 477):

$$G_{\delta g}(\psi) = -\frac{\cos\frac{\psi}{2}}{2\sin^2\frac{\psi}{2}} + 3\sin\psi - \cos\frac{\psi}{2}\frac{1 + 2\sin\frac{\psi}{2}}{2\sin\frac{\psi}{2}\left(1 + \sin\frac{\psi}{2}\right)}, \tag{3.45}$$

$$G_{\Delta g}(\psi) = \operatorname{cosec}\frac{\psi}{2} + \ln\left(\frac{\sin^3\frac{\psi}{2}}{1 + \sin\frac{\psi}{2}}\right). \tag{3.46}$$

3.2 Integral equations for inversion of temporal variations of orbital elements

In this section, our goal is to develop integral equations for inverting orbital perturbations and temporal variations of the orbital elements of a satellite to the gravity disturbance/anomaly at sea level. Obenson (1970) used Gaussian equations to invert these temporal variations, and Eshagh and Ghorbannia (2013, 2014) developed them further and presented different integral equations to recover the gravity anomalies at sea level. In the following discussion, this process is developed further to determine the gravity disturbance/anomaly from along, cross and radial track orbital perturbations as well as the temporal variations of orbital elements and their combinations.

3.2.1 Integral equations for recovering gravity anomaly/disturbance from temporal variations of orbital elements

As was discussed in Chapter 2, Section 2.3.2, perturbations in the terrestrial reference frame (TRF) are used to connect the potential to perturbations in this frame (Eqs. 2.69 or 2.71). According to our definition of the gravity field and Eq. (3.3), total potential V is the summation of disturbing potential T and normal potential U; so, with Eqs. (3.25) and (3.27), we can write:

$$V(r,\theta,\lambda) = U(r,\theta,\lambda) + \begin{cases} \dfrac{R}{4\pi} \displaystyle\iint_\sigma \Delta g(R,\theta',\lambda') S(r,\psi) \mathrm{d}\sigma \\ \dfrac{R}{4\pi} \displaystyle\iint_\sigma \delta g(R,\theta',\lambda') H(r,\psi) \mathrm{d}\sigma \end{cases}, \tag{3.47}$$

which is the mathematical relation connecting the gravity disturbance/anomaly to the Earth's gravitational potential. From the along, cross and radial track components of the perturbations, their temporal variations and accelerations can be determined. According to Eqs. (2.85)−(2.87), their combinations give perturbing accelerations V_u^P, V_v^P and V_w^P, which have a direct relation to the Earth gravitational potential. Thus, we can write:

$$\left.\begin{array}{l} \dfrac{R}{4\pi} \displaystyle\iint_\sigma \Delta g(R,\theta',\lambda') S_r(r,\psi) \mathrm{d}\sigma \\ \dfrac{R}{4\pi} \displaystyle\iint_\sigma \delta g(R,\theta',\lambda') H_r(r,\psi) \mathrm{d}\sigma \end{array}\right\} = V_u^P(r,\theta,\lambda) - U_r(r,\theta), \tag{3.48}$$

$$\left.\begin{array}{l} \dfrac{R}{4\pi r} \displaystyle\iint_\sigma \Delta g(R,\theta',\lambda') S_\psi(r,\psi) \cos(\alpha + \alpha_{\mathrm{Sat}}) \mathrm{d}\sigma \\ \dfrac{R}{4\pi r} \displaystyle\iint_\sigma \delta g(R,\theta',\lambda') H_\psi(r,\psi) \cos(\alpha + \alpha_{\mathrm{Sat}}) \mathrm{d}\sigma \end{array}\right\} = V_v^P(r,\theta,\lambda) - \frac{1}{r} U_\theta(r,\theta) \cos \alpha_{\mathrm{Sat}}, \tag{3.49}$$

$$\left.\begin{array}{l} -\dfrac{R}{4\pi r} \displaystyle\iint_\sigma \Delta g(R,\theta',\lambda') S_\psi(r,\psi) \sin(\alpha + \alpha_{\mathrm{Sat}}) \mathrm{d}\sigma \\ -\dfrac{R}{4\pi r} \displaystyle\iint_\sigma \delta g(R,\theta',\lambda') H_\psi(r,\psi) \sin(\alpha + \alpha_{\mathrm{Sat}}) \mathrm{d}\sigma \end{array}\right\} = V_w^P(r,\theta,\lambda) + \frac{U_\theta(r,\theta) \sin \alpha_{\mathrm{Sat}}}{r}, \tag{3.50}$$

in which α is the forward azimuth, whereas α_{Sat} is the satellite track azimuth (Eqs. 1.77 or 2.31). The kernel functions of Eq. (3.48) are (Heiskanen and Moritz, 1967, p. 233−235, and Kern and Haagmans, 2005):

$$S_r(r,\psi) = -\frac{t^2}{R}\left[\frac{1-t^2}{D^3}+\frac{4}{D}+1-6D-t\widetilde{x}\left(13+6\ln\frac{1-t\widetilde{x}+D}{2}\right)\right], \tag{3.51}$$

$$H_r(r,\psi) = -\frac{2t}{rD}-\frac{2t^2(\widetilde{x}-t)}{rD^3}-\frac{t(\widetilde{x}-t)}{Dr(t-\widetilde{x}+D)}+\frac{t}{r(t-\widetilde{x}+D)}, \tag{3.52}$$

By solving Eqs. (3.48)−(3.50), respectively, for V_u^P, V_v^P and V_w^P and inserting the results into the Gaussian equations (Eqs. 2.58−2.63) and further algebraic manipulations, the following integral equations are derived (Eshagh and Ghorbannia, 2013):

$$\left.\begin{array}{l}\dfrac{R}{4\pi}\displaystyle\iint_\sigma \Delta g(R,\theta',\lambda')K_{\dot{a}}^S(r,\psi,\alpha_{\mathrm{Sat}},\alpha)\mathrm{d}\sigma \\ \dfrac{R}{4\pi}\displaystyle\iint_\sigma \delta g(R,\theta',\lambda')K_{\dot{a}}^H(r,\psi,\alpha_{\mathrm{Sat}},\alpha)\mathrm{d}\sigma\end{array}\right\} = \frac{\dot{a}}{2a\mu} - U_r e\sin f - \frac{U_\theta\cos\alpha_{\mathrm{Sat}}}{r^2}, \tag{3.53}$$

$$\left.\begin{array}{l}\dfrac{R}{4\pi}\displaystyle\iint_\sigma \Delta g(R,\theta',\lambda')K_{\dot{e}}^S(r,\psi,\alpha_{\mathrm{Sat}},\alpha)\mathrm{d}\sigma \\ \dfrac{R}{4\pi}\displaystyle\iint_\sigma \delta g(R,\theta',\lambda')K_{\dot{e}}^H(r,\psi,\alpha_{\mathrm{Sat}},\alpha)\mathrm{d}\sigma\end{array}\right\} = \frac{\dot{e}}{\overline{m}} - \left(U_r\sin f + \frac{\overline{\Theta}}{r}U_\theta\cos\alpha_{\mathrm{Sat}}\right), \tag{3.54}$$

$$\left.\begin{array}{l}\dfrac{R}{4\pi}\displaystyle\iint_\sigma \Delta g(R,\theta',\lambda')K_{\dot{i}}^S(r,\psi,\alpha_{\mathrm{Sat}},\alpha)\mathrm{d}\sigma \\ \dfrac{R}{4\pi}\displaystyle\iint_\sigma \delta g(R,\theta',\lambda')K_{\dot{i}}^H(r,\psi,\alpha_{\mathrm{Sat}},\alpha)\mathrm{d}\sigma\end{array}\right\} = \frac{a}{\cos(\omega+f)\mu}\dot{i} + U_\theta\sin\alpha_{\mathrm{Sat}}, \tag{3.55}$$

$$\left.\begin{array}{l}\dfrac{R}{4\pi}\displaystyle\iint\limits_{\sigma}\Delta g(R,\theta',\lambda')K_{\dot\Omega}^{S}(r,\psi,\alpha_{\text{Sat}},\alpha)\mathrm{d}\sigma\\ \dfrac{R}{4\pi}\displaystyle\iint\limits_{\sigma}\delta g(R,\theta',\lambda')K_{\dot\Omega}^{H}(r,\psi,\alpha_{\text{Sat}},\alpha)\mathrm{d}\sigma\end{array}\right\}=\frac{a\sin i}{\sin(\omega+f)\mu}\dot\Omega+U_\theta\sin\alpha_{\text{Sat}}, \tag{3.56}$$

$$\left.\begin{array}{l}\dfrac{R}{4\pi}\displaystyle\iint\limits_{\sigma}\Delta g(R,\theta',\lambda')K_{\dot\omega}^{S}(r,\psi,\alpha_{\text{Sat}},\alpha')\mathrm{d}\sigma\\ \dfrac{R}{4\pi}\displaystyle\iint\limits_{\sigma}\delta g(R,\theta',\lambda')K_{\dot\omega}^{H}(r,\psi,\alpha_{\text{Sat}},\alpha')\mathrm{d}\sigma\end{array}\right\}=\frac{\dot\omega}{\overline{m}}$$
$$-\left(-\frac{\cos f}{e}U_r+\frac{(r+p)\sin f}{epr}U_\theta\cos\alpha_{\text{Sat}}-\frac{\sin(\omega+f)\cot i}{p}U_\theta\sin\alpha_{\text{Sat}}\right), \tag{3.57}$$

$$\left.\begin{array}{l}-\dfrac{R}{4\pi}\displaystyle\iint\limits_{\sigma}\Delta g(R,\theta',\lambda')K_{\dot M}^{S}(r,\psi,\alpha_{\text{Sat}},\alpha)\mathrm{d}\sigma\\ -\dfrac{R}{4\pi}\displaystyle\iint\limits_{\sigma}\delta g(R,\theta',\lambda')K_{\dot M}^{H}(r,\psi,\alpha_{\text{Sat}},\alpha)\mathrm{d}\sigma\end{array}\right\}=\dot{\overline{M}}-\overline{n}+\Gamma U_r+\frac{\Pi}{r}U_\theta\cos\alpha_{\text{Sat}}, \tag{3.58}$$

where

$$\left\{\begin{array}{l}K_{\dot a}^{S}(r,\psi,\alpha_{\text{Sat}},\alpha)\\ K_{\dot a}^{H}(r,\psi,\alpha_{\text{Sat}},\alpha)\end{array}\right\}=e\left\{\begin{array}{l}S_r(r,\psi)\\ H_r(r,\psi)\end{array}\right\}\sin f+\frac{p}{r^2}\left\{\begin{array}{l}S_\psi(r,\psi)\\ H_\psi(r,\psi)\end{array}\right\}\cos(\alpha+\alpha_{\text{Sat}}), \tag{3.59}$$

$$\left\{\begin{array}{l}K_{\dot e}^{S}(r,\psi,\alpha_{\text{Sat}},\alpha)\\ K_{\dot e}^{H}(r,\psi,\alpha_{\text{Sat}},\alpha)\end{array}\right\}=\left\{\begin{array}{l}S_r(r,\psi)\\ H_r(r,\psi)\end{array}\right\}\sin f+\frac{\overline{\Theta}}{r}\left\{\begin{array}{l}S_\psi(r,\psi)\\ H_\psi(r,\psi)\end{array}\right\}\cos(\alpha+\alpha_{\text{Sat}}), \tag{3.60}$$

$$\left\{\begin{array}{l}K_{\dot i}^{S}(r,\psi,\alpha_{\text{Sat}},\alpha)\\ K_{\dot i}^{H}(r,\psi,\alpha_{\text{Sat}},\alpha)\end{array}\right\}=\left\{\begin{array}{l}K_{\dot\Omega}^{S}(r,\psi,\alpha_{\text{Sat}},\alpha)\\ K_{\dot\Omega}^{S}(r,\psi,\alpha_{\text{Sat}},\alpha)\end{array}\right\}=-\left\{\begin{array}{l}S_\psi(r,\psi)\\ H_\psi(r,\psi)\end{array}\right\}\sin(\alpha+\alpha_{\text{Sat}}), \tag{3.61}$$

$$\left\{\begin{array}{l} K^S_{\dot{\omega}}(r,\psi,\alpha_{\mathrm{Sat}},\alpha) \\ K^H_{\dot{\omega}}(r,\psi,\alpha_{\mathrm{Sat}},\alpha) \end{array}\right\} = -\frac{\cos f}{e}\left\{\begin{array}{l} S_r(r,\psi) \\ H_r(r,\psi) \end{array}\right\} + \frac{1}{p}\left\{\begin{array}{l} S_\psi(r,\psi) \\ H_\psi(r,\psi) \end{array}\right\}$$
$$\times\left(\frac{(r+p)\sin f}{er}\cos(\alpha+\alpha_{\mathrm{Sat}}) + \sin(\omega+f)\cot i\,\sin(\alpha+\alpha_{\mathrm{Sat}})\right), \quad (3.62)$$

$$\left\{\begin{array}{l} K^S_{\dot{M}}(r,\psi,\alpha_{\mathrm{Sat}},\alpha) \\ K^H_{\dot{M}}(r,\psi,\alpha_{\mathrm{Sat}},\alpha) \end{array}\right\} = \Gamma\left\{\begin{array}{l} S_r(r,\psi) \\ H_r(r,\psi) \end{array}\right\} + \frac{\Pi}{r}\left\{\begin{array}{l} S_\psi(r,\psi) \\ H_\psi(r,\psi) \end{array}\right\}\cos(\alpha+\alpha_{\mathrm{Sat}}). \quad (3.63)$$

The rest of parameters were already defined in Section 2.3.2 in Eqs. (2.64)−(2.68).

The kernel functions presented in Eqs. (3.59)−(3.63) are functions of the orbital elements and their temporal changes. To present these kernels, Eshagh and Ghorbannia (2013) simulated a 2-month orbit of a Gravity Field and Steady-State Ocean Circulation Explorer (GOCE)-type satellite with an integration step size of 10 s using the Runge−Kutta integrator of the fourth order and Earth Gravitational Model 96 (Lemoine et al., 1998) to degree and order 360 as the main gravity field. Passes of the satellite over Fennoscandia were selected, along with the position and velocity vectors of the satellite derived by orbit integration in the celestial reference frame and converted to the orbital elements. Their statistics are presented in Table 3.1, which shows that a changes about 15 km over this area with a standard deviation of 3.6 km. No significant variation is seen in the orbital eccentricity (e), and variations of i are less than 1 degree. Large changes are seen in ω and Ω owing to the rotations of the Earth and the satellite's revolutions. Eshagh and Ghorbannia (2013) used the mean values of these orbital elements to present the kernels' behaviour. However, in real applications, the orbital elements are different from one point to another and the kernels should be regenerated point by point. The mean values are only for the purposes of presenting the kernel and its variations with respect to the geocentric angle ψ and the azimuth α between the computation and the integration point.

Fig. 3.3 shows the behaviour of the kernels presented in Eqs. (3.60)−(3.63) with three azimuths of $\alpha' = 45$, 90 and 180 degrees and geocentric angles from $\psi = 0$ to 10 degrees. Fig. 3.3A is the plot of $K^S_{\dot{a}}$; it is a bell-shaped kernel with the maximum at $\psi = 1°$. In addition, the kernel has similar behaviour with different azimuths. Fig. 3.3B is a similar plot except

Table 3.1 Statistics of orbital elements of an imaginary satellite with Gravity Field and Steady-State Ocean Circulation Explorer properties and its azimuth over Fennoscandia in a 2-month revolution.

Orbital element	Maximum	Mean	Minimum	Standard deviation
a (km)	6737.536	6727.653	6722.158	3.600
E	0.016	0.015	0.013	0.000
I	97 degrees47′31″	97 degrees15′13′	96 degrees48′45″	0 degrees16′19″
ω	301 degrees23′30″	270 degrees46′46″	239 degrees55′03″	20 degrees46′36″
Ω	171 degrees09′24″	142 degrees27′50″	114 degrees35′18″	15 degrees23′47″
$\overline{M}$ (m/s^2)	17,175.776	7379.775	2589.608	4919.724
α	−13 degrees36′01″	−18 degrees24′22″	−27 degrees29′33″	03 degrees47′12″

e, the orbit eccentricity; a, the semimajor axis of orbital ellipse; ω, the argument of the perigee; i, inclination; Ω, right ascension of ascending node; $\overline{M}$, the mean anomaly; and α, satellite track azimuth.

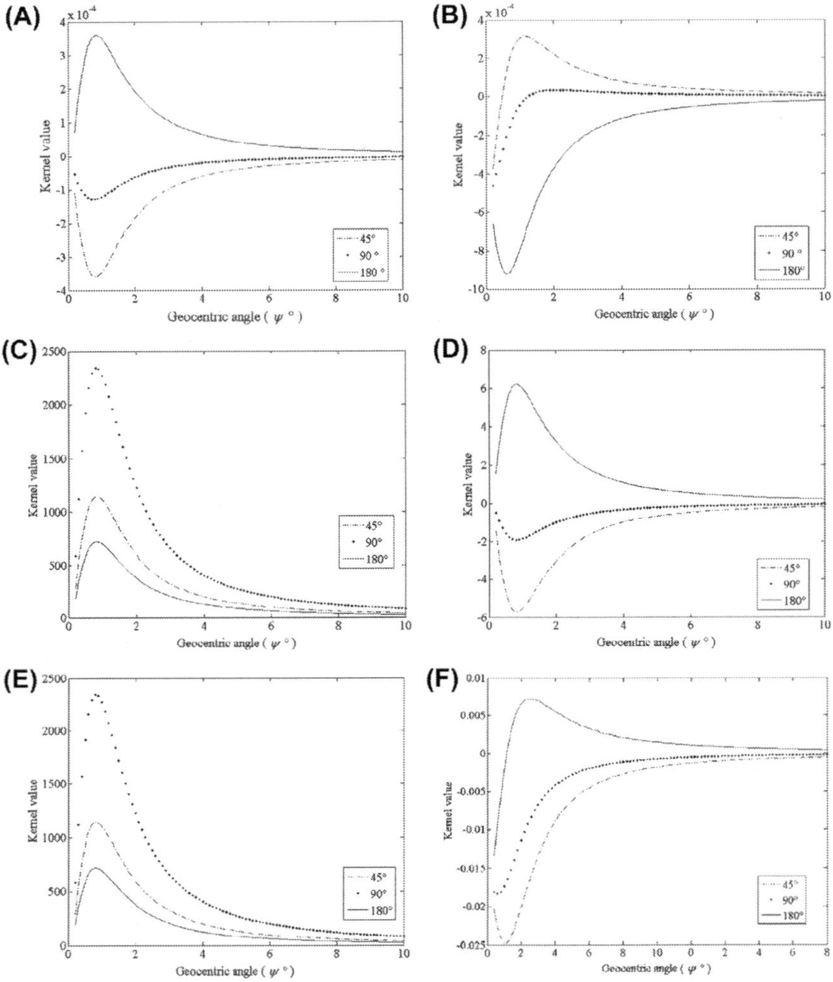

Figure 3.3 Behaviour of (A) $K^S_{\dot{a}}$, (B) $K^S_{\dot{e}}$, (C) $K^S_{\dot{i}}$, (D) $K^S_{\dot{\bar{M}}}$, (E) $K^S_{\dot{\Omega}}$ and (F) $K^S_{\dot{\omega}}$ for three azimuths of 45, 90 and 180 degrees when the satellite track azimuth and mean altitude of a satellite are −18 and 6727.653 km, respectively (Eshagh and Ghorbannia, 2013).

for $K^S_{\dot{e}}$. The kernel behaves differently with the chosen azimuths. For some azimuths, it behaves well, but not for others. Figs. 3.3C and E are the same, as are the mathematical models of $K^S_{\dot{i}}$ and $K^S_{\dot{\Omega}}$; they are of the bell-shaped type. Fig. 3.3D shows the plot of $K^S_{\dot{\bar{M}}}$, which is similar to the behaviour of $K^S_{\dot{a}}$ in Fig. 3.3A; the difference is that this kernel has smaller values than does $K_{\dot{a}}^{S}$. Finally, Fig. 3.3F shows that $K_{\dot{\omega}}^{S}$ behaves well for some azimuths, but not for all.

3.2.2 Integral equations for inverting combination of Gaussian equations for gravity disturbance/anomaly recovery

The kernel functions presented for inverting the temporal variation of each orbital element are complicated because of their dependence on different orbital parameters as well as the azimuth between the computation and the integration points. Eshagh and Ghorbannia (2013) presented some combinations of orbital elements to obtain simpler kernel functions that are easier for integral inversion of the velocity of the orbital elements to gravity anomalies at sea level.

Three systems of equations can be made from the Gaussian equations (Eqs. 2.58, 2.59 and 2.63) with V_u^P and V_v^P as their unknowns. In addition, V_w^P can be obtained from Eq. (2.60) or (2.61) and inserted into Eq. (2.62) to obtain another equation in terms of V_u^P and V_v^P. In this case, V_u^P and V_v^P can be solved from all four equations in a least-squares sense. Moreover, when U_r is subtracted from V_u^P and U_θ^λ from V_v^P, the derivative of the disturbing potential (T) involving gravity anomaly is obtained. Now, we solve these equations for T in terms of V_u^P and V_v^P by solving the equations containing them, two by two. Eq. (3.64), which is derived from this strategy, has three right-hand sides (rhss); the top one is the result of combining Eqs. (2.58) and (2.59), the middle results from Eqs. (2.58) and (2.63) and the lowest one is from Eqs. (2.59) and (2.63):

$$\left.\begin{array}{l}\dfrac{R}{4\pi}\displaystyle\iint_\sigma \Delta g' S_r(r,\psi)\mathrm{d}\sigma \\ \dfrac{R}{4\pi}\displaystyle\iint_\sigma \delta g' H_r(r,\psi)\mathrm{d}\sigma\end{array}\right\} = \left\{\begin{array}{l}\dfrac{1}{\sin f\left(er\overline{\Theta}-p\right)}\left(\dot{a}\dfrac{r\overline{\Theta}}{2a\mu}-\dfrac{\dot{e}p}{\overline{m}}\right)-U_r \\ \dfrac{1}{(p\Gamma - r\Pi e\sin f)}\left(\dot{a}\dfrac{-r\Pi}{2a\mu}-p\left(\dot{\overline{M}}-\overline{n}\right)\right)-U_r \\ \dfrac{1}{\left(\Gamma\overline{\Theta}-\Pi\sin f\right)}\left(\dot{e}\dfrac{-\Pi}{\overline{m}}-\overline{\Theta}\left(\dot{\overline{M}}-\overline{n}\right)\right)-U_r\end{array}\right. . \tag{3.64}$$

Eq. (3.65) is derived similarly. The kernel of this integral is not an isotropic kernel, but it is much simpler than those presented in Eqs. (3.59)–(3.63):

$$\left.\begin{array}{l}\dfrac{R}{4\pi}\displaystyle\iint_{\sigma}\Delta g' S_{\psi}(r,\psi)\cos(\alpha_{\text{Sat}}+\alpha)\mathrm{d}\sigma\\[2ex]\dfrac{R}{4\pi}\displaystyle\iint_{\sigma}\delta g' H_{\psi}(r,\psi)\cos(\alpha_{\text{Sat}}+\alpha)\mathrm{d}\sigma\end{array}\right\}$$

$$=\begin{cases}\dfrac{r^2}{\left(\overline{\Theta}re-p\right)}\left(\dot{e}\dfrac{e}{\overline{m}}-\dfrac{\dot{a}}{2a\mu}\right)-U_{\theta}\cos\alpha_{\text{Sat}}\\[2ex]\dfrac{r^2}{\left(\Gamma\overline{\Theta}-\Pi\sin f\right)}\left(\dot{e}\dfrac{\Gamma}{\overline{m}}+\sin f\left(\dot{\overline{M}}-\overline{n}\right)\right)-U_{\theta}\cos\alpha_{\text{Sat}}\\[2ex]\dfrac{r^2}{\left(\Gamma p-re\Pi\sin f\right)}\left(\dfrac{\dot{a}\Gamma}{2a\mu}+e\sin f\left(\dot{\overline{M}}-\overline{n}\right)\right)-U_{\theta}\cos\alpha_{\text{Sat}}\end{cases}\qquad(3.65)$$

Analogously, another integral equation is derived from Eqs. (2.60) and (2.61). These equations are solved for V_w^P. Eq. (2.62) is used to determine the gravity anomaly using $\dot{\omega}$, V_u^P and V_v^P which can be presented in terms of orbital elements and their velocities as well. As mentioned earlier, the first terms on the rhs of Eqs. (3.64) and (3.65) are equivalent to V_u^P and V_v^P, respectively. The integral of Eq. (3.66) does not have an isotropic kernel, either, like that of Eq. (3.65):

$$\left.\begin{array}{l}\dfrac{R}{4\pi}\displaystyle\iint_{\sigma}\Delta g' S_{\psi}(r,\psi)\sin(\alpha_{\text{Sat}}+\alpha)\mathrm{d}\sigma\\[2ex]\dfrac{R}{4\pi}\displaystyle\iint_{\sigma}\delta g' H_{\psi}(r,\psi)\sin(\alpha_{\text{Sat}}+\alpha)\mathrm{d}\sigma\end{array}\right\}$$

$$=\begin{cases}-U_{\theta}\sin\alpha_{\text{Sat}}-\dfrac{\dot{i}ar}{\mu\cos(\omega+f)}\\[2ex]-U_{\theta}\sin\alpha_{\text{Sat}}-\dfrac{r\dot{\Omega}a\sin i}{\mu\sin(\omega+f)}\\[2ex]-U_{\theta}\sin\alpha_{\text{Sat}}-\dfrac{rp\tan i}{\sin(\omega+f)}\\[2ex]\times\left(\dfrac{r+p}{ep}\sin f\,U_{\theta}\cos\alpha_{\text{Sat}}+\dfrac{\cos f}{e}U_{\theta}\sin\alpha_{\text{Sat}}-\dfrac{\dot{\omega}}{\overline{m}}\right)\end{cases}\qquad(3.66)$$

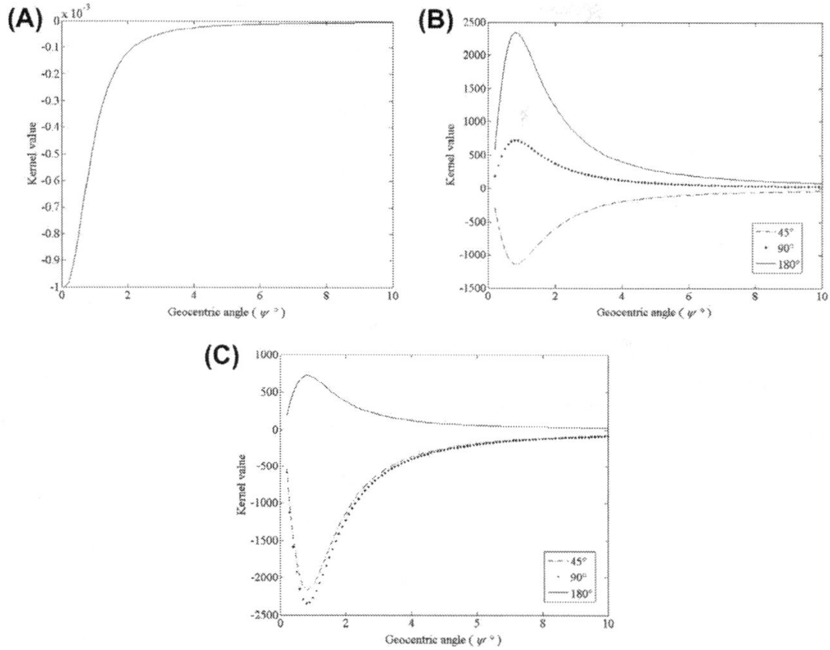

Figure 3.4 Behaviour of (A) $S_r(r,\psi)$, (B) $S_\psi(r,\psi)\cos(\alpha+\alpha')$ and (C) $S_\psi(r,\psi)\sin(\alpha+\alpha')$ at different azimuths (Eshagh and Ghorbannia, 2013).

Fig. 3.4A is the plot of $S_r(r,\psi)$, which is needed for Eq. (3.64). It is well-behaving and suitable for local gravity anomaly recovery. The important issue is to derive the rhs of Eq. (3.64). Fig. 3.4B illustrates that the kernel of Eq. (3.65) is bell-shaped, and so is Fig. 3.4C, which is similar to Fig. 3.4B, except for the kernel of Eq. (3.66), and the same conclusion can be drawn from it.

3.3 Integral inversion of acceleration and velocity of perturbations

It was explained in Section 2.3.3 that Hill equations represent the relation between the acceleration and the velocity of orbital perturbations in the track-oriented frame (TOF): radial, along and cross-track perturbations, and the Earth gravitational potential. The Hill equations were presented in Eqs. (2.85)−(2.87). Here, our goal is to show theoretically how these equations can be applied to recover gravity anomaly or gravity disturbance. We presented these perturbing potentials in terms of the gravity anomaly and normal gravity field. If each of these equations is inserted into the corresponding Hill equation (Eqs. 2.85–2.87) and the result is solved for the integral term, and considering Eqs. (3.48) and (3.50), we obtain integral equations:

$$\left.\begin{array}{l}\dfrac{R}{4\pi}\displaystyle\iint_{\sigma}\Delta g' S_r(r,\psi)\mathrm{d}\sigma \\ \dfrac{R}{4\pi}\displaystyle\iint_{\sigma}\delta g' H_r(r,\psi)\mathrm{d}\sigma\end{array}\right\} = \ddot{\overline{u}} + 2\overline{n}\dot{\overline{v}} - U_r(r,\theta), \tag{3.67}$$

$$\left.\begin{array}{l}\dfrac{R}{4\pi r}\displaystyle\iint_{\sigma}\Delta g' S_\psi(r,\psi)\cos(\alpha_{Sat}+\alpha)\mathrm{d}\sigma \\ \dfrac{R}{4\pi r}\displaystyle\iint_{\sigma}\delta g' H_\psi(r,\psi)\cos(\alpha_{Sat}+\alpha)\mathrm{d}\sigma\end{array}\right\} = \ddot{\overline{v}} - 2\overline{n}\dot{\overline{u}} - 3\overline{n}^2\overline{v} - \frac{1}{r}(U_\theta(r,\theta)\cos\alpha_{Sat}), \tag{3.68}$$

$$\left.\begin{array}{l}-\dfrac{R}{4\pi r}\displaystyle\iint_{\sigma}\Delta g' S_\psi(r,\psi)\sin(\alpha_{\mathrm{Sat}}+\alpha)\mathrm{d}\sigma \\ -\dfrac{R}{4\pi r}\displaystyle\iint_{\sigma}\delta g' H_\psi(r,\psi)\sin(\alpha_{\mathrm{Sat}}+\alpha)\mathrm{d}\sigma\end{array}\right\} = \ddot{\overline{w}} + \overline{n}^2\overline{w} + \frac{1}{r}(U_\theta(r,\theta)\sin\alpha_{\mathrm{Sat}}), \tag{3.69}$$

in which $\overline{u}, \overline{v}$ and $\overline{w}$ are the along, radial and cross-track geopotential perturbations, respectively; the double dots over them mean their acceleration and the single dots mean their velocity. Also, $\Delta g' = \Delta g(R,\theta',\lambda')$ and $\delta g' = \delta g(R,\theta',\lambda')$ to shorten the formulas.

The behaviour of the kernel functions of these integrals was presented in Fig. 3.4. To use these integrals, the accelerations and velocities of the perturbations should be transferred to the TRF.

3.4 Integral equations for inversion of satellite acceleration and velocity

In Chapter 2, the mathematical foundations for satellite acceleration and velocity and their relation to the gravitational potential were presented. In addition, some integral formulae for inverting them to the potential at the surface of the reference sphere were derived. In this section, our goal is to show how to modify those formulae to recover gravity anomaly/ disturbance.

3.4.1 Inversion of satellite acceleration in the terrestrial reference frame

The spherical harmonic expansions of the satellite accelerations in the TRF were presented in Eq. (2.98). The contribution of the normal gravity field to the satellite acceleration can be simply computed by inserting Eq. (3.1) into Eq. (2.98):

$$\ddot{\mathbf{r}}_{\mathrm{TRF}}^{\mathrm{normal}} = \frac{1}{r}\begin{bmatrix} \sum_{n=0}^{\infty}\left(\frac{R}{r}\right)^{n+1} u_n\left[-(n+1)\sin\theta\cos\lambda - \cos\theta\cos\lambda\frac{\mathrm{d}}{\mathrm{d}\theta}\right]P_n(\cos\theta) \\ \sum_{n=0}^{\infty}\left(\frac{R}{r}\right)^{n+1} u_n\left[-(n+1)\sin\theta\sin\lambda - \cos\theta\sin\lambda\frac{\mathrm{d}}{\mathrm{d}\theta}\right]P_n(\cos\theta) \\ \sum_{n=0}^{\infty}\left(\frac{R}{r}\right)^{n+1} u_n\left[-(n+1)\cos\theta + \sin\theta\frac{\mathrm{d}}{\mathrm{d}\theta}\right]P_n(\cos\theta) \end{bmatrix}, \tag{3.70}$$

which is the satellite acceleration vector induced only by the normal gravity potential field.

Now, if the integral equation derived for potential is written for the disturbing potential and normal potential, it will not be difficult to find integral equations for inverting satellite accelerations to gravity anomalies:

$$\left.\begin{aligned} &\frac{R}{4\pi}\iint_{\sigma}\Delta g'\sin\theta\cos\lambda\left(S_r(r,\psi)+\frac{1}{r}S_{\psi}(r,\psi)(-\cot\theta\cos\alpha+\tan\lambda\sin\alpha/\sin\theta)\right)\mathrm{d}\sigma \\ &\frac{R}{4\pi}\iint_{\sigma}\delta g'\sin\theta\cos\lambda\left(H_r(r,\psi)+\frac{1}{r}H_{\psi}(r,\psi)(-\cot\theta\cos\alpha+\tan\lambda\sin\alpha/\sin\theta)\right)\mathrm{d}\sigma \end{aligned}\right\} \\ = \ddot{x}_{\mathrm{TRF}} - \ddot{x}_{\mathrm{TRF}}^{\mathrm{normal}} \tag{3.71}$$

$$\left.\begin{array}{l}\dfrac{R}{4\pi}\displaystyle\iint_{\sigma}\Delta g'\sin\theta\sin\lambda\left(S_r(r,\psi)+\frac{1}{r}S_{\psi}(r,\psi)(-\cot\theta\cos\alpha+\cot\lambda\sin\alpha/\sin\theta)\right)\mathrm{d}\sigma\\[2ex]\dfrac{R}{4\pi}\displaystyle\iint_{\sigma}\delta g'\sin\theta\sin\lambda\left(H_r(r,\psi)+\frac{1}{r}H_{\psi}(r,\psi)(-\cot\theta\cos\alpha+\cot\lambda\sin\alpha/\sin\theta)\right)\mathrm{d}\sigma\end{array}\right\}$$

$$=\ddot{y}_{\mathrm{TRF}}-\ddot{y}^{\,\mathrm{normal}}_{\mathrm{TRF}},\tag{3.72}$$

$$\left.\begin{array}{l}\dfrac{R}{4\pi}\displaystyle\iint_{\sigma}\Delta g'\left(\cos\theta S_r(r,\psi)+\sin\theta\cos\alpha\frac{1}{r}S_{\psi}(r,\psi)\right)\mathrm{d}\sigma\\[2ex]\dfrac{R}{4\pi}\displaystyle\iint_{\sigma}\delta g'\left(\cos\theta H_r(r,\psi)+\sin\theta\cos\alpha\frac{1}{r}H_{\psi}(r,\psi)\right)\mathrm{d}\sigma\end{array}\right\}=\ddot{z}_{\mathrm{TRF}}-\ddot{z}^{\,\mathrm{normal}}_{\mathrm{TRF}}.\tag{3.73}$$

As seen, the kernels of these integral equations are functions of satellite coordinates and not just the geocentric angle between the computation point at the satellite and the integration points at sea level. Therefore, it is better to show their behaviour in a three-dimensional (3D) form. Fig. 3.5A

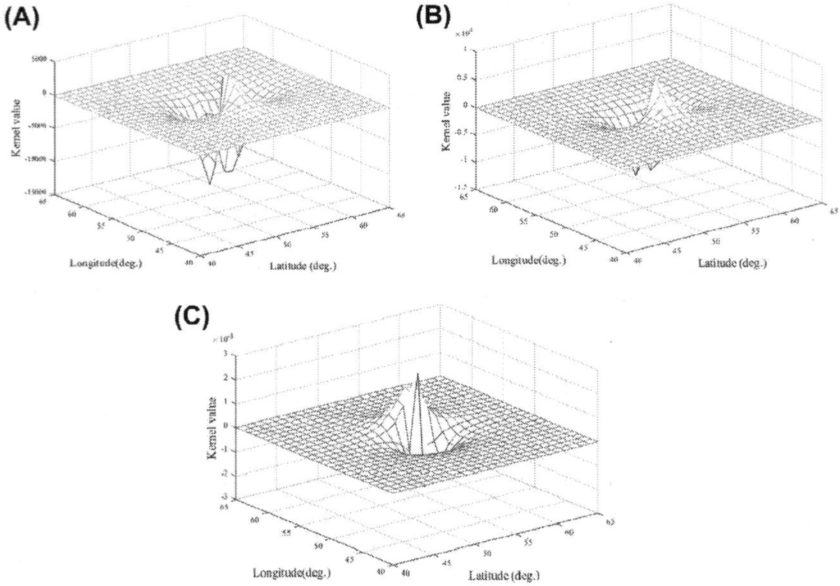

Figure 3.5 Three-dimensional plot of kernels of (A) Eq. (3.71), (B) Eq. (3.72) and (C) Eq. (3.73) for recovering $\Delta g'$.

shows the kernel of Eq. (3.71), which is used to determine $\Delta g'$ over an area limited between latitudes 40°N and 65°N and longitudes 40°E and 65°E and a satellite at 250 km elevation and a latitude of 53.1°N and longitude of 53.1°E. Fig. 3.5B illustrates the corresponding kernel of Eq. (3.72), and Fig. 3.5C that of Eq. (3.73). As Fig. 3.5 shows, the contribution of the far-zone $\Delta g'$is significant in the integral domain. In addition, as observed in Fig. 3.5B and C, the kernels are bell-shaped, but they do not have zero value at the computation point. Fig. 3.5D shows a well-behaving kernel and the contributions of the far-zone data are not as significant as those in the other kernels.

3.4.2 Inversion of satellite velocity in the terrestrial reference frame

Theoretical issues related to the energy integral and its solution were presented in Section 2.5. Here, we reformulate the final formula to relate it to the gravity anomaly. In the potential on the rhs of this formula, Eq. (2.104) is written as the summation of the disturbing and normal potentials, and the contribution related to the normal potential is taken to the left-hand side of Eq. (2.104). Disregarding dissipative acceleration, the integral equations will be:

$$\left.\begin{matrix} -\dfrac{R}{4\pi}\displaystyle\iint\limits_{\sigma} \Delta g(R,\theta',\lambda')S(r,\psi)\mathrm{d}\sigma \\ -\dfrac{R}{4\pi}\displaystyle\iint\limits_{\sigma} \delta g(R,\theta',\lambda')H(r,\psi)\mathrm{d}\sigma \end{matrix}\right\} + H$$

$$= \frac{1}{2}\left(\dot{x}_{\mathrm{TRF}}^2 + \dot{y}_{\mathrm{TRF}}^2 + \dot{z}_{\mathrm{TRF}}^2\right) - \frac{1}{2}\omega_e^2\left(x_{\mathrm{TRF}}^2 + y_{\mathrm{TRF}}^2\right) + U(r,\theta). \qquad (3.74)$$

Note that H is the Hamiltonian, an unknown that should be solved simultaneously with Δg or δg.

The kernel function of this integral equation is the extended Stokes function, which is not as sharp as its derivatives. This means that a large inversion area should be considered for inverting this integral, because the effect of the spatial truncation error (STE) (see Chapter 4) STE will be large and starts from areas close to the central part. Even if the truncation error removal is successful, only a long-wavelength structure of the gravity anomaly, or a considerably smoother solution, will be obtained compared with those derived from other methods.

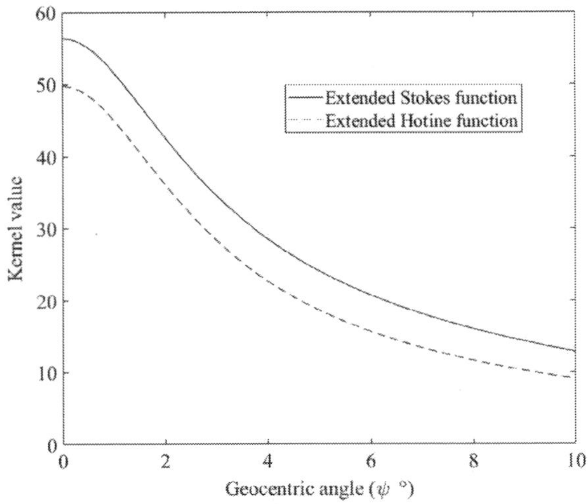

Figure 3.6 Behaviour of the extended Stokes and Hotine functions.

Fig. 3.6 illustrates the behaviour of the extended Stokes and Hotine function when the computation point is at 250 km altitude above sea level. Both kernels are well-behaving but do not decay quickly to zero. This means the contribution of the far-zone data remains significant in the integration domain. As observed, none of the kernels approaches zero, even up to a geocentric angle of 10 degrees. However, the extended Hotine function is slightly better for inversion purposes. Therefore, inverting the integrals in Eq. (3.74) results in a smooth solution of $\Delta g(R, \theta', \lambda')$ or $\delta g(R, \theta', \lambda')$. So, a much larger area should be considered as the inversion area, or recovery area, so that by performing the inversion and selecting the results in the central area, the STE will be reduced.

Fig. 3.6 illustrates the behaviour of the extended Stokes and Hotine functions when the computation point is at 250 km altitude above sea level. Both kernels are well-behaving but do not decay quickly to zero. This means the contribution of the far-zone data remains significant in the integration domain. As observed, none of the kernels approaches zero even to up to a geocentric angle of 10 degrees. However, the extended Hotine function is slightly better for inversion. Therefore, inverting the integrals in Eq. (3.74) results in a smooth solution of $\Delta g(R, \theta', \lambda')$ or $\delta g(R, \theta', \lambda')$. In addition, the STE is significant. Thus, a considerably larger area should be considered as an inversion area, or recovery area, so that by performing the inversion and selecting the results in the central area, the spatial truncation error reduces.

3.5 Integral equations for inversion of low–low tracking data

Here, two methods for deriving integral equations for gravity anomaly recovery are presented. The first uses the average velocity and the velocity difference between twin satellites moving in the same orbit with a specific distance or range between them. The second uses the range acceleration, range rate and range, as well as the satellites' relative positions, which we called it line-of-sight measurements. The mathematical backgrounds of the methods are given in Section 2.7.

3.5.1 Integral equations for inversion of range rates

Eq. (2.108) shows that the velocity difference between the twin satellites has a direct relation to the gravitational potential difference between them. If we write the gravitational potential for each satellite as the summation of the disturbing potential (T) and the normal potential (U), and then replace T with the extended Hotine or Stokes integral, Eqs. (3.25 or 3.27), the following integral equation is derived to invert the range rates ($\dot{\tilde{\rho}}$):

$$\left.\begin{array}{l}\dfrac{R}{4\pi}\displaystyle\iint_{\sigma}\Delta g(R,\theta',\lambda')[S(r_2,\psi_2)-S(r_1,\psi_1)]\mathrm{d}\sigma \\ \dfrac{R}{4\pi}\displaystyle\iint_{\sigma}\delta g(R,\theta',\lambda')[H(r_2,\psi_2)-H(r_1,\psi_1)]\mathrm{d}\sigma\end{array}\right\}=v_0\dot{\tilde{\rho}}-\Upsilon, \tag{3.75}$$

where v_0 stands for the average velocity of satellites and $\Upsilon$is the difference between the normal gravitational potentials for satellite 1 and 2 we the following mathematical formula:

$$\Upsilon=\sum_{n=2}^{\infty}u_n\left(\left(\frac{R}{r_2}\right)^{n+1}P_n(\cos\theta_2)-\left(\frac{R}{r_1}\right)^{n+1}P_n(\cos\theta_1)\right). \tag{3.76}$$

The kernels of the integral formulae (3.75) are isotropic, meaning that they are not functions of the azimuth. However, both are of a bipolar type with two computation points. However, because the satellites are close to each other, the differential kernels, the kernel with satellite 2 as the computation point minus the one with satellite 1, reduce the long-wavelength portion of the resulting kernel, and it will be sharper and able to recover higher frequencies of the gravitational potential. The separation between the satellites is a parameter determining how much low-frequency content should be reduced.

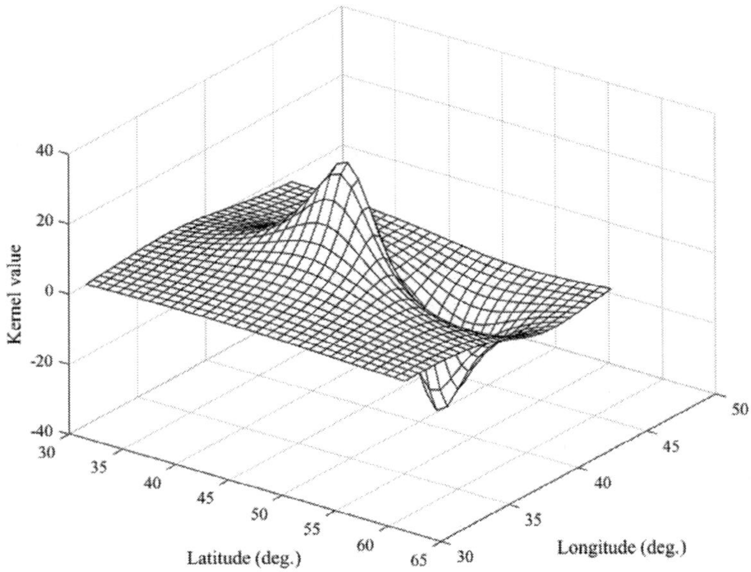

Figure 3.7 $S(r_2, \psi_2) - S(r_1, \psi_1)$.

Fig. 3.7 represents the bipolar kernel $S(r_2, \psi_2) - S(r_1, \psi_1)$ for two imaginary satellites with the following positions: latitude 50°N, longitude 50°E and altitude 250 km, and 55°N, 55°E and 260 km. The inversion area is limited between latitudes 30°N and 60°N and longitudes 30°E and 55°E. As observed, the contribution of the far-zone data is significant, but we should consider that the distance between the satellites is large and more than 500 km in this example for better visualisation of the kernel. If the satellites are closer, the behaviour of the kernel is similar because the contribution of the far-zone data reduces. This means the distance between the satellites has an important role in recovering higher frequencies of the gravity field. For high-resolution gravity field modelling, this distance should be smaller.

3.5.2 Integral equations for inversion of line-of-sight measurements

In a similar manner as that explained in the previous section, and according to Eqs. (2.127) and (2.128), the following integral equations are derived to recover gravity anomaly/disturbance from intersatellite line-of-sight (LOS) measurements:

$$\left.\begin{array}{l}\dfrac{R}{4\pi}\displaystyle\iint_{\sigma}\Delta g(R,\theta',\lambda')S^{\mathrm{B}}(r_1,\psi_1,\alpha_1,r_2,\psi_2,\alpha_2)\mathrm{d}\sigma\\[2ex]\dfrac{R}{4\pi}\displaystyle\iint_{\sigma}\delta g(R,\theta',\lambda')H^{\mathrm{B}}(r_1,\psi_1,\alpha_1,r_2,\psi_2,\alpha_2)\mathrm{d}\sigma\end{array}\right\}=\ddot{\widetilde{\rho}}-\frac{(\delta\dot{\mathbf{r}}_{12})^2-\dot{\widetilde{\rho}}^2}{\widetilde{\rho}}-\Sigma, \tag{3.77}$$

where Σ is a contribution of the normal gravitational potential to the LOS measurements and has the mathematical expression:

$$\Sigma=\sum_{n=2}^{\infty}u_n\left(\frac{n+1}{R}\left(\left(\frac{R}{r_2}\right)^{n+2}P_n(\cos\theta_2)\mathbf{e}_{z2}-\left(\frac{R}{r_1}\right)^{n+2}P_n(\cos\theta_1)\mathbf{e}_{z2}\right)\right.$$
$$\left.+\frac{1}{R}\left(\left(\frac{R}{r_2}\right)^{n+2}\frac{\mathrm{d}P_n(\cos\theta_2)}{\mathrm{d}\theta_2}\mathbf{e}_{x_2}-\left(\frac{R}{r_1}\right)^{n+2}\frac{\mathrm{d}P_n(\cos\theta_1)\mathbf{e}_{x_1}}{\mathrm{d}\theta_1}\right)\right)\cdot\mathbf{e}_{\mathrm{LOS}}, \tag{3.78}$$

$$\left\{\begin{array}{c}S^{\mathrm{B}}(r_1,\psi_1,\alpha_1,r_2,\psi_2,\alpha_2)\\[1ex]H^{\mathrm{B}}(r_1,\psi_1,\alpha_1,r_2,\psi_2,\alpha_2)\end{array}\right\}$$
$$=\left(\left\{\begin{array}{c}S_{r_2}(r_2,\psi_2)\\[1ex]H_{r_2}(r_2,\psi_2)\end{array}\right\}\mathbf{e}_{z_2}+\frac{1}{r_2}\left\{\begin{array}{c}S_{\psi_2}(r_2,\psi_2)\\[1ex]H_{\psi_2}(r_2,\psi_2)\end{array}\right\}(\cos\alpha_2\mathbf{e}_{x_2}-\sin\alpha_2\mathbf{e}_{y_2})\right)\cdot\mathbf{e}_{\mathrm{LOS}}$$
$$-\left(\left\{\begin{array}{c}S_{r_1}(r_1,\psi_1)\\[1ex]H_{r_1}(r_1,\psi_1)\end{array}\right\}\mathbf{e}_{z_1}+\frac{1}{r_1}\left\{\begin{array}{c}S_{\psi_1}(r_1,\psi_1)\\[1ex]H_{\psi_1}(r_1,\psi_1)\end{array}\right\}(\cos\alpha_1\mathbf{e}_{x_1}-\sin\alpha_1\mathbf{e}_{y_1})\right)\cdot\mathbf{e}_{\mathrm{LOS}}, \tag{3.79}$$

All parameters of Eqs. (3.78) and (3.79) were presented in Eqs. (2.126) and (2.127).

As seen, the kernel functions of the integral (Eq. 3.79) contain derivatives of the extended Stokes/Hotine formula for each satellite, which are more sensitive to the higher frequencies than the extended Stokes/Hotine functions. In addition, subtracting these derivatives for Satellite 2 from those at Satellite 1 amplifies the sensitivity of the resulting kernel even more to higher frequencies of the gravity anomalies/disturbance. Fig. 3.8

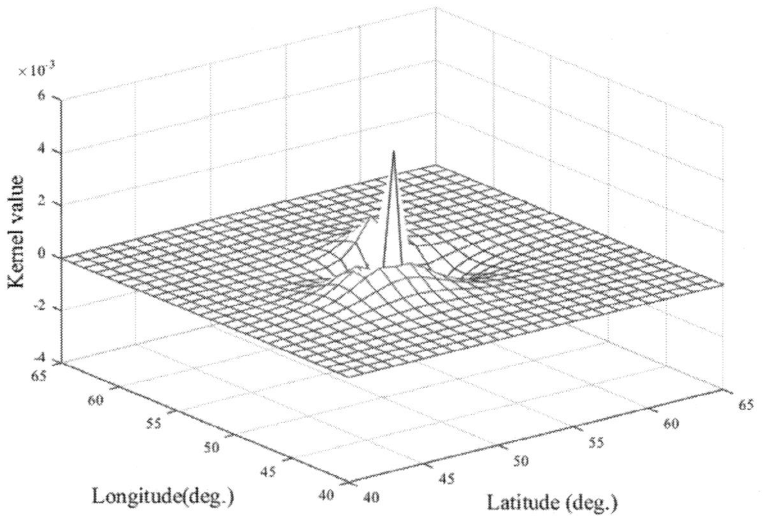

Figure 3.8 $S^B(r_1, \psi_1, \alpha_1, r_2, \psi_2, \alpha_2)$

shows the bipolar kernel $S^B(r_1, \psi_1, \alpha_1, r_2, \psi_2, \alpha_2)$, which can be used to recover $\Delta g(R, \theta', \lambda')$. Again, two imaginary satellites with a latitude, longitude and altitude of 55°N, 55°E and 250 km and 53.3°N, 53.3°E and 260 km, respectively, are considered over an area limited between the latitudes of 40°N and 65°N and longitudes 40°E and 65°E, respectively.

3.6 Integral equations for inversion of satellite gradiometry data

Satellite gradiometry data are more suitable than the rest of satellite gravimetry data for local gravity field recovery owing to their sensitivity to higher frequencies of the gravity field. Here, we consider two frames in which these types of data can be represented; the local north-oriented frame (LNOF) and the TOF.

3.6.1 Integral equations for inversion of gravity gradients in the local north-oriented frame

The satellite gradiometry data in the LNOF have simpler mathematical expressions than the other frames and are easier to use in real applications. However, these data should be transferred from the gradiometer frame to the LNOF. In the following discussion, we explain how gravity gradients are inverted.

The second-order partial derivatives of the disturbing potential have the general form (Reed, 1973):

$$T_i(r,\theta,\lambda) = \frac{R}{4\pi}\begin{cases}\iint\limits_{\sigma} \Delta g(R,\theta',\lambda')S_i(r,\psi)\mathrm{d}\sigma \\ \iint\limits_{\sigma} \delta g(R,\theta',\lambda')H_i(r,\psi)\mathrm{d}\sigma\end{cases}, \tag{3.80}$$

$$i = zz, xx, yy, xy, xz \text{ and } yz.$$

T_i with $i = zz, xx, yy, xy, xz$ and yz is representative partial derivatives of a disturbing potential in the LNOF. These are obtained by subtracting equivalent normal potential partial derivatives from the gravitational tensor measured at satellite level. The rest of parameters used in Eq. (3.80) are the same as those defined and presented in Section 3.1, except the second-order radial derivative of the extended Stokes function $S_{zz}(r,\psi) = S_{rr}(r,\psi)$, which is (Reed, 1973, Eq. 5.35):

$$S_{rr}(r,\psi) = \frac{t^3}{R^2}\left\{(1-t\tilde{x})\left[\frac{3(1-t^2)}{D^5} - \frac{4}{D^3}\right] - \frac{1+t^2}{D^3} - \frac{10}{D} - 18D + 2\right.$$
$$\left. -3t\tilde{x}\left(15 + 6\ln\frac{1-t\tilde{x}+D}{2}\right) + \frac{24}{D} + \frac{4(1-t^2)}{D^3}\right\}, \tag{3.81}$$

The other kernels are:

$$\begin{bmatrix} S_{xx}(r,\psi) \\ S_{yy}(r,\psi) \end{bmatrix} = \frac{S_r(r,\psi)}{r} + \frac{1}{2r^2}\{[S_{\psi\psi}(r,\psi) + \cot\psi S_\psi(r,\psi)] \pm$$
$$\pm [S_{\psi\psi}(r,\psi) - \cot\psi S_\psi(r,\psi)]\cos 2\alpha\}, \tag{3.82}$$

$$S_{xy}(r,\psi) = \frac{-1}{2r^2}[S_{\psi\psi}(r,\psi) + \cot\psi S_\psi(r,\psi)]\sin 2\alpha, \tag{3.83}$$

$$\begin{bmatrix} S_{xz}(r,\psi) \\ S_{yz}(r,\psi) \end{bmatrix} = \frac{1}{r}\left[\frac{1}{r}S_\psi(r,\psi) - S_{r\psi}(r,\psi)\right]\begin{pmatrix}\cos\alpha \\ -\sin\alpha\end{pmatrix}. \tag{3.84}$$

The derivatives $S_r(r,\psi)$ and $S_\psi(r,\psi)$ were presented in Eqs. (3.51) and (3.35), respectively, and in Kern and Haagmans (2005):

$$S_{\psi\psi}(r,\psi) = -t^2\widetilde{x}\left[\frac{2}{D^3}+\frac{6}{D}-8-3\frac{1-t\widetilde{x}-D}{D\sin^2\psi}-3\ln\frac{1-t\widetilde{x}+D}{2}\right]+t^3\sin^2\psi$$
$$\left[\frac{6}{D^5}+\frac{6}{D^3}+3\frac{D-1}{D^2\sin^2\psi}+3\frac{D+1}{D(1-t\widetilde{x}+D)}-3\frac{1-t\widetilde{x}-D}{D\sin^2\psi}\left(\frac{2\widetilde{x}}{t\sin^2\psi}+\frac{1}{D^2}\right)\right], \tag{3.85}$$

$$S_{r\psi}(r,\psi)=\frac{t^3\sin\psi}{R}$$
$$\left[\frac{3(1-t^2)}{D^5}+\frac{4}{D^3}+\frac{6}{D}-13-6\ln\frac{1-t\widetilde{x}+D}{2}+\frac{6t\widetilde{x}(D+1)}{D(1-t\widetilde{x}+D)}\right]. \tag{3.86}$$

Equivalent derivatives for the extended Hotine function are (Kern and Haagmans, 2005):

$$H_{rr}(r,\psi)=\frac{t}{r^2D}\left(4-\frac{2t^2}{D^2}-\frac{t}{g}\right)+\frac{t}{r^2g}\left(\frac{t}{g}-2\right)$$
$$+\frac{t(\widetilde{x}-t)}{r^2D}\left(\frac{8t}{D^2}+\frac{6t^2(\widetilde{x}-t)}{D^4}+\frac{t(\widetilde{x}-t)}{D^2g}+\frac{t(\widetilde{x}-t)}{Dg^2}-\frac{2t}{g^2}+\frac{2}{g}\right), \tag{3.87}$$

$$\begin{bmatrix} H_{xx}(r,\psi) \\ H_{yy}(r,\psi) \end{bmatrix} = \frac{H_r(r,\psi)}{r}+\frac{1}{2r^2}\{[H_{\psi\psi}(r,\psi)+\cot\psi H_\psi(r,\psi)]$$
$$\pm[H_{\psi\psi}(r,\psi)-\cot\psi H_\psi(r,\psi)]\cos 2\alpha\}, \tag{3.88}$$

$$H_{xy}(r,\psi)=\frac{-1}{2r^2}[H_{\psi\psi}(r,\psi)+\cot\psi H_\psi(r,\psi)]\sin 2\alpha, \tag{3.89}$$

$$\begin{bmatrix} H_{xz}(r,\psi) \\ H_{yz}(r,\psi) \end{bmatrix}=\frac{1}{r}\left[\frac{1}{r}H_\psi(r,\psi)-H_{r\psi}(r,\psi)\right]\begin{pmatrix}\cos\alpha \\ -\sin\alpha\end{pmatrix}, \tag{3.90}$$

where (Kern and Haagmans, 2005):

$$H_{\psi\psi}(r,\psi)=\frac{-2t^2\widetilde{x}}{D^3}+\frac{6t^3(1-\widetilde{x}^2)}{D^5}-\frac{t\widetilde{x}}{Dg}+\frac{t(1-\widetilde{x}^2)}{Dg}\left[\frac{t}{D^2}+\frac{t}{Dg}+\frac{2}{g}\right]-\frac{\widetilde{x}}{g}$$
$$+\frac{(1-\widetilde{x}^2)}{g^2}-\frac{1}{1-\widetilde{x}}, \tag{3.91}$$

$$H_{r\psi}(r,\psi)=\sqrt{1-\widetilde{x}^2}\left\{\frac{2t^2}{rD^3}\left[2+\frac{3t(\widetilde{x}-t)}{D^2}\right]\right.$$
$$\left.+\frac{t(\widetilde{x}-t)}{rDg}\left[\frac{t}{D}\left(\frac{1}{D}+\frac{1}{g}\right)+\frac{1}{g}\right]-\frac{t}{rg}\left[\frac{1}{D}\left(\frac{t}{g}+1\right)-\frac{1}{g}\right]\right\}. \tag{3.92}$$

where $g=t-\widetilde{x}+D$.

In fact, both types of integral equations presented in Eq. (3.80) are applicable for recovering Δg or δg, depending on the purpose of the inversion. The kernels of both types of integral equations behave similarly. However, slightly better results are obtained from inverting the gradients to δg, because the extended Hotine function decays slightly faster than the extended Stokes function.

Now, only the kernel functions of the integral equations Eq. (3.80) for recovering Δg are considered. The only isotopic kernel is $T_{zz}=T_{uu}$, presented in Eq. (3.81). Fig. 3.9 shows the isotropic parts of all kernels.

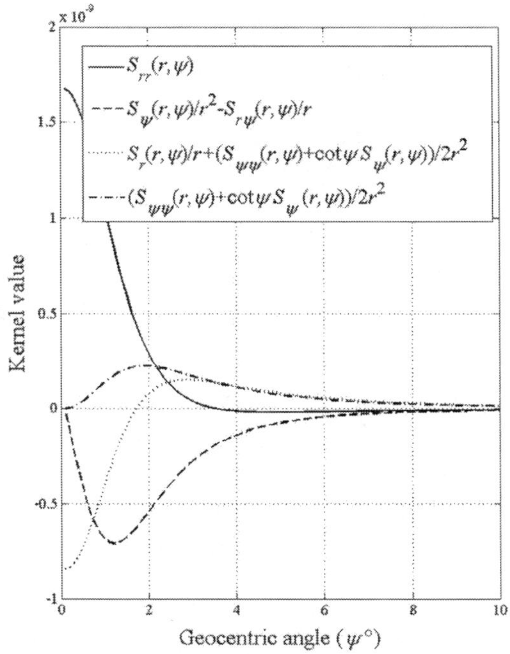

Figure 3.9 Isotropic parts of kernels presented in Eqs. (3.80)–(3.84) (Eshagh, 2011).

3.6.2 Integral equations for inversion of gravity gradients in the track-oriented frame

The second-order derivatives of the Earth's gravitational potential can also be expressed in a TOF, which is a moving frame, like the LNOF on the orbit, but it rotates around their common axes (the u-axis coincides with the z-axis of the local frame) based on the satellite track azimuth. In such a frame, the second-order derivatives of the disturbing potential can be presented by the integral formulas:

$$T_i(r,\theta,\lambda) = \begin{cases} \dfrac{R}{4\pi}\displaystyle\iint_\sigma \Delta g(R,\theta',\lambda')S_i(r,\psi)\mathrm{d}\sigma \\ \dfrac{R}{4\pi}\displaystyle\iint_\sigma \delta g(R,\theta',\lambda')H_i(r,\psi)\mathrm{d}\sigma \end{cases}, \tag{3.93}$$

$$i = ww, uu, vv, uv, uw \text{ and } vw,$$

where

$$\begin{Bmatrix} S_{uu}(r,\psi) \\ H_{uu}(r,\psi) \end{Bmatrix} = \begin{Bmatrix} S_{zz}(r,\psi) \\ H_{zz}(r,\psi) \end{Bmatrix}, \tag{3.94}$$

$$\begin{Bmatrix} S_{vv}(r,\psi,\alpha_{\mathrm{Sat}}) \\ H_{vv}(r,\psi,\alpha_{\mathrm{Sat}}) \end{Bmatrix} = \begin{Bmatrix} S_{xx}(r,\psi) \\ H_{xx}(r,\psi) \end{Bmatrix}\cos^2\alpha_{\mathrm{Sat}} + \begin{Bmatrix} S_{yy}(r,\psi) \\ H_{yy}(r,\psi) \end{Bmatrix}\sin^2\alpha_{\mathrm{Sat}} + \begin{Bmatrix} S_{xy}(r,\psi) \\ H_{xy}(r,\psi) \end{Bmatrix}\sin 2\alpha_{\mathrm{Sat}}, \tag{3.95}$$

$$\begin{Bmatrix} S_{ww}(r,\psi,\alpha_{\mathrm{Sat}}) \\ H_{ww}(r,\psi,\alpha_{\mathrm{Sat}}) \end{Bmatrix} = \begin{Bmatrix} S_{xx}(r,\psi) \\ H_{xx}(r,\psi) \end{Bmatrix}\sin^2\alpha_{\mathrm{Sat}} + \begin{Bmatrix} S_{yy}(r,\psi) \\ H_{yy}(r,\psi) \end{Bmatrix}\cos^2\alpha_{\mathrm{Sat}} - \begin{Bmatrix} S_{xy}(r,\psi) \\ H_{xy}(r,\psi) \end{Bmatrix}\sin 2\alpha_{\mathrm{Sat}}, \tag{3.96}$$

$$\begin{Bmatrix} S_{vw}(r,\psi,\alpha_{\mathrm{Sat}}) \\ H_{vw}(r,\psi,\alpha_{\mathrm{Sat}}) \end{Bmatrix} = \frac{\sin 2\alpha_{\mathrm{Sat}}}{2}\left[\begin{Bmatrix} S_{xx}(r,\psi) \\ H_{xx}(r,\psi) \end{Bmatrix} - \begin{Bmatrix} S_{yy}(r,\psi) \\ H_{yy}(r,\psi) \end{Bmatrix}\right] + \begin{Bmatrix} S_{xy}(r,\psi) \\ H_{xy}(r,\psi) \end{Bmatrix}\cos 2\alpha_{\mathrm{Sat}}, \tag{3.97}$$

$$\left\{\begin{array}{l} S_{uw}(r,\psi,\alpha_{\text{Sat}}) \\ H_{uw}(r,\psi,\alpha_{\text{Sat}}) \end{array}\right\} = \left\{\begin{array}{l} S_{xz}(r,\psi) \\ H_{xz}(r,\psi) \end{array}\right\}\cos\alpha_{\text{Sat}} + \left\{\begin{array}{l} S_{yz}(r,\psi) \\ H_{yz}(r,\psi) \end{array}\right\}\sin\alpha_{\text{Sat}}, \tag{3.98}$$

$$\left\{\begin{array}{l} S_{uv}(r,\psi,\alpha_{\text{Sat}}) \\ H_{uv}(r,\psi,\alpha_{\text{Sat}}) \end{array}\right\} = -\left\{\begin{array}{l} S_{xz}(r,\psi) \\ H_{xz}(r,\psi) \end{array}\right\}\sin\alpha_{\text{Sat}} + \left\{\begin{array}{l} S_{yz}(r,\psi) \\ H_{yz}(r,\psi) \end{array}\right\}\cos\alpha_{\text{Sat}}. \tag{3.99}$$

Some of these kernels are non-isotropic and satellite track azimuth dependent. However, as we can observe, the kernel (Eq. 3.94) is the simplest and is not involved in the transformation between the LNOF and the TOF.

If we add the kernels Eqs. (3.95) and (3.96), we obtain:

$$\begin{aligned} \left\{\begin{array}{l} S_{vv}(r,\psi,\alpha_{\text{Sat}}) + S_{uw}(r,\psi,\alpha_{\text{Sat}}) \\ H_{vv}(r,\psi,\alpha_{\text{Sat}}) + H_{uw}(r,\psi,\alpha_{\text{Sat}}) \end{array}\right\} &= \left\{\begin{array}{l} S_{xx}(r,\psi) + S_{yy}(r,\psi) \\ H_{xx}(r,\psi) + H_{yy}(r,\psi) \end{array}\right\} \\ &= \left\{\begin{array}{l} S_{vvuw}(r,\psi) \\ H_{vvuw}(r,\psi) \end{array}\right\}. \end{aligned} \tag{3.100}$$

This means that summation $T_{vv}(r,\theta,\lambda) + T_{uw}(r,\theta,\lambda)$ can be inverted using a simpler kernel:

$$\left.\begin{array}{l} \dfrac{R}{4\pi}\displaystyle\iint_{\sigma} \Delta g(R,\theta',\lambda') S_{vvuw}(r,\psi)\mathrm{d}\sigma \\ \dfrac{R}{4\pi}\displaystyle\iint_{\sigma} \delta g(R,\theta',\lambda') H_{vvuw}(r,\psi)\mathrm{d}\sigma \end{array}\right\} = T_{vv}(r,\theta,\lambda) + T_{uw}(r,\theta,\lambda), \tag{3.101}$$

where

$$\left\{\begin{array}{l} S_{vvuw}(r,\psi) \\ H_{vvuw}(r,\psi) \end{array}\right\} = \frac{1}{r}\left\{\begin{array}{l} S_{r}(r,\psi) \\ H_{r}(r,\psi) \end{array}\right\} + \frac{1}{2r^2}\left\{\begin{array}{l} S_{\psi\psi}(r,\psi) + \cot\psi S_{r\psi}(r,\psi) \\ H_{\psi\psi}(r,\psi) + \cot\psi H_{r\psi}(r,\psi) \end{array}\right\}. \tag{3.102}$$

The behaviour of the kernels (3.94)–(3.99) is presented in Fig. 3.10A–C. Because the kernels depend on the satellite track azimuths, an average latitude of 62 degrees and inclination of 96.6 degrees are considered for generating the kernels. Fig. 3.10A and B shows the kernels Eqs. (3.94)–(3.99) in two azimuths of 0 degrees and 180 degrees. They have different behaviour with respect to the azimuth changes and both have

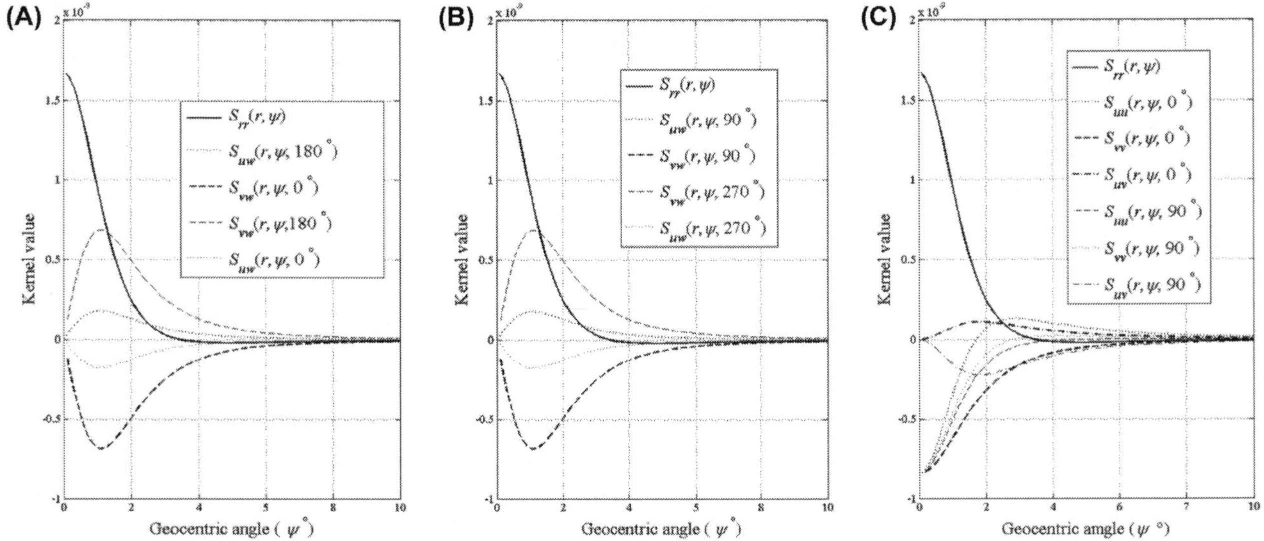

Figure 3.10 Kernels for inverting satellite gradiometry data in TOF with $\varphi^{\circ} = 90 - \theta = 62^{\circ}$ and $I = 96.6$ degrees in different azimuths (Eshagh, 2011).

a bell-shaped form. Fig. 3.10B is the plot of these kernels for the azimuths of 90 and 270 degrees. Fig. 3.10C shows that the kernels (3.95), (3.96) and (3.97) are well-behaving; therefore, the quality of gravity anomaly recovery from T_{vw} and $T_{xx} + T_{yy} = T_{uw} + T_{vv}$ should be acceptable, but the kernel $S_{vv}(r, \psi)$ has a bell-shaped form and is poor for inversion purposes.

3.7 Integral inversion of satellite altimetry data

The satellite altimetry data, defined in Section 2.9, are convertible to the geoid height or deflections of the vertical. The principle and integral transformations amongst these quantities were presented in Section 3.1. Now, our goal is to look more closely at the mathematical problems for gravity field recovery over oceans.

3.7.1 Determination of gravity anomaly/disturbance from altimetry geoid heights and deflections of the vertical

Let us start with a determination of gravity disturbance from the altimetry geoid height (see Section 2.9). The integral formula for such a purpose was presented in Eq. (3.17). One problem with performing the integration is that the kernel is singular at the computation point where $\psi = 0$; this has been discussed in numerous scientific books and papers in geodesy (Heiskanen and Moritz, 1967, Martinec, 1998, Hwang, 1998). Here, we explain how to solve this problem. Let us write the integral (3.17) as:

$$\delta g(R, \theta, \lambda) = I_1 + I_2, \tag{3.103}$$

where

$$I_1 = \frac{\gamma}{4\pi R} \iint_\sigma [N(R, \theta', \lambda') - N(R, \theta, \lambda)] P'_r(\psi) d\sigma, \tag{3.104}$$

$$I_2 = \frac{\gamma}{4\pi R} \iint_\sigma N(R, \theta, \lambda) P'_r(\psi) d\sigma. \tag{3.105}$$

where I_1 means the integral of differences between all geoid heights at the integration points and the one at the computation point all over the unit sphere and I_2, this integral is the integral of the constant value of $N(R, \theta, \lambda)$ over the sphere. Writing Eq. (3.17) in the form of Eq. (3.103) means nothing more than adding and subtracting I_2 to Eq. (3.103). Because

$N(R, \theta, \lambda)$ is constant in the integration domain of the integral in Eq. (3.105) we can write:

$$I_2 = \frac{\gamma}{4\pi R} N(R, \theta, \lambda) \int_0^{2\pi} \int_0^{\pi} P_r'(\psi) \sin\psi d\psi d\alpha. \tag{3.106}$$

The kernel $P_r'(\psi)$ is isotropic and does not depend on the azimuth; thus, after performing the integration over azimuth, Eq. (3.106) will change to:

$$I_2 = \frac{\gamma}{2R} N(R, \theta, \lambda) \int_0^{\pi} P_r'(\psi) \sin\psi d\psi. \tag{3.107}$$

By inserting the spectral form of the kernel, presented in Eq. (3.18), into Eq. (3.107), we obtain:

$$\int_0^{\pi} P_r'(\psi) \sin\psi d\psi = \sum_{n=0}^{\infty} (2n+1)(n+1) \int_0^{\pi} P_n(\cos\psi) \sin\psi d\psi. \tag{3.108}$$

The solution of the integral on the rhs of Eq. (3.108) is (Ilk, 1983):

$$\int_0^{\pi} P_n(\cos\psi) \sin\psi d\psi = \frac{2\delta_{n0}}{2n+1}. \tag{3.109}$$

By inserting Eq. (3.109) into Eq. (3.108), we obtain:

$$I_2 = \frac{\gamma}{R} N(R, \theta, \lambda). \tag{3.110}$$

Now we try to simplify the first integral, I_1. Let us write $N = N(R, \theta, \lambda)$ and $N' = N(R, \theta', \lambda')$ to shorten the formulas and derivations. The integration domain of I_1 can be partitioned into two parts: the distant zone minus the parts in the vicinity of the computation point or the innermost zone $\sigma_{\odot}$, $\sigma - \sigma_{\odot}$. Note that $\sigma_{\odot}$ is close to the computation point and should not be considered the integration cap or area. In this case, I_1 can be mathematically rewritten:

$$I_1 = J_1 + J_2, \tag{3.111}$$

where

$$J_1 = \frac{\gamma}{4\pi R} \iint_{\sigma - \sigma_{\odot}} (N' - N) P_r'(\psi) d\sigma, \tag{3.112}$$

$$J_2 = \frac{\gamma}{4\pi R} \iint_{\sigma_\odot} (N' - N) P_r'(\psi) d\sigma. \quad (3.113)$$

J_1 is an integral over the whole domain except the area around the computation point. To solve J_2, assume a local 2D coordinate system, with a centre at the centre of $\sigma_\odot$, an x-axis towards the east and a y-axis pointing to the north (Fig. 3.11).

Any point in this circular area has the Cartesian coordinates of x and y, and polar coordinates s and α. The relation between these two types of coordinates is well-known:

$$x = s \cos \alpha \text{ and } y = s \sin \alpha. \quad (3.114)$$

If we expand the geoid height at the computation point by a Taylor series around the computation point with coordinates x and y (Fig. 3.11):

Now, insert Eq. (3.114) into Eq. (3.115) and the result into Eq. (3.115) to obtain:

$$N' = N + x\frac{\partial N}{\partial x} + y\frac{\partial N}{\partial y}. \quad (3.115)$$

$$J_2 = \frac{\gamma}{4\pi R} \iint_{\sigma_\odot} \left[s \sin \alpha \frac{\partial N}{\partial x} + s \cos \alpha \frac{\partial N}{\partial y} \right] P_r'(\psi) d\sigma. \quad (3.116)$$

The kernel of this integral formula was presented in Eq. (3.18). It can be approximated in the near zone by:

$$P_r'(\psi) = \frac{-\cos\frac{\psi}{2}}{2 \sin^2\frac{\psi}{2}} - \sin \psi - 6 \cos \psi \sin \psi \approx -\frac{2}{\psi^2} \approx -\frac{2}{s^2}. \quad (3.117)$$

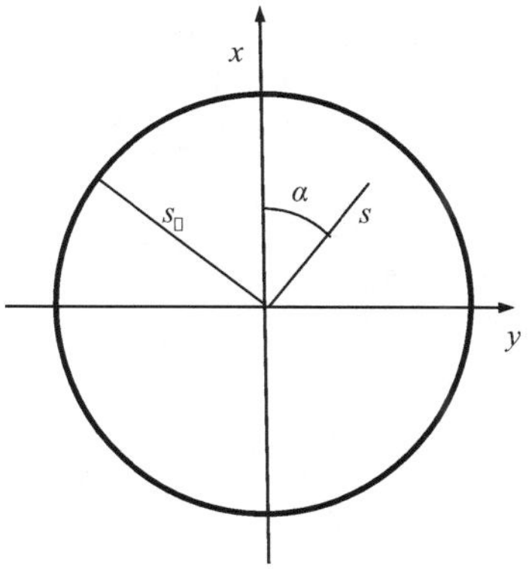

Figure 3.11 Local coordinate system at the vicinity of computation point $\sigma_\odot$.

Because $\frac{\partial N}{\partial x}$ and $\frac{\partial N}{\partial y}$ are the derivative of the geoid at the computation point, they remain constant over the integration domain; therefore, by inserting Eq. (3.114) into Eq. (3.113), we obtain:

$$J_2 = -\frac{\gamma s_{\odot}}{2\pi R}\left\{\frac{\partial N}{\partial x}\int_0^{2\pi}\sin\alpha d\alpha + \frac{\partial N}{\partial y}\int_0^{2\pi}\cos\alpha d\alpha\right\} = 0. \tag{3.118}$$

This means that J_2 is zero by the linear approximation of Eq. (3.115). Therefore, it is enough to subtract N at the computation point from all integration points and perform the integral over them. The integral formula for recovering gravity disturbance from the altimetry geoid height will be:

$$\left\{\begin{matrix}\delta g(R,\theta,\lambda)\\ \Delta g(R,\theta,\lambda)\end{matrix}\right\} = \frac{\gamma}{4\pi R}\iint_{\sigma-\sigma_{\odot}}[N(R,\theta',\lambda') - N(R,\theta,\lambda)]\left\{\begin{matrix}P_r'(\psi)\\ K''(\psi)\end{matrix}\right\}d\sigma + \frac{\gamma}{R}N(R,\theta,\lambda), \tag{3.119}$$

where$K''(\psi)$ was presented in Eq. (3.24).

Eq. (3.119) is valid when $\sigma_{\odot}$ is small enough that the linear approximation, Eq. (3.115), can be applied, if the integrand is not smooth like the geoid height or $\sigma_{\odot}$ is a large higher-order approximation that needs to be considered. Heiskanen and Moritz (1967) and Sjöberg (2009) show some examples.

3.7.2 Determination of gravity anomaly/disturbance from altimetry-derived deflections of the vertical

The integral formulas for computing δg and Δg from ξ and η were presented, respectively, in Eq. (3.42). The kernels of these integral formulae are also singular when the computation and the integration points coincide. To solve these integrals, we use a method similar to the one we used in the previous section. Let us partition the integral (Eq. 3.42) into two:

$$\left\{\begin{matrix}\delta g\\ \Delta g\end{matrix}\right\} = I_1' + I_2', \tag{3.120}$$

where

$$I_1' = \frac{R^2\gamma}{4\pi} \iint_{\sigma} [(\xi' - \xi)\cos\alpha' - (\eta' - \eta)\sin\alpha'] \left\{ \begin{array}{c} G_{\delta g}(\psi) \\ G_{\Delta g}(\psi) \end{array} \right\} d\sigma, \quad (3.121)$$

$$I_2' = \frac{R^2\gamma}{4\pi} \iint_{\sigma} [\xi \cos\alpha' - \eta \sin\alpha'] \left\{ \begin{array}{c} G_{\delta g}(\psi) \\ G_{\Delta g}(\psi) \end{array} \right\} d\sigma. \quad (3.122)$$

This means that we add and subtract ξ and η at the computation point from the integration points inside the integral. ξ and η will be constant in the integration domain of I_2'; therefore, I_2' can be simplified to:

$$\begin{aligned} I_2' = & \frac{R^2\gamma}{4\pi}\xi \int_0^{2\pi} \cos\alpha' d\alpha' \sum_{n=2}^{\infty} \frac{2n+1}{n} \left\{ \begin{array}{c} 1 \\ \dfrac{n-1}{n+1} \end{array} \right\} R_{n1} \\ & - \frac{R^2\gamma}{4\pi}\eta \int_0^{2\pi} \sin\alpha' d\alpha' \sum_{n=2}^{\infty} \frac{2n+1}{n} \left\{ \begin{array}{c} 1 \\ \dfrac{n-1}{n+1} \end{array} \right\} R_{n1} \\ = & \, 0, \end{aligned} \quad (3.123)$$

where

$$R_{n1} = \int_0^{\pi} P_{n1}(\cos\psi)\sin\psi d\psi. \quad (3.124)$$

As observed, the solution of the second integral, I_2', is zero.

The solution of the first integral, I_1', by partitioning it into two:

$$I_1' = J_1' + J_2', \quad (3.125)$$

where

$$J_1' = \frac{R^2\gamma}{4\pi} \iint_{\sigma-\sigma_\odot} [(\xi' - \xi)\cos\alpha' - (\eta' - \eta)\sin\alpha'] \left\{ \begin{array}{c} G_{\delta g}(\psi) \\ G_{\Delta g}(\psi) \end{array} \right\} d\sigma, \quad (3.126)$$

$$J_2' = \frac{R^2\gamma}{4\pi} \iint_{\sigma_\odot} [(\xi' - \xi)\cos\alpha' - (\eta' - \eta)\sin\alpha'] \left\{ \begin{array}{c} G_{\delta g}(\psi) \\ G_{\Delta g}(\psi) \end{array} \right\} d\sigma. \quad (3.127)$$

The goal is to solve J_2', the domain of which is the innermost zone area, $\sigma_{\odot}$. Let us approximate ξ and η at the computation point by the Taylor series to the linear term around the computation point (Fig. 3.11):

$$\xi' = \xi + s\cos\alpha\frac{\partial\xi}{\partial x} + s\sin\alpha\frac{\partial\xi}{\partial y}, \tag{3.128}$$

$$\eta' = \eta + s\cos\alpha\frac{\partial\eta}{\partial x} + s\sin\alpha\frac{\partial\eta}{\partial y}. \tag{3.129}$$

By inserting these equations into Eq. (3.127), we obtain:

$$J_2' = \frac{R^2\gamma}{4\pi}\int_0^{2\pi}\int_0^{s_\odot} s\left(\cos\alpha\frac{\partial\xi}{\partial x} + \sin\alpha\frac{\partial\xi}{\partial y}\right)\cos\alpha'\begin{Bmatrix} G_{\delta g}(\psi) \\ G_{\Delta g}(\psi) \end{Bmatrix} s ds d\alpha$$
$$- \frac{R^2\gamma}{4\pi}\int_0^{2\pi}\int_0^{s_\odot} s\left(\cos\alpha\frac{\partial\eta}{\partial x} + \sin\alpha\frac{\partial\eta}{\partial y}\right)\sin\alpha'\begin{Bmatrix} G_{\delta g}(\psi) \\ G_{\Delta g}(\psi) \end{Bmatrix} s ds d\alpha. \tag{3.130}$$

Note that α' is the backward azimuth or the azimuth from the integration point to the computation point. Because the convergence of meridian is negligible over the small area, we can assume that $\alpha' = \alpha + \pi$. Therefore, by inserting it into Eq. (3.130), and with further simplifications, we obtain:

$$J_2' = \frac{R^2\gamma}{4\pi}\int_0^{2\pi}\int_0^{s_\odot} s\left(\cos\alpha\cos(\pi+\alpha)\frac{\partial\xi}{\partial x} + \sin\alpha\cos(\pi+\alpha)\frac{\partial\xi}{\partial y}\right)\begin{Bmatrix} G_{\delta g}(\psi) \\ G_{\Delta g}(\psi) \end{Bmatrix} s ds d\alpha$$
$$- \frac{R^2\gamma}{4\pi}\int_0^{2\pi}\int_0^{s_\odot} s\left(\cos\alpha\sin(\pi+\alpha)\frac{\partial\eta}{\partial x} + \sin\alpha\sin(\pi+\alpha)\frac{\partial\eta}{\partial y}\right)\begin{Bmatrix} G_{\delta g}(\psi) \\ G_{\Delta g}(\psi) \end{Bmatrix} s ds d\alpha \tag{3.131}$$

Because $\cos(\pi+\alpha) = -\cos\alpha$, $\sin(\pi+\alpha) = \sin\alpha$ and derivatives of $\frac{\partial\xi}{\partial x}, \frac{\partial\xi}{\partial y}, \frac{\partial\eta}{\partial x}, \frac{\partial\eta}{\partial y}$ are related to the computation point,

$$J_2' = -\frac{R^2\gamma}{4\pi}\int_0^{s_\odot}\left[\frac{\partial\xi}{\partial x}\int_0^{2\pi}\cos^2\alpha d\alpha + \frac{\partial\xi}{\partial y}\int_0^{2\pi}\sin\alpha\cos\alpha d\alpha\right]\left\{\begin{matrix} G_{\delta g}(\psi) \\ G_{\Delta g}(\psi)\end{matrix}\right\}s^2 ds$$

$$\frac{R^2\gamma}{4\pi}\int_0^{s_\odot}\left[\frac{\partial\eta}{\partial x}\int_0^{2\pi}\sin\alpha\cos\alpha d\alpha - \frac{\partial\eta}{\partial y}\int_0^{2\pi}\sin^2\alpha d\alpha\right]\left\{\begin{matrix} G_{\delta g}(\psi) \\ G_{\Delta g}(\psi)\end{matrix}\right\}s^2 ds \tag{3.132}$$

The solution of the integrals in Eq. (3.132) over α yields:

$$J_2' = -\frac{R^2\gamma}{4}\left(\frac{\partial\xi}{\partial x}+\frac{\partial\eta}{\partial y}\right)\int_0^{s_\odot} s^2\left\{\begin{matrix} G_{\delta g}(\psi) \\ G_{\Delta g}(\psi)\end{matrix}\right\}ds. \tag{3.133}$$

In the small $\sigma_\odot$, the kernel $G_{\delta g}(\psi)$ and $G_{\Delta g}(\psi)$, presented in Eqs. (3.43) and (3.44), can be approximated to:

$$G_{\delta g}(\psi) \approx G_{\Delta g}(\psi) \approx -\frac{1}{2\left(\frac{\psi}{2}\right)^2}, \tag{3.134}$$

where s is the distance from any point in the integration domain and its centre, at which the kernel is singular (Fig. 3.11). After inserting Eq. (3.134) into Eq. (3.133) and solving the integral, we obtain (Hwang, 1998):

$$J_2' = \frac{R^2\gamma s_\odot}{2}\left(\frac{\partial\xi}{\partial x}+\frac{\partial\eta}{\partial y}\right). \tag{3.135}$$

Finally, Eq. (3.120) can be rewritten as:

$$\left\{\begin{matrix}\delta g \\ \Delta g\end{matrix}\right\} = \frac{R^2\gamma}{4\pi}\iint_{\sigma-\sigma_\odot}[(\xi'-\xi)\cos\alpha' - (\eta'-\eta)\sin\alpha']\left\{\begin{matrix} G_{\delta g}(\psi) \\ G_{\Delta g}(\psi)\end{matrix}\right\}d\sigma$$

$$+\frac{R^2\gamma s_\odot}{2}\left(\frac{\partial\xi}{\partial x}+\frac{\partial\eta}{\partial y}\right). \tag{3.136}$$

3.7.3 Integral inversion of geoid and deflections of the vertical

In the numerical inversion of the Hotine, Stokes and Vening Meinesz integrals, the kernel singularity occurs when satellite altimetry data are in a gridded form. In practice, these types of data are given along the satellite ground track at a certain interval. Therefore, it is possible to use the mentioned integral formulae directly determining the gravity anomaly/disturbance on regular grids. In such a case, singularity seldom occurs. However, if the satellite altimetry data are dense, there is a low singularity risk. In this case, the following method for those singular points can be applied.

3.7.3.1 Integral equations for inversion of altimetry geoid heights

Let us separate the area around the singularity point from the Hotine and Stokes integral:

$$N(R,\theta,\lambda) = I_1 + I_2, \tag{3.137}$$

where

$$I_1 = \begin{cases} \dfrac{R}{4\pi\gamma}\displaystyle\iint_{\sigma-\sigma_\odot} \delta g(R,\theta',\lambda')H(\psi)\mathrm{d}\sigma \\ \dfrac{R}{4\pi\gamma}\displaystyle\iint_{\sigma-\sigma_\odot} \Delta g(R,\theta',\lambda')S(\psi)\mathrm{d}\sigma \end{cases} \quad \text{and} \quad I_2 = \begin{cases} \dfrac{R}{4\pi\gamma}\displaystyle\iint_{\sigma_\odot} \delta g(R,\theta',\lambda')H(\psi)\mathrm{d}\sigma \\ \dfrac{R}{4\pi\gamma}\displaystyle\iint_{\sigma_\odot} \Delta g(R,\theta',\lambda')S(\psi)\mathrm{d}\sigma \end{cases}. \tag{3.138}$$

I_2 has a small integration domain around the singular point, which can be simplified further to:

$$I_2 = \frac{R}{4\pi\gamma}\begin{Bmatrix} \delta g' \displaystyle\int_0^{2\pi}\int_0^{\psi_\odot} H(\psi)\sin\psi\mathrm{d}\psi\mathrm{d}\alpha \\ \Delta g' \displaystyle\int_0^{2\pi}\int_0^{\psi_\odot} S(\psi)\sin\psi\mathrm{d}\psi\mathrm{d}\alpha \end{Bmatrix} = \frac{R}{2\gamma}\begin{Bmatrix} \delta g' \displaystyle\int_0^{\psi_\odot} H(\psi)\sin\psi\mathrm{d}\psi \\ \Delta g' \displaystyle\int_0^{\psi_\odot} S(\psi)\sin\psi\mathrm{d}\psi \end{Bmatrix}, \tag{3.139}$$

where $\psi_\odot$ is the radius of the circular cap $\sigma_\odot$.

The Hotine and Stokes kernels can be approximated in the small integration domain, and when ψ is small, to:

$$H(\psi) \approx S(\psi) \approx \frac{1}{\sin\frac{\psi}{2}} \approx \frac{2}{\psi}. \tag{3.140}$$

By inserting this kernel into Eq. (3.139) and solving for the integral, I_2 will be simplified to:

$$I_2 = \frac{R}{\gamma}\sin\frac{\psi_{\odot}}{2}\left\{\begin{array}{c}\delta g(R,\theta,\lambda)\\ \Delta g(R,\theta,\lambda)\end{array}\right\}, \tag{3.141}$$

and finally, Eq. (3.137) will change to:

$$N(R,\theta,\lambda) = \left\{\begin{array}{c}\frac{R}{4\pi\gamma}\iint\limits_{\sigma-\sigma_{\odot}}\delta g(R,\theta',\lambda')H(\psi)\mathrm{d}\sigma\\ \frac{R}{4\pi\gamma}\iint\limits_{\sigma-\sigma_{\odot}}\Delta g(R,\theta',\lambda')S(\psi)\mathrm{d}\sigma\end{array}\right\} + \frac{R}{\gamma}\sin\frac{\psi_{\odot}}{2}\left\{\begin{array}{c}\delta g(R,\theta,\lambda)\\ \Delta g(R,\theta,\lambda)\end{array}\right\}. \tag{3.142}$$

Alternatively, we can simplify I_2 another way. We know the geocentric angle ψ has the following relation to the length of the spherical curve in front of it in the sphere $\psi = s/R$. Now, we try to rewrite I_2 based on s, but first, we have to find the integration element. The differential of this element is $\mathrm{d}\psi = \mathrm{d}s/R$. Therefore:

$$\sin\psi\mathrm{d}\psi\mathrm{d}\alpha = \frac{s\mathrm{d}s\mathrm{d}\alpha}{R^2}. \tag{3.143}$$

Inserting Eq. (3.143) into Eq. (3.139) reads:

$$I_2 = \left\{\begin{array}{c}\delta g(R,\theta',\lambda')\\ \Delta g(R,\theta',\lambda')\end{array}\right\}\frac{1}{2\pi\gamma}\iint\limits_{\sigma_{\odot}}\frac{1}{s}s\mathrm{d}s\mathrm{d}\alpha. \tag{3.144}$$

Further simplification of Eq. (3.144) leads to:

$$I_2 = \left\{\begin{array}{c}\delta g(R,\theta',\lambda')\\ \Delta g(R,\theta',\lambda')\end{array}\right\}\frac{1}{4\pi\gamma}\int_0^{2\alpha}\int_0^{s_{\odot}}\mathrm{d}s\mathrm{d}\alpha = \left\{\begin{array}{c}\delta g(R,\theta',\lambda')\\ \Delta g(R,\theta',\lambda')\end{array}\right\}\frac{s_{\odot}}{\gamma}. \tag{3.145}$$

Eq. (3.145) is equivalent to Eq. (3.143) as $s_{\odot} = R\sin(\psi_{\odot}/2)$ over the sphere. However, writing the formula in this form is helpful for writing the planar approximation of the Hotine or Stokes integral, which is good for a small area around $\sigma_{\odot}$, the singular point.

So far, we have assumed that the area around the singular point is a small spherical cap. However, when the data are gridded, the cells are in rectangular forms and not caps. Therefore, the planar approximation for I_2 is:

$$I_2 = \left\{ \begin{array}{c} \delta g(R, \theta', \lambda') \\ \Delta g(R, \theta', \lambda') \end{array} \right\} \frac{1}{2\pi\gamma} \int_{y_1}^{y_2} \int_{x_1}^{x_2} \frac{1}{\sqrt{x^2 + y^2}} \mathrm{d}x\mathrm{d}y. \tag{3.146}$$

Let us consider a local projection for this small area with the centre as a singular point with coordinates θ and λ:

$$x = R(\theta' - \theta), \tag{3.147}$$

$$y = R\sin\theta'(\lambda' - \lambda). \tag{3.148}$$

The upper and lower bounds of this rectangle will be (Hirt et al., 2011):

$$x_1 = R\left(\theta' - \theta - \frac{\Delta\theta}{2}\right) \text{ and } y_1 = R\sin\theta'\left(\lambda' - \lambda - \frac{\Delta\lambda}{2}\right), \tag{3.149}$$

$$x_2 = R\left(\theta' - \theta + \frac{\Delta\theta}{2}\right) \text{ and } y_1 = R\sin\theta'\left(\lambda' - \lambda + \frac{\Delta\lambda}{2}\right), \tag{3.150}$$

where $\Delta\theta$ and $\Delta\lambda$ are the cell sizes along meridians and parallels.

The solution of the integral part of Eq. (3.146) is:

$$\int_{y_1}^{y_2} \int_{x_1}^{x_2} \frac{1}{\sqrt{x^2 + y^2}} \mathrm{d}x\mathrm{d}y = F(x, y) \bigg|_{x_1}^{x_2} \bigg|_{y_1}^{y_2} = W, \tag{3.151}$$

where

$$F(x, y) = 2x\ln\left(y + \sqrt{x^2 + y^2}\right) - 2y + 2y\ln\left(x + \sqrt{x^2 + y^2}\right). \tag{3.152}$$

Finally,

$$I_2 = \left\{ \begin{array}{c} \delta g(R, \theta', \lambda') \\ \Delta g(R, \theta', \lambda') \end{array} \right\} \frac{W}{2\pi\gamma}. \tag{3.153}$$

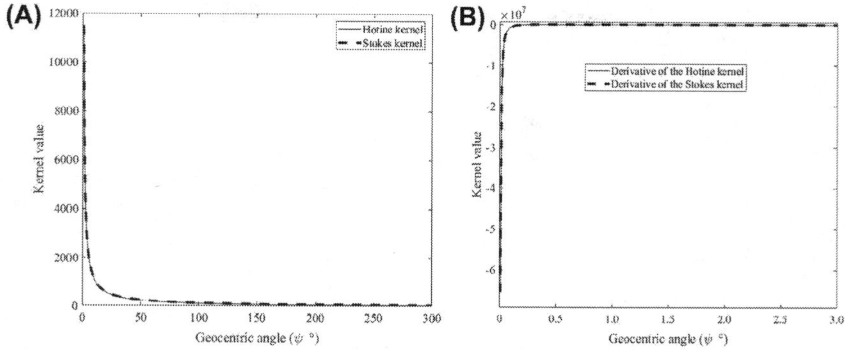

Figure 3.12 Behaviour of (A) the Stokes and Hotine functions and (B) derivative of the Stokes and Hotine functions.

One of the three expressions for I_2 can be applied when the kernel becomes singular instead of the Hotine or Stokes kernel. The discretised form of Eq. (3.137) will be:

$$N_i = \sum_{j=1}^{N} \begin{Bmatrix} \delta g_j \\ \Delta g_j \end{Bmatrix} \times \left(\frac{R}{4\pi\gamma} \begin{Bmatrix} H(\psi_{ij}) \\ S(\psi_{ij}) \end{Bmatrix} \sin\theta_j \Delta\theta_j \Delta\lambda_j (1-\delta_{ij}) + \frac{\delta_{ij}}{\gamma} \begin{Bmatrix} R\sin(\psi_{\odot}/2) \\ s_{\odot} \\ W/(2\pi) \end{Bmatrix} \right), \tag{3.154}$$

where δ_{ij} is the Kronecker delta, i stands for the computation points and j stands for the integration points.

Fig. 3.12A shows the behaviour of the Stokes and Hotine functions, and we can see that they behave well and are applicable to integral inversion of satellite altimetry data, but care should be taken for the singularity of their kernels.

3.7.3.2 Integral equations for inversion of altimetry deflections of the vertical

The satellite altimetry-derived geoid heights can be converted to components of the deflections of the vertical by numerical differentiation. Because Δg or δg are unknown parameters, we cannot use the Taylor expansion of them around the vicinity of the computation point ($\sigma_{\odot}$) to solve the singularity problem as we do for forward integrals. Therefore, another technique should be applied.

The Vening Meinesz formula can be directly applied to invert these components to Δg or δg. The Vening Meinesz formula is:

$$\left.\begin{array}{l}\dfrac{R}{4\pi\gamma}\displaystyle\iint_{\sigma}\delta g(R,\theta',\lambda')H_{\psi}(\psi)\begin{pmatrix}\cos\alpha\\ -\sin\alpha\end{pmatrix}d\sigma\\ \dfrac{R}{4\pi\gamma}\displaystyle\iint_{\sigma}\Delta g(R,\theta',\lambda')S_{\psi}(\psi)\begin{pmatrix}\cos\alpha\\ -\sin\alpha\end{pmatrix}d\sigma\end{array}\right\}=\begin{pmatrix}\xi(R,\theta,\lambda)\\ \eta(R,\theta,\lambda)\end{pmatrix}. \quad (3.155)$$

Fig. 3.12B represents the behaviour of $H_{\psi}(\psi)$ and $S_{\psi}(\psi)$ and shows that they are well-behaving but singular at the computation point and have large values around it. Over $\sigma_{\odot}$, with the rectangular coordinate system defined in Eqs. (3.113) and (3.114), we can use the planar approximation of the Hotine or Stokes formula, which means (Eqs. 3.149 and 3.150):

$$\left.\begin{array}{l}\dfrac{R}{4\pi\gamma}\displaystyle\iint_{\sigma_{\odot}}\delta g(R,\theta',\lambda')H_{\psi}(\psi)\begin{pmatrix}\cos\alpha\\ -\sin\alpha\end{pmatrix}d\sigma\\ \dfrac{R}{4\pi\gamma}\displaystyle\iint_{\sigma_{\odot}}\Delta g(R,\theta',\lambda')S_{\psi}(\psi)\begin{pmatrix}\cos\alpha\\ -\sin\alpha\end{pmatrix}d\sigma\end{array}\right\}=I_2$$

$$=\left\{\begin{array}{l}\delta g(R,\theta',\lambda')\\ \Delta g(R,\theta',\lambda')\end{array}\right\}\frac{1}{2\pi\gamma}\int_{y_1}^{y_2}\int_{x_1}^{x_2}\frac{-1}{(x^2+y^2)^{3/2}}\begin{pmatrix}x\\ y\end{pmatrix}dxdy \quad (3.156)$$

Let us consider the solution of the integral on the rhs of Eq. (3.156):

$$\int_{y_1}^{y_2}\int_{x_1}^{x_2}\frac{-1}{(x^2+y^2)^{3/2}}\begin{pmatrix}x\\ y\end{pmatrix}dxdy=\begin{pmatrix}W_{\xi}\\ W_{\eta}\end{pmatrix}, \quad (3.157)$$

where

$$W_{\xi}=\ln\left(y+\sqrt{x^2+y^2}\right)\Big|_{x_1}^{x_2}\Big|_{y_1}^{y_2} \text{ and } W_{\eta}=\ln\left(x+\sqrt{x^2+y^2}\right)\Big|_{x_1}^{x_2}\Big|_{y_1}^{y_2}. \quad (3.158)$$

finally,

$$I_2 = \frac{1}{2\pi\gamma}\begin{cases} \delta g(R, \theta', \lambda')\begin{pmatrix} W_\xi \\ W_\eta \end{pmatrix} \\ \Delta g(R, \theta', \lambda')\begin{pmatrix} W_\xi \\ W_\eta \end{pmatrix} \end{cases}, \tag{3.159}$$

$$\begin{pmatrix} \xi(R, \theta, \lambda) \\ \eta(R, \theta, \lambda) \end{pmatrix} = \begin{cases} \dfrac{R}{4\pi\gamma}\iint\limits_{\sigma-\sigma_\odot} \delta g(R, \theta', \lambda') H_\psi(\psi)\begin{pmatrix} \cos\alpha \\ -\sin\alpha \end{pmatrix} d\sigma \\ +\dfrac{1}{2\pi\gamma}\delta g(R, \theta, \lambda)\begin{pmatrix} W_\xi \\ W_\eta \end{pmatrix} \\ \dfrac{R}{4\pi\gamma}\iint\limits_{\sigma-\sigma_\odot} \Delta g(R, \theta', \lambda') S_\psi(\psi)\begin{pmatrix} \cos\alpha \\ -\sin\alpha \end{pmatrix} d\sigma \\ +\dfrac{1}{2\pi\gamma}\Delta g(R, \theta, \lambda)\begin{pmatrix} W_\xi \\ W_\eta \end{pmatrix} \end{cases}, \tag{3.160}$$

with the discretised form:

$$\begin{pmatrix} \xi(R, \theta, \lambda) \\ \eta(R, \theta, \lambda) \end{pmatrix} = \begin{cases} \sum\limits_{j=1}^{N} \delta g_j \left(\dfrac{R}{4\pi\gamma} H_\psi(\psi_{ij}) \sin\theta_j \Delta\theta_j \Delta\lambda_j (1 - \delta_{ij}) + \dfrac{\delta_{ij}}{2\pi\gamma}\begin{pmatrix} W_\xi \\ W_\eta \end{pmatrix} \right) \\ \sum\limits_{j=1}^{N} \Delta g_j \left(\dfrac{R}{4\pi\gamma} S_\psi(\psi_{ij}) \sin\theta_j \Delta\theta_j \Delta\lambda_j (1 - \delta_{ij}) + \dfrac{\delta_{ij}}{2\pi\gamma}\begin{pmatrix} W_\xi \\ W_\eta \end{pmatrix} \right) \end{cases}. \tag{3.161}$$

References

Eshagh, M., 2011a. Inversion of satellite gradiometry data using statistically modified integral formulas for local gravity field recovery. Adv. Space Res. 47 (1), 74–85.

Eshagh, M., 2011b. The effect of spatial truncation error on integral inversion of satellite gravity gradiometry data. Adv. Space Res. 47, 1238–1247.

Eshagh, M., Ghorbannia, M., 2013. The use of Gaussian equations of motions of a satellite for local gravity anomaly recovery. Adv. Space Res. 52 (1), 30–38.

Eshagh, M., Ghorbannia, M., 2014. The effect of spatial truncation error on variance of gravity anomalies aderived from inversion of satellite orbital and gradiometric data. Adv. Space Res. 54, 261–271.

Heiskanen, W., Moritz, H., 1967. Physical Geodesy. W.H. Freeman and company, San Francisco and London.

Hirt, C., Featherstone, W.E., Claessens, S.J., 2011. On the accurate numerical evaluation of geodetic convolution integrals. J Geod. 85, 519–538.

Hwang, C., 1998. Inverse Vening Meinesz formula and deflection-geoid formula: applications to the predictions of gravity and geoid over the South China Sea. J. Geodes. 72, 304–312.

Ilk, K.H., 1983. Ein Eitrag zür Dynamik Ausgedehnter Körper-Gravitations-Wechselwirkung (A contribution to the dynamics of extended-body gravitational interaction). Deutsche Geodätische Kommission, Reihe C, Heft 288, München.

Kern, M., Haagmans, R., 2005. Determination of gravity gradients from terrestrial gravity data for calibration and validation of gradiometric GOCE data. In: Proc. Gravity, Geoid and Space Missions, GGSM 2004, IAG International Symposium, Portugal, August 30-September 3, pp. 95–100, 2004.

Lemoine, F.G., Kenyon, S.C., Factor, J.K., Trimmer, R.G., Pavlis, N.K., Chinn, D., Cox, C.M., Klosko, S.M., Luthcke, S.B., Torrence, M.H., Wang, Y.M., Williamson, R.G., Pavlis, E.C., Rapp, R.H., Olson, T.R., 1998. Geopotential Model EGM96. NASA/TP-1998-206861. Goddard Space Flight Center, Greenbelt.

Martinec, Z., 1998. Boundary-Value Problems for Gravimetric Determination of a Precise Geoid. Lecture notes in Earth Sciences, Springer.

Obenson, G., 1970. Direct Evaluation of the Earth's Gravity Anomaly Field from Orbital Analysis of Artificial Earth Satellites, Report No. 3. Department of Geodetic Science, The Ohio State University (OSU), Columbus, Ohio.

Pick, M., Jan, P., Vyskočil, V., Tauer, J., 1973. Theory of the Earth's Gravity Field. Elsevier, p. 538.

Reed, G.B., 1973. Application of Kinematical Geodesy for Determining the Shorts Wavelength Component of the Gravity Field by Satellite Gradiometry. Ohio state University, Dept. of Geod. Science, Rep. No. 201, Columbus, Ohio.

Sjöberg, L.E., 2009. Solving Vening Meinesz-Moritz inverse problem in isostasy. Geophysical Journal International 179 (3), 1527–1536.

Torge, W., 2001. Geodesy. Walter de Gruyter, 2001 - Science - 416 pages.

Vaníček, P., Krakiwsky, E.J., 1982. Geodesy, the Concepts. North-Holland Pub. Co., p. 691

CHAPTER 4

Numerical inversion of satellite gravimetry data

4.1 Discretisation of integral formulae

So far, a lot of discussion has gone on about integral equations. In this section, our goal is to show how this process is done in practice. Consider the following general form of a spherical integral formula:

$$A(r,\theta,\lambda) = \frac{1}{4\pi}\iint\limits_{\sigma} B(R,\theta',\lambda')\Pi(r,\psi)\mathrm{d}\sigma \quad \text{with}$$
$$\Pi(r,\psi) = \sum_{n=0}^{\infty}(2n+1)\pi_n(r)P_n(\cos\psi), \tag{4.1}$$

where $A(r,\theta,\lambda)$ is the solution of the integral at the computation point (r,θ,λ) from $B(R,\theta',\lambda')$ at integration points (R,θ',λ') at the surface of a sphere with radius R with the spherical kernel $\Pi(r,\psi)$ in the forward computation process. $\pi_n(r)$ is the spectra of the kernal $\Pi(r,\psi)$. In the inversion of the integral (Eq. 4.1), $A(r,\theta,\lambda)$ is known but $B(R,\theta',\lambda')$ unknown. In the following, we explain forward computation from which $A(r,\theta,\lambda)$ is computed from $B(R,\theta',\lambda')$ using the spherical integral. Later, the method for the inverse solution of this integral, or determining $B(R,\theta',\lambda')$ from $A(r,\theta,\lambda)$, will be discussed.

4.1.1 Numerical solution of integrals

According to Eq. (4.1), $B(R,\theta',\lambda')$ is integrated to obtain a single value for $A(r,\theta,\lambda)$. In practice, the gravimetric data are measured point by point, and for solving integrals, they normally need to be gridded; therefore, $B(R,\theta',\lambda')$ should be in gridded form. A schematic figure of this grid is presented in Fig. 4.1, containing, e.g. 16 values of $B(R,\theta',\lambda')$. To take the integral means to obtain a weighted mean of these values, and the kernel function $\Pi(r,\psi)$ plays the role of weight for this process. This is the reason

Satellite Gravimetry and the Solid Earth
ISBN 978-0-12-816936-0
https://doi.org/10.1016/B978-0-12-816936-0.00004-9

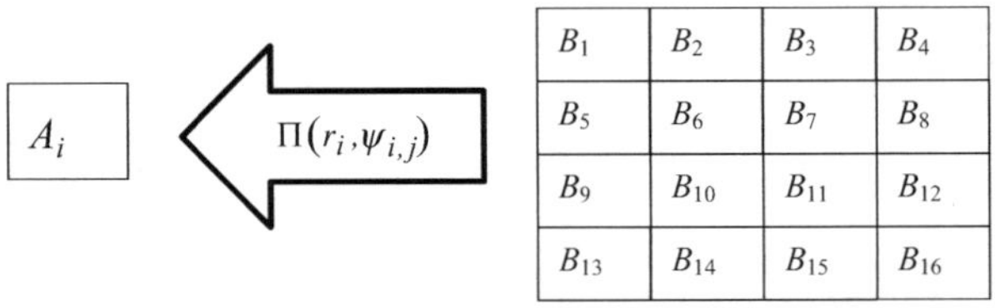

Figure 4.1 Discretisation for solving integrals.

for presenting the plots of the kernel functions of the integral formulae whenever we present them. These kernels show the significance and the weighting scheme of the data in the solution. In addition, the kernel acts as a converter to change one quantity to another while weighting them. We will come back to the kernel functions later, but for now, we know that a kernel is a function of two points, the computation point at which the solution $A(r, \theta, \lambda)$ is desired and the integration point $B(R, \theta', \lambda')$ at which the data are given. Therefore, in Eq. (4.1) $B(R, \theta', \lambda')$ are integrated to deliver only one value for $A(r, \theta, \lambda)$. This process is shown in Fig. 4.1 as well.

The discretised form of Eq. (4.1), based on the grid presented in Fig. 4.1, is:

$$A_i = \frac{1}{4\pi} \sum_{j=1}^{16} B_j \Pi(r_i, \psi_{ij}) \sin \theta_j \Delta\theta \Delta\lambda, \tag{4.2}$$

where $\Delta\theta$ and $\Delta\lambda$ are the size of each grid cell along the meridian and parallel, respectively. The terms i and j stand for the computation and integration points.

4.1.2 Kernels of integrals

As explained, the integration is similar to a weighting process with a weight equal to the kernel values. Generally, the kernel functions are divided into two groups according to Eshagh (2011a): well-behaving and bell-shaped kernels which we explain in the following.

4.1.2.1 Well-behaving kernels

Fig. 4.2A shows a well-behaving kernel, having its maximum value at the computation point, where the geocentric angle is $\psi = 0°$ and decays fast to zero. If we rotate the kernel plot around the vertical axis in Fig. 4.2A and B is obtained, which is the 3D form of this kernel. The integration domain will be a circle and the value of the kernel shows the weighting scheme in this circular domain.

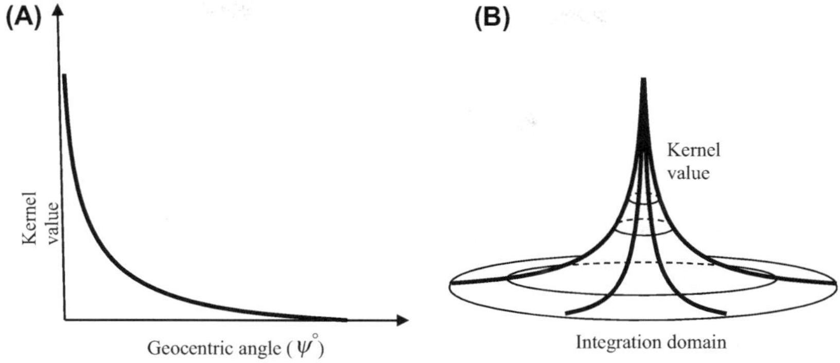

Figure 4.2 A well-behaving kernel. (A) 2D plot and (B) 3D plot.

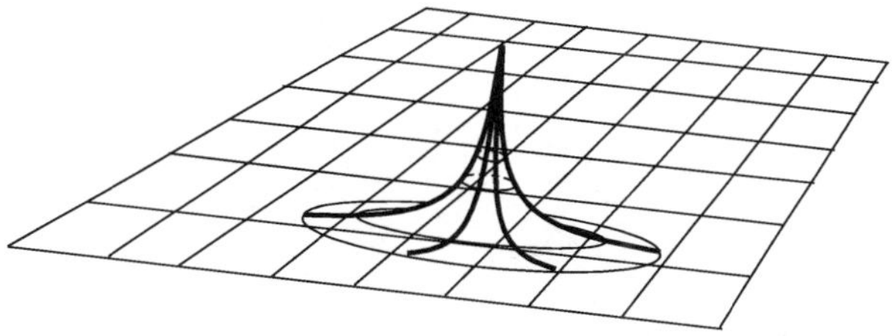

Figure 4.3 Integration inside the area.

To explain how this kernel acts on the data, Fig. 4.3 is presented. The integration is performed on gridded data, and this kernel function will be placed on the grid, and its integration domain covers some parts of the data, depending on the kernel. The kernel function weights the data and when it is well-behaving the size of its integration domain will be small, and therefore, those data close to the centre of integration are required. This kernel with its domain moves step by step over the grid of data with an overlap depending on the grid resolution, and the integral delivers one value for each integration. However, when the computation points get closer to the margins and corners of the grid, the situation presented in Fig. 4.4 will occur. As we can observe, half of the integration domain is outside the grid at the margins and three quarters at the corners. This means that when the integration is performed in such situation, the integration domain is not covered by data, and the solution of the integral is based on an incomplete integration domain. Such a problem leads to spatial truncation error (STE) if the integration is done all over the area. To avoid this problem, the integration should not be done all over the area; see Fig. 4.5 showing that the area of the results should be smaller than the data area to

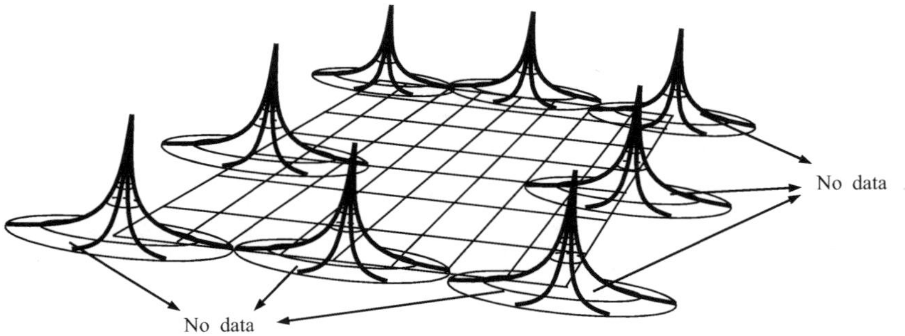

Figure 4.4 Well-behaving kernels close to the margins and corners of the area.

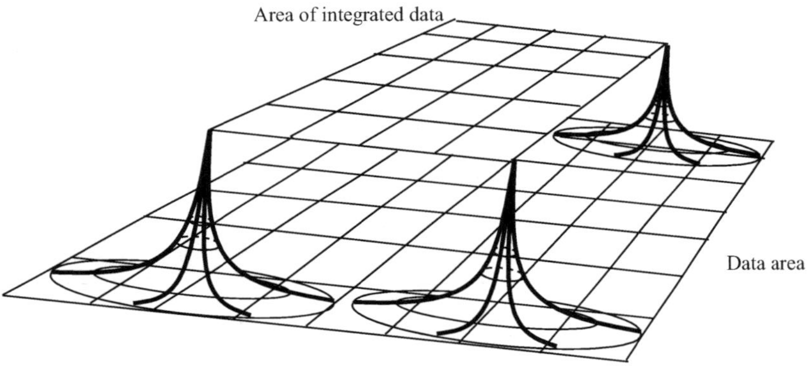

Figure 4.5 Data area and area of the integrated data.

avoid STE. In addition, one can perform the integration all over the area but select the results in the central area as the solution.

4.1.2.2 Bell-shaped kernels

There are kernels with small values at the computation point where $\psi = 0^\circ$, and when this angle increases, the function reaches its maximal value and then decays and gets closer to zero; see Fig. 4.6A. This type of kernel, which normally includes the derivative of the Legendre polynomials in their spectral form, was named bell-shaped (Eshagh, 2011a). The plot of a bell-shaped kernel is presented in Fig. 4.6A, and if rotated around the vertical axis, in this case Fig. 4.6B, the 3D plot of this kernel is obtained.

This kernel plays the role of a weighting function for the data, but it does not give the highest weight to the data at the computation point, but rather to those data which are placed around this point at a certain distance from the centre of integration domain. The procedure for calculating the

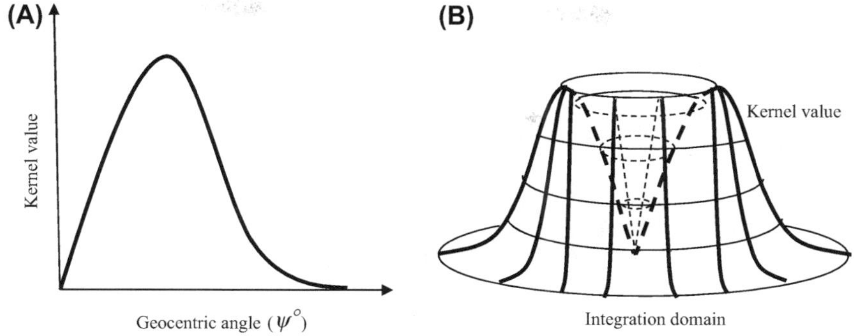

Figure 4.6 A bell-shaped kernel. (A) 2D plot and (B) 3D plot.

integral with such kernel is similar to that presented in Section 4.1.2.1. As long as the integration domain remains on the grid of the data, there will not be any problem with integration. However, when the computation points start approaching the margins and corners of the area, STE occurs. Fig. 4.7 shows the kernel function at the margins and corners of the data area and, as observed, at the margins, half of the integration domain goes outside the grid, and at the corners, three-quarters of the domain. These missing data cause the solution of the integral to contain truncation bias due to loss of the data in its integration domain. In comparison to the well-behaving kernels (see Fig. 4.4), we observe in Fig. 4.7 that the some parts of the maximum value of the kernel are outside the grid, whilst for the

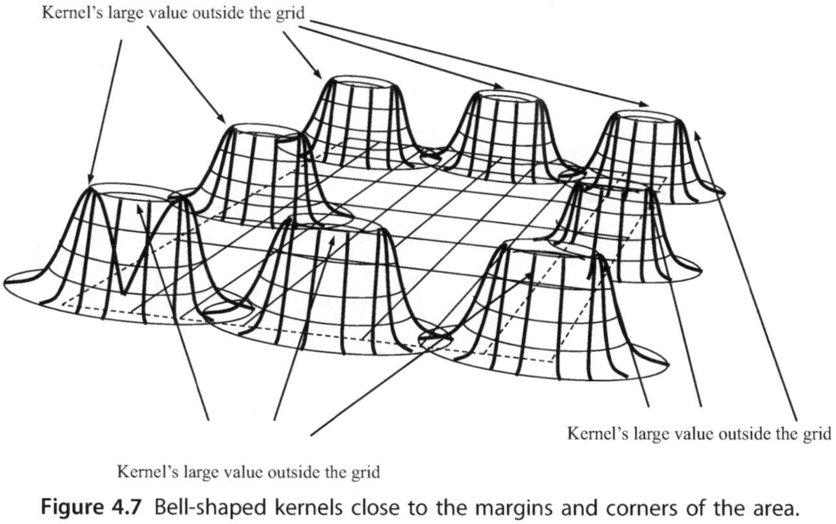

Figure 4.7 Bell-shaped kernels close to the margins and corners of the area.

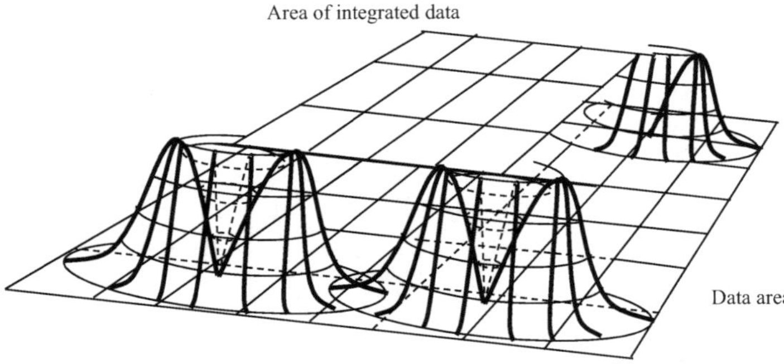

Figure 4.8 Data area and area of the integrated data.

well-behaving kernels the maximum value is still inside the area. When this large value of the kernel is placed outside, it means that it gives a very high weight to non-existing data. Therefore, it will not be difficult to conclude that the STE is more serious for integrating with bell-shaped kernels than with well-behaving ones.

For solving integrals, or direct solution of integrals, it would be enough to perform the integration point to point and with overlapping integration areas to cover the data region as long as the integration domain is inside the area. Therefore, the result of the integrations in the central area is free of such error. This means that we always need to have an area with data that is larger than our desired area for performing the integration. Fig. 4.8 shows this principle and the data area and the area of the integrated data, or the results' area.

4.1.3 Numerical inverse solution of integrals

For solving integrals numerically, the data inside the integration domain are our measurement, and the quantity which is sought comes from solution of the integral. Consider again Eq. (4.1). This time $A(r,\theta,\lambda)$ is known and $B(R,\theta',\lambda')$ is the unknown. We should solve this integral in reverse, to get $B(R,\theta',\lambda')$. See Fig. 4.9 showing the principle of this process. Again, the integral should be discretised according to the desired resolution. Here again we consider a 4-by-4 grid of $B(R,\theta',\lambda')$, shown in Fig. 4.9, right side. This time these $B(R,\theta',\lambda')$ are our unknowns, and in this example, we have 16 unknowns. Here, Eq. (4.1) is considered as an observation equation connecting each measured $A(r,\theta,\lambda)$ to all data on the regular grid of $B(R,\theta',\lambda')$.

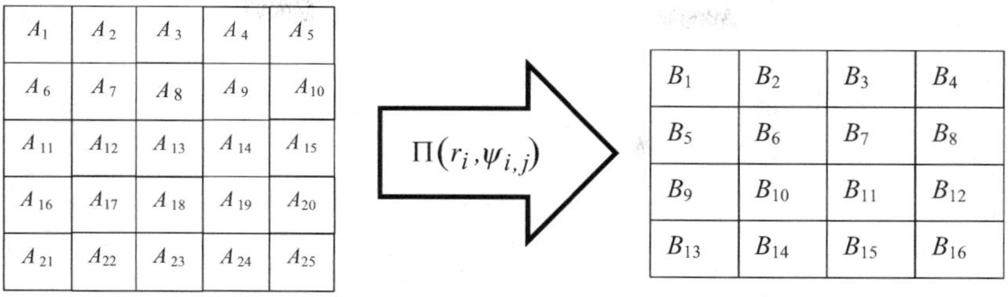

Figure 4.9 Discretisation for inversion of integrals.

Therefore, for each value of $A(r, \theta, \lambda)$ we write observation equations:

$$A_1 = \frac{\sin\theta_1 \Delta\theta\Delta\lambda}{4\pi} B_1 \Pi(r_1, \psi_{1,1}) + \frac{\sin\theta_2 \Delta\theta\Delta\lambda}{4\pi} B_2 \Pi(r_1, \psi_{1,2}) + \cdots + \frac{\sin\theta_{16} \Delta\theta\Delta\lambda}{4\pi} B_{16} \Pi(r_1, \psi_{1,16})$$

$$A_2 = \frac{\sin\theta_1 \Delta\theta\Delta\lambda}{4\pi} B_1 \Pi(r_2, \psi_{2,1}) + \frac{\sin\theta_2 \Delta\theta\Delta\lambda}{4\pi} B_2 \Pi(r_2, \psi_{2,2}) + \cdots + \frac{\sin\theta_{16} \Delta\theta\Delta\lambda}{4\pi} B_{16} \Pi(r_2, \psi_{2,16}).$$

$$\vdots$$

$$A_{25} = \frac{\sin\theta_1 \Delta\theta\Delta\lambda}{4\pi} B_1 \Pi(r_{25}, \psi_{25,1}) + \frac{\sin\theta_2 \Delta\theta\Delta\lambda}{4\pi} B_2 \Pi(r_{25}, \psi_{25,2}) + \cdots + \frac{\sin\theta_{16} \Delta\theta\Delta\lambda}{4\pi} B_{16} \Pi(r_{25}, \psi_{25,16})$$

This is a system of equations with 25 observations and 16 unknowns, which can be written in the following matrix form:

$$\mathbf{Ax} = \mathbf{L} - \boldsymbol{\varepsilon},$$

where $\mathbf{A}$ is a 25×16 coefficients matrix, $\mathbf{L}$ is a 25×1 vector of measurement, $\mathbf{x}$ is the 16×1 vector of unknowns and $\boldsymbol{\varepsilon}$ is the vector of random errors:

$$\mathbf{A} = \frac{\Delta\theta\Delta\lambda}{4\pi} \begin{bmatrix} \sin\theta_1 \Pi(r_1, \psi_{1,1}) & \sin\theta_2 \Pi(r_1, \psi_{1,2}) & \cdots & \sin\theta_{16} \Pi(r_1, \psi_{1,16}) \\ \sin\theta_1 \Pi(r_2, \psi_{2,1}) & \sin\theta_1 \Pi(r_2, \psi_{2,2}) & \cdots & \sin\theta_{16} \Pi(r_2, \psi_{2,16}) \\ \vdots & \vdots & \ddots & \vdots \\ \sin\theta_1 \Pi(r_{25}, \psi_{25,1}) & \sin\theta_1 \Pi(r_{25}, \psi_{25,2}) & \cdots & \sin\theta_{16} \Pi(r_{25}, \psi_{25,16}) \end{bmatrix},$$

$\mathbf{L} = [A_1, A_2, \ldots, A_{25}]^T$, and $\mathbf{x} = [B_1, B_2, \ldots, B_{16}]^T$.

The system of equations has a larger number of observations than the unknowns, and therefore, it needs to be solved by the least-squares method. The important issue for solving this system is the size of the discretisation; when the resolution is high the elements of matrix $\mathbf{A}$ are closer to each other. This causes the determinant of the matrix becomes close to zero. By inverting the normal equations, we place the determinant in the denominator, and therefore, since it is close to zero, the inverted system will amplify the high frequencies of the solution and any small error in the data will lead to a large value of changes in the obtained results. Regularisation is a method to obtain a stable solution and control this error amplification.

4.2 Handling spatial truncation error

When gravity field determination is done locally by integral inversion, the problem of STE always occurs. This is because truncating the integration domain of the integral to a small part of the sphere and definitely integrating limited data gives limited information about the desired quantity. The difference between what a quantity is and what is obtained by the limited coverage of the same quantity is a measure of the STE. This systematic error is inevitable in integral inversion and should be controlled so that the results are not significantly influenced by that. In the following, two methods are presented for estimating this error.

4.2.1 Estimation of spatial truncation error based on the integral and spherical harmonics

To estimate the STE on a quantity and see whether the loss of data in the corners and margins of the area is significant, the quantity is generated using its spherical harmonic expansion and its integral formula. Theoretically, the integral formula and this expansion should deliver the same results, if the data are dense enough to reduce the discretisation errors and the integration domain is the whole sphere, for example:

$$A(r,\theta,\lambda)=\sum_{n=0}^{\infty}\sum_{m=-n}^{n}A_{nm}Y_{nm}(\theta,\lambda)=\frac{1}{4\pi}\iint_{\sigma}B(R,\theta',\lambda')\Pi(r,\psi)d\sigma. \tag{4.3}$$

As the discretisation error is assumed to be small and our data are dense, but the integration domain is $\sigma_0 \subset \sigma$, this causes the STE around the data area. Mathematically, we can write:

$$\delta A(r,\theta,\lambda)=\sum_{n=0}^{\infty}\sum_{m=-n}^{n}A_{nm}Y_{nm}(\theta,\lambda)-\frac{1}{4\pi}\iint_{\sigma_0\subset\sigma}B(R,\theta',\lambda')\Pi(r,\psi)d\sigma. \tag{4.4}$$

This is the simplest method for estimating the STE.

4.2.2 Estimation of spatial truncation error based on spherical harmonics

To find a spherical harmonic expression for the truncation error, we try to transfer back the integral part of Eq. (4.4) into the spectral domain:

$$J = \frac{1}{4\pi} \iint_{\sigma_0 \subset \sigma} \Pi(r, \psi) B(R, \theta', \lambda') \mathrm{d}\sigma$$
$$= \frac{1}{4\pi} \sum_{n=0}^{\infty} (2n+1) \pi_n(r) P_n(\cos\psi) \sum_{n'=0}^{\infty} \sum_{m'=-n'}^{n'} B_{n'm'} Y_{n'm'}(\theta', \lambda') \mathrm{d}\sigma. \tag{4.5}$$

By considering a spherical harmonic expansion for $B(R, \theta', \lambda')$, and the kernel of Eq. (4.1), using the addition theorem of spherical harmonics, Eq. (1.9), we arrive at:

$$J = \frac{1}{4\pi} \sum_{n=0}^{\infty} \pi_n(r) \sum_{m'=-n'}^{n'} Y_{n'm'}(\theta', \lambda') Y_{nm}(\theta, \lambda) \sum_{n'=0}^{\infty} \sum_{m'=-n'}^{n'} B_{n'm'} Y_{n'm'}(\theta', \lambda') \mathrm{d}\sigma$$
$$= \frac{1}{4\pi} \sum_{n=0}^{\infty} \sum_{m=-n}^{n} \sum_{n'=0}^{\infty} \sum_{m'=-n'}^{n'} \pi_n(r) B_{n'm'} I_{nmn'm'} Y_{nm}(\theta, \lambda), \tag{4.6}$$

where

$$I_{nmn'm'} = \iint_{\sigma_0 \subset \sigma} Y_{nm}(\theta', \lambda') Y_{n'm'}(\theta', \lambda') \mathrm{d}\sigma. \tag{4.7}$$

Inserting Eq. (4.6) back into Eq. (4.4) and further simplifying lead to:

$$\delta A_{nm} = A_{nm} - \frac{1}{4\pi} \sum_{n'=0}^{\infty} \sum_{m'=-n'}^{n'} \pi_n(r) B_{n'm'} I_{nmn'm'}, \tag{4.8}$$

which is the spherical harmonic coefficient (SHC) of the STE. Also, the product of the spherical harmonics can be written by (see Eq. 1.143):

$$Y_{nm}(\theta, \lambda) Y_{n'm'}(\theta, \lambda) = \sum_{n''=|n-n'|}^{n+n'} \sum_{m''=-n''}^{n''} Q_{nmn'm'}^{n''m''} Y_{n''m''}(\theta, \lambda). \tag{4.9}$$

Therefore, Eq. (4.7) can be rewritten as:

$$I_{nmn'm'} = \sum_{n''=|n-n'|}^{n+n'} \sum_{m''=-n''}^{n''} Q_{nmn'm'}^{n''m''} \iint_{\sigma_0 \subset \sigma} Y_{n''m''}(\theta', \lambda') \mathrm{d}\sigma, \tag{4.10}$$

where $Q_{nmn'm'}^{n''m''}$ is the Gaunt coefficient. Note that in computations of the Gaunt coefficients instability may happen. The estimated STE by this method should be removed from the data prior to inverting them. Solution of the integral of Eq. (4.10) and the problem of determining the gravity field over the ocean using satellite altimetry data have been discussed in Section 2.9.

4.2.3 Size of the inversion area and spatial truncation error

We observed that for solving integrals the integration should be performed over the data grid, and the behaviour of the kernel of the integral shows how large the central area should be. However, in inversion of the integrals, they should be discretised all over the area and not only over the integration domain, especially for inverting satellite gravimetry data. One important issue is that the satellite data being inverted should have larger coverage, even if the data are dense and our system has high redundancy, than the recovery area; therefore, the size of inversion and recovery area is of vital importance for integral inversion. Here, we consider two cases, in which the data area is larger than or equal to the recovery area. Schematically, these two situations are presented in Fig. 4.10.

Fig. 4.10A shows the case in which the data area or the coverage of the satellite data is larger than the recovery area. In contrast to the common belief about having more redundancy to reduce the error, this area is not proper for integral inversion. For each data point in the upper data area, a linear equation is written for the desired data in the lower grid, and since this area is smaller than the data area, the truncation error occurs more in the central part. Fig. 4.10B is the case in which the two areas are equal; in

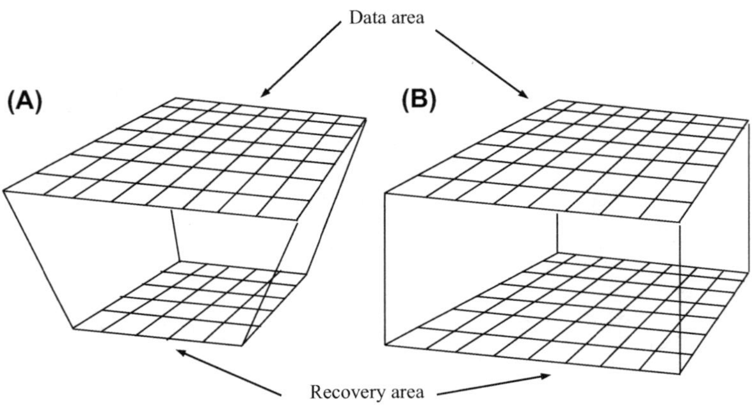

Figure 4.10 (A) data area is larger than recovery area, (B) data and recovery area are equal.

this case the truncation error occurs closer to the margins and corners than in the case seen in Fig. 4.10A. Note that the data do not need to be in grid form and over a constant elevation in practice. This is not an important issue when the satellite data have the same coverage as the recovery area.

4.2.4 Example: spatial truncation error of satellite gradiometry data

Here, the goal is to present a part of the Eshagh (2011b) study for determining STE. Gravity anomalies with a resolution of $1^\circ \times 1^\circ$ were generated in Fennoscandia and a grid of second-order radial gradients was generated from them at a constant elevation of 250 km and with a resolution of $0.5^\circ \times 0.5^\circ$. To estimate the STE of the integral formula, the principle presented in Section 4.2.1 was used. The difference between the gradients generated by Earth Gravitational Model (EGM08) and those with the anomalies through the internal formula is nothing other than the STE. Fig. 4.11A shows this truncation error. Fig. 4.11B is the case in which the recovery area is smaller than the data coverage and, as we can observe, the STEs are more inside the central part of the area compared with Fig. 4.11A, where the error is closer to the border of the area.

Fig. 4.12A and B show the STE of all the gravity gradients over Fennoscandia. We can observe that the errors of T_{xx} and T_{yy} are on the same order and the largest errors are located in the northern and eastern parts of the region. Fig. 4.12C shows that many errors are related to T_{zz} but the errors are in the edges and boundaries of the region and the majority of them are small. T_{xy} and T_{xz} contain large truncation errors, presented in Fig. 4.12D and E, respectively. The STE of T_{yz} is seen in Fig. 4.12F.

4.2.5 Example: spatial truncation error in the results of inversion of satellite inter-satellite tracking data

The result of the integral inversion of the other satellite gravimetry data suffers from STE as well. Similar to the direct integration, the effect of STE will show up in the margins and corners of the area. This effect can be estimated by comparing the result of inversion with the corresponding ones generated by a spherical harmonic series. Fig. 4.13 is an example of a sub-lithospheric stress-generating function recovered from the GRACE (Gravity Recovery and Climate Experiment) range-rate data over Himalaya. It shows very well that this effect is considerably large in the marginal

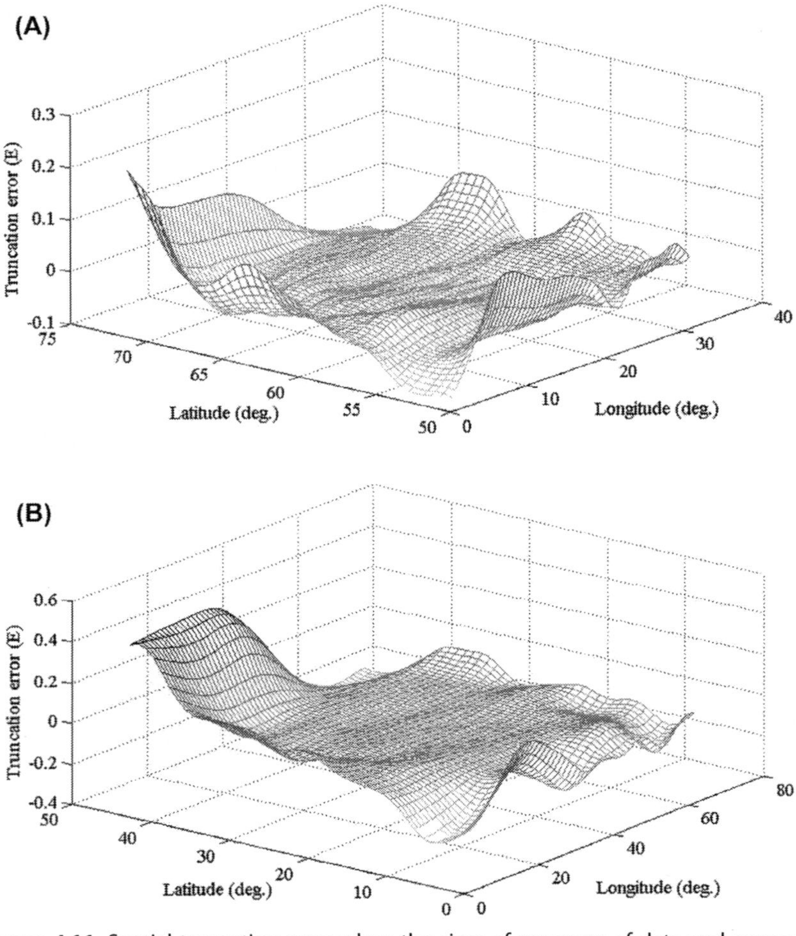

Figure 4.11 Spatial truncation error when the sizes of coverage of data and recovery area are equal and (B) when recovery area is smaller than data coverage (Eshagh, 2011b).

areas and corners. Note that when considering a smaller recovery area the effect of this STE will be more towards the centre of the area. In this example, the sizes of the data coverage and recovery areas are the same, and this truncation error can be removed or at least reduced by performing the inversion all over the area and selecting the central part as the result of the inversion, which is not affected by the STE. The size of the central area depends on the behaviour of the kernel function, which is used for data inversion.

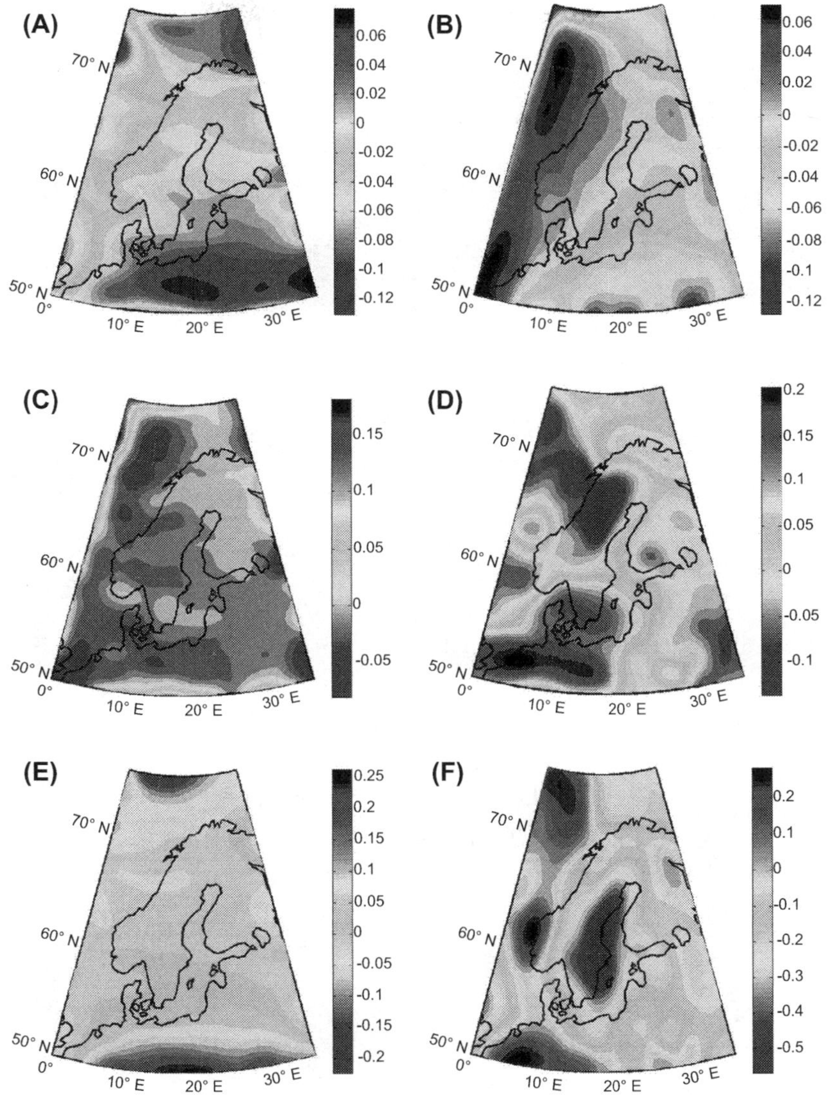

Figure 4.12 STE of (A) T_{xx}, (B) T_{yy}, (C) T_{zz}, (D) T_{xy}, (E) T_{xz} and (F) T_{yz}. Unit: 1 E (from Eshagh 2011a).

4.2.6 Example: inversion of T_{rr} to Δg

Here, Δg is recovered over Fennoscandia from the second-order radial derivative of disturbing potential. The Tikhonov regularisation with L-curve (which will be discussed in Section 4.3) method is used for solving

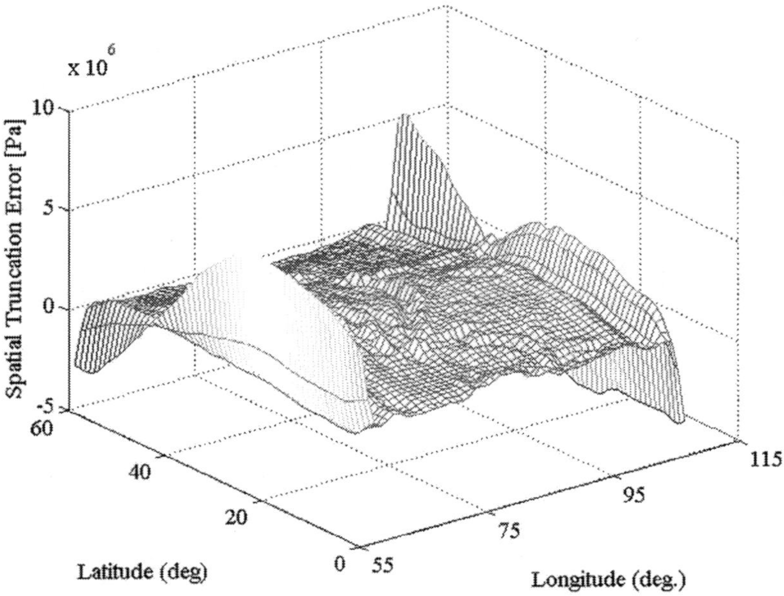

Figure 4.13 The effect of spatial truncation error on the stress-generating function recovered from the GRACE range-rate data (Sprlak and Eshagh, 2016).

the integral equations. Two sets of Δg, one from EGM96 to degree and order 360 and one from inversion, assess the quality of the inversion process. Such an integral is spherical, and the result of its inversions is always contaminated with STE, which is reduced by those recovered Δg located in the central part of the recovery area. Therefore, the size of this central area and the resolution of recovery should be tested for a successful inversion. Fig. 4.14A illustrates the STE of the satellite gradients, estimated by the method presented in Section 4.2.1, and by cutting the marginal areas 3° inwards from each side Fig. 4.14B is obtained, which shows a significant reduction in the estimated error.

4.2.7 Example: inversion of orbital data to gravity anomaly

Here, the study done by Eshagh and Ghorbannia (2014) is presented. The region of interest is Fennoscandia and the goal is to test different resolutions and sizes of the STE for inversion of orbital data using Eq. (4.4). Fig. 4.15A shows the truncation error computed for the orbital data, the right-hand side (rhs) of Eq. (4.4), and this error is large in the marginal parts of the area. Fig. 4.15B shows the case in which the central area, which is smaller

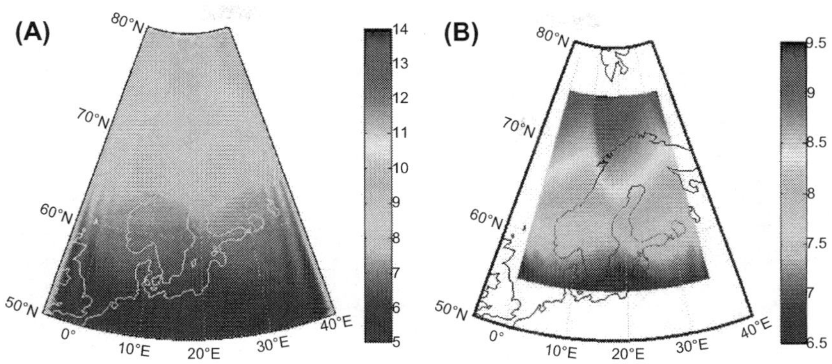

Figure 4.14 Spatial truncation error of second-order radial derivative of disturbing potential (A) before windowing and (B) after windowing (mE) by cutting 5 degrees from each side (Eshagh and Ghorbannia, 2014).

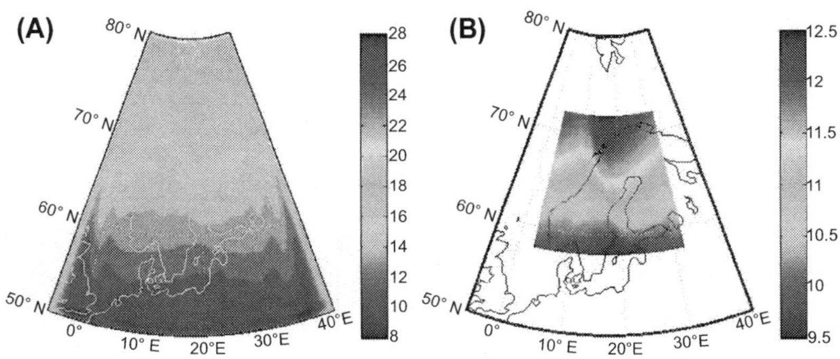

Figure 4.15 Spatial truncation error of orbital data (A) over the whole area and (B) over the central area (mE) (Eshagh and Ghorbannia, 2014).

by 5° from each side, is selected as the truncation error-free data, and we can observe a significant reduction by removing those parts around the area.

Fig. 4.16A shows the map of the orbital data, which should be inverted, and Fig. 4.16B the map of recovered gravity anomalies after removing the effect of the STE by selecting those anomalies in a central area smaller by 8°. Table 4.2 presents the statistics of the differences between the recovered anomalies from orbital data and the anomalies derived from the EGM96 gravity model. Column A presents the size of truncation; for example, when it is 8° it means that the results are from a central area which is 8° smaller than the recovery area. The table suggests that the size of the central area should be smaller than the recovery area by more than 7° and the resolution should be higher than $1° \times 1°$, which is reasonable for the orbital data inversion.

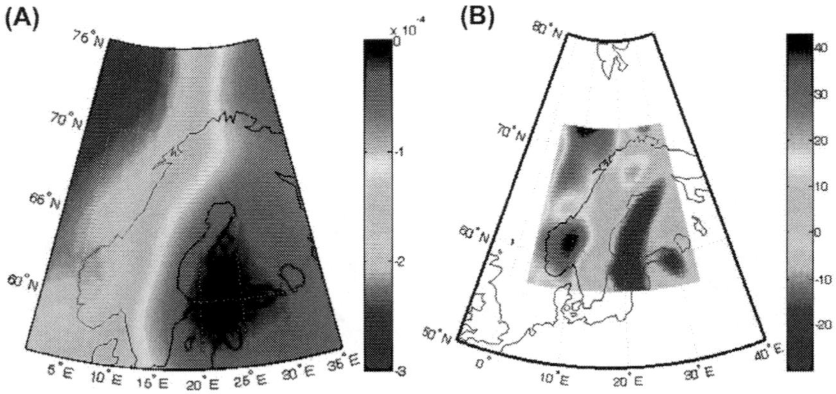

Figure 4.16 Maps of (A) orbital data (m/s^2) and (B) gravity anomalies derived from orbital data inversion in a central area smaller by 8° (mGal) (Eshagh and Ghorbannia 2014).

Table 4.1 Statistics of differences between Δg generated by EGM96 and recovered from second-order radial derivatives of disturbing potential, based on different resolutions and size of cutting the results contaminated with the spatial truncation error.

A	Resolution	Max.	Mean	Min.	Std	RMSE
5°	1.5° × 1.5°	13.8	−6.2	−27.0	7.1	9.2
	1.0° × 1.0°	16.7	−6.0	−26.9	7.6	9.7
	0.5° × 0.5°	33.0	−6.2	−42.1	7.9	10.1
4°	1.5° × 1.5°	23.6	−6.2	−41.2	7.5	9.7
	1.0° × 1.0°	24.6	−5.8	−46.8	7.9	9.8
	0.5° × 0.5°	33.0	−6.3	−50.2	8.1	10.3
3°	1.5° × 1.5°	31.7	−6.0	−45.0	8.7	10.6
	1.0° × 1.0°	24.6	−6.2	−63.5	8.8	10.7
	0.5° × 0.5°	42.0	−6.4	−55.8	8.9	11.0
2°	1.5° × 1.5°	24.5	−6.6	−45.7	9.3	11.4
	1.0° × 1.0°	30.4	−6.3	−64.9	10.3	12.0
	0.5° × 0.5°	42.1	−5.7	−59.3	11.1	12.5

Unit: 1 mGal. *RMSE*, root mean squared error; *Std*, standard deviation (Eshagh and Ghorbannia, 2014). Column A in Table 4.1 shows the size of the inversion area; for example, 5° means that the inversion area is larger than the target area by 5° but the results in the central area are selected. The table shows that the gravity anomaly is recoverable with 0.5° × 0.5° resolution but the recovery area should be considered at least 3° larger than the desired area for controlling the STE.

Table 4.2 Statistics of differences between gravity anomalies generated by EGM96 and those recovered from orbital data (Eshagh and Ghorbannia 2014).

A	Resolution	Max.	Mean	Min.	Std	RMSE
8°	1.5° × 1.5°	38.9	5.1	−30.0	9.9	11.1
	1.0° × 1.0°	52.7	4.1	−32.0	10.3	11.1
	0.5° × 0.5°	46.6	5.2	−31.4	10.9	12.1
7°	1.5° × 1.5°	38.9	5.3	−30.0	10.4	11.7
	1.0° × 1.0°	59.0	4.6	−32.0	10.9	11.8
	0.5° × 0.5°	46.6	5.5	−40.4	11.0	12.3
6°	1.5° × 1.5°	43.1	5.3	−50.2	11.6	12.8
	1.0° × 1.0°	65.6	5.1	−33.7	11.8	12.9
	0.5° × 0.5°	47.3	5.6	−58.2	12.0	13.2
5°	1.5° × 1.5°	43.3	5.2	−50.2	11.5	12.6
	1.0° × 1.0°	64.4	5.8	−51.2	12.8	14.1
	0.5° × 0.5°	47.3	4.9	−66.9	13.4	14.3

RMSE, root mean squared error; *Std*, standard deviation.

4.3 Regularisation methods

The mathematical relations between the satellite gravimetry observations and the SHCs of the Earth's gravitational field showed that a system of observation equations needs to be organised for determining these coefficients. The least-squares method is used for such over-determined systems. However, the problem of such systems is that they are not well-conditioned in practice and their solutions are very sensitive to the noise of observations. Therefore, another way should be considered so that stable solutions are achieved. Such a method is called regularisation. Here some of the well-known regularisation techniques are briefly presented and discussed.

4.3.1 Gauss–Markov model

All mathematical formulae relating the satellite gravimetry observables are observation equations with gravitational potential or its SHCs (v_{nm}) as their unknowns. For solving such over-determined systems, having millions of measurements, the well-known least-squares method should be applied. Fortunately such systems are linear with respect to the unknowns, and they can be presented in a general matrix form known as the Gauss–Markov model (see e.g. Koch, 1999):

$$\mathbf{Ax} = \mathbf{L} - \boldsymbol{\varepsilon}, \mathrm{E}\{\boldsymbol{\varepsilon}\boldsymbol{\varepsilon}^{\mathrm{T}}\} = \sigma_0^2\mathbf{Q}, \text{and } \mathrm{E}\{\boldsymbol{\varepsilon}\} = 0, \tag{4.11}$$

where $\mathbf{A}$ is the $n \times m$ coefficient matrix, where n and m stand for the numbers of measurements and unknowns, respectively. $\mathbf{L}$ is the $n \times 1$ vector of observations, $\boldsymbol{\varepsilon}$ the $n \times 1$ vector of the random error of $\mathbf{L}$ and $\mathbf{x}$ the $m \times 1$ vector of unknown parameters. In Eq. (4.11) E stands for the statistical expectation operator; σ_0^2 is the a priori variance factor, which is often equal to 1, and $\mathbf{Q}$ is the variance—covariance matrix of observations.

Let us solve Eq. (4.11) for $\boldsymbol{\varepsilon}$:

$$\boldsymbol{\varepsilon} = \mathbf{L} - \mathbf{A}\mathbf{x}, \tag{4.12}$$

which is the error vector. The aim of the least-squares method is to minimise the norm of this vector; i.e.

$$\|\boldsymbol{\varepsilon}\|^2_{\mathbf{Q}^{-1}} = \sigma_0^{-2}\boldsymbol{\varepsilon}^{\mathrm{T}}\mathbf{Q}^{-1}\boldsymbol{\varepsilon}. \tag{4.13}$$

Eq. (4.13) is an objective function being minimised. Inserting Eq. (4.12) into Eq. (4.13) and taking the derivative of the result with respect to $\mathbf{x}$ and solving it for $\mathbf{x}$ yields:

$$\widehat{\mathbf{x}} = \mathbf{A}^{-}\mathbf{L} \quad \text{where} \quad \mathbf{A}^{-} = \left(\mathbf{A}^{\mathrm{T}}\mathbf{Q}^{-1}\mathbf{A}\right)^{-1}\mathbf{A}^{\mathrm{T}}\mathbf{Q}^{-1}\mathbf{L}, \tag{4.14}$$

which is an unbiased and best estimation of $\mathbf{x}$. $\mathbf{A}^{-}$ is the generalised matrix of A (Bjerhammar, 1973), or the left inverse.

Being unbiased means that by increasing the number of measurements the probability of reaching the true value of $\mathbf{x}$ increases. Having the minimum variance for the solution means the estimator is best compared with others. Based on the error propagation law of random errors, the variance—covariance matrix of the estimated parameters is:

$$\begin{aligned}\mathbf{C}_{\widehat{\mathbf{x}}} &= \mathrm{E}\left\{(\mathbf{x}-\widehat{\mathbf{x}})(\mathbf{x}-\widehat{\mathbf{x}})^{\mathrm{T}}\right\} = \mathrm{E}\left\{(\mathbf{x}-\mathbf{A}^{-}(\mathbf{A}\mathbf{x}+\boldsymbol{\varepsilon}))(\mathbf{x}-\mathbf{A}^{-}(\mathbf{A}\mathbf{x}+\boldsymbol{\varepsilon}))^{\mathrm{T}}\right\} = \\ &= \mathrm{E}\left\{\mathbf{A}^{-}\boldsymbol{\varepsilon}\boldsymbol{\varepsilon}^{\mathrm{T}}(\mathbf{A}^{-})^{\mathrm{T}}\right\} = \mathbf{A}^{-}\mathrm{E}\{\boldsymbol{\varepsilon}\boldsymbol{\varepsilon}^{\mathrm{T}}\}(\mathbf{A}^{-})^{\mathrm{T}} = \sigma_0^2\mathbf{A}^{-}\mathbf{Q}(\mathbf{A}^{-})^{\mathrm{T}} = \\ &= \sigma_0^2\left(\mathbf{A}^{\mathrm{T}}\mathbf{Q}^{-1}\mathbf{A}\right)^{-1},\end{aligned} \tag{4.15}$$

where σ_0^2 is called the a priori variance factor or variance of unit weight. Normally, it is assumed equal to 1. This means that we assume that errors of measurements are in balance with the estimated residual:

$$\widehat{\boldsymbol{\varepsilon}} = \mathbf{L} - \mathbf{A}\widehat{\mathbf{x}} = (\mathbf{I}_n - \mathbf{A}\mathbf{A}^{-})\mathbf{L}, \tag{4.16}$$

In is an $n \times n$ identity matrix.

Solving Eq. (4.11) for $\mathbf{L}$, inserting the result into Eq. (4.16) and performing further simplifications yields a mathematical relation between $\widehat{\boldsymbol{\varepsilon}}$, and the true errors of the measurements $\boldsymbol{\varepsilon}$ are derived:

$$\widehat{\boldsymbol{\varepsilon}} = (\mathbf{I}_n - \mathbf{A}\mathbf{A}^-)\boldsymbol{\varepsilon}. \tag{4.17}$$

The a priori variance factor σ_0^2 should be estimated and tested statistically after solving $\mathbf{x}$. Such an estimate is called the a posteriori variance. To estimate it we write:

$$\begin{aligned} \mathrm{E}\{\widehat{\boldsymbol{\varepsilon}}^{\mathrm{T}}\mathbf{Q}^{-1}\widehat{\boldsymbol{\varepsilon}}\} &= \mathrm{E}\{(\mathbf{I}_n - \mathbf{A}\mathbf{A}^-)^{\mathrm{T}}\boldsymbol{\varepsilon}^{\mathrm{T}}\mathbf{Q}^{-1}\boldsymbol{\varepsilon}(\mathbf{I}_n - (\mathbf{A}^-)^{\mathrm{T}}\mathbf{A}^{\mathrm{T}})\} = \\ &= \mathrm{trace}\left[\mathrm{E}(\mathbf{Q}^{-1}\boldsymbol{\varepsilon}^{\mathrm{T}}\boldsymbol{\varepsilon})(\mathbf{I}_n - (\mathbf{A}^-)^{\mathrm{T}}\mathbf{A}^{\mathrm{T}})(\mathbf{I}_n - \mathbf{A}\mathbf{A}^-)\right] \\ &= \sigma_0^2\{\mathrm{trace}(\mathbf{I}_n - \mathbf{I}_m)\} \\ &= \sigma_0^2(n-m) \end{aligned} \tag{4.18}$$

where $\mathbf{I}_m$ is an identity matrix with size m.

The solution of Eq. (4.18) for σ_0^2 gives the a posteriori variance factor estimator:

$$\widehat{\sigma}_0^2 = \frac{\widehat{\boldsymbol{\varepsilon}}^{\mathrm{T}}\mathbf{Q}^{-1}\widehat{\boldsymbol{\varepsilon}}}{n-m} \tag{4.19}$$

Theoretically the a priori and a posteriori variance factors should be equal, $\sigma_0^2 \approx \widehat{\sigma}_0^2$, if the errors of measurements are realistic and the mathematical models are precise and no gross error exists in the measurements. In the case in which the mathematical models are rigorous, with the absence of gross errors, the deviation of $\widehat{\sigma}_0^2$ from σ_0^2 is due to the improper choice of a priori errors of the data. In this case, the variance—covariance matrix of the estimated parameters (see Eq. 4.15) should be scaled by replacing σ_0^2 with $\widehat{\sigma}_0^2$:

$$\widehat{\mathbf{C}}_{\hat{x}} = \widehat{\sigma}_0^2(\mathbf{A}^{\mathrm{T}}\mathbf{Q}^{-1}\mathbf{A})^{-1}. \tag{4.20}$$

The Gauss—Markov model (Eq. 4.11) is solvable when matrix $\mathbf{A}$ is well-conditioned. In this case, the normal equation:

$$(\mathbf{A}^{\mathrm{T}}\mathbf{Q}^{-1}\mathbf{A})\mathbf{x} = \mathbf{A}^{\mathrm{T}}\mathbf{Q}^{-1}\mathbf{L}, \tag{4.21}$$

is invertible and the least-squares solution is obtained. But when Eq. (4.21) is ill-conditioned, it means that the determinant of $(\mathbf{A}^{\mathrm{T}}\mathbf{Q}^{-1}\mathbf{A})$ is very close to zero (but not equal) and the system is unstable. It is solvable, but any small errors in the data lead to large variations in the solution. Regularisation is a method for controlling the effect of the noise of the data on the solution.

4.3.2 A conceptual overview of regularisation methods

Generally, there are two types of regularisation, direct and iterative. In the iterative ones the direct inversion of the system of equations is avoided, which is an advantage for solving large systems. However, in the direct methods the system is solved by inverting its coefficients matrix, which is rather complicated, especially when the system is ill-conditioned. In this section, some of the well-known direct and iterative regularisation methods are presented.

4.3.2.1 Iterative methods

The iterative methods are of two types: (1) the classical iterative methods and (2) the methods based on Krylov subspaces. Some examples of the former are the ν method (Brakhage, 1987) and the algebraic reconstruction technique (ART) (Kaczmarz, 1937), which are presented in this section, and some of the latter are the range-restricted generalised minimum residual (RRGMRES) and conjugate gradient (CG).

One of the important issues in the iterative methods is related to the estimation of the errors of the estimated parameters, as they have very complicated spectral properties. Therefore, if the goal is only to solve a large system of equations, the use of iterative methods is recommended, otherwise, the direct methods should be applied.

4.3.2.1.1 Classical iterative methods

The ν method and ART are two examples of the classical iterative regularisation methods. Here, only a brief review of them will be presented without details.

4.3.2.1.1.1 The ν method

This method is known as an extension of the Landweber method (Brakhage, 1987), the solution of which has the following structure:

$$\mathbf{x}^{(k)} = \mathbf{x}^{(k-1)} + \omega^{*} \mathbf{A}^{\mathrm{T}} \left(\mathbf{A}\mathbf{x}^{(k-1)} - \mathbf{L} \right), \tag{4.22}$$

where k is the iteration number.

The key issue is to select ω^*, which in the classical Landweber method is $\omega^* = F(\mathbf{A}^T\mathbf{Q}^{-1}\mathbf{A})$, where F is a rational function of $\mathbf{A}^T\mathbf{Q}^{-1}\mathbf{A}$; for example, if $\omega^* = (\mathbf{A}^T\mathbf{Q}^{-1}\mathbf{A} + \alpha^2\mathbf{I})^{-1}$ the iterative Tikhonov regularisation is derived. In the ν method, a weighted average of some of the last iterations is considered as $\mathbf{x}^{(k-1)}$, and ω^* is allowed to depend on k. This method has two steps (Hansen, 1998):

$$\mathbf{x}^{(k)} = \mu_k\mathbf{x}^{(k-1)} + (1-\mu_k)\mathbf{x}^{(k-2)} + \omega_k^*\mathbf{A}^T\left(\mathbf{A}\mathbf{x}^{(k-1)} - \mathbf{L}\right), \tag{4.23}$$

where

$$\mu_k = 1 + \frac{(k-1)(2k-3)(2k+2\nu-1)}{(k+2\nu-1)(2k+4\nu-1)(2k+2\nu-3)}, \tag{4.24}$$

$$\omega_k^* = \frac{4(2k+2\nu-1)(k+\nu-1)}{(k+2\nu-1)(2k+4\nu-1)} \tag{4.25}$$

and $0 < \nu < 1$.

The best value of ν can be chosen empirically.

4.3.2.1.1.2 Algebraic reconstruction technique In the ART each row of matrix $\mathbf{A}$ is used, one at a time, for solving the system, which is very suitable when $\mathbf{A}$ is large. Each iteration involves a step through the rows of $\mathbf{A}$ (Hansen, 1998):

$$\mathbf{x}^{(k)} = \mathbf{x}^{(k-1)} + \frac{\mathbf{L}_i - \mathbf{a}_i^T\mathbf{x}^{(k-1)}}{\|\mathbf{a}_i\|_2^2}\mathbf{a}_i, \tag{4.26}$$

where $\mathbf{L}_i$ is the ith component of $\mathbf{L}$ and $\mathbf{a}_i$ is the ith row of $\mathbf{A}$.

4.3.2.1.2 Krylov subspaces-based methods

In the Krylov subspaces-based methods (Hansen, 1998), the goal is to minimise the following objective function:

$$\min_{\mathbf{x}\in K_k}\|\mathbf{L} - \mathbf{A}\mathbf{x}\|_2. \tag{4.27}$$

Eq. (4.27) means that the residuals of the system are minimised so that the solution remains in the k-dimensional Krylov subspaces K_k. A Krylov space of order k, generated by an $n \times n$ matrix $\mathbf{A}$ and a vector $\mathbf{L}$ of dimension n, is the linear subspace spanned by the images of $\mathbf{L}$ under the first k powers of $\mathbf{A}$. The general iterative solution of the Gauss–Markov model (Eq. 4.11) is based on this method (Jensen and Hansen, 2007):

$$\mathbf{x}^{(k)} = \Phi_k(\mathbf{B})\mathbf{B}'\mathbf{L}, \tag{4.28}$$

where $\Phi_k(\mathbf{B})$ is a polynomial of degree $k-1$ with matrix $\mathbf{B}$. $\mathbf{B}$ and $\mathbf{B}'$ are defined based on each method and its corresponding Krylov subspaces. The term $\mathbf{x}^{(k)}$ stands for the kth iterated solution. Eq. (4.28) illustrates that the solution is a combination of $\mathbf{B}$ and $\mathbf{B}'\mathbf{L}$, or in Krylov subspaces:

$$\mathbf{x} \in K_k(\mathbf{B}, \mathbf{B}'\mathbf{L}) = \mathrm{span}\{\mathbf{B}'\mathbf{L}, \mathbf{B}\mathbf{B}'\mathbf{L}, \ldots, \mathbf{B}^{k-1}\mathbf{B}'\mathbf{L}\}. \tag{4.29}$$

4.3.2.1.2.1 Range-restricted generalised minimum residual method

The RRGMRES (Calvetti et al., 2000) is designed for solving systems having a square coefficient matrix. It is a modified form of the generalised minimum residual (GMRES) method (Saad and Schultz, 1986) that restricts the solution to the range of the coefficient matrix of the system. Calvetti et al. (2000) proved that RRGMRES has an intrinsic regularisation property and the vector $\mathbf{L}$ is damped by the ill-conditioned matrix $\mathbf{A}$. If these methods are applied for solving ill-conditioned systems the iterative solutions will converge to the minimum-norm least-squares solution. However, the first few iterated solutions can be considered as a regularised solution. In other words, the long wavelength structure of the solution is constructed by the first iterations and iteration adds higher frequencies to it. Jensen and Hansen (2007) studied this method in the spectral domain and mentioned that the first large singular value components contribute more strongly to the solution than small singular values. Another study of Hansen and Jensen (2006) showed that such methods are superior to minimum residual and GMRES owing to their noise suppression.

To develop the RRGMRES method the Arnoldi decomposition (Saad, 1996) was applied by Calvetti et al. (2000). This decomposition is:

$$\mathbf{A}\mathbf{V}_k = \mathbf{V}_{k+1}\overline{\mathbf{H}}_k, \tag{4.30}$$

where $\mathbf{V}_{k+1} = [\mathbf{v}_1\mathbf{v}_2\ldots\mathbf{v}_{k+1}]_{n\times(k+1)}$ has orthogonal columns, i.e. $\mathbf{V}_{k+1}^{\mathrm{T}}\mathbf{V}_{k+1} = \mathbf{I}_{k+1}$, which span the Krylov subspaces $K_k(\mathbf{A}, \mathbf{L})$, and $(\overline{\mathbf{H}}_k)_{(k+1)\times k}$ is of the Hessenberg type, which is an almost triangular matrix. The upper and lower triangular Hessenberg matrices have zero elements below the first sub-diagonal and above the first super-diagonal elements, respectively (see Eshagh, 2011c). Calvetti et al. (2000) selected $\mathbf{x} = \mathbf{V}_k\mathbf{y}$ and inserted it into Eq. (4.27), then:

$$\min_{\mathbf{x}\in K_k(\mathbf{A},\mathbf{L})} \|\mathbf{L} - \mathbf{A}\mathbf{x}\|_2 = \min_{\mathbf{y}\in R^k} \left\|\mathbf{L} - \mathbf{V}_{k+1}\overline{\mathbf{H}}_k\mathbf{y}\right\|_2. \tag{4.31}$$

In Eq. (4.31) the Arnoldi decomposition transfers the solution space from $K_k(\mathbf{A}, \mathbf{AL})$ to $\mathbf{y} \in R^k$, which is simpler to solve, as the problem is converted to an ordinary minimisation. Once $\mathbf{y}$ is estimated $\mathbf{x}$ is obtained from it by $\mathbf{x} = \mathbf{V}_k \mathbf{y}$. As observed, this method has a regularisation property due to multiplication of $\mathbf{A}$ to $\mathbf{L}$ in the subspaces. According to Calvetti et al. (2000) the regularisation philosophy behind this method is that $\overline{\mathbf{H}}_k \in \mathrm{R}^{(k+1)\times k}$ is not ill-conditioned if the Arnoldi decomposition starts with $\mathbf{AL}$. Jensen and Hansen (2007) stated that the role of polynomial $\Phi_k(\mathbf{B})$ is to reduce the large components of $\mathbf{U}^{\mathrm{T}}\mathbf{L}$, where $\mathbf{U}$ is the eigenvector of $\mathbf{A}$, by being small for these large components.

4.3.2.1.2.2 Conjugate gradient

The CG method was designed for solving systems with sparse coefficient matrices. It is applicable to any type of system, but as a least-squares solution is sought, it should be applied to the normal equations of the Gauss–Markov model (Eq. 4.11), Eq. (4.21). Now, consider $\mathbf{B} = \mathbf{A}^{\mathrm{T}}\mathbf{Q}^{-1}\mathbf{A}$ and $\mathbf{B}' = \mathbf{A}^{\mathrm{T}}\mathbf{Q}^{-1}$ in Eq. (4.28). In CG the components related to the large singular values converge faster than those involved with small ones, which is its intrinsic regularisation property (Hansen, 2007). From a geometrical perspective, the minimisation problem creates concentric ellipses around the solution and in each iteration a gradient to that ellipse passing from the iterative solution is generated towards the final solution (Bouman, 1998) so that the gradient is orthogonal to all previous ones.

Being conjugate means that two vectors are orthogonal with respect to the inner product. If $\mathbf{p}_k$ is the sequence of orthogonal directions, then they play the role of base functions and the solution can be presented as a series of these bases:

$$\mathbf{x} = \sum_{i=1}^{n'} \beta_i \mathbf{p}_i, \tag{4.32}$$

where n' stands for iteration and β_i is the coefficient of the expansions. Substituting Eq. (4.32) into Eq. (4.21) yields:

$$\mathbf{A}^{\mathrm{T}}\mathbf{Q}^{-1}\mathbf{A}\mathbf{x} = \sum_{i=1}^{n'} \beta_i \mathbf{A}^{\mathrm{T}}\mathbf{Q}^{-1}\mathbf{A}\mathbf{p}_i = \mathbf{A}^{\mathrm{T}}\mathbf{Q}^{-1}\mathbf{L}. \tag{4.33}$$

Pre-multiplying Eq. (4.33) by $\mathbf{p}_k^{\mathrm{T}}$ and further simplifications read:

$$\beta_k = \frac{\mathbf{p}_k^{\mathrm{T}}\mathbf{A}^{\mathrm{T}}\mathbf{Q}^{-1}\mathbf{L}}{\mathbf{p}_k^{\mathrm{T}}\mathbf{A}^{\mathrm{T}}\mathbf{Q}^{-1}\mathbf{A}\mathbf{p}_k}. \tag{4.34}$$

CG needs an initial value $\mathbf{x}_0$, which can be considered as zero without loss of generality. However, the solution is sought in such a way that the following function is minimised:

$$F(\mathbf{x}) = \frac{1}{2}\mathbf{x}^{\mathrm{T}}\mathbf{A}^{\mathrm{T}}\mathbf{Q}^{-1}\mathbf{A}\mathbf{x} - \mathbf{x}^{\mathrm{T}}\mathbf{A}^{\mathrm{T}}\mathbf{Q}^{-1}\mathbf{L}. \tag{4.35}$$

Since $F(\mathbf{x})$ ought to be a minimiser, its gradient should be $\mathbf{p}_0 = \mathbf{A}\mathbf{x}_0 - \mathbf{L}$ and the other base vectors should be conjugate to it. By having this gradient one can obtain β_1 using Eq. (4.34) and

$$\mathbf{p}_{k+1} = \mathbf{r}_k - \sum_{i \le k} \frac{\mathbf{p}_i^{\mathrm{T}}\mathbf{A}^{\mathrm{T}}\mathbf{Q}^{-1}\mathbf{A}\mathbf{r}_i}{\mathbf{p}_i^{\mathrm{T}}\mathbf{A}^{\mathrm{T}}\mathbf{Q}^{-1}\mathbf{A}\mathbf{p}_i}\mathbf{p}_i \text{ where } \mathbf{r}_j = \mathbf{L} - \mathbf{A}\mathbf{x}^{(j)}, \text{ j} = \text{i or k} \tag{4.36}$$

and

$$\mathbf{x}^{(k+1)} = \mathbf{x}^{(k)} + \beta_k \mathbf{p}_k. \tag{4.37}$$

Eq. (4.32) is a series in which the higher frequencies of the solution are added step by step during iteration. The first iterations create its low frequencies, and the high frequencies, infected by the observation noise, are added by further iterations. Truncating this series of Eq. (3.32) to k means to stop iteration before convergence of the solution and remove the infected high frequencies.

4.3.2.2 Estimation of the optimal iteration number by L-curve

As mentioned, in the iterative methods the solution is constructed by a series to which each iteration adds more frequencies of the solution. If the iteration number goes to infinity the solution will approach the least-squares solution, which is extremely sensitive to the errors of data. This causes a significant amplification of the values of the estimated parameters. On the other hand, by increasing the number of iterations, the residuals $\mathbf{A}\mathbf{x} - \mathbf{L}$ will reduce. Therefore, on one hand, we have an increasing parameter, which is the norm of the estimated parameters $\|\mathbf{x}\|$, and on the other hand, we have a decreasing one, which is the norm of the residuals $\|\mathbf{A}\mathbf{x} - \mathbf{L}\|$. Hansen (1998) suggested a method for finding a balance between this increase and decrease and he named it the L-curve. According to his idea, the logarithm of the norm of residuals is considered as the horizontal axis of a 2D coordinate system and the vertical axis represents the logarithm of $\|\mathbf{x}\|$. When these two parameters are plotted based on different values of the iterations, an L-shaped curve is obtained, which was

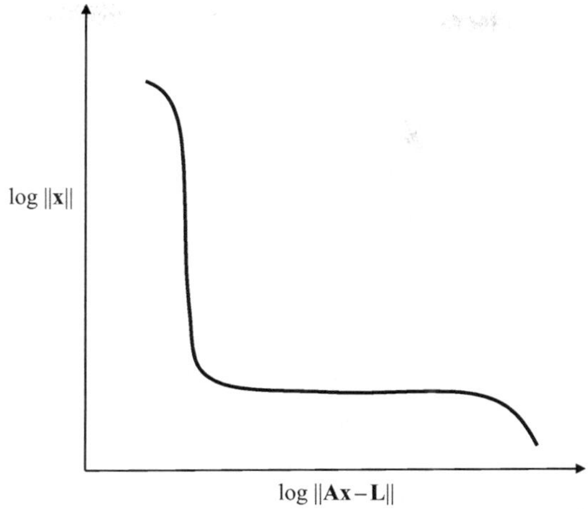

Figure 4.17 An L-curve.

named the L-curve; see Fig. 4.17. The corner of this curve shows the balancing iteration number between $\log\|\mathbf{Ax} - \mathbf{L}\|$ and $\log\|\mathbf{x}\|$. This means that the optimal iteration number does not allow $\|\mathbf{Ax} - \mathbf{L}\|$ to be very small or $\|\mathbf{x}\|$ to be very large. To find this iteration number, the solutions for a large number of iterations are derived and after that the L-curve is plotted for them. After finding the corner of the L-curve, the corresponding iteration number is selected as the optimal iteration.

4.3.2.3 Example: application of the iterative methods for determining equivalent water height from the GRACE mission

Here, we present one part of the Eshagh et al. (2013) study for regularisation of the GRACE range-rate data and estimating equivalent water height (EWH), of which its principle and solution are described. Two periods, 31 December 2009 to 29 February 2010 and 1 March 2010 to 30 April 2010, corresponding to January—February 2010 and March—April 2010, are selected for calculating the temporal changes in the gravity field. These two 60-day periods correspond to six 10-day GRACE normal equations. Two sets of v_{nm} are derived for these periods and they are converted into geoid height differences; the EWH is computed by (cf. Lemoine et al., 2007):

$$\Delta h(R, \theta, \lambda) = \frac{g}{4\pi G R \rho} \sum_{n=2}^{\infty} \frac{2n+1}{1+k_n} \sum_{m=-n}^{n} \Delta N_{nm} Y_{nm}(\theta, \lambda),$$

where $\Delta N_{nm} = \frac{v_{nm}^{(2)} - v_{nm}^{(1)}}{\gamma}$ and $g = 9.8\ \text{m/s}^2$ is the mean gravity attraction, $G = 6.67 \times 10^{-11}\ \text{m}^3/\text{kgs}^2$ is the general gravitational constant, $R = 6{,}378{,}137$ m is the radius of the spherical Earth and $\rho = 1000\ \text{kg/m}^3$ is the density of water. The term ΔN_{nm} stands for the harmonics of geoid change, $v_{nm}^{(1)}$ and $v_{nm}^{(2)}$ are the SHCs derived from each period and γ is the normal gravity. Finally k_n is the load Love number of degree n. A mean gravity field of a longer period, 4 months, was considered as a reference field, and the goal was to compare EWHs derived from the regularised solutions of v_{nm} degree and order 50.

Fig. 4.18 illustrates their computed EWH without applying any regularisation for the period January–February 2010. The numerical errors due to the instability of the system are visible in the inter-tropical zone, where the GRACE tracks are sparse. A regularisation should be used to smooth the gravity field and reduce these errors. Statistics of the generated EWHs over oceans are regarded as a criterion for the smoothness of the solutions, due to the presence of small dynamics of the gravitational signal there (together with the deserts).

Figure 4.18 Equivalent water height estimated without regularisation. Comparison of January–February 2010 with January–April 2010. Unit: m. *(From Eshagh, M., Lemoine, J.M., Gegout P., Biancale, R., 2013. On regularized time varying gravity field models based on GRACE data and their comparisons with hydrological models. Acta Geophys. 61 (1), 1–17, with permission from Springer.)*

After the normal equations, obtained from the GRACE data, were solved by the iterative methods of ν, ART, CG and RRGMRES, smooth solutions for the EWH were derived. As mentioned before, such methods have a regularisation property when their iteration is stopped before the convergence of the solution. In this case, finding the optimum number of iterations (Kilmer and O'Leary, 2001) is very important. The discrepancy principle (cf. Hansen, 2007) and L-curve are two methods for this purpose, but here Eshagh et al. (2013) applied the L-curve, presented in Section 4.3.2.2, to find the optimal iteration number amongst 100 iterated solutions.

The statistics of the smoothed EWHs in oceans are presented in Table 4.3 In the table 'I' stands for the number of iterations; e.g. the ν method should be iterated 10 times to obtain a regularised solution. Logically, the first iteration yields the smoothest solution. The statistics of the EWHs after the first iteration are also presented in the table. It shows that the first iterations of ν, CG and RRGMRES deliver very similar results, which is a normal phenomenon as the non-linearity of each method is sensed more in recovering the higher frequencies of the solution. As seen in Table 4.3, RRGMRES needs 50 iterations based on the L-curve criterion without delivering a small standard deviation of oceanic signal. The table suggests using CG, as it provides the smallest error after five iterations and is considerably faster.

4.3.3 Direct methods

This subsection presents two well-known methods amongst many, truncated singular value decomposition (TSVD) and Tikhonov regularisation, for solving ill-conditioned systems; see e.g. Xu (1992, 1998). Both of these methods have a shortcoming for working with large matrices, but the quality of the parameters is estimable, unlike the iterative methods.

4.3.3.1 Truncated singular value decomposition

The main idea of TSVD comes from the fact that the small eigenvalues of the coefficients matrix of a system are removed and those parts of the solution relating to these small eigenvalues are neglected in the final solution. In fact, the high frequencies of the solutions are removed and a smooth solution is sought.

Table 4.3 Statistics of estimated equivalent water height over the oceans derived from regularised gravity fields by iterative regularisation.

	January–February 2010					March–April 2010				
	I	Max.	Mean	Min.	Std	I	Max.	Mean	Min.	Std
Non-regularised	—	601.5	2.2	−647.7	113.5	—	421.8	3.0	−537	92.9
V	10	305.3	3.1	−500.1	44.3	10	309.2	3.2	−484.6	42.4
	1	306.9	3.2	−493.9	43.2	1	307.2	3.2	−492.8	43.1
ART	47	614.2	2.3	−672.0	104.6	50	412.1	3.2	−489.2	84.4
	1	298.7	2.7	−457.4	68.9	1	340.1	3.3	−405.9	61.2
CG	5	306.1	3.1	−500.8	44.0	5	307.5	3.3	−483.3	42.6
	1	306.8	3.2	−494.2	43.2	1	307.2	3.2	−492.5	43.1
RRGMRES	50	381.1	1.9	−558.1	69.1	51	316.4	3.1	−545.6	59.5
	1	306.8	3.2	−494.2	43.2	1	307.2	3.2	−492.5	43.1

ART, algebraic reconstruction technique; *CG*, conjugate gradient; *RRGMRES*, range-restricted generalised minimum residual. "I" means the number of iterations. Unit: mm.

From Eshagh, M., Lemoine, J.M., Gegout P., Biancale, R., 2013. On regularized time varying gravity field models based on GRACE data and their comparisons with hydrological models. Acta Geophys. 61 (1), 1–17.

Let us assume that the weight matrix is positive-definite and symmetric; in this case, it is straightforward to write:

$$\sigma_0^{-2}\mathbf{Q}^{-1} = \mathbf{G}^{\mathrm{T}}\mathbf{G}, \tag{4.38}$$

where $\mathbf{G}$ can be obtained by the Cholesky decomposition. Let us consider:

$$\mathbf{A}' = \mathbf{GA} \text{ and } \mathbf{L}' = \mathbf{GL}, \boldsymbol{\varepsilon}' = \mathbf{G}\boldsymbol{\varepsilon}, \tag{4.39}$$

where the matrices $\mathbf{A}'$ and $\mathbf{L}'$ contain the weight of observations. The idea of definition of these matrices is to simplify our mathematical derivations using TSVD. According to Eq. (4.39), Eq. (4.11) will change to:

$$\mathbf{A}'\mathbf{x} = \mathbf{L}' - \boldsymbol{\varepsilon}', \ \mathrm{E}(\boldsymbol{\varepsilon}'\boldsymbol{\varepsilon}'^{T}) = \sigma_0^2\mathbf{I}. \tag{4.40}$$

Now, let us write $\mathbf{A}'$ in the following spectral form:

$$\mathbf{A}' = \mathbf{U}\boldsymbol{\Lambda}\mathbf{V}^{\mathrm{T}}, \tag{4.41}$$

where $\mathbf{U}$ and $\mathbf{V}$ are $n \times n$ and $m \times m$ matrices, which are the eigenvectors of $\mathbf{A}'$ with the orthogonality property $\mathbf{U}\mathbf{U}^{\mathrm{T}} = \mathbf{I}$ and $\mathbf{V}^{\mathrm{T}}\mathbf{V} = \mathbf{I}$. These vectors play the role of base functions for $\mathbf{A}'$. $\boldsymbol{\Lambda}$ is an $n \times m$ matrix containing the eigenvalues sorted in decreasing order. The matrix $\mathbf{A}'$ is assumed positive-definite, which means that all eigenvalues are positive, but some of them are very small and close to zero (but not exactly zero). The existence of these small values in the least-squares solution of Eq. (4.40) amplifies the errors of $\mathbf{L}'$ considerably. By inserting Eq. (4.41) into Eq. (4.40) and solving it for $\mathbf{x}$, the spectral form of the least-squares solution of the system is derived:

$$\widehat{\mathbf{x}} = \mathbf{V}\boldsymbol{\Sigma}^{+}\boldsymbol{\Lambda}^{\mathrm{T}}\mathbf{U}^{\mathrm{T}}\mathbf{L}' = \sum_{i=1}^{n} \frac{\mathbf{u}_{\mathrm{i}}'^{\mathrm{T}}\mathbf{L}'}{\lambda_i}\mathbf{v}_i \text{ where } \boldsymbol{\Sigma} = \boldsymbol{\Lambda}^{\mathrm{T}}\boldsymbol{\Lambda}, \tag{4.42}$$

and $\boldsymbol{\Sigma}^{+}$ is the Moore–Penrose inverse of $\boldsymbol{\Sigma}$, i is the number of singular values λ_i, $\mathbf{u}_i$ and $\mathbf{v}_i$ are the eigenvectors of $\mathbf{A}'$ which are in $\mathbf{U}$ and $\mathbf{V}$, respectively, and n is the number of λ_i equivalent to the number of observations. The singular value λ_i in the denominator will amplify the solution, as a small denominator increases the corresponding ratio. Eq. (4.42) considers all λ_i, even the smallest one; the idea of TSVD is to remove the small singular values and truncate the series in Eq. (4.42) to a certain number $k < m$. In such a case, the solution will be biased, and this is the penalty we have to pay for stabilising the system. The main issue in TSVD is the proper selection of k. Different methods have been presented for estimating k, such as generalised cross validation (Wahba, 1976), the L-curve (Hansen, 1998), the quasi-optimal method (Hansen, 2007), the discrepancy principle (cf. Scherzer, 1993), the monotone error rule (Hämarik and Taytenhahn, 2001) and the normalised cumulative periodogram (cf. Diggle, 1991, or Mojabi and LoVetri, 2008).

The L-curve could be a method for such purpose and it was presented in Section 4.3.2.2 for finding the optimal value of k. Here, k starts from 1 to the number of eigenvalues of the system, and for each number the norm of residuals and estimated parameter is estimated and plotted. The results for all selected truncation numbers will be represented by an L-curve plot with a corner at the optimal truncation number.

One can also mention that the minimisation problem of the TSVD is:

$$\min_{\mathbf{x}\in \mathrm{span}\{\mathbf{v}_1,\mathbf{v}_2,\ldots,\mathbf{v}_k\}} \|\mathbf{L}' - \mathbf{A}'\mathbf{x}\|_2, \tag{4.43}$$

where $\|\bullet\|_2$ stands for the L_2-norm. Eq. (4.43) means that $\mathbf{x}$ is obtained from a combination of the bases $\mathbf{v}_1, \mathbf{v}_2, \ldots, \mathbf{v}_k$ and k means that the solution is constructed by k bases. Therefore, the regularised solution for $\mathbf{x}_{\mathrm{reg}}$ is obtained as:

$$\mathbf{x}_{\mathrm{reg}} = \mathbf{V}\boldsymbol{\Sigma}_k^+\boldsymbol{\Lambda}^{\mathrm{T}}\mathbf{U}^{\mathrm{T}}\mathbf{L}', \tag{4.44}$$

where $\boldsymbol{\Sigma}_k^+$ is the pseudo-inverse of $\boldsymbol{\Sigma}_k$, which is similar to $\boldsymbol{\Sigma}$, but the first k diagonal elements are kept and the rest of them switched to zero. To find the bias due to this regularisation we write:

$$\mathrm{Bias}(\mathbf{x}_{\mathrm{reg}}) = \mathrm{E}\{\mathbf{x}_{\mathrm{reg}} - \mathbf{x}\} = \mathrm{E}\left\{\mathbf{V}\boldsymbol{\Sigma}_k^+\boldsymbol{\Lambda}^{\mathrm{T}}\mathbf{U}^{\mathrm{T}}\mathbf{L}' - \mathbf{x}\right\}. \tag{4.45}$$

By solving Eq. (4.40) for $\mathbf{L}'$ and using Eq. (4.41) and properties of the eigenvector $\mathbf{U}$, we arrive at:

$$\mathrm{Bias}(\mathbf{x}_{\mathrm{reg}}) = \left(\mathbf{V}\boldsymbol{\Sigma}_k^+\boldsymbol{\Sigma}\mathbf{V}^{\mathrm{T}} - \mathbf{I}\right)\mathbf{x}. \tag{4.46}$$

The product $\boldsymbol{\Sigma} = \boldsymbol{\Lambda}^{\mathrm{T}}\boldsymbol{\Lambda}$ will be an $m \times m$ diagonal matrix. This matrix is positive-semi-definite and contains zero diagonal elements, and this is why we use the Moore–Penrose inverse of this matrix in the least-squares solution, Eq. (4.42). TSVD puts zero values for the very small singular values for stabilising the system of equations. The product $\boldsymbol{\Sigma}\boldsymbol{\Sigma}_k^+$, which is the product of two diagonal matrices, will be an identity matrix in which the non-zero diagonal elements are replaced by 1:

$$\mathrm{Bias}(\mathbf{x}_{\mathrm{reg}}) = \left(\mathbf{V}\mathbf{I}_u^{1k}\mathbf{V}^{\mathrm{T}} - \mathbf{I}\right)\mathbf{x}, \tag{4.47}$$

where $\mathbf{I}_m^{gh}$ is called the generalised unitary matrix (Eshagh, 2010), an $m \times m$ unitary matrix whose non-zero diagonal elements are between the gth and the hth columns or rows.

Truncation of the spectral form (Eq. 4.42) leads to a bias in estimating unknown parameters; therefore, the mean squared error (MSE) of the parameters should be estimated, as our estimator is not unbiased anymore. To do so write:

$$\begin{aligned}\mathrm{MSE}_{\mathbf{x}_{\mathrm{reg}}} &= \mathrm{E}\{(\mathbf{x}-\mathbf{x}_{\mathrm{reg}})(\mathbf{x}-\mathbf{x}_{\mathrm{reg}})^{\mathrm{T}}\} \\ &= \mathrm{E}\left\{\left(\mathbf{x}-\mathbf{V}\boldsymbol{\Sigma}_k^{+}\boldsymbol{\Lambda}^{\mathrm{T}}\mathbf{U}^{\mathrm{T}}(\mathbf{A}'\mathbf{x}+\boldsymbol{\varepsilon}')\right)\left(\mathbf{x}-\mathbf{V}\boldsymbol{\Sigma}_k^{+}\boldsymbol{\Lambda}^{\mathrm{T}}\mathbf{U}^{\mathrm{T}}(\mathbf{A}'\mathbf{x}+\boldsymbol{\varepsilon}')\right)^{\mathrm{T}}\right\} = \\ &= \left(\mathbf{I}-\mathbf{V}\boldsymbol{\Sigma}_k^{+}\boldsymbol{\Sigma}_k\mathbf{V}^{\mathrm{T}}\right)\mathbf{x}\mathbf{x}^{\mathrm{T}}\left(\mathbf{I}-\mathbf{V}\boldsymbol{\Sigma}_k^{+}\boldsymbol{\Sigma}_k\mathbf{V}^{\mathrm{T}}\right)+\mathbf{V}\boldsymbol{\Sigma}_k^{+}\boldsymbol{\Lambda}^{\mathrm{T}}\mathbf{U}^{\mathrm{T}}\mathrm{E}\{\boldsymbol{\varepsilon}'\boldsymbol{\varepsilon}'^{\mathrm{T}}\}\mathbf{U}\boldsymbol{\Lambda}\boldsymbol{\Sigma}_k^{+}\mathbf{V}^{\mathrm{T}} = \\ &= \left(\mathbf{I}-\mathbf{V}\boldsymbol{\Sigma}_k^{+}\boldsymbol{\Sigma}_k\mathbf{V}^{\mathrm{T}}\right)\mathbf{x}\mathbf{x}^{\mathrm{T}}\left(\mathbf{I}-\mathbf{V}\boldsymbol{\Sigma}_k^{+}\boldsymbol{\Sigma}_k\mathbf{V}^{\mathrm{T}}\right)+\sigma_0^2\mathbf{V}\boldsymbol{\Sigma}_k^{+}\mathbf{V}^{\mathrm{T}}. \end{aligned} \tag{4.48}$$

The first term on the rhs of Eq. (4.48) is due to the truncation bias and the second term to the contribution of the random errors.

However, estimation of the variance factor σ_0^2 is not straightforward as was presented in Eq. (4.19). If one uses the standard formula of the a posteriori variance factor, which is suitable for an ordinary adjustment problem, the estimate will not be correct, because the residuals, obtained from the estimated parameters, contain the truncation bias. Let us express the residual vector by:

$$\boldsymbol{\varepsilon}'_{\mathrm{reg}} = \mathbf{L}' - \mathbf{A}'\mathbf{x}_{\mathrm{reg}} = \mathbf{U}\boldsymbol{\Lambda}\mathbf{V}^{\mathrm{T}}\mathbf{x}+\boldsymbol{\varepsilon}' - \mathbf{U}\boldsymbol{\Lambda}\boldsymbol{\Sigma}_k^{+}\boldsymbol{\Lambda}^{\mathrm{T}}\mathbf{U}^{\mathrm{T}}(\mathbf{U}\boldsymbol{\Lambda}\mathbf{V}^{\mathrm{T}}\mathbf{x}+\boldsymbol{\varepsilon}'). \tag{4.49}$$

Simplification of Eq. (4.49) reads:

$$\boldsymbol{\varepsilon}'_{\mathrm{reg}} = \mathbf{U}\boldsymbol{\Lambda}\mathbf{I}_m^{1,k}\mathbf{V}^{\mathrm{T}}\mathbf{x} + \left(\mathbf{I}_n - \mathbf{U}\boldsymbol{\Lambda}\boldsymbol{\Sigma}_k^{+}\boldsymbol{\Lambda}^{\mathrm{T}}\mathbf{U}^{\mathrm{T}}\right)\boldsymbol{\varepsilon}' \text{ where } \mathbf{I}_m^{1,k} = \mathbf{I}_m - \boldsymbol{\Sigma}_k^{+}\boldsymbol{\Sigma}_k. \tag{4.50}$$

The first term on the rhs of Eq. (4.50) is the bias of the estimated residuals.

Squaring Eq. (4.50) yields:

$$\begin{aligned}\boldsymbol{\varepsilon}'^{\mathrm{T}}_{\mathrm{reg}}\boldsymbol{\varepsilon}'_{\mathrm{reg}} &= \boldsymbol{\varepsilon}'^{\mathrm{T}}\left(\mathbf{I}_n - \mathbf{U}\boldsymbol{\Lambda}\boldsymbol{\Sigma}_k^{+}\boldsymbol{\Lambda}^{\mathrm{T}}\mathbf{U}^{\mathrm{T}}\right)^{\mathrm{T}}\left(\mathbf{I}_n - \mathbf{U}\boldsymbol{\Lambda}\boldsymbol{\Sigma}_k^{+}\boldsymbol{\Lambda}^{\mathrm{T}}\mathbf{U}^{\mathrm{T}}\right)\boldsymbol{\varepsilon}' \\ &\quad + \mathbf{x}^{\mathrm{T}}\mathbf{V}\mathbf{I}_m^{1k}\boldsymbol{\Sigma}\mathbf{I}_m^{1k}\mathbf{V}^{\mathrm{T}}\mathbf{x}.\end{aligned} \tag{4.51}$$

Taking the second term on the rhs into the left-hand side (lhs) of Eq. (4.51) and taking statistical expectation, assuming that $\boldsymbol{\varepsilon}'$ and $\mathbf{x}$ are independent, reads:

$$\boldsymbol{\varepsilon}'^{\mathrm{T}}_{\mathrm{reg}}\boldsymbol{\varepsilon}'_{\mathrm{reg}} - \mathbf{x}^{\mathrm{T}}\mathbf{V}\mathbf{I}^{1k}_{m}\boldsymbol{\Sigma}\mathbf{I}^{1k}_{m}\mathbf{V}^{\mathrm{T}}\mathbf{x} = \sigma_0^2\left[\mathrm{trace}(\mathbf{I}_n) - \mathrm{trace}\left(\mathbf{I}^{1k}_{m}\right)\right]. \tag{4.52}$$

Solving the equation for σ_0^2 gives:

$$\widehat{\sigma}_0^2 = \frac{\boldsymbol{\varepsilon}'^{\mathrm{T}}_{\mathrm{reg}}\boldsymbol{\varepsilon}'_{\mathrm{reg}} - \mathbf{x}^{\mathrm{T}}\mathbf{V}\mathbf{I}^{1k}_{u}\boldsymbol{\Sigma}\mathbf{I}^{1k}_{u}\mathbf{V}^{\mathrm{T}}\mathbf{x}}{n-k}. \tag{4.53}$$

It should be noted that there is no guarantee that this estimator, Eq. (4.53), delivers a non-negative variance factor. If the second term in the numerator of this estimator is larger than the first, a negative variance factor will come out, which is not meaningful. Therefore, a bias-corrected and non-negative form for this factor is needed. If the first term on the rhs of Eq. (4.50) is taken to the lhs the bias-corrected residual will be obtained:

$$\overline{\boldsymbol{\varepsilon}}'_{\mathrm{reg}} = \boldsymbol{\varepsilon}'_{\mathrm{reg}} - \mathbf{U}\boldsymbol{\Lambda}\mathbf{I}^{1,k}_{m}\mathbf{V}^{\mathrm{T}}\mathbf{x} = \left(\mathbf{I}_n - \mathbf{U}\boldsymbol{\Lambda}\boldsymbol{\Sigma}^{+}_{k}\boldsymbol{\Lambda}^{\mathrm{T}}\mathbf{U}^{\mathrm{T}}\right)\boldsymbol{\varepsilon}'. \tag{4.54}$$

Squaring Eq. (4.54) and taking the expectation from both sides of the result yields:

$$\overline{\boldsymbol{\varepsilon}}'^{\mathrm{T}}_{\mathrm{reg}}\overline{\boldsymbol{\varepsilon}}'_{\mathrm{reg}} = \mathrm{E}\left\{\boldsymbol{\varepsilon}'^{\mathrm{T}}\left(\mathbf{I}_n - \mathbf{U}\boldsymbol{\Lambda}\boldsymbol{\Sigma}^{+}_{k}\boldsymbol{\Lambda}^{\mathrm{T}}\mathbf{U}^{\mathrm{T}}\right)^{\mathrm{T}}\left(\mathbf{I}_n - \mathbf{U}\boldsymbol{\Lambda}\boldsymbol{\Sigma}^{+}_{k}\boldsymbol{\Lambda}^{\mathrm{T}}\mathbf{U}^{\mathrm{T}}\right)\boldsymbol{\varepsilon}'\right\} \tag{4.55}$$

or

$$\overline{\boldsymbol{\varepsilon}}'^{\mathrm{T}}_{\mathrm{reg}}\overline{\boldsymbol{\varepsilon}}'_{\mathrm{reg}} = \mathrm{trace}\left[\left(\mathbf{I}_n - \mathbf{U}\boldsymbol{\Lambda}\boldsymbol{\Sigma}^{+}_{k}\boldsymbol{\Lambda}^{\mathrm{T}}\mathbf{U}^{\mathrm{T}}\right)^{\mathrm{T}}\left(\mathbf{I}_n - \mathbf{U}\boldsymbol{\Lambda}\boldsymbol{\Sigma}^{+}_{k}\boldsymbol{\Lambda}^{\mathrm{T}}\mathbf{U}^{\mathrm{T}}\right)\mathrm{E}\{\boldsymbol{\varepsilon}'\boldsymbol{\varepsilon}'^{\mathrm{T}}\}\right]. \tag{4.56}$$

According to Eq. (4.40), Eq. (4.56) will be simplified to:

$$\overline{\boldsymbol{\varepsilon}}'^{\mathrm{T}}_{\mathrm{reg}}\overline{\boldsymbol{\varepsilon}}'_{\mathrm{reg}} = \sigma_0^2\mathrm{trace}\left[\left(\mathbf{I}_n - \mathbf{U}\boldsymbol{\Lambda}\mathbf{I}^{1k}_{m}\boldsymbol{\Sigma}^{+}_{k}\boldsymbol{\Lambda}^{\mathrm{T}}\mathbf{U}^{\mathrm{T}}\right)\left(\mathbf{I}_n - \mathbf{U}\boldsymbol{\Lambda}\mathbf{I}^{1k}_{m}\boldsymbol{\Sigma}^{+}_{k}\boldsymbol{\Lambda}^{\mathrm{T}}\mathbf{U}^{\mathrm{T}}\right)\right]. \tag{4.57}$$

Based on the idempotency (e.g. for the matric $\mathbf{A}$, we have $\mathbf{AA} = \mathbf{A}$) of $\left(\mathbf{I}_n - \mathbf{U}\boldsymbol{\Lambda}\mathbf{I}^{1k}_{m}\boldsymbol{\Sigma}^{+}_{k}\boldsymbol{\Lambda}^{\mathrm{T}}\mathbf{U}^{\mathrm{T}}\right)$ it is straightforward to show that:

$$\widetilde{\sigma}_0^2 = \frac{\overline{\boldsymbol{\varepsilon}}'^{\mathrm{T}}_{\mathrm{reg}}\overline{\boldsymbol{\varepsilon}}'_{\mathrm{reg}}}{\mathrm{trace}(\mathbf{I}_n) - \mathrm{trace}\left(\mathbf{I}^{1k}_{u}\boldsymbol{\Sigma}^{+}_{k}\boldsymbol{\Sigma}\right)} = \frac{\overline{\boldsymbol{\varepsilon}}'^{\mathrm{T}}_{\mathrm{reg}}\overline{\boldsymbol{\varepsilon}}'_{\mathrm{reg}}}{n-k}. \tag{4.58}$$

The estimator presented in Eq. (4.58) is non-negative and the true value of $\mathbf{x}$ or some approximation of it is required for computing the bias-corrected residuals using Eq. (4.54).

4.3.3.2 Tikhonov regularisation

Tikhonov (1963) was one of the earliest persons to begin solving ill-posed problems. His method is summarised in the following minimisation problem:

$$\min\left(\|\mathbf{Ax}-\mathbf{L}\|^2_{\mathbf{Q}^{-1}}+\alpha^2\|\mathbf{x}\|^2\right), \tag{4.59}$$

where $\|\bullet\|_{\mathbf{Q}^{-1}}$ means L_2-norm and $\mathbf{Q}^{-1}$ is the weight matrix, and α^2 is the regularisation parameter. The solution of the above minimisation problem is obtained by solving the following system of equations:

$$\mathbf{Nx}=\mathbf{A}^{\mathrm{T}}\mathbf{Q}^{-1}\mathbf{L}, \quad \text{where } \mathbf{N}=\mathbf{A}^{\mathrm{T}}\mathbf{Q}^{-1}\mathbf{A}+\alpha^2\mathbf{I}_m, \tag{4.60}$$

where $\mathbf{I}_m$ is an $m \times m$ identity matrix.

Tikhonov's idea is to add a small positive number α^2 to the diagonal elements of the coefficients matrix of the normal system of equations to stabilise it. Therefore, the regularised solution of Eq. (4.60) will be:

$$\mathbf{x}_{\mathrm{reg}}=\mathbf{N}^{-1}\mathbf{A}^{\mathrm{T}}\mathbf{Q}^{-1}\mathbf{L}. \tag{4.61}$$

By adding α^2 the solution will be biased; this is a penalty to pay for stabilising the solution. Now, we estimate this bias (see e.g. Xu et al., 2006; Eshagh, 2009):

$$\mathrm{Bias}(\mathbf{x}_{\mathrm{reg}})=\mathrm{E}\{\mathbf{x}_{\mathrm{reg}}-\mathbf{x}\}=\mathrm{E}\{\mathbf{N}^{-1}\mathbf{A}^{\mathrm{T}}\mathbf{Q}^{-1}\mathbf{L}-\mathbf{x}\}. \tag{4.62}$$

Let us solve the Gauss–Markov model (Eq. 4.11) for $\mathbf{L}$ and insert the result into Eq. (4.62); then:

$$\begin{aligned}\mathrm{Bias}(\mathbf{x}_{\mathrm{reg}})&=\mathrm{E}\{\mathbf{N}^{-1}\mathbf{A}^{\mathrm{T}}\mathbf{Q}^{-1}(\mathbf{Ax}+\boldsymbol{\varepsilon})-\mathbf{x}\}\\&=\mathrm{E}\{\mathbf{N}^{-1}\mathbf{A}^{\mathrm{T}}\mathbf{Q}^{-1}\mathbf{Ax}-\mathbf{x}+\mathbf{N}^{-1}\mathbf{A}^{\mathrm{T}}\mathbf{Q}^{-1}\boldsymbol{\varepsilon}\}.\end{aligned} \tag{4.63}$$

After taking the statistical expectation and considering that $\mathrm{E}\{\boldsymbol{\varepsilon}\}=0$, the bias of the Tikhonov regularisation will be:

$$\mathrm{Bias}(\mathbf{x}_{\mathrm{reg}})=(\mathbf{N}^{-1}\mathbf{A}^{\mathrm{T}}\mathbf{Q}^{-1}\mathbf{A}-\mathbf{I}_m)\mathbf{x}. \tag{4.64}$$

It will not be difficult to show that

$$\mathbf{N}^{-1}\mathbf{A}^{\mathrm{T}}\mathbf{Q}^{-1}\mathbf{A}=\mathbf{I}_m-\alpha^2\mathbf{N}^{-1}. \tag{4.65}$$

Inserting Eq. (4.65) into Eq. (4.64) leads to:

$$\text{Bias}(\mathbf{x}_{\text{reg}}) = -\alpha^2 \mathbf{N}^{-1}\mathbf{x}, \tag{4.66}$$

which is the bias of the Tikhonov regularisation. As observed, it is a function of the true value of **x**, which is not available, but the estimated parameter from Eq. (4.61) can be used as an approximation to **x** according to Xu et al. (2006). We commit a second-order bias by this approximation, but it is always smaller than the first-order approximation. Therefore, this bias is somehow estimable and can be removed from the regularised solution to make it closer to the true value **x**.

To estimate the quality of the regularised solution, we write:

$$\text{MSE}_{\mathbf{x}_{\text{reg}}} = \text{E}\left\{(\mathbf{x} - \mathbf{x}_{\text{reg}})(\mathbf{x} - \mathbf{x}_{\text{reg}})^{\text{T}}\right\}. \tag{4.67}$$

By inserting Eq. (4.61) into Eq. (4.67), solving Eq. (4.11) for **L** and substituting it into the results we obtain:

$$\begin{aligned}\text{MSE}_{\mathbf{x}_{\text{reg}}} &= (\mathbf{I}_m - \mathbf{N}^{-1}\mathbf{A}^{\text{T}}\mathbf{Q}^{-1}\mathbf{A})\mathbf{x}\mathbf{x}^{\text{T}}(\mathbf{I}_m - \mathbf{N}^{-1}\mathbf{A}^{\text{T}}\mathbf{Q}^{-1}\mathbf{A})^{\text{T}} \\ &+ \mathbf{N}^{-1}\mathbf{A}^{\text{T}}\mathbf{Q}^{-1}\text{E}\{\boldsymbol{\varepsilon}\boldsymbol{\varepsilon}^{\text{T}}\}\mathbf{Q}^{-1}\mathbf{A}\mathbf{N}^{\text{T}}.\end{aligned} \tag{4.68}$$

According to Eq. (4.11) and the identity (Eq. 4.65), Eq. (4.68) will change to:

$$\text{MSE}_{\mathbf{x}_{\text{reg}}} = \alpha^4 \mathbf{N}^{-1}\mathbf{x}\mathbf{x}^{\text{T}}\mathbf{N}^{-1} + \sigma_0^2 \mathbf{N}^{-1}\mathbf{A}^{\text{T}}\mathbf{Q}^{-1}\mathbf{A}\mathbf{N}^{-1}. \tag{4.69}$$

The first-term on the rhs of Eq. (4.69) is due to the regularisation bias and the second term the propagated random error.

For estimation of the a posteriori variance factor, the estimated residuals vector are required:

$$\boldsymbol{\varepsilon}_{\text{reg}} = \mathbf{L} - \mathbf{A}\mathbf{x}_{\text{reg}} = \mathbf{A}\mathbf{x} + \boldsymbol{\varepsilon} - \mathbf{A}\mathbf{N}^{-1}\mathbf{A}^{\text{T}}\mathbf{Q}^{-1}(\mathbf{A}\mathbf{x} + \boldsymbol{\varepsilon}). \tag{4.70}$$

According to Eq. (4.65) we can rewrite it:

$$\boldsymbol{\varepsilon}_{\text{reg}} = \mathbf{L} - \mathbf{A}\mathbf{x}_{\text{reg}} = \alpha^2 \mathbf{A}\mathbf{N}^{-1}\mathbf{x} + (\mathbf{I}_n - \mathbf{A}\mathbf{N}^{-1}\mathbf{A}^{\text{T}}\mathbf{Q}^{-1})\boldsymbol{\varepsilon}. \tag{4.71}$$

By taking the expectation of the quadratic form of this residual, we obtain:

$$\begin{aligned}\text{E}\left\{\boldsymbol{\varepsilon}_{\text{reg}}^{\text{T}}\mathbf{Q}^{-1}\boldsymbol{\varepsilon}_{\text{reg}}\right\} &= \alpha^4 \mathbf{x}^{\text{T}}\mathbf{N}^{-1}\mathbf{A}^{\text{T}}\mathbf{Q}^{-1}\mathbf{A}\mathbf{N}^{-1}\mathbf{x} \\ &+ \text{E}\{\boldsymbol{\varepsilon}^{\text{T}}(\mathbf{I}_n - \mathbf{Q}^{-1}\mathbf{A}\mathbf{N}^{-1}\mathbf{A}^{\text{T}})\mathbf{Q}^{-1}(\mathbf{I}_n - \mathbf{A}\mathbf{N}^{-1}\mathbf{A}^{\text{T}}\mathbf{Q}^{-1})\boldsymbol{\varepsilon}\}.\end{aligned} \tag{4.72}$$

The second term on the rhs of Eq. (4.72) can be simplified according to Eq. (4.11) and Eq. (4.65) further to:

$$
\begin{aligned}
&\mathrm{E}\{\boldsymbol{\varepsilon}^{\mathrm{T}}(\mathbf{I}_n - \mathbf{Q}^{-1}\mathbf{A}\mathbf{N}^{-1}\mathbf{A}^{\mathrm{T}})\mathbf{Q}^{-1}(\mathbf{I}_n - \mathbf{A}\mathbf{N}^{-1}\mathbf{A}^{\mathrm{T}}\mathbf{Q}^{-1})\boldsymbol{\varepsilon}\} \\
&= \mathrm{trace}(\mathbf{Q}^{-1} - \mathbf{Q}^{-1}\mathbf{A}\mathbf{N}^{-1}\mathbf{A}^{\mathrm{T}}\mathbf{Q}^{-1} - \alpha^2\mathbf{Q}^{-1}\mathbf{A}\mathbf{N}^{-2}\mathbf{A}^{\mathrm{T}}\mathbf{Q}^{-1})\mathrm{E}\{\boldsymbol{\varepsilon}\boldsymbol{\varepsilon}^{\mathrm{T}}\} \\
&= \sigma_0^2 \mathrm{trace}(\mathbf{I}_n - \mathbf{Q}^{-1}\mathbf{A}\mathbf{N}^{-1}\mathbf{A}^{\mathrm{T}} - \alpha^2\mathbf{Q}^{-1}\mathbf{A}\mathbf{N}^{-2}\mathbf{A}^{\mathrm{T}}) \\
&= \sigma_0^2(n - \mathrm{trace}(\mathbf{I}_m - \alpha^2\mathbf{N}^{-1}) - \alpha^2\mathrm{trace}\mathbf{N}^{-1}(\mathbf{I}_m - \alpha^2\mathbf{N}^{-1})) \\
&= \sigma_0^2(n - m + \alpha^4\mathrm{trace}\mathbf{N}^{-2}).
\end{aligned}
\tag{4.73}
$$

Substituting Eq. (4.73) into Eq. (4.72) yields:

$$
\mathrm{E}\left\{\boldsymbol{\varepsilon}_{\mathrm{reg}}^{\mathrm{T}}\mathbf{Q}^{-1}\boldsymbol{\varepsilon}_{\mathrm{reg}}\right\} = \alpha^4\mathbf{x}^{\mathrm{T}}\mathbf{N}^{-1}\mathbf{A}^{\mathrm{T}}\mathbf{Q}^{-1}\mathbf{A}\mathbf{N}^{-1}\mathbf{x} + \sigma_0^2(n - m + \alpha^4\mathrm{trace}\mathbf{N}^{-2}).
\tag{4.74}
$$

Solution of Eq. (4.74) gives an estimator for a posteriori variance factor

$$
\tilde{\sigma}_0^2 = \frac{\boldsymbol{\varepsilon}_{\mathrm{reg}}^{\mathrm{T}}\mathbf{Q}^{-1}\boldsymbol{\varepsilon}_{\mathrm{reg}} - \alpha^4\mathbf{x}^{\mathrm{T}}\mathbf{N}^{-1}\mathbf{A}^{\mathrm{T}}\mathbf{Q}^{-1}\mathbf{A}\mathbf{N}^{-1}\mathbf{x}}{n - m + \alpha^4\mathrm{trace}\ \mathbf{N}^{-2}}.
\tag{4.75}
$$

Normally, since α^4 is very small, the regularisation bias will be very small, and in some cases, one can assume $\alpha^4 \approx 0$. In this case, it is enough to apply Eq. (4.19) for the estimated residuals. However, if the bias is large the estimator (Eq. 4.75) should be used, but in some cases, the variance factor may come out negative. In such a case, the bias-corrected residuals should be applied for estimating this factor. Let us take the first term on the rhs of Eq. (4.71) to the lhs to obtain the bias-corrected residuals by inserting the regularised solution (Eq. 4.61) into Eq. (4.70) and performing further simplifications:

$$
\bar{\boldsymbol{\varepsilon}}_{\mathrm{reg}} = \boldsymbol{\varepsilon}_{\mathrm{reg}} - \alpha^2\mathbf{A}\mathbf{N}^{-1}(\mathbf{I}_m - \alpha^2\mathbf{N}^{-1})\mathbf{x}
\tag{4.76}
$$

or

$$
\bar{\boldsymbol{\varepsilon}}_{\mathrm{reg}} = (\mathbf{I}_n - \mathbf{A}\mathbf{N}^{-1}\mathbf{A}^{\mathrm{T}}\mathbf{Q}^{-1} + \alpha^2\mathbf{A}\mathbf{N}^{-2}\mathbf{A}^{\mathrm{T}}\mathbf{Q}^{-1})\boldsymbol{\varepsilon}.
\tag{4.77}
$$

The quadratic form of Eq. (4.77) with the weight matrix $\mathbf{Q}^{-1}$ is:

$$
\begin{aligned}
\mathrm{E}\left\{\bar{\boldsymbol{\varepsilon}}_{\mathrm{reg}}^{\mathrm{T}}\mathbf{Q}^{-1}\bar{\boldsymbol{\varepsilon}}_{\mathrm{reg}}\right\} = \mathrm{E}\Big\{&\boldsymbol{\varepsilon}^{\mathrm{T}}(\mathbf{I}_n - \mathbf{A}\mathbf{N}^{-1}\mathbf{A}^{\mathrm{T}}\mathbf{Q}^{-1} + \alpha^2\mathbf{A}\mathbf{N}^{-2}\mathbf{A}^{\mathrm{T}}\mathbf{Q}^{-1})^{\mathrm{T}} \\
&\times \mathbf{Q}^{-1}(\mathbf{I}_n - \mathbf{A}\mathbf{N}^{-1}\mathbf{A}^{\mathrm{T}}\mathbf{Q}^{-1} + \alpha^2\mathbf{A}\mathbf{N}^{-2}\mathbf{A}^{\mathrm{T}}\mathbf{Q}^{-1})\boldsymbol{\varepsilon}\Big\}
\end{aligned}
\tag{4.78}
$$

or

$$E\left\{\bar{\boldsymbol{\varepsilon}}_{\text{reg}}^{\text{T}}\mathbf{Q}^{-1}\bar{\boldsymbol{\varepsilon}}_{\text{reg}}\right\} = \text{trace}\left[(\mathbf{I}_n - \mathbf{A}\mathbf{N}^{-1}\mathbf{A}^{\text{T}}\mathbf{Q}^{-1} + \alpha^2\mathbf{A}\mathbf{N}^{-2}\mathbf{A}^{\text{T}}\mathbf{Q}^{-1})^{\text{T}} \times \mathbf{Q}^{-1}(\mathbf{I}_n - \mathbf{A}\mathbf{N}^{-1}\mathbf{A}^{\text{T}}\mathbf{Q}^{-1} + \alpha^2\mathbf{A}\mathbf{N}^{-2}\mathbf{A}^{\text{T}}\mathbf{Q}^{-1})E\right]\{\boldsymbol{\varepsilon}\boldsymbol{\varepsilon}^{\text{T}}\}. \tag{4.79}$$

After a rather lengthy derivation and solving the results for σ_0^2 we obtain (see also Xu et al., 2006):

$$\tilde{\sigma}_0^2 = \frac{\bar{\boldsymbol{\varepsilon}}_{\text{reg}}^{\text{T}}\mathbf{Q}^{-1}\bar{\boldsymbol{\varepsilon}}_{\text{reg}}}{n - m + \alpha^8 \text{trace } \mathbf{N}^{-4}}, \tag{4.80}$$

which is a non-negative reduced-bias estimator of the a posteriori variance factor.

4.3.3.3 Generalised cross validation

Generalised cross validation is another idea for estimating the regularisation parameter for the Tikhonov regularisation. The philosophy of cross validation is based on removing an observation and trying to estimate it using the rest of the observations. A good regularisation parameter will produce solutions where the difference between the removed and the estimated values of the parameters in an average sense will be minimal. Mathematically, this idea can be written as (cf. Kusche and Klees, 2002):

$$G(\alpha^2) = \frac{\|\mathbf{A}\mathbf{x}_{\text{reg}} - \mathbf{L}\|^2}{[\text{trace}(\mathbf{I}_n - \mathbf{A}\mathbf{N}^{-1}\mathbf{A}^{\text{T}}\mathbf{Q}^{-1})]^2} = \frac{\|\mathbf{A}\mathbf{x}_{\text{reg}} - \mathbf{L}\|^2}{[n - m + \alpha^2\text{trace}(\mathbf{N}^{-1})]^2}, \tag{4.81}$$

and the regularisation parameter will be found by:

$$\alpha^2 = \arg\ \min G(\alpha^2). \tag{4.82}$$

This means that different values for α^2 are selected and inserted into Eq. (4.81) and $G(\alpha^2)$ is computed for them. Normally, a mathematical model is fitted to $G(\alpha^2)$ and then the minimum of this function will be obtained, and the corresponding α^2, which minimises $G(\alpha^2)$, is selected as the optimal value.

Generalised cross validation can also be applied for estimating the truncation number in the TSVD method.

4.3.3.4 Example: application of the TSVD and Tikhonov regularisation methods for determining equivalent water height from the GRACE mission

Here, another part of the Eshagh et al. (2013) study is considered, and the system of equations constructed from the GRACE range-rate data is solved by the TSVD and Tikhonov regularisation methods. Their regularisation parameter or truncation number is estimated by L-curve and generalised cross validation, to see which case delivers the better results. Table 4.4 shows the statistics of the EWH over the oceans estimated by these methods, which we expect to be zero.

Fig. 4.19 shows the EWH derived by the regularised gravity field by Tikhonov's regularisation. As we can observe, it shows a significant reduction of artefact over ocean.

4.3.3.5 Example: inversion of on-orbit satellite gradiometry data in the local north-oriented frame and orbital reference frame

A 3-month orbit of the Gravity Field and Steady-State Ocean Circulation Explorer (GOCE) with a rate of 10 s was simulated by Eshagh (2010) using EGM96 (Lemoine et al., 1998) to degree and order 250 and the Runge–Kutta integrator of fourth order. The initial state vector for the numerical integration was provided in a personal communication from Plank (2005). The estimated positions and velocities of the satellite were converted to co-latitude and longitudes and afterwards to the Keplerian elements (see e.g. Seeber, 2003, pp. 69–70) including the satellite orbit inclination i. The satellite gradiometry data in the LNOF and TOF were also simulated from EGM08 (Pavlis et al., 2008) to degree and order 360. A coloured noise is generated by filtering a random noise with a standard deviation of 0.02 E using an auto-regressive moving average filter of degree 1 and order 2 (ARMA (1,2)); see Klees et al. (2003). The on-orbit satellite gradiometry data are infected by the generated noise prior to their inversions and the Tikhonov regularisation and L-curve were used for solving the discretised integral equations.

Table 4.5 represents the statistics of the recovered Δg from the on-orbit satellite gradiometry data in the LNOF and ORF. It shows unsatisfactory recoveries from T_{vv}, T_{uv}, and T_{uw}. Similarly, worse results were derived from inverting T_{uv}, T_{uw} and T_{vw}. In addition, Table 4.5 shows the largest error for T_{vw}, which is even larger than the error for T_{uw}.

Table 4.4 Statistics of estimated equivalent water height over oceans un-regularised and by direct regularisation.

		January–February 2010				March–April 2010			
		Max.	**Mean**	**Min.**	**Std**	**Max.**	**Mean**	**Min.**	**Std**
Non-regularised		601.5	2.2	−647.7	113.5	421.8	3.0	−537.7	92.9
TR	LC	304.2	2.8	−478.9	44.0	309.8	3.4	−454.1	41.7
	GCV	370.4	2.1	−546.4	63.9	313.2	3.1	−537.4	58.3
TSVD	LC	329.9	2.8	−474.1	43.8	329.2	3.2	−501. 9	48.1
	GCV	319.2	2.9	−516.5	45.6	317.2	3.2	−535.9	58.7

GCV, generalised cross validation; *LC*, L-curve; *TR*, Tikhonov regularisation; *TSVD*, truncated singular value decomposition. Unit: mm.
From Eshagh, M., Lemoine, J.M., Gegout P., Biancale, R., 2013. On regularized time varying gravity field models based on GRACE data and their comparisons with hydrological models. Acta Geophys. 61 (1), 1–17, with permission from Springer.

Figure 4.19 Equivalent water height computed from regularised gravity field by Tikhonov's regularisation. Comparison of January—February 2010 with January—April 2010. Unit: m. *(From Eshagh, M., Lemoine, J.M., Gegout P., Biancale, R., 2013. On regularized time varying gravity field models based on GRACE data and their comparisons with hydrological models. Acta Geophys. 61 (1), 1—17, with permission from Springer.)*

$T_{zz} = T_{uu}$ and $T_{vv} + T_{ww}$ can also be inverted to the gravity anomaly. The idea is that the signal in $T_{vv} + T_{ww}$ should be equivalent to that of $T_{zz} = T_{uu}$; but the problem is that the noise of $T_{vv} + T_{ww}$ is larger than that of $T_{zz} = T_{uu}$, as T_{vv} and T_{ww} are measured independently and they have their own noise. Considering a noise twice as large as that of each one seems to be reasonable. Table 4.6 shows the statistics of the errors of the recovered gravity anomalies using these two types of satellite gradiometry data. It shows that the anomalies can be recovered with an error of 6.3 mGal from the on-orbit $S_{xy}(r, \psi)$. The quality of inversion of $T_{vv} + T_{ww}$ is fine as well, and the error will be 7.0 mGal.

4.3.3.6 Example for bias corrections and estimation of a posteriori variance factor

This section deals with numerical studies on downward continuation of satellite gradiometry data to gravity anomalies at sea level, which was a part of the Eshagh and Sjöberg (2011) study. They generated a regular grid of gravity anomalies using the global gravity model EGM96 (Lemoine et al., 1998) to degree and order 360 at the surface of the spherical Earth in a regional which is called Fennoscandia, limited to between latitudes 55°N and 70°N and longitudes 5°E and 30°E. The maximum, minimum, mean and standard deviations of the gravity anomalies were 57.5, −51.8, 0.1 and

Table 4.5 Statistics of errors of recovered Δg from on-orbit satellite gradiometry data (mGal) (Eshagh, 2011b).

	Recovery from data in the LNOF						Recovery from data in the TOF				
	Min.	**Mean**	**Max.**	**Std**	**rms**		**Min.**	**Mean**	**Max.**	**Std**	**rms**
T_{xz}	−46.1	−13.4	11.5	8.6	15.9	T_{uv}	−171.4	0.4	179.4	49.6	49.6
T_{yz}	−116.4	5.5	118.5	48.2	48.5	T_{uw}	−110.0	−4.8	95.7	28.8	29.1
T_{xx}	−31.3	−5.4	28.4	9.1	10. 6	T_{vv}	−40.3	0.2	38.3	9.7	9.7
T_{yy}	−27.9	−3.0	22.0	8.0	8.6	T_{ww}	−43.0	0.2	38.7	9.7	9.7
T_{xy}	−90.6	0.7	78.7	35.4	35.4	T_{vw}	−178.0	2.3	178.5	65.2	65.2

LNOF, local north-oriented frame; *TOF*, track-oriented frame.

Table 4.6 Statistics of errors of recovered gravity anomalies from on-orbit $T_{zz} = T_{uu}$ and $T_{vv} + T_{ww}$ (mGal) (Eshagh, 2011b).

	Min.	Mean	Max.	Std	rms
$T_{zz} = T_{uu}$	−19.2	−0.5	22.5	6.3	6.3
$T_{vv} + T_{vv}$	−27.5	−0.9	27.6	6.9	7.0

17.8 mGal, respectively. Later they used these anomalies and regenerated six second-order derivatives, gravity gradients, of the gravity field at an altitude of 250 km over the study area, similar to Eq. (3.80); see Eshagh and Sjöberg (2011). A Gaussian noise of 10 mE was added to the generated derivatives. Their goal was to invert these gradients by discretising the integral and invert them numerically. The system of equations created by a discretised integral is ill-conditioned, and regularisation should be applied to control the noise of the data and obtain a smoother solution. The stability of the system has a direct relation to the resolution of the parameters, in this example, gravity anomalies. When the resolution is higher and a denser grid of anomalies is desired, the system will be more unstable, because the values of elements of the coefficients matrix will be closer to one another. This causes the determinant of the coefficients matrix of the normal equations to become close to zero. In the Eshagh and Sjöberg (2011) study, three cases were considered based on the resolution of recovering anomalies:

Case 1. recovery of $1° \times 1°$ gravity anomaly from a grid of $0.5° \times 0.5°$ gradients,

Case 2. recovery of $0.5° \times 0.5°$ gravity anomaly from a grid of $0.25° \times 0.25°$ gradients,

Case 3. recovery of $1° \times 1°$ gravity anomaly from a grid of $0.25° \times 0.25°$ gradients.

The condition numbers of the coefficient matrices related to the discretised integral for inverting T_{xx}, T_{yy}, T_{zz}, T_{xy}, T_{xz} and T_{yz} are 2.85×10^7, 1.46×10^6, 5.57×10^6, 1.49×10^3, 5.83×10^7 and 2.56×10^3, respectively, for Case 1. In Case 2, they are 1.10×10^8, 7.10×10^6, 9.68×10^{13}, 1.19×10^6, 2.51×10^{14} and 3.63×10^6, and finally, 5.49×10^6, 2.24×10^6, 6.12×10^6, 1.88×10^3, 7.61×10^7 and 3.09×10^3 in Case 3. The condition number is a measure of instability of the system of equations; when it is large the solution will be more sensitive to the noise and the system is more unstable. For example, the condition numbers of the systems of equations related to the inversion of T_{xy} and T_{yz}

data are smaller than those of the others. The condition numbers related to the inversions of T_{zz} and T_{xz} are large in Case 2 when the resolution of the gradients is higher than the ones in Cases 1 and 3. In fact, however, in Case 3 the resolution of the gradients is the same as in Case 2.

These systems, six in each case, were solved using the Tikhonov regularisation in combination with the generalised cross-validation method for estimating its regularisation parameter. Table 4.7 shows the statistics of the differences between the generated anomalies from EGM96 and those recovered from the gradients. It shows two cases where the bias of regularisation is not removed from the recovered anomalies and the case where the bias is removed. This bias is computed by Eq. (4.66) and removed from the recovered anomalies. As we can observe, the smallest root mean squared error (RMSE) of the differences is related to the anomalies recovered from T_{zz} and the largest from T_{xy}. The rhs of the table shows that the bias-correction step can improve the results significantly.

In addition, Eshagh and Sjöberg (2011) estimate the a posteriori variance factor of each system by using Eq. (4.75) and assuming the estimated anomalies are considered as true values in this equation. The estimated a posteriori variance factors are presented in Table 4.8 in addition to their bias. In Case 3, they had problems using this equation for estimating the bias due to very large matrix operations; therefore, they applied the bias-corrected estimator (Eq. 4.81) instead. The estimated biases in Cases 1 and 2 are small and can be simply subtracted from the estimated variances to make them bias corrected.

4.4 Sequential Tikhonov regularisation

The Tikhonov regularisation can also be applied sequentially for solving large systems of equations. Different solutions to this problem have been proposed, such as the Arnoldi–Tikhonov regularisation (Lewis and Reichel, 2009) and Lanczos bidiagonalisation of Tikhonov (Björck, 1988; Calvetti et al., 2000; Calvetti and Reichel, 2003; Golub and von Matt, 1997; Kilmer and O'Leary, 2001; O'Leary and Simmons, 1981). These variants of the Tikhonov regularisation are based on some matrix factorisations like the Lanczos bidiagonalisation (Björck, 1988) and Arnoldi decomposition (Saad, 1996), which are applied to the structure of the system. The idea of applying the Tikhonov method to the normal equations is not relevant due to:

Table 4.7 Error of recovered gravity anomalies from satellite gradiometry data contaminated with 10 mE noise.

		Biased						Bias-corrected					
	T_{xx}	T_{yy}	T_{zz}	T_{xy}	T_{xz}	T_{yz}		T_{xx}	T_{yy}	T_{zz}	T_{xy}	T_{xz}	T_{yz}
Case 1	Min.	−26.7	−23.9	−22.9	−28.6	−27.4	−28.5	−13.7	−7.8	−9.1	−14.2	−14.3	−11.1
	Mean	−0.1	−0.1	0.1	−0.6	0.0	0.0	−0.1	−0.1	0.1	−0.2	0.0	0.0
	Max.	27.7	28.4	24.3	31.3	22.4	35.9	13.8	7.9	8.0	14.3	16.4	12.5
	rms	**7.9**	**7.3**	**6.1**	**10.0**	**7.4**	**7.9**	**4.3**	**2.9**	**3.0**	**5.1**	**4.7**	**3.7**
Case 2	Min.	−26.5	−23.9	−19.7	−31.3	−26.3	−23.6	−12.5	−8.3	−10.4	−13.8	−15.1	−10.8
	Mean	0.0	0.0	0.0	−0.2	−0.2	0.0	0.0	0.0	0.0	0.1	−0.1	0.0
	Max.	29.1	23.6	26.8	33.3	34.6	27.8	22.1	10.7	7.9	17.1	13.2	10.5
	rms	**7.3**	**6.7**	**5.7**	**8.7**	**6.9**	**7.2**	**4.6**	**3.3**	**2.7**	**5.3**	**4.3**	**3.8**
Case 3	Min.	−23.6	−23.0	−15.7	−22.3	−18.8	−16.3	−14.3	−10.0	−11.7	−17.6	−17.0	−15.7
	Mean	0.1	0.0	0.0	−0.2	−0.2	0.0	0.0	0.0	0.0	0.0	−0.1	0.0
	Max.	23.19	19.8	23.1	29.3	21.9	22.3	24.9	12.2	8.8	20.6	15.8	10.7
	rms	**6.5**	**6.5**	**5.2**	**8.3**	**6.4**	**6.7**	**4.7**	**3.3**	**2.7**	**5.4**	**4.2**	**3.9**

rms, root mean square. Unit: mGal.

Adapted from Eshagh, M., Sjöberg, L.E., 2011. Determination of gravity anomaly at sea level from inversion of satellite gravity gradiometric data. J. Geodyn. 51, 366–377.

Table 4.8 Variance of unit weight and bias due to regularisation in recovering gravity anomalies.

	Case 1		Case 2		Case 3
	$\hat{\sigma}_0^2$	Bias	$\hat{\sigma}_0^2$	Bias	$\hat{\sigma}_0^2$
T_{xx}	1.64	0.26	1.12	0.05	1.94
T_{yy}	1.67	0.27	1.12	0.05	1.95
T_{zz}	1.55	0.24	1.10	0.04	1.91
T_{xy}	1.55	0.25	1.10	0.04	1.89
T_{xz}	1.57	0.24	1.11	0.04	1.92
T_{yz}	1.48	0.23	1.08	0.04	1.85

(a) increase in the system condition number by a power of 2 making the normal system even more unstable;

(b) loss of some physical properties of **A** and **L** due to being multiplied by $\mathbf{A}^{\mathrm{T}}$ (cf. Maitre and Levy, 1983).

Solving the ill-posed problems using Tikhonov regularisation is possible in another simple equation when the number of unknown parameters is not very large. In this case, the method is applied sequentially, so that working with a large number of observations is possible. Eshagh (2011a) named this method sequential Tikhonov regularisation (STR). The principle is the same as with the sequential least-squares adjustment (see e.g. Cooper, 1987) and the system of equations is partitioned into small subsystems. The first subsystem is solved and the improvements due to the involvement of the other subsystems are computed and implied on the former solution sequentially. The advantage of this method is its ability to work with large systems without estimating the regularisation parameter from the normal equation. Let us assume that our Gauss–Markov model is divided into two subsystems:

$$\begin{bmatrix}\mathbf{A}_1\\ \mathbf{A}_2\end{bmatrix}\mathbf{x}=\begin{bmatrix}\mathbf{L}_1\\ \mathbf{L}_2\end{bmatrix}-\begin{bmatrix}\boldsymbol{\varepsilon}_1\\ \boldsymbol{\varepsilon}_2\end{bmatrix},\ \text{where } \mathrm{E}\{\boldsymbol{\varepsilon}_1\}=\mathrm{E}\{\boldsymbol{\varepsilon}_2\}=0,$$
$$\mathrm{E}\{\boldsymbol{\varepsilon}_1\boldsymbol{\varepsilon}_1^{\mathrm{T}}\}=\sigma_1^2\mathbf{Q}_1,\ \mathrm{E}\{\boldsymbol{\varepsilon}_2\boldsymbol{\varepsilon}_2^{\mathrm{T}}\}=\sigma_2^2\mathbf{Q}_2, \tag{4.83}$$

where $\mathbf{A}_1$ and $\mathbf{A}_2$ are the coefficient matrices of the first and the second subsystems. The vector of observation **L** is divided into $\mathbf{L}_1$ and $\mathbf{L}_2$ with corresponding errors $\boldsymbol{\varepsilon}_1$ and $\boldsymbol{\varepsilon}_2$.

The solution of the first system by the Tikhonov regularisation is:

$$\mathbf{x}_{\mathrm{reg}}^{1}=\mathbf{N}_1^{-1}\mathbf{A}_1^{\mathrm{T}}\mathbf{Q}_1^{-1}\mathbf{L}_1 \text{ where } \mathbf{N}_1=\mathbf{A}_1^{\mathrm{T}}\mathbf{Q}_1^{-1}\mathbf{A}_1+\alpha^2\mathbf{I}. \tag{4.84}$$

Now, assume that the regularisation parameter was already estimated from the first system. Therefore, the solution of both systems together will be:

$$\mathbf{x}_{\text{reg}}^2 = \mathbf{N}_{12}^{-1}\left(\mathbf{A}_1^{\mathrm{T}}\mathbf{Q}_1^{-1}\mathbf{L}_1 + \mathbf{A}_2^{\mathrm{T}}\mathbf{Q}_2^{-1}\mathbf{L}_2\right) \text{ with}$$
$$\mathbf{N}_{12} = \mathbf{A}_1^{\mathrm{T}}\mathbf{Q}_1^{-1}\mathbf{A}_1 + \mathbf{A}_2^{\mathrm{T}}\mathbf{Q}_2^{-1}\mathbf{A}_2 + \alpha^2\mathbf{I}. \tag{4.85}$$

This means that the same regularisation parameter that was used in the first systems (Eq. 4.84) is used to estimate $\mathbf{x}_{\text{reg}}^2$. In this case, the system of equations will not be ill-conditioned but the regularisation parameter is not optimal. Now we have to find the relation between the solutions $\mathbf{x}_{\text{reg}}^1$ and $\mathbf{x}_{\text{reg}}^2$. It is not difficult to show that:

$$\mathbf{N}_{12}^{-1}\mathbf{N}_1 = \mathbf{I} - \mathbf{N}_{12}^{-1}\mathbf{A}_2^{\mathrm{T}}\mathbf{Q}_2^{-1}\mathbf{A}_2. \tag{4.86}$$

If Eq. (4.85) is expanded and further simplifications considering Eq. (4.86) are performed, this leads to:

$$\mathbf{x}_{\text{reg}}^2 = \mathbf{x}_{\text{reg}}^1 + \mathbf{N}_{12}^{-1}\mathbf{A}_2^{\mathrm{T}}\mathbf{Q}_2^{-1}\left(\mathbf{L}_2 - \mathbf{A}_2\mathbf{x}_{\text{reg}}^1\right). \tag{4.87}$$

The solution $\mathbf{x}_{\text{reg}}^1$ contains the regularisation bias, which is estimable by Eq. (4.66) and approximating the true value by $\mathbf{x}_{\text{reg}}^1$. Therefore, the estimator of $\mathbf{x}_{\text{reg}}^2$ will be:

$$\mathbf{x}_{\text{reg}}^2 = \mathbf{x}_{\text{reg}}^1 - \text{Bias}\left(\mathbf{x}_{\text{reg}}^1\right) + \mathbf{N}_{12}^{-1}\mathbf{A}_2^{\mathrm{T}}\mathbf{Q}_2^{-1}\left(\mathbf{L}_2 - \mathbf{A}_2^{\mathrm{T}}\left(\mathbf{x}_{\text{reg}}^1 - \text{Bias}\left(\mathbf{x}_{\text{reg}}^1\right)\right)\right) \tag{4.88}$$

and

$$\text{Bias}\left(\mathbf{x}_{\text{reg}}^2\right) = -\alpha^2\mathbf{N}_{12}^{-1}\mathbf{x}. \tag{4.89}$$

Generalisation of this idea is:

$$\begin{aligned}\mathbf{x}_{\text{reg}}^{(k+1)} &= \mathbf{x}_{\text{reg}}^{(k)} - \text{Bias}\left(\mathbf{x}_{\text{reg}}^{(k)}\right)\\ &+ \widetilde{\mathbf{N}}_{k+1}^{-1}\mathbf{A}_{k+1}^{\mathrm{T}}\mathbf{Q}_{k+1}^{-1}\left(\mathbf{L}_{k+1} - \mathbf{A}_{k+1}^{\mathrm{T}}\left(\mathbf{x}_{\text{reg}}^{(k)} - \text{Bias}\left(\mathbf{x}_{\text{reg}}^{(k)}\right)\right)\right)\end{aligned} \tag{4.90}$$

and

$$\text{Bias}\left(\mathbf{x}_{\text{reg}}^{k+1}\right) = -\alpha^2\left(\sum_{i=1}^{k+1}\mathbf{A}_i^{\mathrm{T}}\mathbf{Q}_i^{-1}\mathbf{A}_i + \alpha^2\mathbf{I}\right)^{-1}\mathbf{x}, \tag{4.91}$$

where

$$\widetilde{\mathbf{N}}_{k+1} = \sum_{i=1}^{k+1} \mathbf{A}_i^{\mathrm{T}} \mathbf{Q}_i^{-1} \mathbf{A}_i + \alpha^2 \mathbf{I}. \tag{4.92}$$

This means that the regularisation bias should be estimated and removed from the estimate at each sequence.

One drawback of this method is the successive approximation errors due to each bias correction. This can be problematic when the number of subsystems is large. The MSE and the a posterior variance factor can be estimated in the same way as was presented for the Tikhonov regularisation.

4.4.1 Example: application of sequential Tikhonov regularisation

Now we present the Eshagh (2011c) study's application of the STR and its comparison with other regularisation methods. In that study, he used the global gravity model EGM08 (Pavlis et al., 2008) to generate gravity anomalies at sea level and the second-order radial derivative of the disturbing potential on a 3-month simulated orbit of GOCE over Fennoscandia. The goal is to see whether the sequential Tikhonov method can recover the gravity anomaly with acceptable error. An area larger by 5 degrees, the study area, is selected for inverting the gradients to reduce the effect of the STE of the integral formula. Fig. 4.20A and B shows respectively the map of the gradients and the gravity anomalies.

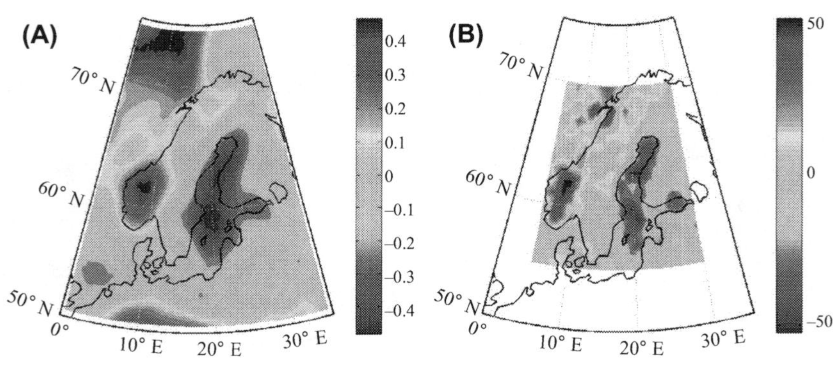

Figure 4.20 (A) Gradients, second-order radial derivative of disturbing potential. Unit 1 E. (B) Gravity anomalies. Unit: 1 mGal. *(From Eshagh M., 2011c. Sequential Tikhonov regularization: an alternative method for inversion of satellite gravity gradiometry data. ZfV. 136, 113–121.)*

Table 4.9 Error of recovered gravity anomalies from the on-orbit gradient data.

	1.0° × 1.0°						0.5° × 0.5°					
Method	**Min.**	**Mean**	**Max.**	**Std**	**rms**	**T (s)**	**Min.**	**Mean**	**Max.**	**Std**	**rms**	**T (s)**
ν	−35.9	−1.2	33.0	7.9	8.0	2	−39.8	−1.1	35.2	8.4	8.5	2
ART	−21.5	−0.2	22.7	7.2	7.2	30	−22.2	−0.1	24.4	1.0	7.0	53
RRGMRES	−18.7	−0.6	23.0	6.6	6.6	1	−22.2	−0.6	24.1	7.1	7.1	7
CG	−25.3	−0.5	19.6	6.6	6.6	4	−25.4	−0.4	23.5	6.5	6.5	13
TSVD	−23.4	−0.6	28.9	7.9	7.9	10	−30.1	−0.71	31.8	8.2	8.2	240
TR	−18.2	−0.6	20.6	6.4	6.5	10	—	—	—	—	—	—

ART, algebraic reconstruction technique; *CG*, conjugate gradient; *RRGMRES*, range-restricted generalised minimum residual; *TR*, Tikhonov regularisation; *TSVD*, truncated singular value decomposition. Unit: mGal.

From Eshagh, M., 2011a. Sequential Tikhonov regularization: an alternative method for inversion of satellite gravity gradiometry data. ZfV 136, 113−121.

In the Eshagh (2011a) study, two resolutions of 1° × 1° and 0.5° × 0.5° were considered, and a coloured noise was generated for the gradients by passing a white noise with a standard deviation of 0.02 E (Eötvös) through an autoregressive moving average filter of degree 2 and order 1; i.e. ARMA (2,1) (Kless et al., 2003). The maximum, mean, minimum and standard deviation of the gradient are 0.53, 0.04, −0.05 and 0.2 in units of E, respectively, and the corresponding statistics for the gravity anomalies are 69.0, 3.4, −55.1 and 17.4 in units of mGal, respectively.

The MATLAB package Regularization Tools (Hansen, 2007) was used by Eshagh (2011a) for regularising the systems of equations. Table 4.9 shows the summary of his study comparing the regularisation method to recover the anomalies based on the two aforementioned resolutions.

It seems that ART works better than ν in this example, but it is the slowest iterative method according to this table. In the ν method, $\nu = 0.75$ was selected as the best value for ν. Amongst the Krylov subspaces-based iterative methods the CG is better, but slightly slower, than RRGMRES. Tikhonov regularisation could not be as good as the CG is in this example. Table 4.10 presents the results of using the STR. The resolution 1° × 1° was also considered for comparisons. The 3-month orbit of GOCE is divided into three parts and the recovery is done month by month. In Table 4.10, '1' means that only the first subsystem is inverted. '1 + 2' means that the second subsystem is sequentially added to the result of the first one, and similarly for '1 + 2 + 3'.

Now the 3-month simulated orbit of GOCE is divided into two parts, as the number of unknowns in this case is four times larger than that in recovering the 1° × 1° anomaly; therefore, more observations are required. So, the recovery is done in two steps of 1.5 month. The recovery is done with an error of 6.6 mGal when the first set of observations is used. Including the second set of equations sequentially reduces the error of recovered gravity anomalies down to 6.2 mGal, which is even smaller than that of the CG method.

Table 4.10 Statistics of errors of recovered gravity anomalies using sequential Tikhonov regularisation method.

Resolution	Subsystem	Min.	Mean	Max.	Std	rms
1.0° × 1.0°	1	−26.2	−0.6	22.0	6.9	7.0
	1 + 2	−17.0	−0.6	20.3	6.4	6.4
	1 + 2 + 3	−20.2	−0.6	19.6	6.3	6.4
0.5° × 0.5°	1	−27.3	−0.7	25.8	6.5	6.6
	1 + 2	−25.6	−0.6	23.7	6.2	6.2

Unit: mGal.

4.5 Variance component estimation in ill-conditioned systems

Estimation of one a posteriori variance factor when there are different types of data in a system of equations is not very correct, as each type of observation has its own nature and noise property. In this case, it is better to estimate one variance factor for each group of measurements of the same type. This process is a generalised form of estimation of variance factor and is known as a variance component estimation. Numerous methods have been presented for this purpose, but in this section we use the best quadratic unbiased estimator of the variance components and present how it should be used when a system of equation is ill-conditioned. The method will be presented for the cases where the system is solved by TSVD and Tikhonov regularisation; see e.g. Koch and Kusche (2002), Xu et al. (2006), Xu (2009), Eshagh and Sjöberg (2011), Eshagh (2010, 2011c).

4.5.1 Best quadratic unbiased estimator of variance component in ordinary systems

Consider the Gauss–Markov model presented in Eq. (4.11), but with the following property:

$$\mathrm{E}(\boldsymbol{\varepsilon}\boldsymbol{\varepsilon}^{\mathrm{T}}) = \sigma_0^2\mathbf{Q} = \sum_{i=1}^{q}\sigma_i^2\mathbf{Q}_i, \tag{4.93}$$

where $\mathbf{Q}_i$ is the co-factor matrix of each group of observations and σ_i^2 is the variance components of that group. q is the number of variance components, which is equivalent to the number of the groups.

The best quadratic unbiased estimation of these variance components is well-known and accessible in much of the geodetic literature and many text books in statistics, and the variance components are derived by solving the following system of equations (Koch, 1999; Xu et al., 2006):

$$\mathbf{S}\boldsymbol{\sigma} = \mathbf{u}, \tag{4.94}$$

where $\boldsymbol{\sigma}$ is the vector of variance components, and $\mathbf{S}$ and $\mathbf{u}$ are a matrix and a vector with the following elements:

$$s_{ij} = \mathrm{trace}[(\mathbf{I}_n - \mathbf{A}\mathbf{A}^-)\mathbf{Q}_i(\mathbf{I}_n - \mathbf{A}\mathbf{A}^-)\mathbf{Q}_j] \tag{4.95}$$

and

$$u_i = \widehat{\boldsymbol{\varepsilon}}^{\mathrm{T}}\mathbf{Q}^{-1}\mathbf{Q}_i\mathbf{Q}^{-1}\widehat{\boldsymbol{\varepsilon}}, \tag{4.96}$$

and $\widehat{\boldsymbol{\varepsilon}}$ is the residual vector; see Eq. (4.16).

The solution of the system (Eq. 4.94) gives the best quadratic unbiased estimation of the variance components. Once these components are estimated the co-factors matrix $\mathbf{Q}$ is updated by multiplying the components of the corresponding sub-co-factor matrices. After that, the process is repeated until the variance components ratios between the two successive iteration converge to 1.

As we can see, $\widehat{\boldsymbol{\varepsilon}}$, resulting from a least-squares solution for $\mathbf{x}$, is used in Eq. (4.96). In this case, the regular inverse can be applied and there will be no bias in the solution. However, when any regularisation is used for solving the Gauss–Markov model (Eq. 4.11), the solution will be biased. This means that no regular matrix inversion can be applied and the estimated residuals are biased, which depends on the type of regularisation method. In the following, we will investigate the case in which an ill-conditioned system is solved by TSVD.

4.5.2 Best quadratic unbiased estimator of variance component in a system solved by truncated singular value decomposition

As discussed in Section 4.3.3.1, the Gauss–Markov model (Eq. 4.11) is converted to the corresponding one in Eq. (4.40), in which:

$$\mathrm{E}(\boldsymbol{\varepsilon}'\boldsymbol{\varepsilon}'^{\mathrm{T}}) = \sigma_0^2\mathbf{I} = \sum_{i=1}^{q}\sigma_i^2\mathbf{I}_i. \tag{4.97}$$

Considering Eq. (4.97) and the residual vector, presented in Eq. (4.49), and inserting the result into the quadratic form of Eq. (4.96), we obtain:

$$\begin{aligned}\mathrm{E}\left\{\boldsymbol{\varepsilon}'^{\mathrm{T}}_{\mathrm{reg}}\mathbf{I}_i\boldsymbol{\varepsilon}'_{\mathrm{reg}}\right\} &= \mathbf{x}^{\mathrm{T}}\mathbf{V}\left(\mathbf{I}_m^{k+1,m}\right)\mathbf{I}_i\mathbf{V}^{\mathrm{T}}\mathbf{x} \\ &\quad + \mathrm{E}\left\{\boldsymbol{\varepsilon}'^{\mathrm{T}}\left(\mathbf{I}_n - \mathbf{U}\boldsymbol{\Lambda}\mathbf{I}_m^{1k}\boldsymbol{\Sigma}_k^{+}\boldsymbol{\Lambda}^{\mathrm{T}}\mathbf{U}^{\mathrm{T}}\right)^{\mathrm{T}}\mathbf{I}_i\left(\mathbf{I}_n - \mathbf{U}\boldsymbol{\Lambda}\mathbf{I}_m^{1k}\boldsymbol{\Sigma}_k^{+}\boldsymbol{\Lambda}^{\mathrm{T}}\mathbf{U}^{\mathrm{T}}\right)\boldsymbol{\varepsilon}'\right\} \\ &= \mathbf{x}^{\mathrm{T}}\mathbf{V}\left(\mathbf{I}_m^{k+1,m}\right)\mathbf{I}_i\mathbf{V}^{\mathrm{T}}\mathbf{x} \\ &\quad + \sum_{j=1}^{q}\sigma_j^2\,\mathrm{trace}\left[\left(\mathbf{I}_n - \mathbf{U}\boldsymbol{\Lambda}\mathbf{I}_m^{1k}\boldsymbol{\Sigma}_k^{+}\boldsymbol{\Lambda}^{\mathrm{T}}\mathbf{U}^{\mathrm{T}}\right)^{\mathrm{T}}\mathbf{I}_i\left(\mathbf{I}_n - \mathbf{U}\boldsymbol{\Lambda}\mathbf{I}_m^{1k}\boldsymbol{\Sigma}_k^{+}\boldsymbol{\Lambda}^{\mathrm{T}}\mathbf{U}^{\mathrm{T}}\right)\mathbf{I}_j\right].\end{aligned} \tag{4.98}$$

If we consider

$$u_i = \boldsymbol{\varepsilon}'^{\mathrm{T}}_{\mathrm{reg}}\mathbf{I}_i\boldsymbol{\varepsilon}'_{\mathrm{reg}} \tag{4.99a}$$

$$du_i = \mathbf{x}^{\mathrm{T}}\mathbf{V}\left(\mathbf{I}_m^{k+1,m}\right)\mathbf{I}_i\mathbf{V}^{\mathrm{T}}\mathbf{x} \tag{4.99b}$$

$$s_{ij} = \mathrm{trace}\left[\left(\mathbf{I}_n - \mathbf{U}\boldsymbol{\Lambda}\mathbf{I}_m^{1k}\boldsymbol{\Sigma}_k^{+}\boldsymbol{\Lambda}^{\mathrm{T}}\mathbf{U}^{\mathrm{T}}\right)^{\mathrm{T}}\mathbf{I}_i\left(\mathbf{I}_n - \mathbf{U}\boldsymbol{\Lambda}\mathbf{I}_m^{1k}\boldsymbol{\Sigma}_k^{+}\boldsymbol{\Lambda}^{\mathrm{T}}\mathbf{U}^{\mathrm{T}}\right)\mathbf{I}_j\right] \tag{4.99c}$$

We can write

$$\bar{u}_i - d\bar{u}_i = \sum_{j=1}^{q} \sigma_j^2 \bar{s}_{ij}, \tag{4.100}$$

with the following matrix form:

$$\mathbf{u}' - d\mathbf{u}' = \mathbf{S}'\boldsymbol{\sigma}, \tag{4.101}$$

and the solution:

$$\boldsymbol{\sigma} = \mathbf{S}'^{-1}\mathbf{u}' - \mathbf{S}'^{-1}\mathbf{du}'. \tag{4.102}$$

The second term on the rhs of Eq. (4.102) is nothing other than the bias of the variance components due to truncation of singular values. However, it may happen that this term comes out larger than the first one and in this case the variance components will be negative. To avoid this problem, the bias-corrected form of the variance component should be developed as well. To do so, we consider the estimated parameter $\mathbf{x}_{\text{reg}}$ from Eq. (4.45) and insert it into the equation of the estimated residuals, Eq. (4.54). Taking the expectation of the quadratic form of this equation and further implications leads to:

$$\bar{\boldsymbol{\varepsilon}}'^{\mathrm{T}}_{\text{reg}} \mathbf{I}_i \bar{\boldsymbol{\varepsilon}}'_{\text{reg}} = \sum_{j=1}^{q} \sigma_j^2 \operatorname{trace}\left(\left(\mathbf{I}_n - \mathbf{U}\boldsymbol{\Lambda}\mathbf{I}_u^{1k}\boldsymbol{\Sigma}_k^{+}\boldsymbol{\Lambda}^{\mathrm{T}}\mathbf{U}^{\mathrm{T}}\right)\left(\mathbf{I}_n - \mathbf{U}\boldsymbol{\Lambda}\mathbf{I}_u^{1k}\boldsymbol{\Sigma}_k^{+}\boldsymbol{\Lambda}^{\mathrm{T}}\mathbf{U}^{\mathrm{T}}\right)^{\mathrm{T}}\mathbf{I}_j\right), \tag{4.103}$$

or in matrix form:

$$\boldsymbol{\sigma} = \bar{\mathbf{S}}^{-1}\bar{\mathbf{u}}, \tag{4.104}$$

where elements of $\bar{\mathbf{u}}$ and $\bar{\mathbf{S}}$ are:

$$\bar{u}'_i = \bar{\boldsymbol{\varepsilon}}'^{\mathrm{T}}_{\text{reg}} \mathbf{I}_i \bar{\boldsymbol{\varepsilon}}'_{\text{reg}}, \tag{4.105}$$

$$\bar{s}'_{ij} = \operatorname{trace}\left(\left(\mathbf{I}_n - \mathbf{U}\boldsymbol{\Lambda}\mathbf{I}_u^{1k}\boldsymbol{\Sigma}_k^{+}\boldsymbol{\Lambda}^{\mathrm{T}}\mathbf{U}^{\mathrm{T}}\right)\mathbf{I}_i\left(\mathbf{I}_n - \mathbf{U}\boldsymbol{\Lambda}\mathbf{I}_u^{1k}\boldsymbol{\Sigma}_k^{+}\boldsymbol{\Lambda}^{\mathrm{T}}\mathbf{U}^{\mathrm{T}}\right)^{\mathrm{T}}\mathbf{I}_j\right). \tag{4.106}$$

4.5.3 Best quadratic unbiased estimator of the variance components in systems solved by Tikhonov regularisation

Now, consider that the Gauss–Markov model (Eq. 4.11) is ill-conditioned with the property Eq. (4.93) for its errors. The residual vector after solving this system by the Tikhonov approach was presented in Eq. (4.71). To compute the variance components we use Eq. (4.96), but we replace $\widehat{\boldsymbol{\varepsilon}}$ with $\boldsymbol{\varepsilon}_{\text{reg}}$, as the system has not been solved by the least-squares method:

$$u_{\text{reg},i} = \boldsymbol{\varepsilon}_{\text{reg}}^{\text{T}}\mathbf{Q}^{-1}\mathbf{Q}_i\mathbf{Q}^{-1}\boldsymbol{\varepsilon}_{\text{reg}} = \alpha^4\mathbf{x}^{\text{T}}\mathbf{N}^{-1}\mathbf{A}^{\text{T}}\mathbf{Q}^{-1}\mathbf{Q}_i\mathbf{Q}^{-1}\mathbf{A}\mathbf{N}^{-1}\mathbf{x} + \text{trace}\left((\mathbf{I}_n - \mathbf{A}\mathbf{N}^{-1}\mathbf{A}^{\text{T}}\mathbf{Q}^{-1})^{\text{T}}\mathbf{Q}^{-1}\mathbf{Q}_i\mathbf{Q}^{-1}(\mathbf{I}_n - \mathbf{A}\mathbf{N}^{-1}\mathbf{A}^{\text{T}}\mathbf{Q}^{-1})\text{E}\{\boldsymbol{\varepsilon}\boldsymbol{\varepsilon}^{\text{T}}\}\right). \tag{4.107}$$

Inserting Eq. (4.93) into Eq. (4.107) yields:

$$u_{\text{reg},i} = \alpha^4\mathbf{x}^{\text{T}}\mathbf{N}^{-1}\mathbf{A}^{\text{T}}\mathbf{Q}^{-1}\mathbf{Q}_i\mathbf{Q}^{-1}\mathbf{A}\mathbf{N}^{-1}\mathbf{x} + \sum_{j=1}^{q}\sigma_j^2\text{trace}\left((\mathbf{I}_n - \mathbf{A}\mathbf{N}^{-1}\mathbf{A}^{\text{T}}\mathbf{Q}^{-1})^{\text{T}}\mathbf{Q}^{-1}\mathbf{Q}_i\mathbf{Q}^{-1}(\mathbf{I}_n - \mathbf{A}\mathbf{N}^{-1}\mathbf{A}^{\text{T}}\mathbf{Q}^{-1})\mathbf{Q}_j\right). \tag{4.108}$$

The first term appears to be due to the regularisation bias. If it is taken into the lhs of Eq. (4.108) and the result is written in a matrix form we have:

$$\mathbf{u}_{\text{reg}} - \mathbf{du}_{\text{reg}} = \mathbf{S}_{\text{reg}}\boldsymbol{\sigma}, \tag{4.109}$$

where the elements of the vectors and matrices will be:

$$u_{\text{reg},i} = \boldsymbol{\varepsilon}_{\text{reg}}^{\text{T}}\mathbf{Q}^{-1}\mathbf{Q}_i\mathbf{Q}^{-1}\boldsymbol{\varepsilon}_{\text{reg}}, \tag{4.110}$$

$$du_{\text{reg},i} = \alpha^4\mathbf{x}^{\text{T}}\mathbf{N}^{-1}\mathbf{A}^{\text{T}}\mathbf{Q}^{-1}\mathbf{Q}_i\mathbf{Q}^{-1}\mathbf{A}\mathbf{N}^{-1}\mathbf{x}, \tag{4.111}$$

$$s_{\text{reg},ij} = \text{trace}\left((\mathbf{I}_n - \mathbf{A}\mathbf{N}^{-1}\mathbf{A}^{\text{T}}\mathbf{Q}^{-1})^{\text{T}}\mathbf{Q}^{-1}\mathbf{Q}_i\mathbf{Q}^{-1}(\mathbf{I}_n - \mathbf{A}\mathbf{N}^{-1}\mathbf{A}^{\text{T}}\mathbf{Q}^{-1})\mathbf{Q}_j\right). \tag{4.112}$$

The solution of the system (Eq. 4.109) is:

$$\boldsymbol{\sigma} = \mathbf{S}_{\text{reg}}^{-1}\mathbf{u}_{\text{reg}} - \mathbf{S}_{\text{reg}}^{-1}\mathbf{du}_{\text{reg}}. \tag{4.113}$$

The second term on the rhs of Eq. (4.113) is the regularisation bias on the variance components.

To obtain the bias-corrected estimator, we use the bias-corrected residual $\bar{\boldsymbol{\varepsilon}}_{\mathrm{reg}}$ (Eq. 4.76) and use the quadratic form of Eq. (4.96), but with $\bar{\boldsymbol{\varepsilon}}_{\mathrm{reg}}$ instead of $\widehat{\boldsymbol{\varepsilon}}$:

$$\mathrm{E}\left\{\bar{\boldsymbol{\varepsilon}}_{\mathrm{reg}}^{\mathrm{T}}\mathbf{Q}^{-1}\mathbf{Q}_i\mathbf{Q}^{-1}\bar{\boldsymbol{\varepsilon}}_{\mathrm{reg}}\right\}=\mathrm{trace}\left[(\mathbf{I}_n-\mathbf{A}\mathbf{N}^{-1}\mathbf{A}^{\mathrm{T}}\mathbf{Q}^{-1}+\alpha^2\mathbf{A}\mathbf{N}^{-2}\mathbf{A}^{\mathrm{T}}\mathbf{Q}^{-1})^{\mathrm{T}}\right.$$
$$\left.\mathbf{Q}^{-1}\mathbf{Q}_i\mathbf{Q}^{-1}\times(\mathbf{I}_n-\mathbf{A}\mathbf{N}^{-1}\mathbf{A}^{\mathrm{T}}\mathbf{Q}^{-1}+\alpha^2\mathbf{A}\mathbf{N}^{-2}\mathbf{A}^{\mathrm{T}}\mathbf{Q}^{-1})\mathrm{E}\{\boldsymbol{\varepsilon}\boldsymbol{\varepsilon}^{\mathrm{T}}\}\right]. \tag{4.114}$$

By using Eq. (4.93) in Eq. (4.114) the following system of equations can be created for solving the variance components:

$$\bar{\mathbf{S}}_{\mathrm{reg}}\boldsymbol{\sigma}=\bar{\mathbf{u}}_{\mathrm{reg}}, \tag{4.115}$$

where

$$\bar{u}_{\mathrm{reg},i}=\bar{\boldsymbol{\varepsilon}}_{\mathrm{reg}}^{\mathrm{T}}\mathbf{Q}^{-1}\mathbf{Q}_i\mathbf{Q}^{-1}\bar{\boldsymbol{\varepsilon}}_{\mathrm{reg}}, \tag{4.116}$$

$$\bar{s}_{\mathrm{reg},ij}=\mathrm{trace}\left[(\mathbf{I}_n-\mathbf{A}\mathbf{N}^{-1}\mathbf{A}^{\mathrm{T}}\mathbf{Q}^{-1}+\alpha^2\mathbf{A}\mathbf{N}^{-2}\mathbf{A}^{\mathrm{T}}\mathbf{Q}^{-1})^{\mathrm{T}}\right.$$
$$\left.\mathbf{Q}^{-1}\mathbf{Q}_i\mathbf{Q}^{-1}\times(\mathbf{I}_n-\mathbf{A}\mathbf{N}^{-1}\mathbf{A}^{\mathrm{T}}\mathbf{Q}^{-1}+\alpha^2\mathbf{A}\mathbf{N}^{-2}\mathbf{A}^{\mathrm{T}}\mathbf{Q}^{-1})\mathbf{Q}_j\right]. \tag{4.117}$$

4.5.4 Example: variance component estimation for inverting satellite gravity gradients

Here, another part of the numerical studies of Eshagh and Sjöberg (2011) is presented as an example of variance component estimation in joint inversion of gravity gradients. They organised six integral equations connecting gravity anomalies with a resolution of 0.5° × 0.5° at sea level over Fennoscandia and used them for generating the gravity gradients at a constant level of 250 km above the spherical Earth with a resolution of 0.25° × 0.25°. The discretised integrals organised a system of equations for solving the gravity anomalies from all six gravity gradients. Since the system was ill-conditioned the Tikhonov regularisation with the generalised cross-validation method was applied to solve it. A Gaussian noise of 10 mE was generated and added to the simulated gravity gradients. The important issue for estimation of the variance components is the choice of the stochastic model, Eq. (4.93). In their example, two stochastic models were used, but here we present one of them:

$$\mathbf{Q} = \sigma_{zz}^2 \mathbf{Q}'_{zz} + \sigma_{xx}^2 \mathbf{Q}'_{xx} + \sigma_{yy}^2 \mathbf{Q}'_{yy} + \sigma_{xy}^2 \mathbf{Q}'_{xy} + \sigma_{xz}^2 \mathbf{Q}'_{xz} + \sigma_{yz}^2 \mathbf{Q}'_{yz},$$

where

$$\begin{aligned}
\mathbf{Q}'_{zz} &= \mathrm{diag}(\mathbf{Q}_{zz} \;\; 0 \;\; 0 \;\; 0 \;\; 0 \;\; 0), \\
\mathbf{Q}'_{xx} &= \mathrm{diag}(0 \;\; \mathbf{Q}_{xx} \;\; 0 \;\; 0 \;\; 0 \;\; 0), \\
\mathbf{Q}'_{yy} &= \mathrm{diag}(0 \;\; 0 \;\; \mathbf{Q}_{yy} \;\; 0 \;\; 0 \;\; 0), \\
\mathbf{Q}'_{xy} &= \mathrm{diag}(0 \;\; 0 \;\; 0 \;\; \mathbf{Q}_{xy} \;\; 0 \;\; 0), \\
\mathbf{Q}'_{xz} &= \mathrm{diag}(0 \;\; 0 \;\; 0 \;\; 0 \;\; \mathbf{Q}_{xz} \;\; 0), \\
\mathbf{Q}'_{yz} &= \mathrm{diag}(0 \;\; 0 \;\; 0 \;\; 0 \;\; 0 \;\; \mathbf{Q}_{yz}),
\end{aligned}$$

where σ_i^2 and $\mathbf{Q}_i$, $i = zz, xx, yy, xy, xz$ and yz, are the variance component and co-factor matrix of each gradient.

Table 4.11 shows the estimated variance components and their biases and, as we can observe, the estimated biases are unrealistically large and cannot be removed from the estimated components as the results will be negative and meaningless. The bias-corrected estimator delivered positive numbers and, as seen, the values are the same as the estimated variance components with bias. Therefore, in the simulated example by Eshagh and Sjöberg (2011), the bias of the variance components is very small so that their biased and bias-corrected estimates are the same. However, when the bias is estimated directly the values are large. In practice, the bias-corrected estimates are preferred, but in some cases, the biased estimates work fine if the regularisation parameter is very small.

Table 4.11 Variance components, their biases due to regularisation and their bias-corrected variance components for gravity gradients.

	Variance component	Bias	Bias-corrected variance component
$\widehat{\sigma}_{zz}^2$	1.01	50.43	1.01
$\widehat{\sigma}_{xx}^2$	0.98	17.49	0.98
$\widehat{\sigma}_{yy}^2$	1.01	15.60	1.00
$\widehat{\sigma}_{xy}^2$	0.97	18.32	0.97
$\widehat{\sigma}_{xz}^2$	0.97	27.82	0.97
$\widehat{\sigma}_{yz}^2$	0.97	71.73	0.98

Adapted from Eshagh, M., Sjöberg, L.E., 2011. Determination of gravity anomaly at sea level from inversion of satellite gravity gradiometric data. J. Geodyn. 51, 366–377.

4.6 Quality of integral inversion in the presence of spatial truncation error

Since the system of equations derived from the discretisation of an integral formula is ill-conditioned, regularisation methods should be applied to solve it and obtain a smooth solution. Regularisation is a way to control the effect of noise of data, which is drastically amplified through the inversion process. So far, the principle to estimate the error and bias of solutions of such systems was presented and discussed. However, one issue in estimating the quality of integral inversion of the satellite gravimetry data is the effect of STE on the variance of the estimates. This section deals with this problem and presents a method for reducing this effect from the quality of the estimated a posteriori variance factor and variance components.

4.6.1 Reduction of spatial truncation error from the a posteriori variance factor

As discussed before, to reduce the effect of STE, the inversion should be performed all over the recovery area, but the central area should be chosen. Mathematically, this idea can be expressed for a system which is solved by Tikhonov regularisation:

$$\mathbf{x}_{\mathrm{W}} = \mathbf{W}_m \mathbf{x}_{\mathrm{reg}}, \mathbf{x}_{\mathrm{reg}} = \mathbf{N}^{-1}\mathbf{A}^{\mathrm{T}}\mathbf{Q}^{-1}\mathbf{L}, \tag{4.118}$$

where $\mathbf{x}_{\mathrm{reg}}$ is an $m \times 1$ vector of the solution over the recovery area; see Eq. (4.61). $\mathbf{W}_m$ is an identity matrix having diagonal elements of 0 and 1, designed in such a way that by its multiplication to $\mathbf{x}_{\mathrm{reg}}$, those values in the marginal areas become zero. In other words, $\mathbf{W}_m$ is a window function giving zero weight to those values of $\mathbf{x}_{\mathrm{reg}}$ in the marginal areas. Similar to Eq. (4.69), the MSE of the windowed parameters will be:

$$\mathrm{MSE}_{\mathbf{x}_{\mathrm{w}}} = \widehat{\sigma}_{\mathrm{w}}^2 \mathrm{diag}(\mathbf{W}_m \mathbf{N}^{-1}\mathbf{A}^{\mathrm{T}}\mathbf{Q}^{-1}\mathbf{A}\mathbf{N}^{-1}\mathbf{W}_m) + \alpha^4 \mathbf{W}_m \mathbf{N}^{-1}\mathbf{x}\mathbf{x}^{\mathrm{T}}\mathbf{N}^{-1}\mathbf{W}_m. \tag{4.119}$$

The main issue in estimating $\mathrm{MSE}_{\mathbf{x}_{\mathrm{w}}}$ is the estimation of the a posteriori variance factor after windowing, $\widehat{\sigma}_{\mathrm{w}}^2$, or the a posteriori variance factor less affected by the STE. To find the estimator, the STE should be reduced from the residual vectors as well, because they are used for estimating this variance factor. From $\mathbf{x}_{\mathrm{W}}$ we can derive the windowed residual vector $\boldsymbol{\varepsilon}_{\mathrm{W}}$:

$$\boldsymbol{\varepsilon}_{\mathrm{W}} = \mathbf{L} - \mathbf{A}\mathbf{W}_m \mathbf{N}^{-1}\mathbf{A}^{\mathrm{T}}\mathbf{Q}^{-1}\mathbf{L}. \tag{4.120}$$

Now, according to the solution of Eq. (4.11) (the Gauss–Markov model) for **L** and inserting the result into Eq. (4.116), we reach:

$$\boldsymbol{\varepsilon}_{\mathrm{W}} = \mathbf{Ax} + \boldsymbol{\varepsilon} - \mathbf{AW}_m\mathbf{N}^{-1}\mathbf{A}^{\mathrm{T}}\mathbf{Q}^{-1}(\mathbf{Ax} + \boldsymbol{\varepsilon}). \tag{4.121}$$

By considering that $\mathbf{N}^{-1}\mathbf{A}^{\mathrm{T}}\mathbf{Q}^{-1}\mathbf{A} = \mathbf{I} - \alpha^2\mathbf{N}^{-1}$, Eq. (4.121) will change to:

$$\boldsymbol{\varepsilon}_{\mathrm{W}} = \mathbf{Mx} + \mathbf{K}\boldsymbol{\varepsilon}, \tag{4.122}$$

where

$$\mathbf{M} = \mathbf{A} - \mathbf{AW}_m + \alpha^2\mathbf{W}_m\mathbf{N}^{-1} \text{ and } \mathbf{K} = \mathbf{I} - \mathbf{AW}_m\mathbf{N}^{-1}\mathbf{A}^{\mathrm{T}}\mathbf{Q}^{-1}. \tag{4.123}$$

Squaring and taking the statistical expectation reads:

$$\begin{aligned}\mathrm{E}\{\boldsymbol{\varepsilon}_{\mathrm{W}}^{\mathrm{T}}\mathbf{W}_n\mathbf{Q}^{-1}\mathbf{W}_n\boldsymbol{\varepsilon}_{\mathrm{W}}\} &= \mathrm{E}\{\mathbf{x}^{\mathrm{T}}\mathbf{M}^{\mathrm{T}}\mathbf{W}_n\mathbf{Q}^{-1}\mathbf{W}_n\mathbf{Mx}\}\\ &\quad + \mathrm{E}\{\boldsymbol{\varepsilon}^{\mathrm{T}}\mathbf{K}^{\mathrm{T}}\mathbf{W}_n\mathbf{Q}^{-1}\mathbf{W}_n\mathbf{K}\boldsymbol{\varepsilon}\},\end{aligned} \tag{4.124}$$

where $\mathbf{W}_n$ is another identity windowing matrix with the diagonal elements of 0 and 1. Zero is considered for those residuals which are around the area and closer to the margins.

Further simplification of Eq. (4.124) reads:

$$\begin{aligned}\mathrm{E}\{\boldsymbol{\varepsilon}_{\mathrm{W}}^{\mathrm{T}}\mathbf{W}_n\mathbf{Q}^{-1}\mathbf{W}_n\boldsymbol{\varepsilon}_{\mathrm{W}}\} &= \mathrm{E}\{\mathbf{x}^{\mathrm{T}}\mathbf{M}^{\mathrm{T}}\mathbf{W}_n\mathbf{Q}^{-1}\mathbf{W}_n\mathbf{Mx}\}\\ &\quad + \mathrm{trace}(\mathbf{K}^{\mathrm{T}}\mathbf{W}_n\mathbf{Q}^{-1}\mathbf{W}_n\mathbf{K}\mathrm{E}\{\boldsymbol{\varepsilon}\boldsymbol{\varepsilon}^{\mathrm{T}}\}).\end{aligned} \tag{4.125}$$

As seen in Eq. (4.11), $\mathrm{E}\{\boldsymbol{\varepsilon}\boldsymbol{\varepsilon}^{\mathrm{T}}\} = \sigma_0^2\mathbf{Q}$ by inserting it into Eq. (4.125) and solving the result for σ_0^2:

$$\widehat{\sigma}_0^2 = \widehat{\sigma}_{\mathbf{W}}^2 = \frac{\boldsymbol{\varepsilon}_{\mathrm{W}}^{\mathrm{T}}\mathbf{W}_n\mathbf{Q}^{-1}\mathbf{W}_n\boldsymbol{\varepsilon}_{\mathrm{W}} - \mathbf{x}^{\mathrm{T}}\mathbf{M}^{\mathrm{T}}\mathbf{W}_n\mathbf{Q}^{-1}\mathbf{W}_n\mathbf{Mx}}{\mathrm{trace}(\mathbf{K}^{\mathrm{T}}\mathbf{W}_n\mathbf{K})}. \tag{4.126}$$

However, if the second term of the numerator of Eq. (4.126) is larger than the first term, $\widehat{\sigma}_{\mathbf{W}}^2$ will come out negative. A non-negative estimator is achievable by taking the first term on the rhs of Eq. (4.125), which is the bias of the residuals, to the lhs to obtain the bias-corrected residuals. After constructing the following quadratic form, we arrive at:

$$(\boldsymbol{\varepsilon}_{\mathrm{W}} - \mathbf{Mx})^{\mathrm{T}}\mathbf{W}_n\mathbf{Q}^{-1}\mathbf{W}_n(\boldsymbol{\varepsilon}_{\mathrm{W}} - \mathbf{Mx}) = \boldsymbol{\varepsilon}^{\mathrm{T}}\mathbf{K}^{\mathrm{T}}\mathbf{W}_n\mathbf{Q}^{-1}\mathbf{W}_n\mathbf{K}\boldsymbol{\varepsilon}. \tag{4.127}$$

Since $\mathrm{E}\{\boldsymbol{\varepsilon}\boldsymbol{\varepsilon}^{\mathrm{T}}\} = \sigma_0^2\mathbf{Q}$, the rhs can be written in trace from, and finally, by solving the result for σ_0^2, the non-negative estimator will be achieved:

$$\widehat{\widehat{\sigma}}_0^2 = \widehat{\widehat{\sigma}}_{\mathbf{W}}^2 = \frac{(\boldsymbol{\varepsilon}_{\mathrm{W}} - \mathbf{Mx})^{\mathrm{T}}\mathbf{W}_n\mathbf{Q}^{-1}\mathbf{W}_n(\boldsymbol{\varepsilon}_{\mathrm{W}} - \mathbf{Mx})}{\mathrm{trace}(\mathbf{K}^{\mathrm{T}}\mathbf{W}_n\mathbf{K})}. \tag{4.128}$$

Here, one part of the Eshagh and Ghorbannia (2014) study, about the estimation of the a posteriori variance factor, is presented. They inverted the orbital elements of a GOCE-type satellite as well as the on-orbit GOCE second-order radial derivative (T_{rr}) to Δg at sea level over Fennoscandia. As was already discussed, the central area should be 8 degrees smaller than each side of the recovery area for the successful reduction of the STE. This value is 3° for inverting T_{rr}. Eshagh and Ghorbannia (2014) estimated the bias and propagated errors of the estimated Δg and studied the differences in estimated variance factors with and without reduction of the STE. Table 4.12 summarises their results. In the first step, they considered $\sigma_0^2 = 1$ for both inversion processes. The mean values of errors are 17.5 and 14.5 mGal, respectively, from the orbital and gradiometry data. In the second case, $\widehat{\sigma}_0^2$ was estimated without considering any reduction in the STE and values of 6.7×10^{-7} for the error from the orbital data and 3.9×10^{-4} from the T_{rr}. Significant reductions are seen in the mean value of the errors. Finally, these mean values reduce down to 11.0 and 8.5 mGal, respectively. The table also shows the bias and RMSE of Δg, but the propagated errors are affected by the a posteriori variance factor and not the bias. Therefore, if any reduction is seen in the RMSE it is related to the reduction of the propagated errors (Table 4.12).

4.6.2 Reduction of spatial truncation error from variance components

In some cases, more than one type of satellite gravimetry data needs to be inverted. Each type of data has its own noise property and even the STE will be different for each type. So the data should be properly weighted using variance component estimation. This estimation process was already discussed for ill-conditioned systems being solved by TSVD or Tikhonov regularisation. In this section, reduction of the STE from the variance component is discussed.

Let us start with windowed residual vector $\boldsymbol{\varepsilon}_{\mathrm{W}}$, Eq. (4.120), and write the following quadratic form for it:

$$\mathrm{E}\{\boldsymbol{\varepsilon}_{\mathrm{W}}^{\mathrm{T}}\mathbf{Q}^{-1}\mathbf{W}_n\mathbf{Q}_i\mathbf{W}_n\mathbf{Q}^{-1}\boldsymbol{\varepsilon}_{\mathrm{W}}\} = \mathrm{E}\{\mathbf{x}^{\mathrm{T}}\mathbf{M}^{\mathrm{T}}\mathbf{Q}^{-1}\mathbf{W}_n\mathbf{Q}_i\mathbf{W}_n\mathbf{Q}^{-1}\mathbf{M}\mathbf{x}\} + \mathrm{E}\{\boldsymbol{\varepsilon}^{\mathrm{T}}\mathbf{K}^{\mathrm{T}}\mathbf{Q}^{-1}\mathbf{W}_n\mathbf{Q}_i\mathbf{W}_n\mathbf{Q}^{-1}\mathbf{K}\boldsymbol{\varepsilon}\}. \tag{4.129}$$

The first term on the rhs of Eq. (4.129) is the bias of this quadratic form and the second one the stochastic part. By writing the second term on the

Table 4.12 Statistics of biases, errors and root mean squared error of recovered gravity anomalies from orbital elements and satellite gradiometry data before and after considering the a posteriori variance factor.

From orbital data						From gradiometry data					
σ_0^2		Max.	Mean	Min.	Std	σ_0^2		Max.	Mean	Min.	Std
$\widehat{\sigma}_0^2 = 1$	Bias	62.1	0.0	−44.6	3.7	$\widehat{\sigma}_0^2 = 1$	Bias	30.1	−0.1	−51.8	5.8
	Error	34.5	17.5	9.1	3.4		Error	27.9	13.9	7.9	2.7
	RMSE	68.9	17.8	9.7	3.7		RMSE	55.1	14.9	8.9	3.3
$\widehat{\sigma}_0^2 = 6.7 \times 10^{-7}$	Error	28.2	14.3	7.4	2.8	$\widehat{\sigma}_0^2 = 2.7 \times 10^{-9}$	Error	24.4	12.2	6.9	2.3
	RMSE	66.7	14.7	8.0	3.3		RMSE	54.4	13.3	7.8	3.2
$\widehat{\sigma}_W^2 = 3.9 \times 10^{-4}$	Error	21.6	11.0	5.7	2.1	$\widehat{\sigma}_W^2 = 3.7 \times 10^{-7}$	Error	17.1	8.5	4.8	1.6
	RMSE	64.9	11.4	6.2	2.9		RMSE	53.1	9.9	5.5	3.1

RMSE, root mean squared error; *Std*, standard deviation. Unit: mGal.
From Eshagh, M., Ghorbannia, M., 2014. The effect of spatial truncation error on variance of gravity anomalies derived from inversion of satellite orbital and gradiometric data. Adv. Space Res. 54, 261–271.

rhs of Eq. (4.129) in trace form and considering that $\mathrm{E}\{\boldsymbol{\varepsilon}\boldsymbol{\varepsilon}^{\mathrm{T}}\} = \sum_{j=1}^{q} \sigma_j^2 \mathbf{Q}_j$ we obtain:

$$\mathrm{E}\{\boldsymbol{\varepsilon}_{\mathrm{W}}^{\mathrm{T}}\mathbf{Q}^{-1}\mathbf{W}_n\mathbf{Q}_i\mathbf{W}_n\mathbf{Q}^{-1}\boldsymbol{\varepsilon}_{\mathrm{W}}\} = \mathrm{E}\{\mathbf{x}^{\mathrm{T}}\mathbf{M}^{\mathrm{T}}\mathbf{Q}^{-1}\mathbf{W}_n\mathbf{Q}_i\mathbf{W}_n\mathbf{Q}^{-1}\mathbf{M}\mathbf{x}\} + \sum_{j=1}^{q} \sigma_j^2 \mathrm{trace}\{\mathbf{K}^{\mathrm{T}}\mathbf{Q}^{-1}\mathbf{W}_n\mathbf{Q}_i\mathbf{W}_n\mathbf{Q}^{-1}\mathbf{K}\mathbf{Q}_j\}. \tag{4.130}$$

Therefore, the windowed variance components will be estimated by solving the following system of equations:

$$\mathbf{S}\boldsymbol{\sigma}_{\mathrm{W}} = \mathbf{u} - \boldsymbol{\delta}\mathbf{u}, \tag{4.131}$$

where the elements of the matrix $\mathbf{S}$ and the vectors $\mathbf{u}$ and $\delta\mathbf{u}$ are:

$$s_{ij} = \sigma_j^2 \mathrm{trace}\{\mathbf{K}^{\mathrm{T}}\mathbf{Q}^{-1}\mathbf{W}_n\mathbf{Q}_i\mathbf{W}_n\mathbf{Q}^{-1}\mathbf{K}\mathbf{Q}_j\}, \tag{4.132}$$

$$u_i = \boldsymbol{\varepsilon}_{\mathrm{W}}^{\mathrm{T}}\mathbf{Q}^{-1}\mathbf{W}_n\mathbf{Q}_i\mathbf{W}_n\mathbf{Q}^{-1}\boldsymbol{\varepsilon}_{\mathrm{W}}, \tag{4.133}$$

$$\delta u_i = \mathbf{x}^{\mathrm{T}}\mathbf{M}^{\mathrm{T}}\mathbf{Q}^{-1}\mathbf{W}_n\mathbf{Q}_i\mathbf{W}_n\mathbf{Q}^{-1}\mathbf{M}\mathbf{x}. \tag{4.134}$$

The non-negative estimate of the variance components is obtained by taking the bias term in the residual vector in Eq. (4.131) to the lhs to obtain the bias-corrected windowed residuals. After a process similar to what was done to obtain Eqs. (4.116) and (4.117) is carried out, the following system of equations is obtained for estimating the bias-corrected windowed variance components for the STE:

$$\overline{\mathbf{S}}\overline{\boldsymbol{\sigma}}_{\mathrm{W}} = \overline{\mathbf{u}}, \tag{4.135}$$

where the elements of $\overline{\mathbf{S}}$ and $\overline{\mathbf{u}}$ are:

$$\bar{s}_{ij} = \mathrm{trace}(\mathbf{K}^{\mathrm{T}}\mathbf{Q}^{-1}\mathbf{W}_n\mathbf{Q}_i\mathbf{W}_n\mathbf{Q}^{-1}\mathbf{K}\mathbf{Q}_j), \tag{4.136}$$

$$\bar{u}_j = (\boldsymbol{\varepsilon}_{\mathrm{W}} - \mathbf{M}\mathbf{x})^{\mathrm{T}}\mathbf{Q}^{-1}\mathbf{W}_n\mathbf{Q}_i\mathbf{W}_n\mathbf{Q}^{-1}(\boldsymbol{\varepsilon}_{\mathrm{W}} - \mathbf{M}\mathbf{x}). \tag{4.137}$$

Eshagh and Ghorbannia (2014) simulated an orbit similar to that of GOCE using EGM96 (Lemoine et al., 1998) and a Runge–Kutta integrator of the fourth order for a period of 3 months with a step size of 10 s over Fennoscandia. Five hundred fifty passes of the imaginary satellite over this region were selected and the integrated position and velocity vectors were converted to their equivalent orbital elements using Eqs. (2.18)–(2.24) (see Chapter 2). The velocities of the orbital elements were

Table 4.13 Statistics of biases, errors and root mean squared error of recovered gravity anomalies from joint inversion of orbital and gradiometry data before and after considering variance component estimation and its windowed version, and statistics of differences between the recovered anomalies from orbital elements and satellite gradiometry data and computed ones from the Earth Gravitational Model.

		From joint inversion				Compared with EGM				
VCs		Max.	Mean	Min.	Std	Max.	Mean	Min.	Std	RMSE
$\sigma_O^2 = 1$	Bias	31.9	−0.7	−109.8	10.5	62.0	4.8	−31.1	10.5	11.5
$\sigma_G^2 = 1$	Error	20.5	8.7	5.8	1.7					
	RMSE	111.7	11.7	5.9	7.2					
$\widehat{\sigma}_O^2 = 41.9$	Bias	100.6	−0.1	−139.4	15.5	20.0	−6.2	−31.1	6.6	9.1
$\widehat{\sigma}_G^2 = 0.02$	Error	19.4	6.9	4.4	1.4					
	RMSE	89.8	10.2	4.9	6.1					
$\widehat{\sigma}_{OW}^2 = 10.2$	Bias	48.6	0.3	−65.7	9.7	25.1	−5.8	−24.2	6.1	8.4
$\widehat{\sigma}_{GW}^2 = 1.0$	Error	9.6	6.7	4.5	1.0					
	RMSE	46.9	8.6	5.3	4.6					

EGM, Earth Gravitational Model; *RMSE*, root mean squared error; *Std*, standard deviation; *VC*, variance component. Unit: mGal.
From Eshagh, M., Ghorbannia, M., 2014. The effect of spatial truncation error on variance of gravity anomalies derived from inversion of satellite orbital and gradiometric data. Adv. Space Res. 54, 261–271.

estimated for this purpose, and therefore, the Newtonian polynomial of order 9 (Reubelt et al., 2003) was fitted to each orbital element, but in an area larger by 5 degrees to reduce the interpolation error. The studies of Földváry (2007) showed that choosing order 7 is more effective. Furthermore, the gradiometry data were generated on the orbit. The goal is to invert these two types of data to Δg at sea level with a resolution of $0.5^\circ \times 0.5^\circ$ over the study area. In addition, two types of white noise were considered in these data, with a mean value of 0 and Std of 0.01, as the main assumption in their mathematical derivations is to have this type of noise.

The statistics of the differences between the recovered Δg and those generated by EGM96 are presented in Table 4.13. σ_O^2 and σ_G^2 stand for the a priori variance factor of the orbital and gradiometry data, and when they are both 1, they are equally weighted in the inversion process. The Std and RMSE of the differences between the recovered and the generated Δg, respectively, are 10.5 and 11.7 mGal. By using the variance component estimation process and reweighting the data without considering the STE, these values reduce down to 6.6 and 9.1 mGal, meaning the process is meaningful. It is also observed that the RMSE of the differences reduces by 1.5 mGal and the Std by about 4 mGal, which is quite significant. When the windowed variance component estimation is applied, the RMSE reduced to 8.4 mGal and the Std to 6.1 mGal. The improvements are not very significant but the values of the variance components are more reasonable, as they are 10.2 and 1.0 for the orbital and gradiometry data, which seems to be more reasonable than those values obtained from non-windowed variance components estimation.

References

Bjerhammer, A., 1973. Theory of Errors and Generalized Matrix Inverses. Elsevier, p. 420.

Björck, Å., 1988. A bidiagonalization algorithm for solving large and sparse ill-posed systems of linear equations. BIT Numer. Math. 28, 656–670.

Bouman, J., 1998. Quality of Regularization Methods, DEOS Report 98.2. Delft University Press, Delft, The Netherlands.

Brakhage, H., 1987. On ill-posed problems and the method of conjugate gradient. In: Engl, H.W., Groetsche, C.W. (Eds.), Inverse and Ill-Posed Problems. Academic Press, London.

Calvetti, D., Reichel, L., 2003. Tikhonov regularization of large linear problems. BIT Numer. Math. 43, 263–283.

Calvetti, D., Morigi, S., Reichel, L., Sgallari, F., 2000. Tikhonov regularization and the L-curve for large, discrete ill-posed problems. J. Comput. Appl. Math. 123, 423–446.

Cooper, M.R.A., April 1987. Control Surveys in Civil Engineering. Nichols Pub Co, p. 381.
Diggle, P., 1991. Nonparameteric comparison of cumulative periodograms. Appl. Statistics 40 (3), 423–434.
Eshagh, M., 2009. On Satellite Gravity Gradiometry, Doctoral Thesis in Geodesy. Royal Institute of Technology (KTH), Sweden.
Eshagh, M., 2010. Variance component estimation in linear ill-posed problems: TSVD issue. Acta Geod. Geophys. Hung. 45, 184–194.
Eshagh, M., 2011a. The effect of spatial truncation error on integral inversion of satellite gravity gradiometry data. Adv. Space Res. 47, 1238–1247.
Eshagh, M., 2011b. Inversion of satellite gradiometry data using statistically modified integral formulas for local gravity field recovery. Adv. Space Res. 47 (1), 74–85.
Eshagh, M., 2011c. Sequential Tikhonov regularization: an alternative method for inversion of satellite gravity gradiometry data. ZfV 136, 113–121.
Eshagh, M., Sjöberg, L.E., 2011. Determination of gravity anomaly at sea level from inversion of satellite gravity gradiometric data. J. Geodyn. 51, 366–377.
Eshagh, M., Lemoine, J.M., Gegout, P., Biancale, R., 2013. On regularized time varying gravity field models based on GRACE data and their comparisons with hydrological models. Acta Geophys. 61 (1), 1–17.
Eshagh, M., Ghorbannia, M., 2014. The effect of spatial truncation error on variance of gravity anomalies derived from inversion of satellite orbital and gradiometric data. Adv. Space Res. 54 (2), 261–271.
Földvary, R., 2007. Analysis of numerical differentiation methods applied for determination of kinematic velocities for LEOs. Periodica Polytechnica-Civil Eng. 51 (1), 17–24.
Golub, G.H., von Matt, U., 1997. Tikhonov regularization for large scale problems. In: Golub, G.H., Lui, S.H., Luk, F., Plemmons, R. (Eds.), Workshop on Scientific Computing. Springer, New York, pp. 3–26.
Hämarik, U., Tautenhahn, U., 2001. On the monotone error rule parameter choice in iterative and continuous regularization methods. BIT Numer. Math. 41 (5), 1029–1038.
Hansen, P.C., 1998. Rank-deficient and Discrete Ill-Posed Problems: Numerical Aspects of Linear Inversion. SIAM, Philadelphia.
Hansen, P.C., 2007. Regularization tools version 4.0 for Matlab 7.3. Numer. Algorithm. 46, 189–194.
Hansen, P.C., Jensen, T.K., 2006. Smoothing-norm preconditioning for regularizing minimum-residual methods. SIAM J. Matrix Anal. Appl. 29, 1–14.
Jensen, T.K., Hansen, P.C., 2007. Iterative regularization with minimum residual methods. BIT Numer. Math. 47, 103–120.
Kaczmarz, S., 1937. Angenäherte Auflösung von Systemen linearer Gleichungen, Bulletin International de l'Academie Polonaise des Sciences, Cl. d. Sc. Mathém. A, pp. 355–357.
Kilmer, M.E., O'Leary, D.P., 2001. Choosing regularization parameters in iterative methods for ill-posed problems, SIAM. J. Matrix Anal. Appl. 22, 1204–1221.
Klees, R., Ditmar, P., Broersen, P., 2003. How to handle colored observatioin noise in large least-squares problems. J. Geodes. 76, 629–640.
Koch, K.-R., 1999. Parameter Estimation and Hypothesis Testing in Linear Models. Springer.
Koch, K.-R., Kusche, J., 2002. Regularization of geopotential determination from satellite data by variance components. J. Geodesy. 76, 259–268.
Kusche, J., Klees, R., 2002. Regularization of gravity field estimation from satellite gravity gradients. J. Geodes. 76, 359–368.

Lemoine, F.G., Kenyon, S.C., Factor, J.K., Trimmer, R.G., Pavlis, N.K., Chinn, D.S., Cox, C.M., Klosko, S.M., Luthcke, S.B., Torrence, M.H., Wang, Y.M., Williamson, R.G., Pavlis, E.C., Rapp, R.H., Olson, T.R., July 1998. The Development of the Joint NASA GSFC and the National Imagery and Mapping Agency (NIMA) Geopotential Model EGM96. NASA/TP-1998-206861.

Lemoine, J.M., Bruinsma, S., Loyer, S., Biancale, R., Marty, J.C., Perosanz, F., Balmino, G., 2007. Temporal gravity field models inferred from GRACE data. Adv. Space Res. 39, 1620–1629.

Lewis, B., Reichel, L., 2009. Arnoldi-Tikhonov regularization methods. J. Comput. Appl. Math. 226 (1), 92–102.

Maitre, H., Levy, A.J., 1983. The use of normal equations for superresolution problems. J. Optics 14 (4), 205–207.

Mojabi, P., LoVetri, J., 2008. Adapting the normalized commulative periodogram parameter-choice method to the Tikhonov regularization of 2-D/TM electromagnetic inverse scattering using born iterative method. Prog. Electromagn. Res. 1, 111–138.

O'Leary, D.P., Simmons, J.A., 1981. A bidiagonalization-regularization procedure for large-scale discretization of ill-posed problems. SIAM J. Sci. Statist. Comput. 2, 474–489.

Pavlis, N., Holmes, S.A., Kenyon, S.C., Factor, J.K., April 13–18, 2008. An Earth Gravitational Model to Degree 2160: EGM08. Presented at the 2008. General Assembly of the European Geosciences Union, Vienna, Austria.

Reubelt, T., Austen, G., Grafarend, E.W., 2003. Space Gravity Spectroscopy – determination of the Earth's gravitational field by means of Newton interpolated LEO ephemeris case studies on dynamic (CHAMP Rapid Science Orbit) and kinematic orbits. Adv. Geosci. 1, 127–135.

Saad, Y., 1996. Iterative Methods for Sparse Linear Systems. PWS, Boston, MA.

Saad, Y., Schultz, M.H., 1986. GMRES: a generalized minimal residual method for solving nonsymmetric linear systems. SIAM J. Sci. Statist. Comput. 7, 856–869.

Scherzer, O., 1993. The use of Morozov's discrepancy principle for Tikhonov regularization for solving nonlinear ill-posed problems. Computing 51, 45–60.

Seeber, G., 2003. Satellite Geodesy Foundations, Methods, and Applications, De Gruyter.

Sprlak, M., Eshagh, M., 2016. Local Recovery of Sub-Crustal Stress Due to Mantle Convection from Satellite-to-Satellite Tracking Data. Acta Geophys. 64, 904–929.

Tikhonov, A.N., 1963. Solution of incorrectly formulated problems and regularization method. Soviet Math. Dokl. 4, 1035–1038. English translation of Dokl. Akad. Nauk. SSSR, 151:501–504.

Wahba, G., 1976. A survey of some smoothing problems and the methods of generalized cross-validation for solving them. In: Krishnaiah, P.R. (Ed.), Proceedings of the Conference on the Applications of Statistics, Held at Dayton, Ohio, June 14–17, 1976.

Xu, P., 1992. Determination of surface gravity anomalies using gradiometric observables. Geophys. J. Int. 110, 321–332.

Xu, P., 1998. Truncated SVD methods for discrete linear ill-posed problems. Geophys. J. Int. 135, 505–514.

Xu, P., 2009. Iterative generalized cross-validation for fusing heteroscedastic data of inverse ill-posed problems. Geophys. J. Int. 179, 182–200.

Xu, P., Shen, Y., Fukuda, Y., Liu, Y., 2006. Variance component estimation in linear inverse ill-posed models. J Geodesy 80, 69–81.

CHAPTER 5

The effect of mass heterogeneities and structures on satellite gravimetry data

5.1 Gravitational potential of topographic and bathymetric masses

Modelling the gravitational attraction of topographic masses is important in both gravity field recovery and the study of the Earth's interior structure. These masses violate the Laplace condition and are disturbances, which should be removed from the satellite gravimetry data prior to modelling the Earth's gravity field to simplify the computational process. However, their effects should be restored to the results, as the real Earth always contains these masses; see e.g. Martinec and Vaníček (1994), Sjöberg (2000), Tziavos and Featherstone (2000), Makhloof (2007), Makhloof and Ilk (2005). The topographic masses are not considered as disturbances, but by removing them from the satellite gravimetry data the unknown part of the gravitational signal is obtained and studied. For example, in isostasy the topographic masses are loads on the Earth's crust, pushing the lithosphere downward and deforming it. In addition, in modelling the stress due to mantle convection the effects of topographic and crustal masses should be removed from satellite gravimetry data to highlight sub-lithospheric gravitational signals. The effect of the bathymetric masses, or water, is important in gravity field modelling when they are at a higher level than the oceans, like some lakes, rivers and so on (Martinec et al., 1995). For studying the Earth's interior, it is considered a known signal, which should be removed from the data.

5.1.1 Gravitational potential of topographic masses

To determine the gravitational potential of the topographic masses, the Newton integral is applied. This integral expresses the potential of the

Satellite Gravimetry and the Solid Earth
ISBN 978-0-12-816936-0
https://doi.org/10.1016/B978-0-12-816936-0.00005-0

attractive topographic masses at point P with spherical coordinates (r, θ, λ), r geocentric radius, co-latitude θ and longitude λ, outside the masses; see Fig. 5.1. It shows that the topography of the solid Earth varies around the mean sea level; this means that over some places with negative heights, a lack of mass exists or they are filled by water. In gravity field recovery, the masses above sea level should be removed to fulfil Laplace conditions and restored after computations. This means that the topographic effect on the result is always positive, as these masses either exist or do not. The effect of mass loss under sea level or water is more related to geophysical studies.

The Newton integral for modelling the topographic masses is:

$$V^{\mathrm{T}}(r, \theta, \lambda) = G \iint_{\sigma} \int_{R}^{R+H'} \frac{\rho^{\mathrm{T}}(r', \theta', \lambda')}{l} \mathrm{d}\sigma r'^{2} \mathrm{d}r', \tag{5.1}$$

where V^{T} is the gravitational potential of topographic masses; G is the gravitational Newton constant; ρ^{T} is the density of the masses; (r', θ', λ') are the spherical coordinates of the integration point, which is inside the masses; R stands for the radius of the spherical Earth and H the topographic heights above sea level when they are positive or below sea level when negative;

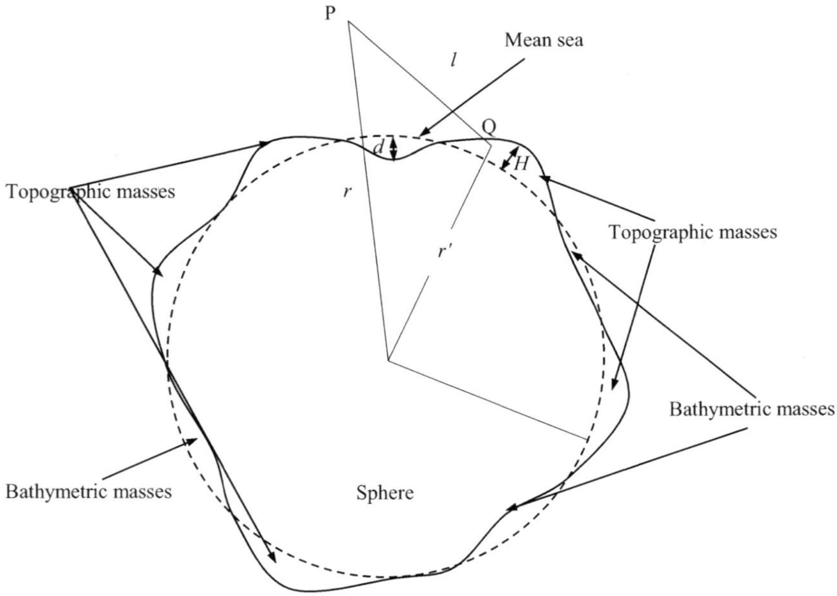

Figure 5.1 Topographic and bathymetric masses.

σ is the sphere at which the integration is taken; $d\sigma$ is the surface integration element and dr' is the radial integration element. l is the distance between the computation and the integration point (see Eq. 1.19); we can write it as:

$$l = \sqrt{r^2 + r'^2 - 2rr' \cos\psi} = rD. \tag{5.2}$$

D has been defined in Eq. (1.19). The Newton integral, Eq. (5.1), can be applied directly for computing the gravitational potential of the topographic masses. Its radial and surface integral can be discretised based on resolution of the available data.

This gravitational potential, Eq. (5.1), can be modelled in terms of spherical harmonics as well. To do so, the Legendre expansion of l^{-1} is applied:

$$\frac{1}{l} = \frac{1}{rD} = \frac{1}{r}\sum_{n=0}^{\infty} t^n P_n(\tilde{x}) \quad \text{where } t = \frac{r'}{r} \text{ and } \tilde{x} = \cos\psi. \tag{5.3}$$

By inserting Eq. (5.3) into Eq. (5.1) and assuming that ρ^{T} is constant radially, Eq. (5.1) changes to:

$$V^{T}(r,\theta,\lambda) = G\iint_{\sigma} \rho^{T}(\theta',\lambda') \sum_{n=0}^{\infty} \frac{1}{r^{n+1}} P_n(\tilde{x}) \int_{R}^{R+H(\theta',\lambda')} r'^{n+2} dr' d\sigma. \tag{5.4}$$

After solving the radial integral Eq. (5.4), it transfers to:

$$V^{T}(r,\theta,\lambda) = G\iint_{\sigma} \rho^{T}(\theta',\lambda') \sum_{n=0}^{\infty} \frac{1}{r^{n+1}} P_n(\tilde{x}) \; \frac{1}{n+3}\left[r'^{n+3}\right]_{R}^{R+H(\theta',\lambda')} d\sigma. \tag{5.5}$$

It can be shown that:

$$\begin{aligned} \frac{1}{n+3}\left[r'^{n+3}\right]_{R}^{R+H(\theta',\lambda')} &= \frac{1}{n+3}\left[(R+H(\theta',\lambda'))^{n+3} - R^{n+3}\right] \\ &= \frac{R^{n+3}}{n+3}\left[\left(1 + \frac{H(\theta',\lambda')}{R}\right)^{n+3} - 1\right]. \end{aligned} \tag{5.6}$$

Now, consider $H' = H(\theta',\lambda')$ to shorten the formulae and approximate the binomial term by the Taylor series up to the fourth term:

$$\begin{aligned} \left(1 + \frac{H'}{R}\right)^{n+3} &\approx 1 + (n+3)\frac{H'}{R} + (n+3)(n+2)\frac{H'^2}{2R^2} \\ &\quad + (n+3)(n+2)(n+1)\frac{H'^3}{6R^3}. \end{aligned} \tag{5.7}$$

According to this approximation, Eq. (5.6) will be simplified further to:

$$\frac{1}{n+3}\left[r'^{n+3}\right]_R^{R+H(\theta',\lambda')} \approx R^{n+3}\left[\frac{H'}{R}+(n+2)\frac{H'^2}{2R^2}+(n+2)(n+1)\frac{H'^3}{6R^3}\right] = R^{n+3}F'. \tag{5.8}$$

The convergence of this approximation has been studied by Sun and Sjöberg (2001), who showed that this approximation is convergent as long as the spherical harmonic series is limited to degree and order 360.

From here the gravitational potential can be modelled in two different ways. It can be done in terms of products of density and topographic heights or based on the spherical harmonic coefficients (SHCs) of density and topographic heights. In the following, these approaches are explained in detail.

5.1.1.1 Approach 1 to consider lateral density variation of topographic masses

As observed in Eq. (5.8), the solution of the radial integral is approximated in terms of topographic heights with powers 1, 2 and 3. Substitution of Eq. (5.8) into Eq. (5.5) yields

$$V^{\mathrm{T}}(r,\theta,\lambda)=G\sum_{n=0}^{\infty}\frac{R^{n+3}}{r^{n+1}}\iint_{\sigma}\left[\frac{\rho'^{\mathrm{T}}H'}{R}+(n+2)\frac{\rho'^{\mathrm{T}}H'^2}{2R^2}+(n+2)(n+1)\frac{\rho'^{\mathrm{T}}H'^3}{6R^3}\right]P_n(\tilde{x})\,\mathrm{d}\sigma. \tag{5.9}$$

According to the addition theorem of spherical harmonics, Eq. (1.9), Eq. (5.9) will be:

$$V^{\mathrm{T}}(r,\theta,\lambda)=GR^2\sum_{n=0}^{\infty}\left(\frac{R}{r}\right)^{n+1}\frac{1}{2n+1}\sum_{m=-n}^{n}\iint_{\sigma}\left[\frac{\rho'^{\mathrm{T}}H'}{R}+(n+2)\frac{\rho'^{\mathrm{T}}H'^2}{2R^2}+(n+2)(n+1)\frac{\rho'^{\mathrm{T}}H'^3}{6R^3}\right] Y_{nm}(\theta',\lambda')Y_{nm}(\theta,\lambda)\,\mathrm{d}\sigma. \tag{5.10}$$

By applying Eq. (1.8), the spherical harmonic expression of the gravitational potential of the topographic masses will be:

$$V^{\mathrm{T}}(r,\theta,\lambda)=\sum_{n=0}^{\infty}\left(\frac{R}{r}\right)^{n+1}\sum_{m=-n}^{n}v_{nm}^{\mathrm{T}}Y_{nm}(\theta,\lambda), \tag{5.11}$$

with the SHCs:

$$v_{nm}^{\mathrm{T}} = \frac{4\pi GR^2}{2n+1}(\rho^{\mathrm{T}}F)_{nm} = \frac{4\pi GR^2}{2n+1}\left[\frac{(\rho^{\mathrm{T}}H)_{nm}}{R} + (n+2)\frac{(\rho^{\mathrm{T}}H^2)_{nm}}{2R^2} + (n+2)(n+1)\frac{(\rho^{\mathrm{T}}H^3)_{nm}}{6R^3}\right], \tag{5.12}$$

where

$$\left(\rho^{\mathrm{T}}H^{i}\right)_{nm} = \frac{1}{4\pi}\iint_{\sigma}\left(\rho^{\mathrm{T}}(\theta',\lambda')H(\theta',\lambda')^{i}\right)Y_{nm}(\theta',\lambda')\mathrm{d}\sigma, \quad i = 1, 2 \text{ and } 3. \tag{5.13}$$

As observed, the SHCs of the potential are functions of the harmonics of products of $(\rho^{\mathrm{T}}H)_{nm}$, $(\rho^{\mathrm{T}}H^2)_{nm}$ and $(\rho^{\mathrm{T}}H^3)_{nm}$.

As mentioned before, the integral (5.1) can be directly applied for determining the topographic potential, which is normally, the method for estimating the topographic effect on terrestrial gravity data in local geoid or gravity field determination (see e.g. Martinec, 1993; Kuehtreiber, 1998; Huang et al., 2001; Hunegnaw, 2001; Sjöberg, 2004, or Kiamehr, 2006). In this case, high frequencies of the topographic potential should be determined as the geoid high frequencies come mainly from them. Topographic effect is an important issue in precise gravity field determination using satellites, but due to the high altitude, these high frequencies are damped and the effect will be considerably smoother than when it is derived for the terrestrial gravimetric data. Therefore, the approximation based on the spherical harmonic series (Eq. 5.11), limited to a specific degree and order depending on the mission, suffices for the satellite gravimetry applications; see e.g. Wild and Heck (2004a,b), Novák and Grafarend (2006), Makhloof (2007), Eshagh and Sjöberg (2008, 2009) or Eshagh (2009a,b).

In many application, a constant density ρ^{T} is considered for the topographic masses; in this case, Eqs. (5.12) and (5.13) can be simplified to:

$$v_{nm}^{\mathrm{T}} = \frac{4\pi GR^2}{2n+1}\rho^{\mathrm{T}}F_{nm}, \tag{5.14}$$

with

$$F_{nm} = \left[\frac{(H)_{nm}}{R} + (n+2)\frac{(H^2)_{nm}}{2R^2} + (n+2)(n+1)\frac{(H^3)_{nm}}{6R^3}\right], \tag{5.15}$$

$$\left(H^{i}\right)_{nm} = \frac{1}{4\pi}\iint_{\sigma} H^{i}(\theta',\lambda')Y_{nm}(\theta',\lambda')\mathrm{d}\sigma, \quad i = 1, 2 \text{ and } 3. \tag{5.16}$$

As Eq. (5.12) shows, the SHCs of the product of the topographic masses and the zeroth, first and third powers of topographic heights are required.

Based on Eq. (5.13), these quantities are required with global coverages, which are not available, at least for the topographic density. In addition, the resolution of the density data and heights should be equal so that the integration can be performed to obtain these SHCs.

Eq. (5.11) is analogous to Eq. (1.3), which was presented for the Earth's gravitational potential. Therefore, by replacing the SHCs v_{nm} with v_{nm}^{T} in the spherical harmonic expansions of the satellite gravimetry data, their corresponding topographic effects can be obtained.

5.1.1.2 Approach 2 to consider lateral density variation of topographic masses

In the case where the density ρ^{T} is considered constant, the situation is simple. However, when it is variable laterally, Eq. (5.12) is applicable if the density data with global coverage are available, but their resolution should be consistent with the topographic heights, so that the SHCs of their products become computable. If the SHCs of the density data are given, then the effect can be computed in two different ways: (1) by synthesising the density data with the same resolution as the topographic heights and applying Eq. (5.12) for computing the SHCs of the gravitational potential or (2) by applying a Clebsch–Gordan series (Loera and McAllister, 2006). The second method is presented in this section.

Novák and Grafarend (2006) derived the topographic effect differently, and Eshagh (2009a) developed it further. If the SHCs of the density are given, F_{nm} can be derived from the topographic heights using Eq. (5.15). Let the spherical harmonic expansion of the density be:

$$\rho^{\mathrm{T}}(\theta,\lambda)=\sum_{i=0}^{\infty}\sum_{j=-i}^{i}\rho_{ij}^{\mathrm{T}}Y_{ij}(\theta,\lambda), \tag{5.17}$$

and that of F:

$$F(\theta,\lambda)=\sum_{k=0}^{\infty}\sum_{l=-k}^{k}F_{kl}Y_{kl}(\theta,\lambda). \tag{5.18}$$

Now, Eqs. (5.17) and (5.18) are inserted into Eq. (5.14), written in the integral form:

$$v_{nm}^{\mathrm{T}}=\frac{GR^2}{2n+1}\iint_{\sigma}\rho^{\mathrm{T}}(\theta',\lambda')F(\theta',\lambda')Y_{nm}(\theta',\lambda')\mathrm{d}\sigma. \tag{5.19}$$

Substitution of Eqs. (5.17) and (5.18) into Eq. (5.19) reads:

$$v_{nm}^{\mathrm{T}}=\frac{GR^2}{2n+1}\sum_{i=0}^{\infty}\sum_{j=-i}^{i}\sum_{k=0}^{\infty}\sum_{l=-k}^{k}F_{kl}\rho_{ij}^{\mathrm{T}}\iint_{\sigma}Y_{ij}(\theta',\lambda')Y_{kl}(\theta',\lambda')\mathrm{d}\sigma. \tag{5.20}$$

Substitution of Eq. (5.20) into Eq. (5.11) yields:

$$V^{\mathrm{T}}(r,\theta,\lambda)=\sum_{n=0}^{\infty}\frac{GR^2}{2n+1}\left(\frac{R}{r}\right)^{n+1}\sum_{m=-n}^{n}\sum_{i=0}^{\infty}\sum_{j=-i}^{i}\sum_{k=0}^{\infty}\sum_{l=-k}^{k}F_{kl}\rho_{ij}^{\mathrm{T}}$$
$$\iint_{\sigma}Y_{ij}(\theta',\lambda')Y_{kl}(\theta',\lambda')\mathrm{d}\sigma Y_{nm}(\theta,\lambda). \tag{5.21}$$

It was presented in Eq. (1.143) that the product of two spherical harmonic functions can be written in terms of the product of the Gaunt coefficient (Xu 1996) and one spherical harmonic. Therefore, Eq. (5.21) will be simplified to:

$$V^{\mathrm{T}}(r,\theta,\lambda)=\sum_{n=0}^{\infty}\frac{GR^2}{2n+1}\left(\frac{R}{r}\right)^{n+1}\sum_{m=-n}^{n}\sum_{i=0}^{\infty}\sum_{j=-i}^{i}\sum_{k=0}^{\infty}\sum_{l=-k}^{k}\sum_{n'=|i-k|}^{i+k}$$
$$\sum_{m'=-n'}^{n'}Q_{ijkl}^{n'm'}F_{kl}\rho_{ij}^{\mathrm{T}}\iint_{\sigma}Y_{n'm'}(\theta',\lambda')Y_{nm}(\theta,\lambda)\mathrm{d}\sigma. \tag{5.22}$$

The spherical harmonics are orthogonal globally over a sphere, according to the orthogonality of these harmonic functions, presented in Eq. (1.6), Eq. (5.22) will change to:

$$V^{\mathrm{T}}(r,\theta,\lambda)=4\pi GR^2\sum_{n=0}^{\infty}\frac{1}{2n+1}\left(\frac{R}{r}\right)^{n+1}\sum_{m=-n}^{n}\sum_{i=0}^{\infty}\sum_{j=-i}^{i}\sum_{k=0}^{\infty}\sum_{l=-k}^{k}\sum_{n=|i-k|}^{i+k}$$
$$\sum_{m=-n}^{n}F_{kl}\rho_{ij}^{\mathrm{T}}Q_{ijkl}^{nm}. \tag{5.23}$$

Therefore, the SHCs of the topographic potential, corresponding to Eq. (5.12), will be:

$$v_{nm}^{\mathrm{T}}=\frac{4\pi GR^2}{2n+1}\sum_{i=0}^{\infty}\sum_{j=-i}^{i}\sum_{k=0}^{\infty}\sum_{l=-k}^{k}F_{kl}\rho_{ij}^{\mathrm{T}}Q_{ijkl}^{nm}. \tag{5.24}$$

According to Eshagh (2009a) the differences between this method and the one presented in the previous section:

1. The density and topographic height data must be given in the same resolution in approach 1, but in approach 2 it is not necessary.
2. Approach 1 is computationally more efficient than approach 2.
3. The number of approximations in approach 2 is larger.
4. In approach 2 the geometry and density data of topography are separately analysed but in approach 1 their products are analysed.

For more details about the numerical applications and detailed discussion, see Eshagh (2009a).

5.1.1.3 Gravitational potential of the bathymetric masses (water)

The Newton integral can also be used for deriving the potential of the oceanic, sea and lake waters. This integral can be written as:

$$V^{\mathrm{B}}(r,\theta,\lambda)=G\iint\limits_{\sigma}\int\limits_{R-d(\theta',\lambda')}^{R}\frac{\rho^{\mathrm{B}}(r',\theta',\lambda')}{l}\mathrm{d}\sigma r'^{2}\mathrm{d}r', \tag{5.25}$$

where ρ^{B} is the density of water and d is the depth of water. Note that the limits of the radial integral are from the bottom of the ocean with depth d to the surface of the sphere R. By assuming that the density of water is constant and according to Eq. (5.3), Eq. (5.25) transfers to:

$$V^{\mathrm{B}}(r,\theta,\lambda)=G\rho^{\mathrm{B}}\iint\limits_{\sigma}\sum_{n=0}^{\infty}\frac{1}{r^{n+1}}P_n(\tilde{x})\int\limits_{R-d(\theta',\lambda')}^{R}r'^{n+2}\mathrm{d}r'\mathrm{d}\sigma. \tag{5.26}$$

The solution of the radial integral of Eq. (5.26) will be:

$$\begin{aligned}\int\limits_{R-d(\theta',\lambda')}^{R}r'^{n+2}\mathrm{d}r'&=\frac{1}{n+3}\left[r'^{n+3}\right]_{R-d(\theta',\lambda')}^{R}\\&=\frac{1}{n+3}\left[R^{n+3}-\left(R-d(\theta',\lambda')\right)^{n+3}\right]\\&=\frac{R^{n+3}}{n+3}\left[1-\left(1-\frac{d(\theta',\lambda')}{R}\right)^{n+3}\right].\end{aligned} \tag{5.27}$$

The Taylor expansion of the binomial term of Eq. (5.27) up to the fourth term is:

$$\begin{aligned}\left(1-\frac{d'}{R}\right)^{n+3}&\approx 1-(n+3)\frac{d'}{R}+(n+3)(n+2)\frac{d'^2}{2R^2}\\&-(n+3)(n+2)(n+1)\frac{d'^3}{6R^3},\end{aligned} \tag{5.28}$$

where $d'=d(\theta',\lambda')$. Now, by inserting Eq. (5.28) into Eq. (5.27) and performing further simplification we arrive at:

$$\frac{1}{n+3}\left[r'^{n+3}\right]_{R-H(\theta',\lambda')}^{R}\approx R^{n+3}\left[\frac{d'}{R}-(n+2)\frac{d'^2}{2R^2}+(n+2)(n+1)\frac{d'^3}{6R^3}\right]. \tag{5.29}$$

After inserting Eq. (5.29) into Eq. (5.26) and applying the addition theorem of spherical harmonics, Eq. (1.9), the spherical harmonic expansion of the gravitational potential of the bathymetric masses will be:

$$V^{\mathrm{B}}(r,\theta,\lambda)=\sum_{n=0}^{\infty}\left(\frac{R}{r}\right)^{n+1}\sum_{m=-n}^{n}v_{nm}^{\mathrm{B}}Y_{nm}(\theta,\lambda), \tag{5.30}$$

where

$$v_{nm}^{\mathrm{B}}=\frac{4\pi GR^2\rho^{\mathrm{B}}}{2n+1}\left[\frac{(d)_{nm}}{R}-(n+2)\frac{(d^2)_{nm}}{2R^2}+(n+2)(n+1)\frac{(d^3)_{nm}}{6R^3}\right] \tag{5.31}$$

and

$$\left(d^i\right)_{nm}=\frac{1}{4\pi}\iint_{\sigma}d(\theta',\lambda')^iY_{nm}(\theta',\lambda')\mathrm{d}\sigma,\quad i=1,2\text{ and }3. \tag{5.32}$$

Fig. 5.2 shows the combined effect of the topographic/bathymetric masses on the second-order radial derivative at an altitude of 250 km above sea level. The ETOPO topographic/bathymetric height model to degree and order 180 corresponding to a resolution of 1° × 1° has been used. In addition, the density of the topographic masses was assumed to be 2670 kg/m^3 and that of water 1000 kg/m^3. The summation of the SHCs presented in Eqs. (5.12) and (5.31) was used to generate the combined

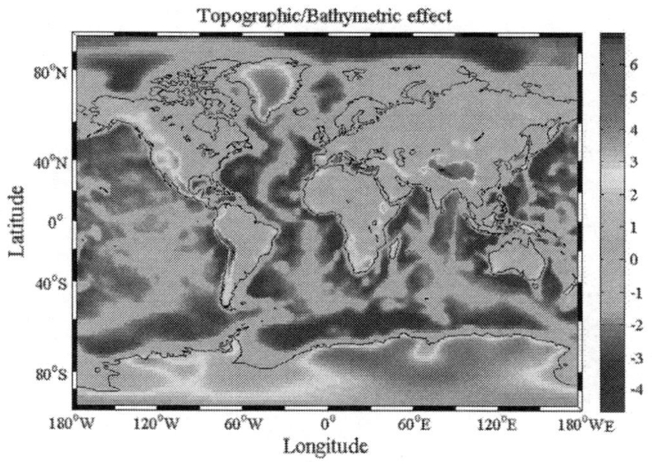

Figure 5.2 The effect of the topographic/bathymetric masses on the second-order radial derivative gradient at 250 km above sea level.

potential and the result was inserted into the spherical harmonic expansion of the second-order radial derivative of the Earth's gravity field, Eq. (2.129). As the figure shows, this effect reaches to a maximum value of about 7E over Himalaya and to -4E over deep parts of the oceans.

5.2 Gravitational potential of crustal layers based on CRUST1.0

The Earth is assumed to have three main layers: crust, mantle and core. The topographic masses are parts of the crust, which has three layers, upper, middle and lower, with different thicknesses and density structures. As mentioned before, in gravity field and geoid determination, the topographic masses play an important role, whilst in studying the Earth's interior structure the densities and thicknesses of these crustal layers are the known part of the gravitational signal, which should be removed from satellite gravimetry data for studying the unknown part. In this section, the goal is to model the gravitational potential of these layers in terms of spherical harmonics and using the CRUST1.0 model (Laske et al., 2013). This model contains eight layers of topographic, bathymetric, upper, middle, and lower layers for the sediments, and crustal crystalline masses as well as ice. The density and thickness of the layers have been estimated from the seismic surveys. Fig. 5.3 shows the crustal layer of CRUST1.0 schematically. In the figure d_{UC}, d_{MC} and d_{LC} are the thicknesses of each crustal layer and ρ^{UC}, ρ^{MC} and ρ^{LC} are the corresponding densities. l stands for the distance between a computation point outside these masses with spherical coordinates (r, θ, λ) and any point inside (integration point) with (r', θ', λ'). Note that the topographic heights are part of the upper crust. The boundary between the bottom of the lower crust and the upper mantle is known as the Moho discontinuity, which is also known as the Moho surface.

The goal is to model the gravitational potential of each layer in terms of spherical harmonics. The radial integral of the Newton formula should now be divided into three parts. One issue that should be considered for the use of the Newton integral in the spherical domain is that the computation and integration points have spherical coordinates, and therefore, the depth of the upper and lower bounds of each layer should be used, and not the thickness. Eq. (5.33) shows the Newton integral for these three crustal layers and as can be seen, the limits of the radial integrals depend on the thicknesses of two or three layer as well as the topographic heights.

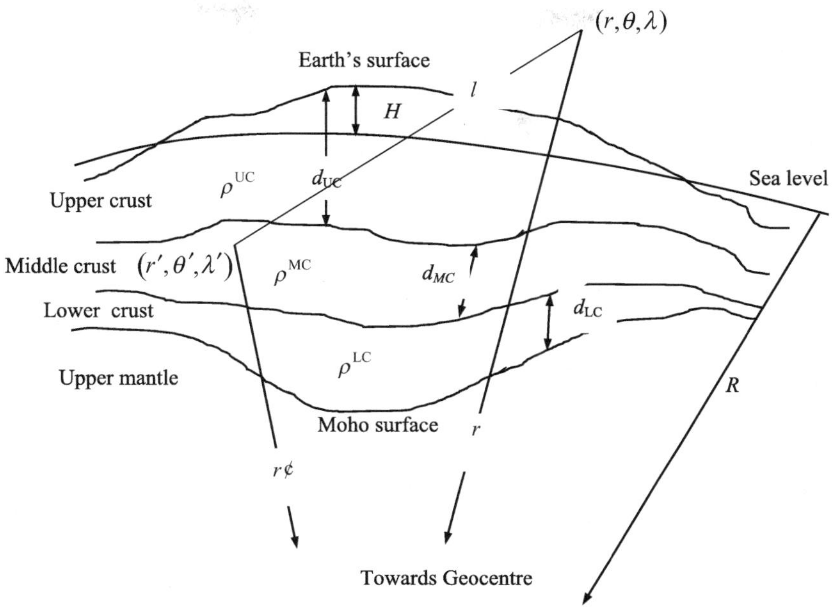

Figure 5.3 Crustal layers.

The thicknesses of the layers have their own uncertainties, therefore the depths computed from them will have larger uncertainties. The prime sign over the parameters of Eq. (5.33) means that they are at integration points:

$$\begin{aligned} V^{\mathrm{C}}(r,\theta,\lambda) = G\iint\limits_{\sigma}\Bigg(& \rho'^{\mathrm{UC}} \int\limits_{R+H(\theta',\lambda')-d_{\mathrm{UC}}(\theta',\lambda')}^{R+H(\theta',\lambda')} \frac{r'^2\mathrm{d}r'}{l} \\ & + \rho'^{\mathrm{MC}} \int\limits_{R+H(\theta',\lambda')-d_{\mathrm{UC}}(\theta',\lambda')-d_{\mathrm{MC}}(\theta',\lambda')}^{R+H(\theta',\lambda')-d_{\mathrm{UC}}(\theta',\lambda')} \frac{r'^2\mathrm{d}r'}{l} \\ & + \rho'^{\mathrm{MC}} \int\limits_{R+H(\theta',\lambda')-d_{\mathrm{UC}}(\theta',\lambda')-d_{\mathrm{MC}}(\theta',\lambda')-d_{\mathrm{LC}}(\theta',\lambda')}^{R+H(\theta',\lambda')-d_{\mathrm{UC}}(\theta',\lambda')-d_{\mathrm{MC}}(\theta',\lambda')} \frac{r'^2\mathrm{d}r'}{l}\Bigg)\mathrm{d}\sigma. \end{aligned} \tag{5.33}$$

The radial integral is divided into three parts, like the crust, to consider the information of each crustal layer separately for deriving the gravitational potential of the crustal masses. Therefore, each radial integral should be solved separately according to its limits. Similar to Eq. (5.4), the integrals can be written in terms of Legendre polynomials:

$$I_1 = \int_{R+H(\theta',\lambda')-d_{UC}(\theta',\lambda')}^{R+H(\theta',\lambda')} \frac{r'^2 dr'}{l} = \sum_{n=0}^{\infty} \frac{1}{r^{n+1}} P_n(\tilde{x}) \int_{R+H(\theta',\lambda')-d_{UC}(\theta',\lambda')}^{R+H(\theta',\lambda')} r'^{n+2} dr'$$
$$= \sum_{n=0}^{\infty} \frac{1}{r^{n+1}} P_n(\tilde{x}) \frac{1}{n+3} [r'^{n+3}]_{R+H(\theta',\lambda')-d_{UC}(\theta',\lambda')}^{R+H(\theta',\lambda')}, \tag{5.34}$$

$$I_2 = \int_{R+H(\theta',\lambda')-d_{UC}(\theta',\lambda')-d_{MC}(\theta',\lambda')}^{R+H(\theta',\lambda')-d_{UC}(\theta',\lambda')} \frac{r'^2 dr'}{l}$$
$$= \sum_{n=0}^{\infty} \frac{1}{r^{n+1}} P_n(\tilde{x}) \int_{R+H(\theta',\lambda')-d_{UC}(\theta',\lambda')-d_{MC}(\theta',\lambda')}^{R+H(\theta',\lambda')-d_{UC}(\theta',\lambda')} r'^{n+2} dr' \tag{5.35}$$
$$= \sum_{n=0}^{\infty} \frac{1}{r^{n+1}} P_n(\tilde{x}) \frac{1}{n+3} [r'^{n+3}]_{R+H(\theta',\lambda')-d_{UC}(\theta',\lambda')-d_{MC}(\theta',\lambda')}^{R+H(\theta',\lambda')-d_{UC}(\theta',\lambda')},$$

$$I_3 = \int_{R+H(\theta',\lambda')-d_{UC}(\theta',\lambda')-d_{MC}(\theta',\lambda')-d_{LC}(\theta',\lambda')}^{R+H(\theta',\lambda')-d_{UC}(\theta',\lambda')-d_{MC}(\theta',\lambda')} \frac{r'^2 dr'}{l}$$
$$= \sum_{n=0}^{\infty} \frac{1}{r^{n+1}} P_n(\tilde{x}) \int_{R+H(\theta',\lambda')-d_{UC}(\theta',\lambda')-d_{MC}(\theta',\lambda')-d_{LC}(\theta',\lambda')}^{R+H(\theta',\lambda')-d_{UC}(\theta',\lambda')-d_{MC}(\theta',\lambda')} r'^{n+2} dr' \tag{5.36}$$
$$= \sum_{n=0}^{\infty} \frac{1}{r^{n+1}} P_n(\tilde{x}) \frac{1}{n+3} [r'^{n+3}]_{R+H(\theta',\lambda')-d_{UC}(\theta',\lambda')-d_{MC}(\theta',\lambda')-d_{LC}(\theta',\lambda')}^{R+H(\theta',\lambda')-d_{UC}(\theta',\lambda')-d_{MC}(\theta',\lambda')}.$$

The solutions of the integrals I_1, I_2 and I_3 are:

$$I_1 = \frac{R^{n+3}}{n+3} \left[\left(1 + \frac{H'}{R}\right)^{n+3} - \left(1 + \frac{H' - d'_{UC}}{R}\right)^{n+3} \right], \tag{5.37}$$

$$I_2 = \frac{R^{n+3}}{n+3} \left[\left(1 + \frac{H' - d'_{UC}}{R}\right)^{n+3} - \left(1 + \frac{H' - d'_{UC-MC}}{R}\right)^{n+3} \right], \tag{5.38}$$

$$I_3 = \frac{R^{n+3}}{n+3}\left[\left(1+\frac{H'-d'_{\mathrm{UC-MC}}}{R}\right)^{n+3} - \left(1+\frac{H'-d'_{\mathrm{UC-MC-LC}}}{R}\right)^{n+3}\right], \tag{5.39}$$

where $d'_{\mathrm{UC-MC}} = d'_{\mathrm{UC}} + d'_{\mathrm{MC}}$ and $d'_{\mathrm{UC-MC-LC}} = d'_{\mathrm{UC}} + d'_{\mathrm{MC}} + d'_{\mathrm{LC}}$.

By expanding each binomial term by the Taylor series to the third order, the solutions of the integral will be:

$$\begin{aligned} I_1 \approx R^{n+3}\Bigg[&\frac{d'_{\mathrm{UC}}}{R} + (n+2)\frac{H'^2 - \left(H' - d'_{\mathrm{UC}}\right)^2}{2R^2} \\ &+ (n+2)(n+1)\frac{H'^3 - \left(H' - d'_{\mathrm{UC}}\right)^3}{6R^3}\Bigg], \end{aligned} \tag{5.40}$$

$$\begin{aligned} I_2 \approx R^{n+3}\Bigg[&\frac{d'_{\mathrm{MC}}}{R} + (n+2)\frac{\left(H' - d'_{\mathrm{UC}}\right)^2 - \left(H' - d'_{\mathrm{UC-MC}}\right)^2}{2R^2} \\ &+ (n+2)(n+1)\frac{\left(H' - d'_{\mathrm{UC}}\right)^3 - \left(H' - d'_{\mathrm{UC-MC}}\right)^3}{6R^3}\Bigg], \end{aligned} \tag{5.41}$$

$$\begin{aligned} I_3 \approx R^{n+3}\Bigg[&\frac{d'_{\mathrm{LC}}}{R} + (n+2)\frac{\left(H' - d'_{\mathrm{UC-MC}}\right)^2 - \left(H' - d'_{\mathrm{UC-MC-LC}}\right)^2}{2R^2} \\ &+ (n+2)(n+1)\frac{\left(H' - d'_{\mathrm{UC-MC}}\right)^3 - \left(H' - d'_{\mathrm{UC-MC-LC}}\right)^3}{6R^3}\Bigg]. \end{aligned} \tag{5.42}$$

Now, these solutions are inserted back into Eqs. (5.34)–(5.36) and the results into Eq. (5.33). After performing the necessary algebraic simplifications, the potential of the crustal masses considering the laterally variable density and thickness of each crustal layer will be:

$$V^{\mathrm{C}}(r,\theta,\lambda) = \sum_{n=0}^{\infty}\left(\frac{R}{r}\right)^{n+1}\sum_{m=-n}^{n}\left(v_{nm}^{\mathrm{UC}} + v_{nm}^{\mathrm{MC}} + v_{nm}^{\mathrm{LC}}\right)Y_{nm}(\theta,\lambda), \tag{5.43}$$

where

$$\begin{aligned} v_{nm}^{\mathrm{UC}} = \frac{4\pi G R^2}{2n+1}\Bigg(&\frac{\left(\rho^{\mathrm{UC}} d_{\mathrm{UC}}\right)_{nm}}{R} + (n+2)\frac{\left(\rho^{\mathrm{UC}} H^2\right)_{nm} - \left(\rho^{\mathrm{UC}}(H - d_{\mathrm{UC}})^2\right)_{nm}}{2R^2} \\ &+ (n+2)(n+1)\frac{\left(\rho^{\mathrm{UC}} H^3\right)_{nm} - \left(\rho^{\mathrm{UC}}(H - d_{\mathrm{UC}})^3\right)_{nm}}{6R^3}\Bigg), \end{aligned} \tag{5.44}$$

$$
\begin{aligned}
v_{nm}^{\mathrm{MC}} = & \frac{4\pi G R^2}{2n+1}\left(\frac{\left(\rho^{\mathrm{LC}} d_{\mathrm{LC}}\right)_{nm}}{R}\right. \\
& + (n+2)\frac{\left(\rho^{\mathrm{MC}}\left(H - d_{\mathrm{UC-MC}}\right)^2\right)_{nm} - \left(\rho^{\mathrm{MC}}\left(Hd'_{\mathrm{UC-MC-LC}}\right)^2\right)_{nm}}{2R^2} \\
& \left. + (n+2)(n+1)\frac{\left(\rho^{\mathrm{MC}}\left(H - d_{\mathrm{UC-MC}}\right)^3\right)_{nm} - \left(\rho^{\mathrm{MC}}\left(H - d'_{\mathrm{UC-MC-LC}}\right)^3\right)_{nm}}{6R^3}\right),
\end{aligned}
\tag{5.45}
$$

$$
\begin{aligned}
v_{nm}^{\mathrm{LC}} = & \frac{4\pi G R^2}{2n+1}\left(\frac{\left(\rho^{\mathrm{LC}} d_{\mathrm{MC}}\right)_{nm} - \left(\rho^{\mathrm{LC}} d_{\mathrm{LC}}\right)_{nm}}{R}\right. \\
& + (n+2)\frac{\left(\rho^{\mathrm{LC}} d_{\mathrm{MC}}^2\right)_{nm} + \left(\rho^{\mathrm{LC}} d_{\mathrm{LC}}^2\right)_{nm}}{2R^2} \\
& \left. + (n+2)(n+1)\frac{\left(\rho^{\mathrm{LC}} d_{\mathrm{MC}}^3\right)_{nm} + \left(\rho^{\mathrm{LC}} d_{\mathrm{LC}}^3\right)_{nm}}{6R^3}\right),
\end{aligned}
\tag{5.46}
$$

where

$$
\begin{gathered}
\left(\rho^i d_j^k\right)_{nm} = \frac{1}{4\pi}\iint_\sigma \rho^i(\theta', \lambda') d_j^k(\theta', \lambda') Y_{nm}(\theta', \lambda') \mathrm{d}\sigma, \\
i,\ j = \mathrm{UC},\ \mathrm{MC}\ \text{and}\ \mathrm{LC};\ k = 1, 2\ \text{and}\ 3,
\end{gathered}
\tag{5.47}
$$

$$
\begin{aligned}
& \left(\rho^i\left(H'^k - \left(H' - d'_{\mathrm{UC}}\right)^k\right)\right)_{nm} \\
& = \frac{1}{4\pi}\iint_\sigma \rho^{i'}\left(H'^k - \left(H' - d'_{\mathrm{UC}}\right)^k\right) Y_{nm}(\theta', \lambda') \mathrm{d}\sigma,
\end{aligned}
\tag{5.48}
$$

$$
\begin{aligned}
& \left(\rho^i((H - d_{\mathrm{UC}})^k - (H - d_{\mathrm{UC-MC}})^k\right)_{nm} \\
& = \frac{1}{4\pi}\iint_\sigma \rho^{i'}\left(\left(H' - d'_{\mathrm{UC}}\right)^k - \left(H' - d'_{\mathrm{UC-MC}}\right)^k\right) Y_{nm}(\theta', \lambda') \mathrm{d}\sigma,
\end{aligned}
\tag{5.49}
$$

$$
\begin{aligned}
& \left(\rho^i((H - d_{\mathrm{UC-MC}})^k - (H - d_{\mathrm{UC-MC-LC}}))^k\right)_{nm} \\
& = \frac{1}{4\pi}\iint_\sigma \rho^{i'}\left(\left(H' - d'_{\mathrm{UC-MC}}\right)^k - \left(H' - d'_{\mathrm{UC-MC-LC}}\right)^k\right) Y_{nm}(\theta', \lambda') \mathrm{d}\sigma.
\end{aligned}
\tag{5.50}
$$

Fig. 5.4 shows the densities and the thicknesses of all layers of the CRUST1.0 model. Those of the upper crust are presented in Fig. 5.4A and B, respectively. Over continents the density is greater than 2.65 gr/cm^3, with the densest part in Antarctica. The less dense masses are located mainly over oceans and along the mid-oceanic ridges. The largest thickness reaches 40 km over Himalaya and the least is observed over oceans. Fig. 5.4C and D are, respectively, the maps of the density and thickness of the middle crust. The former shows that density ranges from 2.7 to 3 gr/cm^3. However, its thickness does not follow the same pattern as the thickness of the upper crust, and large values, reaching 25 km, are observed around the Tibet plateau, Europe and South America. The density and thickness of the lower crust are seen in Fig. 5.4E and F, and according to the former, the density is about 3 gr/cm^3 over oceans and 2.85 gr/cm^3 over Himalaya, Iran and southern Europe, as well as the western part of the American continents. According to Fig. 5.4E the thickest part of the lower crust, reaching about 40 km, is seen in South America and along the Andean belt.

Fig. 5.5 shows the effects of density variation in the crustal layers on the second-order radial derivative of the geopotential at a constant elevation of 250 km. Here, a constant density of 2.670 gr/cm^3 is assumed as a reference value and subtracted from the density of each layer. The SHCs of the gravitational potential of each layer are computed from Eqs. (5.44)–(5.46). These harmonics are, thereafter, inserted into the spherical harmonic expansion of each the second-order radial derivative, Eq. (2.129), for computing the effects. These SHCs can be applied to spherical harmonics expansion of any satellite gravimetry data if these effects are desired on them.

Fig. 5.5A–C shows the effects of the density variation inside the upper, middle and lower crust, respectively. Since the density of the upper crust is close to the assumed reference density the effects come out smaller than those of others. However, as we can observe, the largest value is located in Antarctica based on this model. Fig. 5.5D shows the total effect of crustal density variation and illustrates that it reaches 4E.

5.3 Gravitational potential of sediments

Sediments are located in the upper part of the upper crust and their gravitational potential can be simply determined when their density and thickness are available. In the CRUST1.0 model, the sediment masses are divided into three layers, like the crustal layers, discussed in the previous section. Therefore, the same approach and strategy can be implemented for modelling the

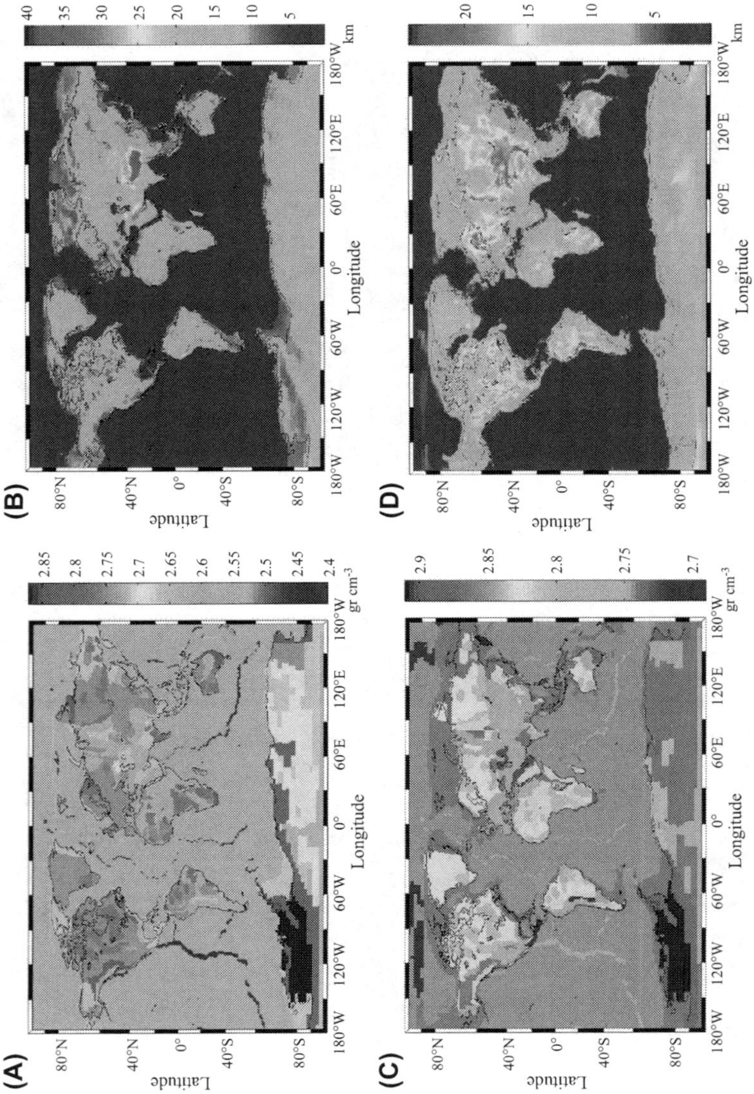
(A)
(B)
(C)
(D)
Latitude
Longitude
80°N
40°N
0°
40°S
80°S
180°W
120°W
60°W
0°
60°E
120°E
180°W
2.85
2.8
2.75
2.7
2.65
2.6
2.55
2.5
2.45
2.4
gr cm-3
40
35
30
25
20
15
10
5
km
2.9
2.85
2.8
2.75
2.7
gr cm-3
20
15
10
5
km

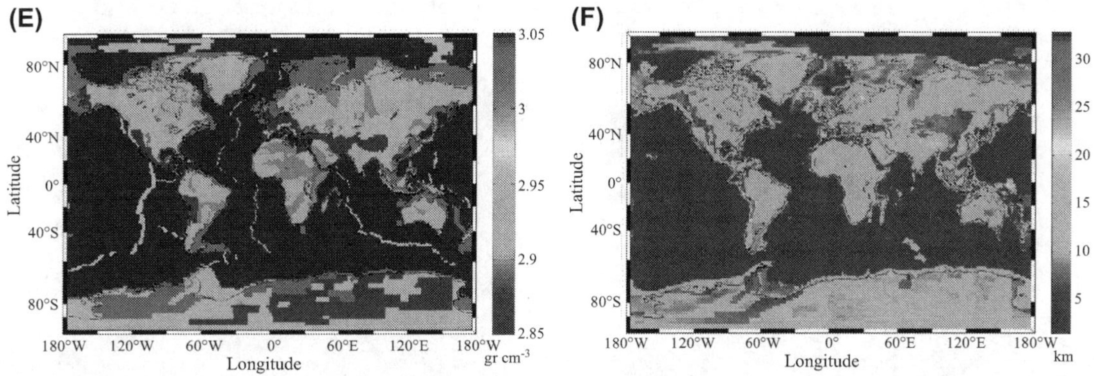

Figure 5.4 Crustal density and thickness according to the CRUST1.0 model. (A) Upper crust density, (B) upper crust thickness, (C) middle crust density, (D) middle crust thickness, (E) lower crust density, (F) lower crust thickness.

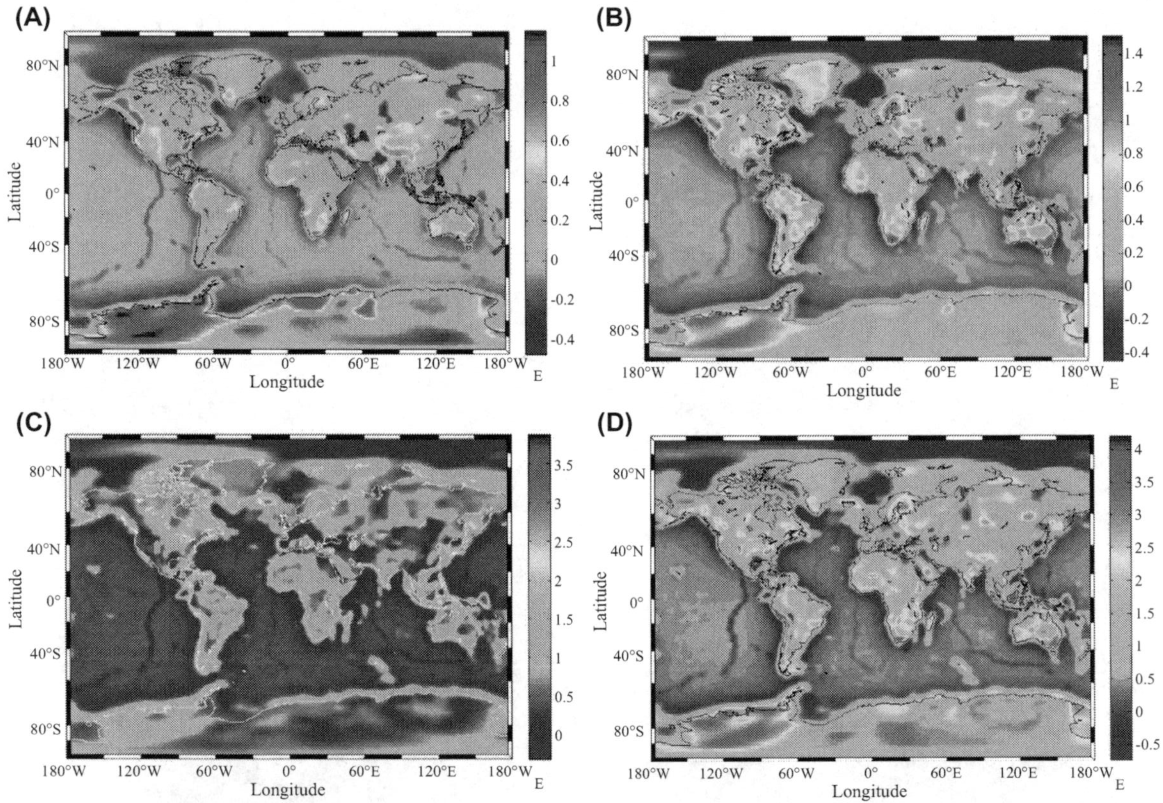

Figure 5.5 The effect of density variation of each crustal layer on the second-order radial derivative of the gravitational potential at 250 km level. (A) Upper crust density and thickness variations, (B) middle crust density and thickness variations, (C) lower crust density and thickness variations, (D) total crust density and thickness variations.

potential of the sediments. In addition, the sediments are compacted with time, meaning that their density increases with depth as well. This makes modelling of the sediment gravitational potential more complicated than that the topographic/bathymetric or crustal layers. So far different mathematical models have been proposed for the density of sediment with respect to the depth, or the density contrast from the upper bound of the sediment layer down to its bottom. Unlike the density, which increases with depth, the density contrast decreases, meaning that the difference between the density at larger depths is closer to the density at the bottom of the sedimentary layers. In this section, some of these density contrast models such as exponential, hyperbolic and exponential compaction models are used for modelling the gravitational potential of sediment masses. In the first step, modelling the sediment potential in terms of spherical harmonics is investigated, and later modelling the SHCs based on different density-contrast or density models will be presented.

5.3.1 Gravitational potential of sediments based on the CRUST1.0 model

Using the CRUST1.0 model, the density and thickness of sediments are presented in three layers of upper, middle and lower by the schematic in Fig. 5.6. Consider that the sediments are the upper part of the solid Earth topographic masses; therefore, the surface of the sediments will be the same as the Earth's surface, when sediments exist.

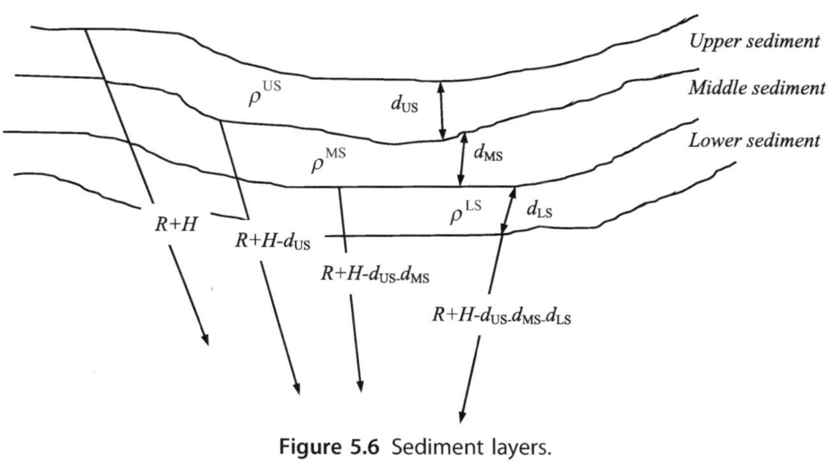

Figure 5.6 Sediment layers.

Fig. 5.6 shows the thickness of each sediment layer and the density using the CRUST1.0 model. Here d_{US}, d_{MS} and d_{LS} are, respectively, the thicknesses of the upper, middle and lower sediments, and ρ^{US}, ρ^{MS} and ρ^{LS} the corresponding density of each layer. The procedure for modelling the gravitational potential of these layers based on such information is very similar for those three layers of the crust. According to the Newton integral, the potential of the sediments is:

$$
\begin{aligned}
V^{S}(r,\theta,\lambda) = G\iint_{\sigma}\Bigg(& \rho'^{US}(\theta',\lambda') \int_{R+H(\theta',\lambda')-d_{US}(\theta',\lambda')}^{R+H(\theta',\lambda')} \frac{r'^2 \mathrm{d}r'}{l} \\
& + \rho'^{MS}(\theta',\lambda') \int_{R+H(\theta',\lambda')-d_{US}(\theta',\lambda')-d_{MS}(\theta',\lambda')}^{R+H(\theta',\lambda')-d_{US}(\theta',\lambda')} \frac{r'^2 \mathrm{d}r'}{l} \\
& + \rho'^{LS}(\theta',\lambda') \int_{R+H(\theta',\lambda')-d_{US}(\theta',\lambda')-d_{MS}(\theta',\lambda')-d_{LS}(\theta',\lambda')}^{R+H(\theta',\lambda')-d_{US}(\theta',\lambda')-d_{MS}(\theta',\lambda')} \frac{r'^2 \mathrm{d}r'}{l} \Bigg) \mathrm{d}\sigma \\
= {} & I_1 + I_2 + I_3.
\end{aligned} \tag{5.51}
$$

The first part of Eq. (5.51) yields the gravitational potential of the first layer of sediments. As the limits of the radial integral show, the lower limit of the integral is the bottom of the upper sediment and the upper one is the solid Earth's surface. By using the Legendre expansion of the reciprocal distance, Eq. (5.3), a formula similar to Eq. (5.4) is, for the first integral of the potential, Eq. (5.51):

$$
\begin{aligned}
V^{US}(r,\theta,\lambda) = GR^2 \iint_{\sigma}\Bigg(& \rho'^{US}(\theta',\lambda') \sum_{n=0}^{\infty} \frac{1}{r^{n+1}} P_n(\tilde{x}) \\
& \int_{R+H(\theta',\lambda')-d_{US}(\theta',\lambda')}^{R+H(\theta',\lambda')} r'^2 \mathrm{d}r' \mathrm{d}\sigma.
\end{aligned} \tag{5.52}
$$

The solution of Eq. (5.52) will be a spherical harmonics series similar to Eq. (5.11), but instead of v^{T}_{nm}, the following SHCs should be used:

$$v^{US}_{nm} = \frac{4\pi GR^2}{2n+1}\left[\frac{(\rho^{US} d_S)_{nm}}{R} + (n+2)\frac{\left(\rho^{US}\left(H^2 - (H-d_S)^2\right)\right)_{nm}}{2R^2} + (n+2)(n+1)\frac{\left(\rho^{US}\left(H^3 - (H-d_S)^3\right)\right)_{nm}}{6R^3}\right]. \tag{5.53}$$

In a very similar manner, the SHCs of the second and third terms of Eq. (5.51) will be:

$$v^{MS}_{nm} \approx \frac{4\pi GR^2}{2n+1}\left(\frac{(\rho^{MS} d_{MS})_{nm}}{R} + (n+2)\frac{\left(\rho^{MS}\left((H-d_{US-MS})^2 - (H-d_{US-MS})^2\right)\right)_{nm}}{2R^2} + (n+2)(n+1)\frac{\left(\rho^{MS}\left((H-d_{US})^3 - (H-d_{US-MS})^3\right)\right)_{nm}}{6R^3}\right) \tag{5.54}$$

and

$$v^{LS}_{nm} \approx \frac{4\pi GR^2}{2n+1}\left(\frac{(\rho^{LS} d_{LS})_{nm}}{R} + (n+2)\frac{\left(\rho^{LS}\left((H-d_{US-MS})^2 - (H-d_{US-MS-LS})^2\right)\right)_{nm}}{2R^2} + (n+2)(n+1)\frac{\left(\rho^{LS}\left((H-d_{US-MS})^3 - (H-d_{US-MS-LS})^3\right)\right)_{nm}}{6R^3}\right), \tag{5.55}$$

where

$$\left(\rho^i d_j^k\right)_{nm} = \frac{1}{4\pi}\iint_\sigma \rho^i(\theta',\lambda') d_j^k(\theta',\lambda') Y_{nm}(\theta',\lambda')\,d\sigma, \tag{5.56}$$

$$\left(\rho^{i}\left(H'^k - (H'-d'_{US})^k\right)\right)_{nm} = \frac{1}{4\pi}\iint_\sigma \rho^{i'}\left(H'^k - (H'-d'_{US})^k\right) Y_{nm}(\theta',\lambda')\,d\sigma, \tag{5.57}$$

$$\left(\rho^i\left((H-d_{US})^k - (H-d_{US-MS})^k\right)\right)_{nm} = \frac{1}{4\pi}\iint_\sigma \rho^{i'}\left((H'-d'_{US})^k - (H'-d'_{US-MS})^k\right) Y_{nm}(\theta',\lambda')\,d\sigma, \tag{5.58}$$

$$\begin{aligned}&\left(\rho^i\left((H-d_{\text{US-MS}})^k-(H-d_{\text{US-MS-LS}})^k\right)\right)_{nm}\\&=\frac{1}{4\pi}\iint\limits_{\sigma}\rho^{i'}\left(\left(H'-d'_{\text{US-MS}}\right)^k-\left(H'-d'_{\text{US-MS-LS}}\right)^k\right)\\&Y_{nm}(\theta',\lambda')\mathrm{d}\sigma\ i,\ j=\text{US, MS and LS};\ \ k=1,2\text{ and }3.\end{aligned}\tag{5.59}$$

The spherical harmonic expansion of all sediment layers will have the same spherical harmonic expansion as Eq. (5.43), but $\upsilon_{nm}^{\text{UC}}$, $\upsilon_{nm}^{\text{MC}}$, $\upsilon_{nm}^{\text{LC}}$ should be replaced by $\upsilon_{nm}^{\text{US}}$, $\upsilon_{nm}^{\text{MS}}$ and $\upsilon_{nm}^{\text{LS}}$, respectively.

Fig. 5.7A and B is the map of the density and thickness of the upper sediments, respectively. The former illustrates that CRUST1.0 gives constant density over large areas, for example, over the American continents, Greenland, a large part of Asia and Antarctica as well as the oceans. This could be due to the lack of seismic data. However, the thickness of this layer, Fig. 5.7B, shows more details about the structure of the masses than the density. According to Fig. 5.7B, the upper sediment thickness is large at the coastal lines and continental margins and close to zero over Fennoscandia, Greenland and the northeast part of the Americas in addition to large places over the oceans. The upper sediment density varies from 1.5 to 2.4 kg/m^3 and thickness reaches about 2.5 km according to CRUST1.0. Fig. 5.7C and D show the density and thickness of the middle sediment layer. The densities range from 2000 to 2500 kg/m^3 and these masses are mainly located over the continents and coastal lines. The thickness of the middle sediments reaches 4 km. The density of the lower sediments reaches about 2600 kg/m^3, these masses have limited coverage compared with the rest of the layers; see Fig. 5.7E. As the map of the lower sediment thickness shows, Fig. 5.7F, the thickness reaches about 16 km in the Caspian Sea.

Here, we assumed a reference density of 2670 kg/m^3 and subtracted it from the density of the sediment of each layer. From Eqs. (5.53), (5.54) and (5.55), the SHCs of the gravitational potential of each sediment layer are obtained. Later they are inserted into Eq. (2.129) to compute the effect of each layer on the second-order radial derivative of the potential. Fig. 5.8A–C are maps of the effect of density variation from the reference value inside upper, middle and lower sediment layers, respectively. The figure shows that the effect varies between 0.8E and 0.2E for the upper sediment, −0.8E to 0.2E for the middle and −0.3E to 0.2E for the lower. Fig. 5.8D is the map of density variations considering all sediment layers. This total effect is between −1.6E and 0.4E on the vertical–vertical gradient.

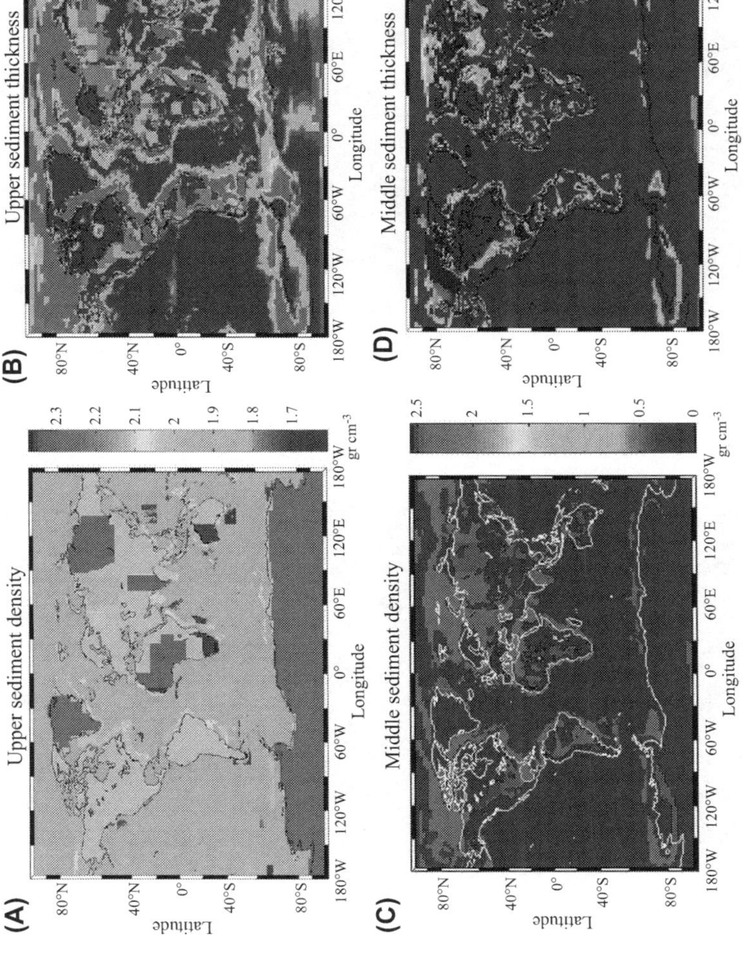
(A)
Upper sediment density
gr cm⁻³
(B)
Upper sediment thickness
km
(C)
Middle sediment density
gr cm⁻³
(D)
Middle sediment thickness
km
Latitude
Longitude

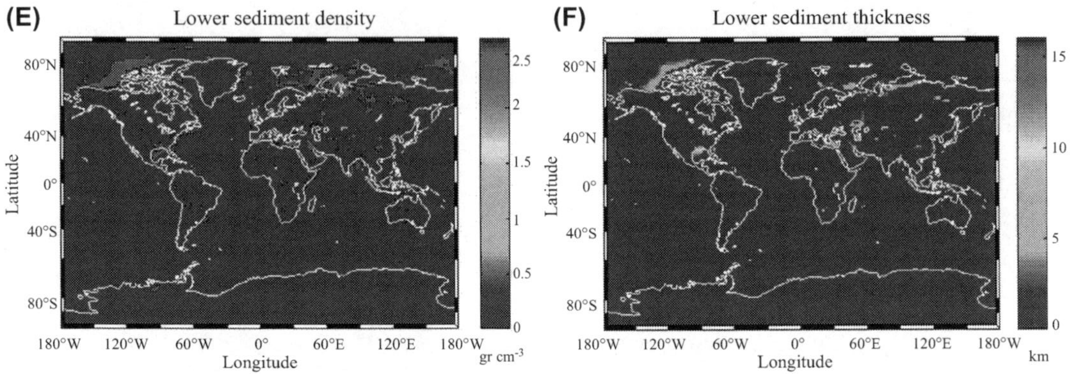

Figure 5.7 (A) Density and (B) thickness for the upper sediment in the CRUST1.0 model, (C) density and (D) thickness of middle sediment and (E) density and (F) thickness of lower sediment.

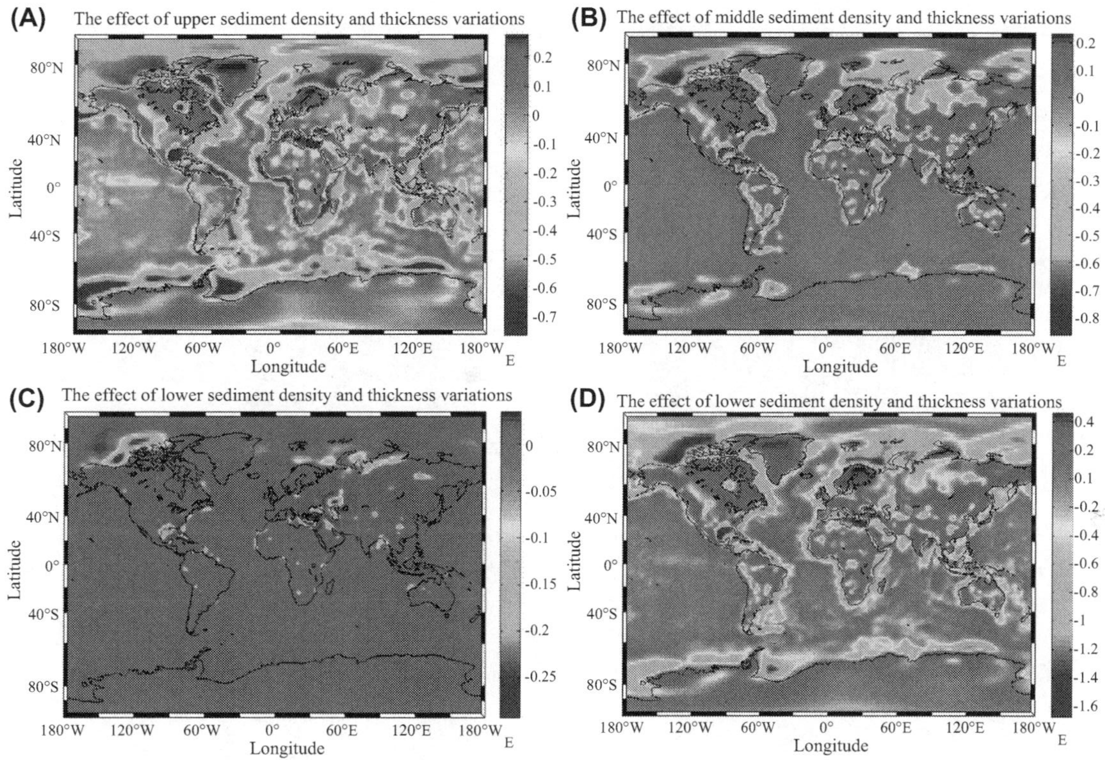

Figure 5.8 The lateral density variation effect of (A) upper, (B) middle, (C) lower and (D) all sediment layers on the vertical–vertical gravity gradient.

5.3.2 Gravitational potential of sediments based on density models

The density of sediment masses increases fast by depth; therefore, a mathematical model can be used for presenting the sediment densities as a function of depth. In such a case, modelling the gravitational potential of the sediment masses will not be simple due to lateral and radial changes of density. In some studies, the density contrast between the upper and the lower parts of a sediment layer is used for modelling the gravitational potential, which is mainly used for determining the thickness of the sediment layers from the gravimetry data. Exponential and hyperbolic density models are two known functions for expressing the radial variation of the contrast. There is also another model for the density of the sediments known as the exponential compact model, which will be discussed later in this section.

Fig. 5.9 shows that the sediment masses are the upper layer of the topographic masses. This principle as well as the density model of the sediments will be applied to determine the gravitational potential or the potential contrast due to sediment masses.

Fig. 5.10 shows the plot of the density contrast of sediments, as presented by Litinsky (1989) for the San Jacinto graven in California. The surface density contrast is assumed as $-0.55\ \mathrm{gr/cm^3}$. Here, an exponential function is fitted to his data and a value of -0.0014 is estimated for its exponent. The constant of the hyperbolic function $\beta = 1368.08$ m is estimated by fitting this model to these data as well. As the figure shows, the hyperbolic model presents the density contrast radial variations slightly better than the exponential one after depths larger than 1 km. In the following, the mathematical models of these density functions are presented and used for modelling the gravitational potential contrast due to the sediment masses.

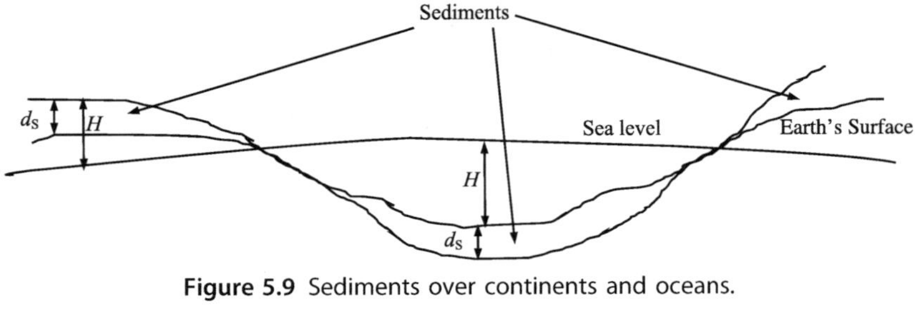

Figure 5.9 Sediments over continents and oceans.

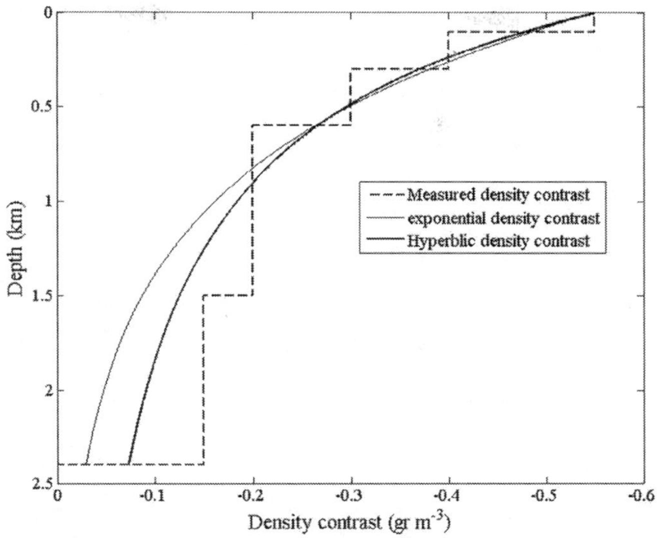

Figure 5.10 Different density contrast models (from Litinsky 1989 in San Jacinto).

5.3.3 Sediments' gravitational potential based on the exponential density contrast model

The exponential density contrast model has the following formula:

$$\Delta\rho^{\mathrm{S}}(r,\theta,\lambda)=\Delta\rho_0^{\mathrm{S}}(\theta,\lambda)e^{-\alpha(R+H(\theta,\lambda)-r)}\quad r\leq R+H, \tag{5.60}$$

where r is the geocentric distance of any point inside the sediment masses, H the topographic heights or bathymetric depths and $\alpha=-0.0014$ is the constant coefficient, which was derived based on the fitting this model to Litinsky's (1989) data. Note that this value is valid for the San Jacinto graven in California and it is not a global value.

Replacing $\rho^{\mathrm{T}}\left(\theta',\lambda'\right)$ by Eq. (5.60) in the Newton integral (Eq. 5.1), applying Eq. (5.3) and simplifying the results yield:

$$\Delta V^{\mathrm{S}}(r,\theta,\lambda)=\mathrm{G}\iint_{\sigma}\sum_{n=0}^{\infty}\frac{1}{r^{n+1}}P_n(\tilde{x})\Delta\rho_0^{\mathrm{S}}\int_{R+H(\theta',\lambda')-d_{\mathrm{S}}(\theta',\lambda')}^{R+H(\theta',\lambda')}e^{-\alpha(R+H(\theta',\lambda')-r)}r'^{n+2}\mathrm{d}r'\mathrm{d}\sigma. \tag{5.61}$$

Solving the radial integral (Eq. 5.61) is rather complicated. Here, we present two approaches for doing so.

5.3.3.1 Approach 1

First, the exponential function (Cordell, 1973) is written in the following form:

$$e^{-\alpha(R+H(\theta',\lambda')-r)} = e^{-\alpha(R+H(\theta',\lambda'))}e^{\alpha r}. \tag{5.62}$$

In this case, those parameters which are radially constant are separated from r. Now, consider the Taylor expansion of the exponential function as:

$$e^{\alpha r} = \sum_{k=0}^{\infty} \frac{\alpha^k r^k}{k!}. \tag{5.63}$$

Substitution of Eq. (5.63) into Eq. (5.62) and the results into Eq. (5.61) gives:

$$\Delta V^{S}(r,\theta,\lambda) = G\iint\limits_{\sigma} \sum_{n=0}^{\infty} \frac{1}{r^{n+1}} P_n(\tilde{x})\Delta\rho_0^{S} e^{-\alpha(R+H(\theta',\lambda'))} \sum_{k=0}^{\infty} \frac{(\alpha)^k}{k!} \int\limits_{R+H(\theta',\lambda')-d_S(\theta',\lambda')}^{R+H(\theta',\lambda')} r'^{n+k+2} \mathrm{d}r' \mathrm{d}\sigma. \tag{5.64}$$

Solving the radial integral and further simplifications lead to:

$$\begin{aligned} \Delta V^{S}(r,\theta,\lambda) = G\iint\limits_{\sigma} \sum_{n=0}^{\infty} \frac{1}{r^{n+1}} P_n(\tilde{x})\Delta\rho_0^{S} e^{-\alpha(R+H(\theta',\lambda'))} \sum_{k=0}^{\infty} \frac{\alpha^k}{k!} \\ \times \frac{1}{n+k+3}\Big[(R+H(\theta',\lambda'))^{n+k+3} \\ -(R+H(\theta',\lambda')-d_S(\theta',\lambda'))^{n+k+3}\Big]\mathrm{d}\sigma. \end{aligned} \tag{5.65}$$

Eq. (5.65) can be rewritten as:

$$\begin{aligned} \Delta V^{S}(r,\theta,\lambda) = G\iint\limits_{\sigma} \sum_{n=0}^{\infty} \frac{1}{r^{n+1}} P_n(x)\Delta\rho_0^{S} e^{-\alpha(R+H(\theta',\lambda'))} \\ \sum_{k=0}^{\infty} \frac{\alpha^k}{k!} \frac{1}{n+k+3}(R+H(\theta',\lambda'))^{n+k+3} \\ \times \left(1-\left(1-\frac{d_S(\theta',\lambda')}{R+H(\theta',\lambda')}\right)^{n+k+3}\right)\mathrm{d}\sigma. \end{aligned} \tag{5.66}$$

By assuming that $d_S(\theta', \lambda') \ll R + H(\theta', \lambda')$ we simplify the binomial term by the Taylor series:

$$\left(1 - \left(1 - \frac{d_S(\theta', \lambda')}{R + H(\theta', \lambda')}\right)^{n+k+3}\right) \approx (n+k+3)\frac{d_S(\theta', \lambda')}{R + H(\theta', \lambda')} - (n+k+3)(n+k+2)\frac{d_S^2(\theta', \lambda')}{2(R + H(\theta', \lambda'))^2}. \tag{5.67}$$

By inserting Eq. (5.67) into Eq. (5.66) we get:

$$\Delta V^S(r, \theta, \lambda) = G\iint_\sigma \sum_{n=0}^{\infty} \frac{1}{r^{n+1}} P_n(\tilde{x})\Delta\rho_0^S e^{-\alpha(R+H(\theta',\lambda'))} \left[\sum_{k=0}^{\infty} \frac{\alpha^k}{k!}(R + H(\theta', \lambda'))^k (R + H(\theta', \lambda'))^{n+3} \left(\frac{d_S(\theta', \lambda')}{R + H(\theta', \lambda')} - (n+k+2)\frac{d_S^2(\theta', \lambda')}{2(R + H(\theta', \lambda'))^2}\right) d\sigma. \tag{5.68}$$

Since

$$\sum_{k=0}^{\infty} \frac{\alpha^k}{k!}(R + H(\theta', \lambda'))^k = e^{R+H(\theta',\lambda')}, \tag{5.69}$$

$$\sum_{k=0}^{\infty} \frac{\alpha^k}{k!}(R + H(\theta', \lambda'))^k (n+2+k) = e^{\alpha(R+H(\theta',\lambda'))}(n+2) + \alpha(R + H(\theta', \lambda'))e^{\alpha(R+H(\theta',\lambda'))}. \tag{5.70}$$

Therefore, Eq. (5.68) will be simplified, after some manipulations, to:

$$\Delta V^S(r, \theta, \lambda) = GR^2\iint_\sigma \sum_{n=0}^{\infty} \frac{R^{n+1}}{r^{n+1}} P_n(\tilde{x})\Delta\rho_0^S \left[\left(1 + \frac{H'}{R}\right)^{n+3} \left(\frac{d'_S}{R + H'} - (n+2+\alpha(R + H'))\frac{d'^2_S}{2(R + H')^2}\right) d\sigma. \tag{5.71}$$

Again, since H is changing laterally, the binomial term involving it should be approximated by the Taylor series, Eq. (5.67). This means that Eq. (5.71) will change to:

$$\Delta V^{\mathrm{S}}(r,\theta,\lambda) = GR^2 \iint_\sigma \sum_{n=0}^{\infty} \frac{R^{n+1}}{r^{n+1}} P_n(\tilde{x}) \Delta\rho_0^{\mathrm{S}} \left[\left(1 + (n+3)\frac{H'}{R} + (n+3)(n+2)\frac{H'^2}{2R^2} \right) \left(\frac{d'_{\mathrm{S}}}{R+H'} - (n+2+\alpha(R+H'))\frac{d'^2_{\mathrm{S}}}{2(R+H')^2} \right) \right] d\sigma. \tag{5.72}$$

The gravitational potential contrast between the surface and the bottom of the sediment layer is derived by applying the addition theorem of spherical harmonics, Eq. (1.9), and performing the horizontal integral:

$$\Delta V^{\mathrm{S}}(r,\theta,\lambda) = \sum_{n=0}^{\infty} \left(\frac{R}{r}\right)^{n+1} \sum_{m=-n}^{n} \Delta v_{nm}^{\mathrm{S}} Y_{nm}(\theta,\lambda), \tag{5.73}$$

where

$$\begin{aligned} \Delta v_{nm}^{\mathrm{S}} = \frac{4\pi GR^2}{2n+1} \Bigg(& \left(\frac{\Delta\rho_0^{\mathrm{S}} d_{\mathrm{S}}}{R+H}\right)_{nm} + \left(\frac{n+3}{R}\right) \left(\frac{\Delta\rho_0^{\mathrm{S}} H d_{\mathrm{S}}}{R+H}\right)_{nm} \\ & + \frac{(n+3)(n+2)}{2R^2} \left(\frac{\Delta\rho_0^{\mathrm{S}} H^2 d_{\mathrm{S}}}{R+H}\right)_{nm} - \frac{(n+2)}{2} \left(\frac{\Delta\rho_0^{\mathrm{S}} d_{\mathrm{S}}^2}{(R+H)^2}\right)_{nm} \\ & - \frac{(n+2)(n+3)}{2R} \left(\frac{\Delta\rho_0^{\mathrm{S}} H d_{\mathrm{S}}^2}{(R+H)^2}\right)_{nm} - \frac{(n+3)(n+2)^2}{4R^2} \left(\frac{\Delta\rho_0^{\mathrm{S}} H^2 d_{\mathrm{S}}^2}{(R+H)^2}\right)_{nm} \\ & - \frac{\alpha}{2} \left(\frac{\Delta\rho_0^{\mathrm{S}} d_{\mathrm{S}}^2}{(R+H)}\right)_{nm} - \frac{\alpha(n+3)}{2R} \left(\frac{\Delta\rho_0^{\mathrm{S}} H d_{\mathrm{S}}^2}{(R+H)}\right)_{nm} \\ & - \frac{\alpha(n+3)(n+2)}{4R^2} \left(\frac{\Delta\rho_0^{\mathrm{S}} H^2 d_{\mathrm{S}}^2}{(R+H)}\right)_{nm} \Bigg), \end{aligned} \tag{5.74}$$

$$\left(\frac{\Delta\rho_0^{\mathrm{S}} H^j d_{\mathrm{S}}^i}{(R+H)^t}\right)_{nm} = \frac{1}{4\pi} \iint_\sigma \left(\frac{\Delta\rho'^{\mathrm{S}}_0 H'^j d'^i_{\mathrm{S}}}{(R+H')^t}\right) Y_{nm}(\theta,\lambda) d\sigma \tag{5.75}$$

$$j = 0,\ 1 \text{ or } 2, \quad i = 0, 1 \text{ or } 2, \quad t = 1 \text{ or } 2.$$

5.3.3.2 Approach 2

In this approach, the radial integral in Eq. (5.61) is solved differently. Let us rewrite this radial integral in the following form:

$$\begin{aligned} I &= \int_{R+H(\theta',\lambda')-d_S(\theta',\lambda')}^{R+H(\theta',\lambda')} e^{-\alpha(R+H(\theta',\lambda')-r)} r'^{n+2} \mathrm{d}r' \\ &= -\int_{-d_S(\theta',\lambda')}^{0} e^{-\alpha z(\theta',\lambda')} (R+H-z(\theta',\lambda'))^{n+2} \mathrm{d}z, \end{aligned} \tag{5.76}$$

where $R+H(\theta',\lambda')-r=z$ and therefore $R+H(\theta',\lambda')-z=r$. Eq. (5.76) will then change to:

$$I=(R+H)^{n+2} \int_{-d_S(\theta',\lambda')}^{0} e^{-\alpha z} \left(1-\frac{z}{R+H}\right)^{n+2} \mathrm{d}z. \tag{5.77}$$

If the binomial term on the right-hand side (rhs) of Eq. (5.77) is approximated up to the second term and the integral is solved, the result will be:

$$\begin{aligned} I = &-R^{n+2}\left(1+\frac{H}{R}\right)^{n+2}\left[\frac{-1}{\alpha}\left(1-e^{\alpha d_S(\theta',\lambda')}\right)\right. \\ &\left.+\alpha^{-2}\frac{n+2}{R+H}\left(1-e^{\alpha d_S(\theta',\lambda')}\left(-d_S(\theta',\lambda')\alpha+1\right)\right)\right]. \end{aligned} \tag{5.78}$$

Now $\left(1+\frac{H}{R}\right)^{n+2}$ is also approximated by the Taylor series up to the third term and inserted back into Eq. (5.78):

$$\begin{aligned} I = &-R^{n+2}\left(1+(n+2)\frac{H}{R}+(n+2)(n+1)\frac{H^2}{2R^2}\right) \\ &\times\left(A+\frac{n+2}{\alpha^2}B\right), \end{aligned} \tag{5.79}$$

where

$$A=\frac{-1}{\alpha}\left(1-e^{\alpha d_S(\theta',\lambda')}\right), \tag{5.80}$$

$$B=\frac{\left(1-e^{\alpha d_S(\theta',\lambda')}\left(-d_S(\theta',\lambda')\alpha+1\right)\right)}{R+H}, \tag{5.81}$$

In this case, Eq. (5.79) will have the expanded form:

$$
\begin{aligned}
I = & -R^{n+2}\left(A+(n+2)\frac{HA}{R}+(n+2)(n+1)\frac{AH^2}{2R^2}\right)\\
& +\frac{1}{\alpha^2}\left(B+(n+2)^2\frac{HB}{R}+(n+2)^2(n+1)\frac{BH^2}{2R^2}\right).
\end{aligned} \tag{5.82}
$$

By inserting Eq. (5.82) back into Eq. (5.61) and using the addition theorem of the spherical harmonics, Eq. (1.9), and performing surface integration over the sphere, the spherical harmonic expansion of the potential contrast will be obtained, Eq. (5.72), but with the following SHCs:

$$
\begin{aligned}
\Delta v_{nm}^{S} = & -\frac{4\pi GR}{2n+1}\left(\left(\Delta\rho_0^S A\right)_{nm}+\left(\frac{n+2}{R}\right)\left(\Delta\rho_0^S HA\right)_{nm}\right.\\
& +\frac{(n+2)(n+1)}{2R^2}\left(\Delta\rho_0^S H^2 A\right)_{nm}\\
& +\frac{\left(\Delta\rho_0^S B\right)_{nm}}{\alpha^2}+\frac{(n+2)^2}{R\alpha^2}\left(\Delta\rho_0^S HB\right)_{nm}\\
& \left.+\frac{(n+2)^2(n+1)}{2\alpha^2R^2}\left(\Delta\rho_0^S BH^2\right)_{nm}\right),
\end{aligned} \tag{5.83}
$$

where

$$
\left(\Delta\rho_0^S H^j A\right)_{nm} = \frac{1}{4\pi}\iint_{\sigma} \Delta\rho_0'^S H'^j A' Y_{nm}(\theta',\lambda')\,d\sigma,\ j=0,\ 1,\ \text{or } 2, \tag{5.84}
$$

$$
\left(\Delta\rho_0^S H^j B\right)_{nm} = \frac{1}{4\pi}\iint_{\sigma} \Delta\rho_0'^S H'^j B' Y_{nm}(\theta',\lambda')\,d\sigma,\ j=0,\ 1,\ \text{or } 2. \tag{5.85}
$$

5.3.4 Sediments' gravitational potential based on the hyperbolic density contrast model

Litinsky (1989) proposed another form of the density contrast model for the sediments, and called it the hyperbolic contrast model in planar form. This model is also a decreasing function with respect to depth inside the sediments, and here a spherical form is written for this model:

$$
\Delta\rho^S(r,\theta,\lambda) = \Delta\rho_0^S(\theta,\lambda)\frac{\beta^2}{\left(R+H(\theta,\lambda)-r+\beta\right)^2}, \tag{5.86}
$$

where $\beta = 1368.08$ m stands for a constant, which should be estimated by fitting the model to the San Jacinto graven in the California sediment data (Cordell, 1973).

Again, to obtain the potential contrast of the sediment masses based on this model, it is inserted into the Newton integral (Eq. 5.1) but with Eq. (5.86) as the density model:

$$\Delta V^{S}(r,\theta,\lambda)=G\iint_{\sigma}\sum_{n=0}^{\infty}\frac{1}{r^{n+1}}P_n(\tilde{x})\Delta\rho_0^{S}(\theta',\lambda')\int_{R+H(\theta',\lambda')-d_S(\theta',\lambda')}^{R+H(\theta',\lambda')}\frac{r'^{n+2}\mathrm{d}r'}{(R+H-r'+\beta)^2}. \tag{5.87}$$

Let us now start simplification of the radial integral prior to solving it:

$$I=-\int_{d_S}^{0}\frac{(R+H-z)^{n+2}\mathrm{d}z}{(z+\beta)^2}=(R+H)^{n+2}\int_{0}^{d_S}\frac{\left(1-\dfrac{z}{R+H}\right)^{n+2}\mathrm{d}z}{(z+\beta)^2}$$
$$=(R+H)^{n+2}\left\{\frac{1}{d_S+\beta}-\frac{1}{\beta}+\frac{n+2}{R+H}\left(\ln(d_S+\beta)+\frac{\beta}{\beta+d_S}-(\ln(\beta)+1)\right)\right\}, \tag{5.88}$$

or

$$I=(R+H)^{n+2}\{A+(n+2)B\}, \tag{5.89}$$

with

$$A=\frac{1}{d_S+\beta}-\frac{1}{\beta},$$
$$B=\frac{1}{R+H}\left(\ln(d_S+\beta)-\frac{\beta}{\beta+d_S}+(\ln(\beta)+1)\right). \tag{5.90}$$

Eq. (5.89) can be rewritten by expanding the binomial term involving the topographic heights or bathymetric depths to the third term by the Taylor series:

$$I\approx R^{n+2}\left(1+(n+2)\frac{H}{R}+(n+2)(n+1)\frac{H^2}{2R^2}\right)\{A+(n+2)B\}. \tag{5.91}$$

Eq. (5.91) is the approximate solution of the radial integral in Eq. (5.87), and by inserting it into Eq. (5.87), applying the addition theorem (Eq. 1.9) and performing the surface integral, the gravitational potential contrast will have the same spherical harmonic expansion as in Eq. (5.73) but with the SHCs:

$$
\begin{aligned}
\Delta v_{nm}^{\mathrm{S}} = \frac{4\pi G R \beta^2}{2n+1} \bigg\{ & \left(\Delta\rho_0^{\mathrm{S}} A\right)_{nm} + (n+2)\left(\Delta\rho_0^{\mathrm{S}} B\right)_{nm} \\
& + \frac{n+2}{R}\left(\left(\Delta\rho_0^{\mathrm{S}} AH\right)_{nm} + (n+2)\left(\Delta\rho_0^{\mathrm{S}} BH\right)_{nm}\right) \\
& + \frac{(n+2)(n+1)}{2R^2}\left(\left(\Delta\rho_0^{\mathrm{S}} AH^2\right)_{nm}\right. \\
& \left. + (n+1)\left(\Delta\rho_0^{\mathrm{S}} BH^2\right)_{nm}\right) \bigg\}.
\end{aligned}
\tag{5.92}
$$

The topographic and bathymetric heights are available globally. If the surface sediment density contrast $\Delta\rho_0^{\mathrm{S}}$ and the sediment thickness d_S are given with a global coverage, the SHCs (Eq. 5.92) can be simply calculated and, after inserting them into Eq. (5.73), the potential of the sediments will be achieved.

5.3.5 Sediments' gravitational potential based on the exponential compaction density model

The difference between the models discussed so far and the exponential compaction density model is that the density inside the sediments is given by this model rather than the density contrast. In addition, another parameter, porosity, is used in this model, which was not applicable in the density contrast model. Generally the exponential compaction density model is (see e.g. Braitenberg et al., 2006):

$$
\begin{aligned}
\rho^{\mathrm{S}}(r,\theta,\lambda) = & \left(1 - \Phi_0(\theta,\lambda) e^{-\frac{R+H(\theta,\lambda)-r}{k}}\right) \rho^{\mathrm{UC}}(\theta,\lambda) \\
& + \Phi_0(\theta,\lambda) e^{-\frac{R+H(\theta,\lambda)-r}{k}} \rho^{\mathrm{B}}(\theta,\lambda),
\end{aligned}
\tag{5.93}
$$

where $\rho^{\mathrm{S}}(r,\theta,\lambda)$ is the density inside the sediment masses at a point with spherical coordinates (r,θ,λ), $\Phi_0(\theta,\lambda)$ is the porosity at the surface of the sediments, $\rho^{\mathrm{UC}}(\theta,\lambda)$ and $\rho^{\mathrm{B}}(\theta,\lambda)$ are the densities of the upper mantle and bathymetric masses (water) and k is a constant, which is called the compaction length.

Eq. (5.93) can also be written in the following form, which is easier for obtaining the gravitational potential of sediments accordingly:

$$
\rho^{\mathrm{S}}(r,\theta,\lambda) = \rho^{\mathrm{UC}}(\theta,\lambda) + \Phi_0(\theta,\lambda) e^{-\frac{R+H(\theta,\lambda)-r}{k}} \left(\rho^{\mathrm{B}}(\theta,\lambda) - \rho^{\mathrm{UC}}(\theta,\lambda)\right). \tag{5.94}
$$

By inserting this model, Eq. (5.94), into the Newton integral (Eq. 5.1) instead of the density $\rho^{\mathrm{T}}(\theta,\lambda)$, the gravitational potential of the sediments will be:

$$V^{\mathrm{S}}(r,\theta,\lambda)=G\iint\limits_{\sigma}\rho^{\mathrm{UC}}(\theta,\lambda)\int\limits_{R+H(\theta,\lambda)-d_{\mathrm{S}}(\theta',\lambda')}^{R+H(\theta,\lambda)}\frac{r'^2\mathrm{d}r'}{l}\mathrm{d}\sigma$$
$$+G\iint\limits_{\sigma}\Phi_0(\theta,\lambda)\left(\rho^{\mathrm{B}}(\theta,\lambda)-\rho^{\mathrm{UC}}(\theta,\lambda)\right)$$
$$\int\limits_{R+H(\theta,\lambda)-d_{\mathrm{S}}(\theta',\lambda')}^{R+H(\theta,\lambda)}\frac{e^{-\frac{R+H(\theta,\lambda)-r'}{k}}}{l}r'^2\mathrm{d}r'\,\mathrm{d}\sigma.$$
$$=I_1+I_2 \tag{5.95}$$

The solution of the first integral on the rhs of Eq. (5.95) is rather simple:

$$I_1=G\iint\limits_{\sigma}\sum_{n=0}^{\infty}\frac{1}{r^{n+1}}P_n(\tilde{x})\rho^{\mathrm{UP}}(\theta',\lambda')\int\limits_{R+H(\theta',\lambda')-d_{\mathrm{S}}(\theta',\lambda')}^{R+H(\theta',\lambda')}r'^{n+2}\mathrm{d}r'. \tag{5.96}$$

After performing the integration, the SHCs of the contribution of the first term of Eq. (5.96) into the total potential of the sediment will be:

$$(I_1)_{nm}=\frac{4\pi GR^2}{2n+1}\left[\frac{\left(\rho^{\mathrm{UC}}d_{\mathrm{S}}\right)_{nm}}{R}+(n+2)\frac{\left(\rho^{\mathrm{UC}}\left(H^2-(H-d_{\mathrm{S}})^2\right)\right)_{nm}}{2R^2}\right.$$
$$\left.+(n+2)(n+1)\frac{\left(\rho^{\mathrm{UC}}\left(H^3-(H-d_{\mathrm{S}})^3\right)\right)_{nm}}{6R^3}\right], \tag{5.97}$$

where

$$\left(\rho^{\mathrm{UC}}d_{\mathrm{S}}\right)_{nm}=\frac{1}{4\pi}\iint\limits_{\sigma}\rho'^{\mathrm{UC}}d'_{\mathrm{S}}Y_{nm}(\theta',\lambda')\mathrm{d}\sigma, \tag{5.98}$$

$$\left(\rho^{\mathrm{UC}}\left(H^i-(H-d_{\mathrm{S}})^i\right)\right)_{nm}=\frac{1}{4\pi}\iint\limits_{\sigma}\left(\rho'^{\mathrm{UC}}\left(H'^i-(H'-d'_{\mathrm{S}})^i\right)\right)$$
$$Y_{nm}(\theta',\lambda')\mathrm{d}\sigma\quad i=2\text{ and }3. \tag{5.99}$$

Now, the second part of Eq. (5.95) is solved. This integral can be simplified to:

$$I_2 = G\iint_\sigma M'(\theta',\lambda')\sum_{n=0}^{\infty}\frac{1}{r^{n+1}}P_n(\tilde{x})\int_{R+H(\theta',\lambda')-d_S(\theta',\lambda')}^{R+H(\theta',\lambda')} e^{-\frac{R+H(\theta',\lambda')-r'}{k}} r'^2\,\mathrm{d}r', \tag{5.100}$$

where

$$M(\theta',\lambda') = \Phi_0(\theta',\lambda')(\rho^{\mathrm{B}}(\theta',\lambda') - \rho^{\mathrm{UC}}(\theta',\lambda')). \tag{5.101}$$

The approximate solution of the radial integral in Eq. (5.100) is:

$$\int_{R+H(\theta',\lambda')-d_S(\theta',\lambda')}^{R+H(\theta',\lambda')} e^{-\frac{R+H(\theta',\lambda')-r'}{k}} r'^{n+2}\,\mathrm{d}r' \approx (R+H)^{n+2}\{A+(n+2)B\}, \tag{5.102}$$

where

$$A = -k\left(1 - e^{-\frac{d_S}{k}}\right), \tag{5.103}$$

$$B = \frac{k}{R+H}\left(1 - e^{-\frac{d_S}{k}}(k + d_S)\right). \tag{5.104}$$

Inserting Eq. (5.102) into Eq. (5.100) and performing the spherical harmonic analysis and taking the advantage of addition theorem, Eq. (1.9), yields:

$$\begin{aligned}(I_2)_{nm} = \frac{4\pi GR}{2n+1}\Bigg\{&(MA)_{nm} + (n+2)(MB)_{nm} \\ &+ \frac{(n+2)}{R}\left((MAH)_{nm} + (n+2)(MBH)_{nm}\right) \\ &+ \frac{(n+2)(n+1)}{2R^2}\left((MAH^2)_{nm} + (n+2)(MBH^2)_{nm}\right)\Bigg\},\end{aligned} \tag{5.105}$$

where

$$\left(MAH^j\right)_{nm} = \frac{1}{4\pi}\iint_\sigma M(\theta',\lambda')A(\theta',\lambda')H^j(\theta',\lambda')Y_{nm}(\theta',\lambda')\mathrm{d}\sigma, \quad j = 0,\ 1 \text{ and } 2, \tag{5.106}$$

$$\left(MBH^j\right)_{nm} = \frac{1}{4\pi}\iint_\sigma M(\theta',\lambda')B(\theta',\lambda')H^j(\theta',\lambda')Y_{nm}(\theta',\lambda')\mathrm{d}\sigma \quad j = 0,\ 1 \text{ and } 2. \tag{5.107}$$

Finally, the gravitational potential of the sediments according to the sediment compact model will be:

$$V^{S}(r,\theta,\lambda)=\sum_{n=0}^{\infty}\left(\frac{R}{r}\right)^{n+1}\sum_{m=-n}^{n}\left((I_1)_{nm}+(I_2)_{nm}\right)Y_{nm}(\theta,\lambda). \qquad (5.108)$$

5.3.6 Example: the effect of upper sediments based on different density models on satellite gradiometry data

To present an application of method, the density and thickness of the upper sediment layer of CRUST1.0 is used. A constant density of 2670 kg/m^3 is considered as a reference density and subtracted from the density values to obtain the density contrast. The exponent $\alpha=-0.0014$ has been obtained by fitting to the density contrast value that Cordell (1973) derived for San Jacinto area. Therefore, the potential that is obtained based on these data will have the best value over this area compared with the rest of the world. In this case, using the aforementioned data of CRUST1.0 and this exponent, the SHCs, presented in Eq. (5.74) are determined and inserted into the spherical harmonic expansion of the second-order radial derivative at 250 km above sea level. The results are illustrated in Fig. 5.11A, showing the global map of the sediment effects. As this map shows, the values are close to those of the case in which the CRUST1.0 model is used. By looking at the mathematical model of the spherical harmonics, it is observed that the last three terms containing α appeared by linearisation of the exponential function. Fig. 5.11A shows that these three terms are not strong enough to reduce the power of the other terms and decrease the potential contrast and impose the exponentiality of the density contrast function. Fig. 5.11B is the map of the gradient contrast based on approach 2. As it shows, the values of the effect are smaller compared with Fig. 5.11A. The reason is that the density contrast decreases fast by depth and this causes the denser part of the sediments to be placed at the lower part of the layer. It is normal to see that gradient contrast will be smaller when the dense part of the layer is at the bottom compared with the case in which the whole layer has a constant density contrast. Therefore, it can be concluded that approach 2 is more appropriate for determining the gravitational potential contrast.

Fig. 5.11C shows the gradient contrast when the hyperbolic density contrast model is applied. The map is very similar to the map presented in Fig. 5.11B, but the gradient contrast reaches to about -0.3E. As Fig. 5.10 showed, the exponential density contrast model yields smaller density contrast for large depths compared with the hyperbolic model. This is the

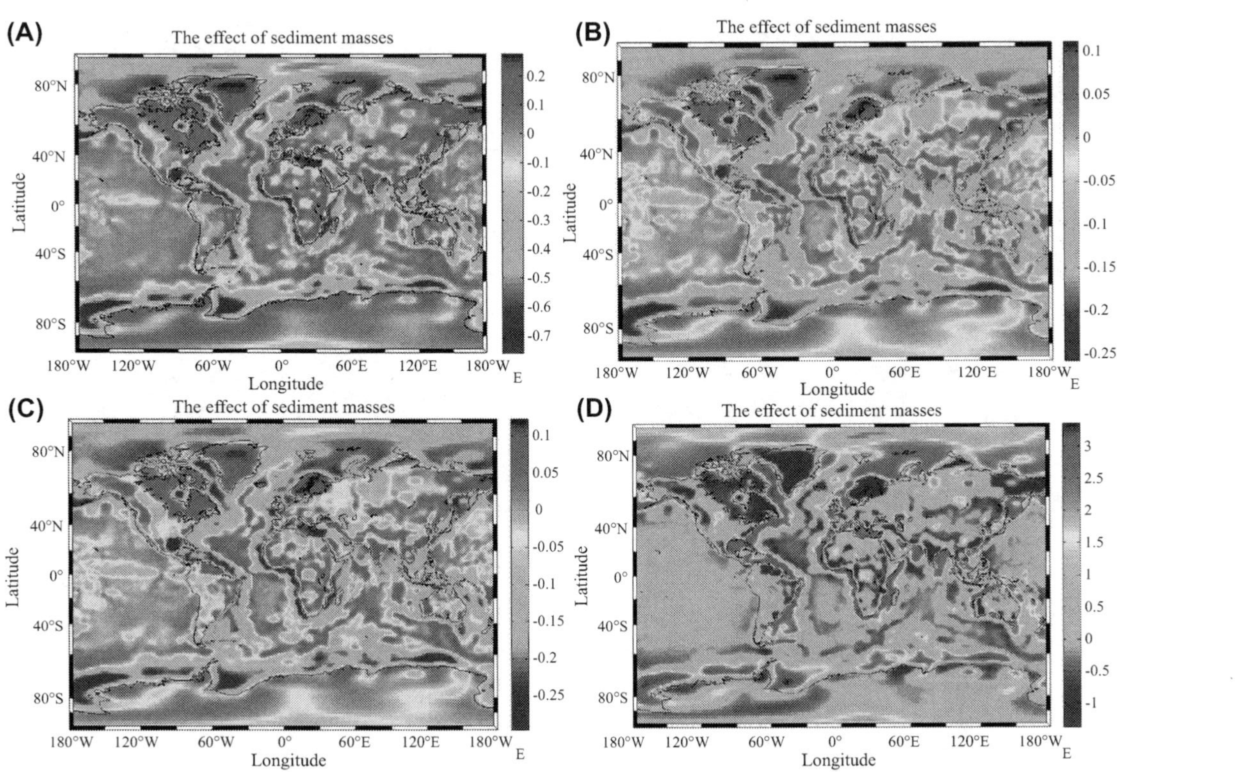

Figure 5.11 (A) The effect of lateral density variation in upper sediments on the second-order radial derivative of the Earth's gravitational potential according to the exponential density contrast model and formulation according to approach 1. (B) Approach 2. (C) Based on the hyperbolic density contrast model. (D) The effect of upper sediment based on the compact density model.

reason that the gradient contrast is slightly larger when the hyperbolic model is applied.

Fig. 5.11D is the map of the upper sediment gravitational potential based on the exponential compact model, where the porosity $\Phi_0 = 0.8$ and $k = 1.5$ km were taken from Braitenberg et al. (2006) for the China Sea. It shows that the effect of the sediment masses based on this model reaches 3E.

5.4 Gravitational potential of atmospheric masses

Modelling the gravitational potential of the atmospheric masses is easier under the assumption that the atmospheric density does not change laterally. The density of the atmospheric masses decreases fast by increasing the distance from the sea level. So far different models have been considered for the relation between the density of atmospheric masses and elevation, such as the exponential model (Lambeck, 1988), the power model (Sjöberg, 1998), the Novák and Grafarened (2006) polynomial and finally the model presented by Eshagh and Sjöberg (2009), which is in fact a combination of the Novák and Grafarend (2006) model and a power model.

Fig. 5.12 shows the atmospheric shell bounded by the Earth's surface and the upper bound of the layer, which is far from sea level at elevation Z.

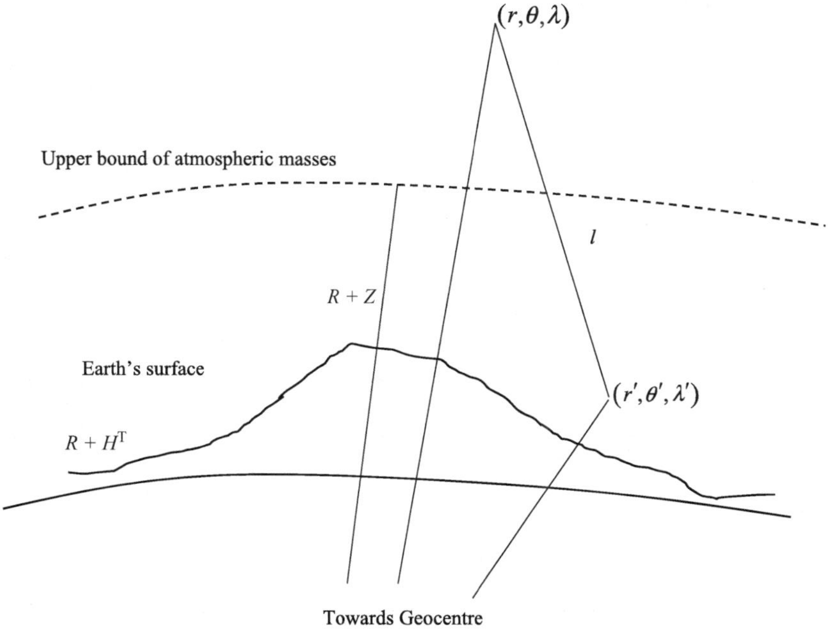

Figure 5.12 Atmospheric masses.

The satellite gravimetry data are collected outside such an atmospheric layer and pulled down slightly. For one point outside this layer with the geocentric distance r, and any point r' inside the atmospheric masses, the Newton integral can be considered again, but the density of the atmosphere changes radially. This means that atmospheric density models should be used inside the Newton integral for modelling the atmospheric gravitational potential.

5.4.1 Gravitational potential of atmospheric masses according to the exponential density model

The density of the atmospheric masses decreases fast by elevation. Therefore, presenting the density of the atmosphere with an exponential function of height will not be far from reality. Such an atmospheric density model can have the following form (Lambeck, 1988; Eshagh, 2009c):

$$\rho^{\mathrm{A}}(r) = \rho_0 e^{-\alpha(r-R)}, \tag{5.109}$$

where $\rho_0 = 1.2227\ \mathrm{kg/m^3}$ is the atmospheric density at sea level, R is the mean radius of the Earth and r is the radial distance of any point inside the atmosphere, and α is a constant. Now, our goal is to determine the gravitational potential of the atmospheric masses at any point outside the atmosphere layer, which is bounded by the Earth's surface and a constant elevation Z above sea level. To do so let us write Eq. (5.109) in the following form:

$$\rho^{\mathrm{A}}(r) = \rho_0 e^{\alpha R} e^{-\alpha r}, \tag{5.110}$$

and thereafter insert it into the Newton integral (Eq. 5.1). In this case, therefore:

$$V^{\mathrm{A}}(r,\theta,\lambda) = G\rho_0 e^{\alpha R} \iint_{\sigma} \sum_{n=0}^{\infty} \frac{1}{r^{n+1}} P_n(\tilde{x}) \int_{R+H(\theta,\lambda)}^{R+Z} e^{-\alpha r'} r'^{n+2} \mathrm{d}r' \mathrm{d}\sigma. \tag{5.111}$$

The radial integral in Eq. (5.111) can be separated into two integrals:

$$\begin{aligned} \int_{R+H(\theta,\lambda)}^{R+Z} e^{-\alpha r'} r'^{n+2} \mathrm{d}r' &= \int_{R}^{R+Z} e^{-\alpha r'} r'^{n+2} \mathrm{d}r' - \int_{R}^{R+H(\theta,\lambda)} e^{-\alpha r'} r'^{n+2} \mathrm{d}r' \\ &= I_n - \int_{R}^{R+H(\theta,\lambda)} e^{-\alpha r'} r'^{n+2} \mathrm{d}r'. \end{aligned} \tag{5.112}$$

The first integral, which is presented by the help variable I_n, is the potential of an atmospheric shell from the surface of the spherical sphere excluding the topographic masses. The second integral on the rhs of Eq. (5.112) is related to the atmospheric potential from the surface of the sphere to the Earth's surface. Subtraction of these two leads to the potential of the atmospheric masses from the Earth's surface up to $R+Z$.

The solution of the first integral is assumed as I_n and, as will be shown later, only I_0 is needed in our derivations and there is no need to perform partial integration to solve the integral. The exponential function in the second term of Eq. (5.112) is expanded by the Taylor series and the integration performs:

$$\sum_{k=0}^{\infty}\frac{(-\alpha)^k}{k!}\int_{R}^{R+H'} r'^{n+k+2}\mathrm{d}r' = \sum_{k=0}^{\infty}\frac{(-\alpha)^k}{k!(n+k+3)}r'^{n+k+3}\Big|_{R}^{R+H'}$$
$$= \sum_{k=0}^{\infty}\frac{(-\alpha)^k R^{n+k+3}}{k!(n+k+3)}\left[\left(1+\frac{H'}{R}\right)^{n+k+3}-1\right], \tag{5.113}$$

where $H' = H(\theta', \lambda')$. The Taylor expansion of the first term in the squared bracket of Eq. (5.113) yields:

$$\sum_{k=0}^{\infty}\frac{(-\alpha)^k}{k!}\int_{R}^{R+H'} r'^{n+k+2}\mathrm{d}r' \approx \sum_{k=0}^{\infty}\frac{(-\alpha)^k R^{n+k+3}}{k!}\left[\frac{H'}{R}+\frac{n+k+2}{2}\left(\frac{H'}{R}\right)^2\right.$$
$$\left.+\frac{(n+k+2)(n+k+1)}{6}\left(\frac{H'}{R}\right)^3\right]. \tag{5.114}$$

Inserting Eq. (5.114) into Eq. (5.112) and substituting the result into Eq. (5.111) leads to:

$$V^{\mathrm{A}}(r,\theta,\lambda) \approx G\rho_0 e^{\alpha R}\sum_{n=0}^{\infty}\frac{1}{r^{n+1}}\iint_{\sigma}$$
$$\left\{I_n - \sum_{k=0}^{\infty}\frac{(-\alpha)^k R_{n+k+3}}{k!}\left[\frac{H'}{R}+\frac{n+k+2}{2R^2}H'^2\right.\right.$$
$$\left.\left.+\frac{(n+k+2)(n+k+1)}{6R^3}H'^3\right]P_n(\tilde{x})\mathrm{d}\sigma. \tag{5.115}$$

Using the addition theorem Eq. (1.9) and after performing the spherical integral, the spherical harmonics expansion of the atmospheric potential will be (Eshagh, 2009c):

$$V^{\mathrm{A}}(r,\theta,\lambda)=\sum_{n=0}^{\infty}\left(\frac{R}{r}\right)^{n+1}\sum_{m=-n}^{n}v_{nm}^{\mathrm{A}}Y_{nm}(\theta,\lambda), \tag{5.116}$$

with the SHCs

$$\begin{aligned}v_{nm}^{\mathrm{A}}&\approx\frac{4\pi GR^2\rho_0}{2n+1}\left\{\frac{I_n e^{\alpha R}}{R^{n+3}}\delta_{n0}-e^{\alpha R}\right.\\&\qquad\sum_{k=0}^{\infty}\frac{(-\alpha)^k R^k}{k!}\left[\frac{H_{nm}}{R}+\frac{n+k+2}{2R^2}(H^2)_{nm}\right.\\&\qquad\left.\left.+\frac{(n+k+2)(n+k+1)}{6R^3}(H^3)_{nm}\right]\right\}\\&\approx\frac{4\pi GR^2\rho_0}{2n+1}\{A\delta_{n0}-B\},\end{aligned} \tag{5.117}$$

where

$$\begin{aligned}A=\frac{I_0e^{\alpha R}}{R^3}=\left\{\left[1-\left(1+\frac{Z}{R}\right)e^{-\alpha Z}\right]\left[\frac{2+\alpha(R+Z)}{\alpha^2R^2}\right]\right.\\\left.+\frac{2(1-e^{-\alpha Z})}{\alpha^3R^3}-\frac{Z}{R^2\alpha}\right\},\end{aligned} \tag{5.118}$$

$$\begin{aligned}B=\left\{\frac{H_{nm}}{R}+\frac{n+2-\alpha R}{2R^2}(H^2)_{nm}\right.\\\left.+\frac{(n+2)(n+1-2\alpha R)+\alpha^2R^2}{6R^3}(H^3)_{nm}\right\},\end{aligned} \tag{5.119}$$

where H_{nm}, $(H^2)_{nm}$ and $(H^3)_{nm}$ are spherical harmonic coefficients H, H^2 and H^3, respectively, which can be derived by using the simple global spherical harmonic analysis and after further simplifications.

5.4.2 Gravitational potential of atmospheric masses according to the power density model

Sjöberg (1998) assumed that the atmospheric density changes only with elevation. This assumption is not so far from reality, as the atmospheric density reduces with increasing height. Based on the results of Ecker and Mittermayer (1969) which were derived from the US standard atmospheric density model presented in 1961 (USSA61, Reference Atmosphere Committee, 1961), Sjöberg (1998) proposed the following density model:

$$\rho^{\mathrm{A}}(r) = \rho_0 \left(\frac{R}{r}\right)^{\nu}, \tag{5.120}$$

where $\rho^{\mathrm{A}}(r)$is the atmospheric density, $R = 6{,}378{,}137$ m is the Earth's mean radius, $R \leq r \leq R + Z$ is the geocentric radius of any point inside the atmosphere, $\rho_0 = 1.2227\ \mathrm{kg/m^3}$ is the atmospheric density at sea level, and $\nu = 850$ (Eshagh, 2009c) is an estimated constant.

Modelling the gravitational potential of the atmospheric masses based on this density model is simpler than the case of applying the exponential model. However, the principle is the same. We insert the density model, Eq. (5.120), into the Newton integral (Eq. 5.1) instead of the density of topographic masses:

$$V^{\mathrm{A}}(r, \theta, \lambda) = G\rho_0 \iint_{\sigma} \sum_{n=0}^{\infty} \frac{R^{\nu}}{r^{n+1}} P_n(\tilde{x}) \int_{R+H(\theta',\lambda')}^{R+Z} r'^{n+2-\nu} \mathrm{d}r' \mathrm{d}\sigma, \tag{5.121}$$

and again the radial integral on the rhs of Eq. (5.121) is divided into two integrals and each one is solved:

$$\begin{aligned} \int_{R+H(\theta',\lambda')}^{R+Z} r'^{n+2-\nu} \mathrm{d}r' &= \int_{R}^{R+Z} r'^{n+2-\nu} \mathrm{d}r' - \int_{R}^{R+H'} r'^{n+2-\nu} \mathrm{d}r' \\ &\approx \frac{R^{n+3-\nu}}{n+3-\nu} \left[\left(1 + \frac{Z}{R}\right)^{n+3-\nu} - 1 \right] \\ &\quad - R^{n+3-\nu} \left[\frac{H'}{R} + \frac{n+2-\nu}{2R^2} H'^2 \right. \\ &\quad \left. + \frac{(n-\nu+2)(n-\nu+1)}{6R^3} H'^3 \right]. \end{aligned} \tag{5.122}$$

By inserting Eq. (5.122) into Eq. (5.121), applying the addition theorem (Eq. 1.9) and performing the integrals, the following SHCs will be obtained for the potential of the atmospheric masses. The potential of the atmosphere, Eq. (5.116), should be applied with the following SHCs:

$$V_{nm}^{\mathrm{A}} \approx \frac{4\pi G\rho_0 R^2}{(2n+1)}\left\{\left[\left(1+\frac{Z}{R}\right)^{3-\nu}-1\right]\frac{\delta_{n0}}{3-\nu}-\frac{H_{nm}}{R}-\frac{n+2-\nu}{2R^2}\left(H^2\right)_{nm}\right.$$
$$\left.-\frac{(n+2-\nu)(n+1-\nu)}{6R^3}\left(H^3\right)_{nm}\right\}. \tag{5.123}$$

5.4.3 Gravitational potential of atmospheric masses according to the polynomial density model

The maximum value of atmospheric density is at sea level and decreases fast with increasing elevation. Novák (2000) proposed the following model to approximate the vertical behaviour of the atmospheric density:

$$\rho^{\mathrm{A}}(r) = \rho_0\left[1+\alpha H+\beta H^2\right], \tag{5.124}$$

where $\alpha = -7.6495 \times 10^{-5}\ \mathrm{m}^{-1}$, $\beta = 2.2781 \times 10^{-9}\ \mathrm{m}^{-2}$ and $H = r - R$ with $0 \leq H < 10$ km.

Novák (2000) considered a second-degree polynomial for approximating the densest part of the atmosphere to an elevation of 10 km. Since 80% of atmospheric masses are below 12 km (Lambeck, 1988), it is reasonable to use a simple polynomial to express the effect of atmospheric roughness, although Wallace and Hobbs (1977) believe that 99% of the masses lie within the lowest 30 km above sea level.

To model the atmospheric potential, the density model (Eq. 5.124) is inserted again into the Newton integral:

$$V^{\mathrm{A}}(r,\theta,\lambda) = G\rho_0 \iint_\sigma \sum_{n=0}^{\infty} \frac{1}{r^{n+1}} P_n(\tilde{x}) I \,\mathrm{d}\sigma, \tag{5.125}$$

where

$$I = \int_{R+H(\theta',\lambda')}^{R+Z} \left[1+\alpha(r'-R)+\beta(r'-R)^2\right]\mathrm{d}r'. \tag{5.126}$$

The solution of the radial integral in Eq. (5.125) is:

$$
\begin{aligned}
I = & (1-\alpha R+\beta R^2)\frac{1}{n+3}\left((R+H_0)^{n+3}-(R+H')^{n+3}\right)\\
& +(\alpha-2R\beta)\frac{1}{n+4}\left((R+H_0)^{n+4}-(R+H')^{n+4}\right)\\
& +\beta\frac{1}{n+5}\left((R+H_0)^{n+5}-(R+H')^{n+5}\right).
\end{aligned}
\tag{5.127}
$$

After factorising R^{n+3} from each term on the rhs of Eq. (5.126), expanding the binomial terms by the Taylor series to the fourth term and further rearrangements and simplifications:

$$
\begin{aligned}
I \approx R^{n+3}\Bigg(&\frac{H_0-H'}{R}+((n+2)+\alpha R)\frac{H_0^2-H'^2}{2R^2}\\
&+[(n+2)(n+1-2\alpha R)+2\beta R^2]\frac{H_0^3-H'^3}{6R^3}\Bigg).
\end{aligned}
\tag{5.128}
$$

After inserting Eq. (5.128) into Eq. (5.125), taking advantage of the addition theorem (Eq. 1.9) and performing the spherical harmonic analysis, the gravitational potential of the atmospheric masses will be obtained. The spherical harmonics expansion of this potential is the same with Eq. (5.116) but with the following SHCs:

$$
\begin{aligned}
v_{nm}^{\mathrm{A}} \approx \frac{4\pi G\rho_0 R^2}{2n+1}\Bigg\{&\frac{H_0\delta_{n0}-H_{nm}}{R}\\
&+((n+2)+\alpha R)\frac{H_0^2\delta_{n0}-(H^2)_{nm}}{2R^2}\\
&+((n+2)(n+1-2\alpha R)+2\beta R^2)\\
&\frac{H_0^3\delta_{n0}-(H^3)_{nm}}{6R^3}\Bigg\}.
\end{aligned}
\tag{5.129}
$$

5.4.4 Gravitational potential of atmospheric masses according to a combination of the polynomial and power density models

In the polynomial model the upper bound of the atmospheric layer is at the level 10 km and it is assumed that the gravitational potential of the masses above this layer is small. Eshagh and Sjöberg (2009) combined this model with the power model so that the polynomial is used for the denser part of the atmospheric masses, which is below 10 km, and for the masses above a power model is considered. This combined model has the following mathematical expression:

$$\rho^{A}(r) = \begin{cases} \rho_0[1 + \alpha H + \beta H^2], & 0 \le H \le H_0 \\ \rho(H_0)\left(\dfrac{R + H_0}{r}\right)^{\nu''}, & H_0 \le H \le Z \end{cases}, \tag{5.130}$$

where $H_0 = 10$ km and $\rho(H_0) = 0.4127$ kg m^3 is based on this model and $\nu'' = 890$ is estimated by Eshagh and Sjöberg (2009).

Modelling the gravitational potential based on this model is rather straightforward. Replacing the topographic density by this density model into the Newton integral (Eq. 5.1) reads:

$$V^{A}(r,\theta,\lambda) = G\iint_{\sigma}\sum_{n=0}^{\infty}\frac{1}{r^{n+1}}P_n(\tilde{x})\left\{\rho_0\int_{R+H(\theta',\lambda')}^{R+H_0}\left[1+\alpha(r'-R)\right.\right.$$

$$\left.\left.+\beta(r'-R)^2\right]r'^{n+2}dr' + \rho^{a}(H_0)(R+H_0)^{\nu''}\int_{R+H_0}^{R+Z} r'^{n-\nu''+2}dr'\right\}d\sigma \tag{5.131}$$

The first radial integral has already been solved, Eq. (5.128). The solution of the second integral is also simple, as its limits are constant. Therefore:

$$\rho(H_0)(R+H_0)^{\nu''}\int_{R+H_0}^{R+Z} r'^{n+2-\nu}dr'$$

$$= \frac{\rho(H_0)R^{n+3}}{n+3-\nu''}\left[\left(1+\frac{H_0}{R}\right)^{\nu''}\left(1+\frac{Z}{R}\right)^{n+3-\nu''} - \left(1+\frac{H_0}{R}\right)^{n+3}\right]. \tag{5.132}$$

The spherical harmonic expansion of the gravitational potential of the atmospheric masses based on this model is the same as Eq. (5.116), but with the following SHCs:

$$v_{nm}^{A} \approx \frac{4\pi GR^2}{2n+1}\left\{\rho_0\left[\frac{H_0\delta_{n0}-H_{nm}}{R}+((n+2)+\alpha R)\frac{H_0^2\delta_{n0}-(H^2)_{nm}}{2R^2}\right.\right.$$
$$\left.+((n+2)(n+1-2\alpha R)+2\beta R^2)\frac{H_0^3\delta_{n0}-(H^3)_{nm}}{6R^3}\right]$$
$$\left.+\frac{\rho(H_0)R^{n+3}}{3-\nu''}\left[\left(1+\frac{H_0}{R}\right)^{\nu''}\left(1+\frac{Z}{R}\right)^{n+3-\nu''}-\left(1+\frac{H_0}{R}\right)^{n+3}\right]\delta_{n0}\right\}. \tag{5.133}$$

5.4.5 Example: atmospheric effect on satellite gradiometry data

Here, the Shuttle Radar Topography Mission (SRTM) topographic model up to degree and order 360 is used to compute the SHCs of the gravitational potential of atmospheric masses based on the exponential, power, polynomial and combination of polynomial and power models. The computed SHCs are inserted into the mathematical model of the second-order radial derivative of the gravitational potential, Eq. (2.129), at the 250 km level. Fig. 5.13A and B are the atmospheric effect based on the exponential and power density models, and as we can observe, they are very similar. The reason is that each one of them represents a line in the plot of height and logarithm of density; see Eshagh (2009c). Fig. 5.13C represents the effect based on the polynomial density model. Eshagh and Sjöberg (2009) showed that the density of the atmospheric masses cannot be represented by the exponential and power models for the low altitudes and these models underestimate the density of this part of the atmospheric masses. A second-degree polynomial can represent this part of the masses better. Therefore, the effect reaches 5 mE, almost twice as large as what the exponential and power models give. However, the polynomial model is valid up to about 10 km and this was the reason Eshagh and Sjöberg (2009) combined it with the power model to estimate the effects of masses at higher elevations. Fig. 5.13D shows that the effect of the atmospheric masses based on this model reaches about 5.5 mE, slightly larger than what the second-degree polynomial yields.

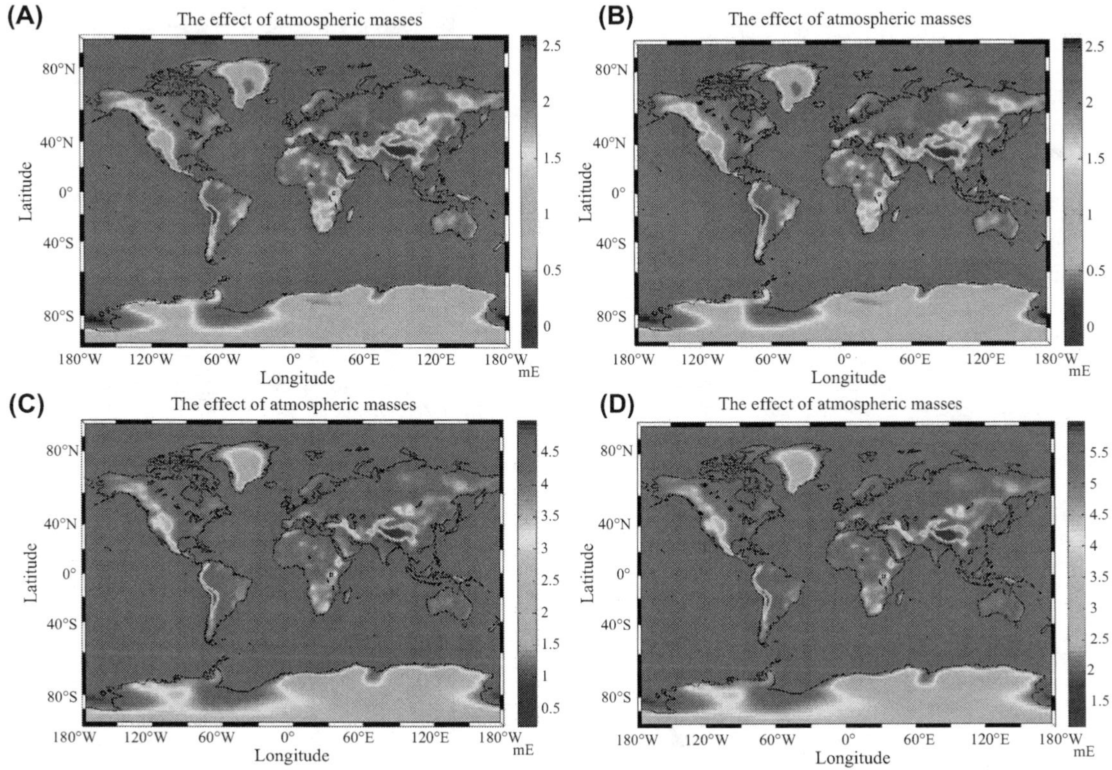

Figure 5.13 The effect of atmospheric masses on the second-order radial derivative of the gravitational potential at the 250 km level based on (A) exponential, (B) power, (C) polynomial and (D) combination of polynomial and power density models.

5.5 Remove–compute–restore model of topographic and atmospheric masses

One reason for determining and removing the effects of topographic and atmospheric masses from satellite gravimetry data is to make the data smooth and simplify the downward continuation of them to sea level; see Eshagh and Bagherbandi (2011). For the gravity field and geoid determination all data should be continued downward to sea level. However, the Laplace condition (Eq. 1.1) is valid when there is no mass outside the geoid or sea level. Therefore, the effects of all disturbing masses, like topographic and atmospheric masses, should be computed and removed from the data to make the computational space harmonic. This is the step which is known as the remove. Later all computations will be performed in such a no-topography and no-atmosphere space, like integration and downward continuation. Later the removed effects should be restored again on the computed quantities. These three steps are known as the remove–compute–restore scheme.

In satellite gravimetry, the goal is to determine the gravity field of the Earth at the surface of a spherical Earth or geoid. This can be done globally or locally. In the global gravity field determination, the effect of topographic and atmospheric masses should be removed from the satellite gravimetry data, and after recovering the gravity field in terms of the spherical harmonics, and the SHCs of the gravity field will be determined. Later the effect of the removed masses should be restored on these determined harmonics as well. In local gravity field recovery the process is rather similar. The effects of the topographic and atmospheric masses are removed from the satellite gravimetry data, but the computations are done locally. Normally, the gravity anomaly, gravity disturbance or geoid height are the quantities which are recovered over a specific area at sea level. The next step is to restore the removed effect on these determined quantities. So far, this chapter has presented the gravitational potentials for removing the effects from the satellite gravimetry data. The methods for gravity field recovery and inversion of the satellite gravimetry data have been presented in Chapter 4. Now, the restore step is discussed.

Fig. 5.14 shows that for restoring the effects of the topographic and atmospheric masses, the computation point with the spherical coordinates (r, θ, λ) is at sea level and below these masses.

5.5.1 Restoring the topographic effect

The satellite gravimetry data are measured outside the topographic masses and far from the Earth's surface. However, when the goal is to restore the

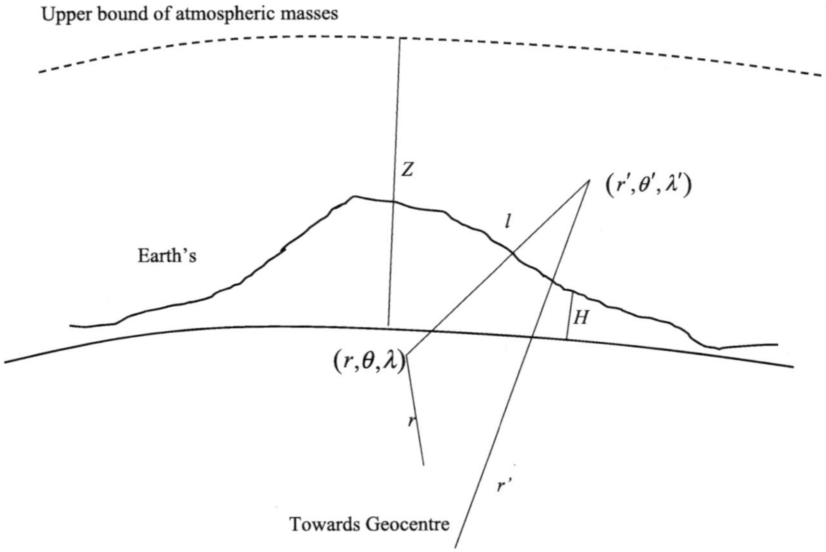

Figure 5.14 Restoration of topographic and atmospheric masses.

effects the computations are done at the surface of these masses and this time it is the topographic masses which are above the computation point. Again the gravitational potential of the topographic masses is computed by the Newton integral (Eq. 5.1) but this time for $r < r'$. In this case, Eq. (5.2) can be rewritten as:

$$l = \sqrt{r^2 + r'^2 - 2rr'\,\tilde{x}} = r'D \quad \text{where } D = \sqrt{1 + t^2 - 2t\tilde{x}} \text{ and } t = \frac{r}{r'}. \tag{5.134}$$

Similar to Eq. (5.3), we can write:

$$\frac{1}{l} = \frac{1}{r'D} = \frac{1}{r'}\sum_{n=0}^{\infty} t^n P_n(\tilde{x}). \tag{5.135}$$

Now, we insert Eq. (5.3) into Eq. (5.1) and assume that the topographic density ρ^T is constant radially, thus:

$$V^T(r,\theta,\lambda) = G\iint_\sigma \rho^T(\theta',\lambda')\sum_{n=0}^{\infty}\frac{r^n}{r'^{n+1}}P_n(\tilde{x})\int_R^{R+H(\theta',\lambda')} r'^2 \mathrm{d}r'\mathrm{d}\sigma. \tag{5.136}$$

After solving the radial integral Eq. (5.4) transfers to:

$$V^{\mathrm{T}}(r,\theta,\lambda)=G\iint\limits_{\sigma}\rho^{\mathrm{T}}(\theta',\lambda')\sum_{n=0}^{\infty}r^{n}P_{n}(\tilde{x})\int\limits_{R}^{R+H(\theta',\lambda')}r'^{-n+1}\mathrm{d}r'\mathrm{d}\sigma, \tag{5.137}$$

and by inserting the limits of the integral into its solution we obtain:

$$\begin{aligned}\int\limits_{R}^{R+H(\theta',\lambda')}r'^{-n+1}\mathrm{d}r'&=\frac{1}{-n+2}\left[r'^{-n+2}\right]_{R}^{R+H(\theta',\lambda')}\\&=\frac{R^{-n+2}}{-n+2}\left[\left(1+\frac{H(\theta',\lambda')}{R}\right)^{-n+2}-1\right].\end{aligned} \tag{5.138}$$

Let us consider $H'=H(\theta',\lambda')$ for shortening our formulae and approximate the binomial term by the Taylor series up to the fourth term; taking advantage of the addition theorem (Eq. 1.9) and performing the global integration, the following spherical harmonic expansion is derived for restoring the topographic effect:

$$V'^{\mathrm{T}}(r,\theta,\lambda)=\sum_{n=0}^{\infty}\left(\frac{r}{R}\right)^{n}\sum_{m=-n}^{n}v'^{\mathrm{T}}_{nm}Y_{nm}(\theta,\lambda)\quad\text{for}\quad r<R, \tag{5.139}$$

with the SHCs of:

$$\begin{aligned}v'^{\mathrm{T}}_{nm}=\frac{4\pi GR^{2}}{2n+1}&\left(\frac{(\rho^{\mathrm{T}}H)_{nm}}{R}+(-n+1)\frac{(\rho^{\mathrm{T}}H^{2})_{nm}}{2R^{2}}\right.\\&\left.+(-n+1)(-n)\frac{(\rho^{\mathrm{T}}H^{3})_{nm}}{6R^{3}}\right).\end{aligned} \tag{5.140}$$

5.5.2 Restoring the atmospheric effect

The principle of determining the gravitational potential of the atmospheric masses at any point below the masses is very similar to what has been done for the topographic effect. However, the density of the atmospheric masses is decreasing radially. The only difference is related to the use of Eq. (5.138) in the Newton integral as well as the density model. If the spherical harmonic expansion of the gravitational potential of the atmospheric masses is:

$$V'^{\mathrm{A}}(r,\theta,\lambda)=\sum_{n=0}^{\infty}\left(\frac{r}{R}\right)^{n}\sum_{m=-n}^{n}v'^{\mathrm{A}}_{nm}Y_{nm}(\theta,\lambda), \tag{5.141}$$

where the SHCs of the series based on the exponential density model are:

$$v'^{A}_{nm} \approx \frac{4\pi GR^2\rho_0}{(2n+1)}\left\{\left\{\frac{1}{\alpha R}\left[1-\left(1+\frac{Z}{R}\right)e^{-\alpha Z}\right]+\frac{(1-e^{-\alpha Z})}{\alpha^2R^2}\right\}\delta_{n0} - \left[\frac{H_{nm}}{R}+\frac{-n+1-\alpha R}{2R^2}H^2_{nm} + \frac{(n-1)(n+2\alpha R)+\alpha^2R^2}{6R^3}(H^3)_{nm}\right]\right\}. \tag{5.142}$$

In the case of using the power density model they are:

$$v'^{A}_{nm} = \frac{4\pi G\rho_0R^2}{2n+1}\left\{\frac{Z\delta_{n0}-H_{nm}}{R} - \frac{(n+\nu-1)}{2R^2}\left(Z^2\delta_{n0}-(H^2)_{nm}\right) + \frac{(n+\nu-1)(n+\nu)}{6R^3}\left(Z^3\delta_{n0}-(H^3)_{nm}\right);\right. \tag{5.143}$$

for the polynomial density model:

$$v'^{A}_{nm} = \frac{4\pi G\rho_0R^2}{2n+1}\left\{\frac{H_0\delta_{n0}-H_{nm}}{R} - (n-1-\alpha R)\frac{H_0^2\delta_{n0}-(H^2)_{nm}}{2R^2} - [(1-n)(n+2\alpha R)-2\beta R^2]\frac{H_0^3\delta_{n0}-(H^3)_{nm}}{6R^3}\right\}; \tag{5.144}$$

and finally based on the combination of polynomial and power density models:

$$v'^{A}_{nm} = \frac{4\pi GR^2}{2n+1}\left\{\rho_0\left[\frac{H_0\delta_{n0}-H_{nm}}{R} - (n-1-\alpha R)\frac{H_0^2\delta_{n0}-H^2_{nm}}{2R^2} - \left[(1-n)(n+2\alpha R)-2\beta R^2\right]\times\frac{H_0^3\delta_{n0}-(H^3)_{nm}}{6R^3}\right] + \frac{\rho(H_0)}{-n-\nu+2}\left[\left(1+\frac{H_0}{R}\right)^{\nu}\left(1+\frac{Z}{R}\right)^{-n-\nu+2} - \left(1+\frac{H_0}{R}\right)^{-n+2}\right]\delta_{n0}\right\}. \tag{5.145}$$

The mathematical proof of these SHCs can be found in Eshagh (2009c,e) and Eshagh and Sjöberg (2009). However, as mentioned before, the principles of determining them are similar.

5.6 Topographic and atmospheric bias

If the gravitational effects of topographic and atmospheric masses are not considered in gravity field determination, the results will contain topographic and atmospheric biases (Sjöberg, 2007; Eshagh, 2009d). They can be computed and removed from the satellite gravimetry data prior to inversion, but in the case of ignoring them, they are continued downward simultaneous with the satellite gravimetry data. Therefore, at this step, our result contains two things, the downward continued gravity field and the topographic and atmospheric effects. Now, these effects are at sea level and they should be restored at this level. Therefore, the bias is defined as the difference between the downward continued topographic or atmospheric gravitational potentials and the restored one at sea level. Mathematically, this idea can be expressed by:

$$V_{\mathrm{b}}^{\mathrm{T,A}}(R,\theta,\lambda)=\left[V^{\mathrm{T,A}}(r,\theta,\lambda)\right]^{*}-V'^{\mathrm{T,A}}(R,\theta,\lambda), \tag{5.146}$$

where $[\,]^{*}$ stands for downward continuation. $\left[V^{\mathrm{T,A}}(r,\theta,\lambda)\right]^{*}$ and $V'^{\mathrm{T,A}}(r,\theta,\lambda)$ are obtained by considering $r=R$ in their corresponding spherical harmonic expansions. Eq. (5.146) can also be expressed by the spherical harmonics series:

$$V_{\mathrm{b}}^{\mathrm{T,A}}(r,\theta,\lambda)=\sum_{n=0}^{\infty}\sum_{m=-n}^{n}\left(v_{\mathrm{b}}^{\mathrm{T,A}}\right)_{nm}Y_{nm}(\theta,\lambda), \tag{5.147}$$

where $\left(v_{\mathrm{b}}^{\mathrm{T,A}}\right)_{nm}=v_{nm}^{\mathrm{T,A}}-v'^{\mathrm{T,A}}_{nm}$ is the SHC of either the topographic or the atmospheric bias.

According to Eqs. (5.12) and (5.140) and the definition of the potential bias in Eq. (5.146) the harmonics of the topographic bias are (see also Ågren, 2004):

$$\left(v_{\mathrm{b}}^{\mathrm{T}}\right)_{nm}\approx 4\pi GR^{2}\left(\frac{\left(\rho^{\mathrm{T}}H^{2}\right)_{nm}}{2R^{2}}+\frac{\left(\rho^{\mathrm{T}}H^{3}\right)_{nm}}{3R^{3}}\right). \tag{5.148}$$

After inserting Eq. (5.148) into Eq. (5.147) and performing the summations, the topographic bias will be:

$$V_{\mathrm{b}}^{\mathrm{T}}(R,\theta,\lambda)=4\pi G\rho^{\mathrm{T}}(\theta,\lambda)\left(\frac{H^{2}(\theta,\lambda)}{2}+\frac{H^{3}(\theta,\lambda)}{3R}\right). \tag{5.149}$$

This means that the gravity field recovery can be performed without considering the topographic effect, but the results will contain topographic bias. This bias can be computed and removed from the results. In the case in which our results are in terms of SHCs of the gravity field, the spectral form of the bias presented in Eq. (5.148) can be applied to estimate the topographic bias on each degree and order.

The SHCs of the atmospheric bias based on the exponential, power and combination of polynomial and power models are, respectively:

$$\left(v_{\mathrm{b}}^{\mathrm{A}}\right)_{nm} \approx 4\pi G R^2 \left\{ \left[(2+Z)\left(1 - Le^{-\alpha' Z}\right) + \left(1 - e^{-\alpha' Z}\right)\left(\frac{2}{\alpha' R} - 1\right) \right] \delta_{n0} - \frac{\left(H^2\right)_{nm}}{2R^2} + (\alpha' R - 1)\frac{\left(H^3\right)_{nm}}{3R^3} \right\}, \tag{5.150}$$

$$\left(v_{\mathrm{b}}^{\mathrm{A}}\right)_{nm} \approx 4\pi G R^2 \left\{ \frac{1}{3-\nu} \left[\frac{ZL^{2-\nu}}{R} + \frac{1 - L^{2-\nu}}{2-\nu} \right] \delta_{n0} - \frac{\left(H^2\right)_{nm}}{2R^2} + (\nu - 2)\frac{\left(H^3\right)_{nm}}{3R^3} \right\}, \tag{5.151}$$

$$\begin{aligned} \left(v_{\mathrm{b}}^{\mathrm{A}}\right)_{nm} \approx 4\pi G R^2 \Bigg\{ & \rho_0 \left[\frac{H_0^2}{2R^2} + (1 - \alpha R)\frac{H_0^3}{3R^3} \right] \delta_{n0} \\ & + \rho(H_0) K^{\nu} \left[L^{2-\nu}\left(\frac{L}{3-\nu} - \frac{1}{2-\nu} \right) \right. \\ & \left. - K^{2-\nu}\left(\frac{K}{3-\nu} + \frac{1}{2-\nu} \right) \right] \delta_{n0} - \frac{\rho_0 \left(H^2\right)_{nm}}{2R^2} \\ & + (\alpha R - 1)\frac{\rho_0 \left(H^3\right)_{nm}}{3R^3} \Bigg\}, \end{aligned} \tag{5.152}$$

where $L = 1 + \frac{Z}{R}$ and $K = 1 + \frac{H_0}{R}$.

In the case of spatial domain, these biases will be:

$$\begin{aligned} V_{\mathrm{b}}^{\mathrm{A}}(r, \theta, \lambda) = & 4\pi G \frac{R^2}{3-\nu} \left[\frac{ZL^{2-\nu}}{R} + \frac{1 - L^{2-\nu}}{2-\nu} \right] \\ & - 4\pi G \left(\frac{H^2(\theta, \lambda)}{2} + (\nu - 2)\frac{H^3(\theta, \lambda)}{3R} \right), \end{aligned} \tag{5.153}$$

$$\begin{aligned} V_{\mathrm{b}}^{\mathrm{A}}(r,\theta,\lambda) = 4\pi GR^2\Bigg[&(2+Z)\left(1-Le^{-\alpha' Z}\right) \\ &+\left(1-e^{-\alpha' Z}\right)\left(\frac{2}{\alpha' R}-1\right)\Bigg] \\ &-4\pi G\left(\frac{H^2(\theta,\lambda)}{2}+(\alpha' R-1)\frac{H^3(\theta,\lambda)}{3R}\right), \end{aligned} \tag{5.154}$$

$$\begin{aligned} V_{\mathrm{b}}^{\mathrm{A}}(r,\theta,\lambda) \approx\ & 4\pi G\rho_0\left[\frac{H_0^2}{2}+(1-\alpha R)\frac{H_0^3}{3R}\right] \\ &+4\pi GR^2\rho(H_0)K^{\nu}\Bigg[L^{2-\nu}\left(\frac{L}{3-\nu}-\frac{1}{2-\nu}\right) \\ &-K^{2-\nu}\left(\frac{K}{3-\nu}+\frac{1}{2-\nu}\right)\Bigg] \\ &+4\pi G\rho_0\left(-\frac{H^2(\theta,\lambda)}{2}+(\alpha R-1)\frac{H^3(\theta,\lambda)}{3R}\right)\Bigg\}, \end{aligned} \tag{5.155}$$

which can be applied directly on the recovered potential for the correcting them.

References

Ågren, J., 2004. Regional Geoid Determination Methods for the Era of Satellite Gravimetry, Numerical Investigations Using Synthetic Earth Gravity Models, Doctoral Thesis in Geodesy. Royal Institute of Technology, Stockholm, Sweden.

Braitenberg, C., Wienecke, S., Wang, Y., 2006. Basement structures from satellite-derived gravity field: South China Sea ridge. J. Geophys. Res. 111, B0540.

Cordell, L., 1973. Gravity analysis using an exponential density-depth function- San Jacinto graben, California. Geophysics 38, 684–690.

Ecker, E., Mittermayer, E., 1969. Gravity corrections for the influence of the atmosphere. Boll. Geofis. Teor. Appl. 11, 70–80.

Eshagh, M., 2009a. Comparison of two approaches for considering laterally varying density in topographic effect on satellite gravity gradiometric data. Acta Geophysica 58 (4), 661–686.

Eshagh, M., 2009b. The effect of lateral density variation of crustal and topographic masses on GOCE gradiometric data: a study in Iran and Fennoscandia. Acta Geod. Geophys. Hung. 44 (4), 399–418.

Eshagh, M., 2009c. Spherical harmonics expansion of the atmospheric gravitational potential based on exponential and power models of atmosphere. Artif. Satell. 43 (1), 26–43.

Eshagh, M., 2009d. On the convergence of spherical harmonic expansion of topographic and atmospheric biases in gradiometry. Contrib. Geophys. Geodes. 39 (4), 273–299.

Eshagh, M., 2009e. Contribution of 1st -3rd order terms of a binomial expansion of topographic heights in topographic and atmospheric effects on satellite gravity gradiometric data. Artif. Satell. 44, 21–31.

Eshagh, M., Bagherbandi, M., 2011. Smoothing impact of isostatic crustal thickness models on local integral inversion of satellite gravity gradiometry data. Acta Geophys. 59 (5), 891–906.

Eshagh, M., Sjöberg, L.E., 2008. Impact of topographic and atmospheric masses over Iran on validation and inversion of GOCE gradiometric data. J. Earth Space Phys. 34, 15–30.

Eshagh, M., Sjöberg, L.E., 2009. Topographic and atmospheric effects on GOCE gradiometric data in local north oriented frame: a case study in Fennoscandia and Iran. Studia Geophys. Geod. 53, 61–80.

Huang, J., Vanicek, P., Pagiatakis, S., Brink, W., 2001. Effect of topographical mass density variation on gravity and geoid on the Canadian Rocky Mountains. J. Geodyn. 74, 805–815.

Hunegnaw, A., 2001. The effect of lateral density variation on local geoid determination. In: Proc. IAG 2001 Scientific Assembly, Budapest, Hungary.

Kiamehr, R., 2006. The impact of lateral density variation model in the determination of precise gravimetric geoid mountainous areas: as case study of Iran. Geophys. J. Int. 167, 521–527.

Kuehtreiber, N., 1998. Precise geoid determination using a density variation model. Phys. Chem. Earth 23, 59–63.

Lambeck, K., 1988. Geophysical Geodesy, the Slow Deformations of the Earth. Clarendon, Oxford University Press, New York.

Laske, G., Masters, G., Ma, Z., Pasyanos, M., 2013. Update on CRUST1.0 – a 1-degree global model of Earth's crust. In: EGU General Assembly Conference Abstracts, vol. 15, p. 2658.

Litinsky, V.A., 1989. Concept of effective density: key to gravity depth determinations for sedimentary basic. Geophysics 54 (11), 1474–1482.

Loera, J.A., McAllister, T.B., 2006. On computation of Clebsch-Gordan coefficients and the dilation effect. Exp. Math. 15, 7–19.

Makhloof, A., 2007. The Use of Topographic-Isostatic Mass Information in Geodetic Applications (Doctoral dissertation). Dept. of Theoretical and Physical Geodesy, Bonn, Germany.

Makhloof, A., Ilk, K.H., 2005. Far-zone topography effects on gravity and geoid heights according to Helmert's methods of condensation and based on Airy-Heiskanen model. In: Proceedings of the 3rd Minia, International Conference for Advanced Trends in Engineering, El-Minia, April 3–5, 2005.

Martinec, Z., 1993. Effect of lateral density variations of topographical masses in view of improving geoid model accuracy over Canada. In: Contract Report for Geodetic Survey of Canada, Ottawa, Canada.

Martinec, Z., Vaníček, P., 1994. Direct topographical effect of Helmert's condensation for a spherical geoid. Manuscripta Geod. 19, 257–268.

Martinec, Z., Vanicek, P., Mainville, A., Veronneau, M., 1995. The effect of lake water on geoidal height. Manuscripta Geod. 20, 193–203.

Novák, P., 2000. Evaluation of Gravity Data for the Stokes-Helmert Solution to the Geodetic Boundary-Value Problem, Technical Report 207. Department of Geodesy and Geomatics Engineering, University of New Brunswick, Fredericton, Canada.

Novák, P., Grafarend, E.W., 2006. The effect of topographical and atmospherical masses on spaceborne gravimetric and gradiomtric data. Studia Geophys. Geod. 50, 549–582.

Reference Atmosphere Committee, 1961. Report of the Preparatory Group for an International Reference Atmosphere Accepted at the COSPAR Meeting in Florance, April 1961. North Holland Publ. Co., Amsterdam.

Sjöberg, L.E., 1998. The atmospheric geoid and gravity corrections. In: Bollettino di geodesia e scienze affini-No.4, pp. 421–435.

Sjöberg, L.E., 2000. Topographic effects by the Stokes-Helmert method of geoid and quasi-geoid determinations. J. Geod. 74, 255–268.

Sjöberg, L.E., 2004. The effect on the geoid of lateral topographic density variation. J. Geod. 78, 34–39.

Sjöberg, L.E., 2007. Topographic bias by analytical continuation in physical geodesy. J. Geod. 81, 345–350.

Sun, W., Sjöberg, L.E., 2001. Convergence and optimal truncation of binomial expansions used in isostatic compensations and terrain corrections. J. Geod. 74, 627–636.

Tziavos, I.N., Featherstone, W.E., 2000. First results of using digital density data in gravimetric geoids computation in Australia. In: IAG Symposia, GGG2000, vol. 123. Springer Verlag, Berlin Heidelberg, pp. 335–340.

Wallace, J.M., Hobbs, P.V., 1977. Atmospheric Science: an Introductory Survey. Academic press, New York.

Wild, F., Heck, B., 2004a. A comparison of different isostatic models applied to satellite gravity gradiometry. In: Gravity, Geoid and Space MissionsGGSM 2004 IAG International Symposium Porto, Portugal August 30–September 3, 2004.

Wild, F., Heck, B., 2004b. Effects of topographic and isostatic masses in satellite gravity gradiometry. In: Proc. Second International GOCE User Workshop GOCE. The Geoid and Oceanography, ESA-ESRIN, Frascati/Italy, March 8-10, 2004 (ESA SP – 569, June 2004), CD-ROM.

Xu, Y.L., 1996. Fast evaluation of the Gaunt coefficients. Math. Com. 65, 1601–1612.

CHAPTER 6
Isostasy

6.1 Isostatic equilibrium

Isostasy is an equilibrium between the Earth's crust and its upper mantle, which properties the crust should have for being in equilibrium. It is an important subject in gravimetry, as the high-frequency portion of the gravity field of the Earth is due to the mass variation and structure of the crust, which includes the topographic and bathymetric (TB) masses. Fig. 6.1A shows a four-layer model of the Earth and there the Earth is divided into crust, upper mantle or asthenosphere, lower mantle and core. The thickness and density of each layer contribute to the gravity field of the Earth. Fig. 6.1A shows that each layer is a spherical shell, whilst in reality this is not the case; at least we are aware of the TB masses, which are parts of the upper crust, see Section 5.2. Also, we are aware of sediments as another part of the upper crust and the crustal crystalline inside the crust. Isostasy represents in fact the relation between the crust and the upper mantle so that the Earth's crust is in an equilibrium state.

Fig. 6.1B shows a closer look at the Earth's crust and upper mantle. It shows that the mountains thrust the mantle downward and they have roots beneath. To have the mountains in isostatic equilibrium they need to have roots inside the upper mantle. This is one of the assumptions of Airy (1855); such roots represent masses, which are known as compensation masses. The geometry and density of such masses play the main role in an isostatic equilibrium. Different geometries and densities of the compensation masses can keep the crust in equilibrium. The geometry of the compensation masses is one of the important parameters in isostasy. It is, in fact, the boundary at which the physical properties of the crust will change. This surface is called Moho discontinuity or simply Moho surface, which is the compensation layer of the topographic and bathymetric masses. The distance from the sea level to this layer is called the Moho depth and from the Earth's surface the crustal thickness. Fig. 6.1B also shows that the mountains can be considered as loads on the lithosphere, which consists of the crust

Satellite Gravimetry and the Solid Earth
ISBN 978-0-12-816936-0
https://doi.org/10.1016/B978-0-12-816936-0.00006-2

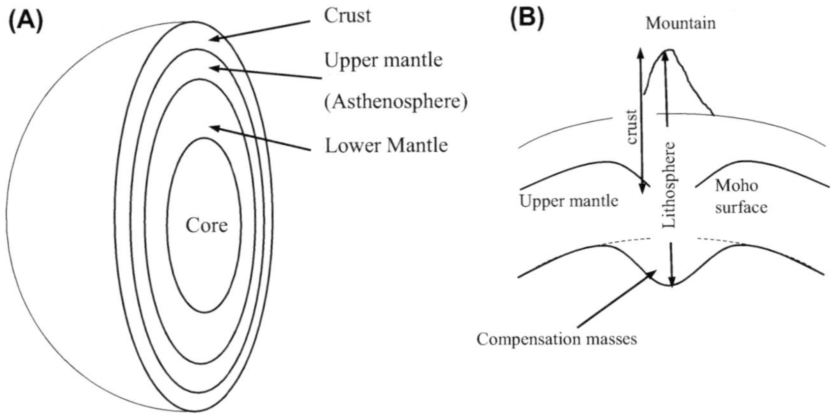

Figure 6.1 (A) Four-layer Earth model. (B) Mountains as a load on the lithosphere.

and upper mantle. Therefore, the physical properties of the upper mantle are key issues in keeping the crust in isostatic balance.

So far isostasy has been looked upon in different ways, according to the stress or the pressure of the mountains due to their weight at the bottom of the roots, the gravitational potential of the masses and the masses as loads on a thin lithospheric shell. These are some of the mechanisms applied to present isostasy. However, isostasy is not valid all over the Earth, and there are places where isostasy is not in agreement with the geological evidence. One example is the orogenic belts, where there are mountains but such mountains do not have deep roots inside the mantle. Also, subduction zones may have no mountains at the border of subduction but there are large roots due to thrusting of the lithosphere inside the mantle. Nevertheless, our goal in this book is to look at isostasy from a theoretical and mathematical point of view based on the gravimetric theories. For details about the geophysical and geological properties the reader is referred to Watts (2001) and Turcotte and Schubert (2014).

In the following, the Pratt—Hayford (Pratt, 1854) and Airy—Heiskanen theories are presented from the gravimetric point of view and potential theories. Later the gravimetric approach to isostasy will be discussed and finally flexure isostasy and its relation with gravimetric isostasy will be presented.

6.2 Pratt–Hayford isostasy model

In the Pratt–Hayford (see e.g. Watts, 2001) isostasy model, the main assumption is that pressure from the mountains is compensated at a constant level with depth $\widetilde{D}_0$, and it is the crustal lateral density variation which compensates for the topographic masses so that the pressure at the compensation level with a depth of $\widetilde{D}_0$ remains constant. In fact, the density variation keeps the mountains and loads in isostatic equilibrium; see Fig. 6.2. According to this assumption, the density under mountains should be smaller than that under valleys. Based on Fig. 6.2 the pressure at the compensation level with a depth of $\widetilde{D}_0$ and with a reference density of ρ_0^{T}, will be:

$$\rho^{\mathrm{T}}\left(H^{\mathrm{T}} + \widetilde{D}_0\right)g = g\rho_0^{\mathrm{T}}\widetilde{D}_0, \tag{6.1}$$

where ρ^{T} and H^{T} are, respectively, density and height of the topographic masses, and g stands for the gravity attraction. The gravity attraction g is

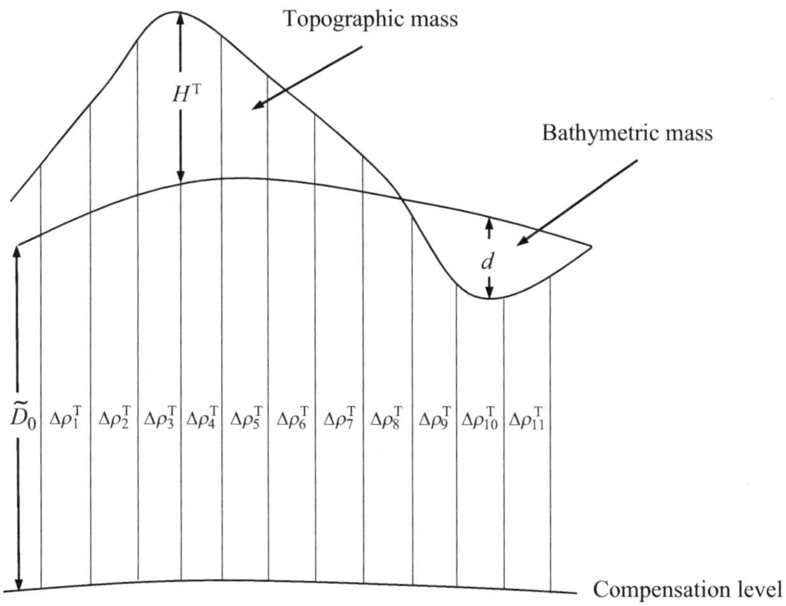

Figure 6.2 Pratt–Hayford isostatic model: $\widetilde{D}_0$ is the mean compensation depth from sea level, H the topographic height, d the bathymetric depth and $\Delta\rho_i^{\mathrm{T}}$, $i = 1, 2, \ldots, 11$ the density contrast from the reference density ρ_0.

considered constant and Eq. (6.1) is solved for density variation from the reference value ρ_0^{T}, $\Delta\rho^{\mathrm{T}} = \rho^{\mathrm{T}} - \rho_0^{\mathrm{T}}$; therefore:

$$\Delta\rho^{\mathrm{T}} = \rho^{\mathrm{T}} - \rho_0^{\mathrm{T}} = -\rho_0^{\mathrm{T}} \frac{H^{\mathrm{T}}}{\widetilde{D}_0 + H^{\mathrm{T}}}. \tag{6.2}$$

In oceans, we can write according to Fig. 6.1:

$$\rho^{\mathrm{T}}\left(\widetilde{D}_0 - d\right) + \rho^{\mathrm{B}} d = \rho_0^{\mathrm{T}} \widetilde{D}_0, \tag{6.3}$$

where d is the depth of the ocean and ρ^{B} the density of water. If the right-hand side (rhs) of Eq. (6.3) is taken to the left-hand side (lhs) and $\rho_0^{\mathrm{T}} d$ is added to and subtracted from the lhs, the following formula for the density contrast from the reference value will be:

$$\Delta\rho^{\mathrm{T}} = -\left(\rho^{\mathrm{B}} - \rho_0^{\mathrm{T}}\right) \frac{d}{\widetilde{D}_0 - d}. \tag{6.4}$$

Therefore, by considering the solid Earth topography, or in other words, topographic/bathymetric (TB) heights together as H, Eqs. (6.2) and (6.4) can be merged to:

$$\Delta\rho^{\mathrm{T}} = -\overline{\rho}_0 \frac{H}{\widetilde{D}_0 + H}, \quad \text{where} \quad \overline{\rho}_0 = \begin{cases} \rho_0^{\mathrm{T}} & \text{continents} \\ \rho^{\mathrm{B}} - \rho_0^{\mathrm{T}} & \text{oceans} \end{cases}. \tag{6.5}$$

Eq. (6.5) represents the mathematical relation between the density contrast from the reference density and the mean compensation depth $\widetilde{D}_0$ to have the TB masses in isostatic equilibrium according to the Pratt–Hayford compensation mechanism. $\widetilde{D}_0$ should be predefined in this model, e.g. from seismic measurements, and its choice has a significant role in determining the value of the density contrast. Reversely, if the density contrast is known, this constant compensation depth can be determined. Note that at least one of these parameters should be known to determine the other. Different combinations of them can satisfy Eq. (6.5), but they should agree with the geological evidence. Therefore, to solve such an inverse problem some geological or geophysical constraints or measurements are necessary.

6.2.1 Pratt–Hayford isostatic model based on gravitational potential in the spherical domain

The Pratt–Hayford isostatic model can also be obtained based on gravitational potential theory. In this case, the gravitational potential of the

TB masses is compensated by the potential variations due to the density contrast beneath them. In the continents the Newton integral can be written for the topographic masses and compensating masses as:

$$G\iint_{\sigma_{\text{continents}}} \rho^{\mathrm{T}}(\theta',\lambda') \int_{R-\widetilde{D}_0}^{R+H'} \frac{r'^2 \mathrm{d}r'}{l} \mathrm{d}\sigma = G\rho_0^{\mathrm{T}} \iint_{\sigma_{\text{continents}}} \int_{R-\widetilde{D}_0}^{R} \frac{r'^2 \mathrm{d}r'}{l} \mathrm{d}\sigma, \tag{6.6}$$

where G stands for the Newtonian gravitational constant, σ is the sphere at which the integration is performed and $\mathrm{d}\sigma$ is the surface integration element. The term r' is the geocentric distance of the integration point inside the topographic and compensation masses, l is the distance between the computation at which the potential is computed and the integration points, $\rho^{\mathrm{T}}(\theta',\lambda')$ stands for the density of the crustal columns at the integration point and finally, ρ_0^{T} is the reference density. Eq. (6.6) is equivalent to Eq. (6.1) but based on the gravitational potential of the topographic and compensation masses. If the rhs of Eq. (6.6) is taken to the lhs and the radial integral is divided into the integral from the compensation level to the sea surface and the sea surface to the Earth's surface, and after adding and subtracting a similar radial integral from the sea to the Earth's surface, but with the reference density, ρ_0^{T}, Eq. (6.6) will change to:

$$G\iint_{\sigma_{\text{continents}}} \Delta\rho'^{\mathrm{T}} \left[\int_{R-\widetilde{D}_0}^{R} \frac{r'^2 \mathrm{d}r'}{l} + \int_{R}^{R+H'} \frac{r'^2 \mathrm{d}r'}{l} \right] \mathrm{d}\sigma = -\, G\rho_0^{\mathrm{T}} \iint_{\sigma_{\text{continents}}} \int_{R}^{R+H'} \frac{r'^2 \mathrm{d}r'}{l} \mathrm{d}\sigma. \tag{6.7}$$

Similarly, the Newton integral can be applied over oceans:

$$G\iint_{\sigma_{\text{ocean}}} \rho^{\mathrm{T}}(\theta',\lambda') \int_{R-\widetilde{D}_0}^{R-d'} \frac{r'^2 \mathrm{d}r'}{l} \mathrm{d}\sigma + G\iint_{\sigma_{\text{ocean}}} \rho^{\mathrm{B}}(\theta',\lambda') \int_{R-d'}^{R} \frac{r'^2 \mathrm{d}r'}{l} \mathrm{d}\sigma$$

$$= G\rho_0^{\mathrm{T}} \iint_{\sigma_{\text{ocean}}} \int_{R-\widetilde{D}_0}^{R} \frac{r'^2 \mathrm{d}r'}{l} \mathrm{d}\sigma, \tag{6.8}$$

which corresponds to Eq. (6.3) but is based on the potentials of the bathymetric and compensation masses. Similar manipulations, as done for deriving Eq. (6.8), can be performed to write this equation in the following form:

$$G\iint\limits_{\sigma_{\text{ocean}}}\Delta\rho'^{\text{T}}\left[\int\limits_{R-\widetilde{D}_0}^{R}\frac{r'^2\text{d}r'}{l}-\int\limits_{R-d'}^{R}\frac{r'^2\text{d}r'}{l}\right]\text{d}\sigma=-G\left(\rho^{\text{B}}-\rho_0^{\text{T}}\right)\iint\limits_{\sigma_{\text{ocean}}}\int\limits_{R-d'}^{R}\frac{r'^2\text{d}r'}{l}\text{d}\sigma. \tag{6.9}$$

Now, Eqs. (6.7) and (6.9) are combined by their addition:

$$G\iint\limits_{\sigma}\Delta\rho'^{\text{T}}\left[\int\limits_{R-\widetilde{D}_0}^{R}\frac{r'^2\text{d}r'}{l}+\int\limits_{R}^{R+H'}\frac{r'^2\text{d}r'}{l}\right]\text{d}\sigma=-G\overline{\rho}_0\iint\limits_{\sigma}\int\limits_{R}^{R+H'}\frac{r'^2\text{d}r'}{l}\text{d}\sigma. \tag{6.10}$$

This represents Eq. (6.5) but is based on the potential of the bathymetric and compensation masses. The parameters $\overline{\rho}_0$ and H (see Eq. 6.5) specify the position of the computation point, whether it is in a continent or in the ocean. Assuming that $\overline{\rho}_0$ and $\widetilde{D}_0$ are known, the density contrast from the reference density can be determined in such a way that the crust remains in isostatic equilibrium. Eq. (6.10) is an integral equation having $\Delta\rho'^{\text{T}}$ as unknown parameters. The value of the two radial integrals on the lhs is computable based on the position of the integral and computation points. The rhs is the known TB gravitational potential. The integral on the lhs should be discretised and solved numerically to obtain $\Delta\rho'^{\text{T}}$. However, the problem is that evaluation of the radial integrals in the bracket on the lhs is very time consuming, as they should be computed for many computation points, depending on the size of the area. Nevertheless, Eq. (6.10) can be simplified further. In the case of expanding $1/l$ by the Legendre series, Eq. (5.3), in all radial integrals on both sides of Eq. (6.10), we arrive at:

$$\begin{aligned}&G\iint\limits_{\sigma}\Delta\rho'^{\text{T}}\sum_{n=0}^{\infty}\frac{1}{r^{n+1}}P_n(\widetilde{x})\left[\int\limits_{R-\widetilde{D}_0}^{R}r'^{n+2}\text{d}r'+\int\limits_{R}^{R+H'}r'^{n+2}\text{d}r'\right]\text{d}\sigma\\&=-G\overline{\rho}_0\iint\limits_{\sigma}\sum_{n=0}^{\infty}\frac{1}{r^{n+1}}P_n(\widetilde{x})\int\limits_{R}^{R+H'}r'^{n+2}\text{d}r'\text{d}\sigma,\end{aligned} \tag{6.11}$$

where $P_n(\widetilde{x})$ is the Legendre polynomial of degree n and $\widetilde{x}=\cos\psi$ is its argument with ψ as the geocentric angle between the computation and the integration points. As shown in Chapter 5, the radial integral on the rhs of Eq. (6.11) can be simplified to:

$$\int_{R}^{R+H'} r'^{n+2}\mathrm{d}r' \approx R^{n+3}\left[\frac{H'}{R}+(n+2)\frac{H'^2}{2R^2}+(n+2)(n+1)\frac{H'^3}{6R^3}\right]=R^{n+3}F'. \tag{6.12}$$

Substitution of Eq. (6.12) into Eq. (6.11) and further manipulations yield:

$$\begin{aligned} GR^2\iint_{\sigma}\Delta\rho'^{\mathrm{T}}\sum_{n=0}^{\infty}\left(\frac{R}{r}\right)^{n+1}\left[\frac{1}{n+3}\left(1-\left(1-\frac{\widetilde{D}_0}{R}\right)^{n+3}\right)+F'\right]P_n(\widetilde{x})\mathrm{d}\sigma \\ = -4\pi GR^2\overline{\rho}_0\sum_{n=0}^{\infty}\frac{1}{2n+1}\left(\frac{R}{r}\right)^{n+1}\sum_{m=-n}^{n}F_{nm}Y_{nm}(\theta,\lambda), \end{aligned} \tag{6.13}$$

where F_{nm} was given in Eq. (5.15). The term $Y_{nm}(\theta, \lambda)$ stands for the fully normalised spherical harmonics of degree n and m with arguments of co-latitude θ and longitude λ.

Eq. (6.13) is in fact an integral equation with respect to $\Delta\rho'^{\mathrm{T}}$, and by solving it numerically, based on one of the methods presented in Chapter 4, the density variation with respect to the reference density ρ_0^{T} is derived. This integral equation gives the density contrast for the compensating masses to keep the loads compensated up to a level with the depth $\widetilde{D}_0$. The kernel function of this integral equation depends on the compensation depth and topographic heights and bathymetric depths of the area in which the inversion performs. Therefore, its behaviour varies from one computation point to another. In addition, writing a closed-form formula for such a kernel function is not possible, and the kernel with the spectral form and truncated to a specified maximum degree can be applied.

6.2.2 Approximate solution in spherical harmonics

Determination of density variation with respect to the reference density can be derived in terms of spherical harmonics. However, such a solution is obtained based on approximating the binomial term involving the compensation depth $\widetilde{D}_0$. It should be mentioned that in the Pratt—Hayford model this depth is about 100 km or larger and the linear approximation of the binomial term may not be enough for presenting this term. If this term in Eq. (6.13) is approximated by the Taylor series to the linear term, the equation will change to

$$GR^2\iint_\sigma \sum_{n=0}^{\infty}\left(\frac{R}{r}\right)^{n+1}\Delta\rho'^{\mathrm{T}}\left(\frac{\widetilde{D}_0}{R}+F'\right)P_n(\widetilde{x})d\sigma$$
$$= -4\pi GR^2\rho_0^{\mathrm{T}}\sum_{n=0}^{\infty}\left(\frac{R}{r}\right)^{n+1}\sum_{m=-n}^{n}\frac{F_{nm}}{2n+1}Y_{nm}(\theta,\lambda). \tag{6.14}$$

Based on the addition theorem of spherical harmonics (Eq. 1.9), it will not be difficult to show that Eq. (6.14) changes to:

$$4\pi GR^2\sum_{n=0}^{\infty}\left(\frac{R}{r}\right)^{n+1}\frac{1}{2n+1}\sum_{m=-n}^{n}\left(\Delta\rho'^{\mathrm{T}}\left(\frac{\widetilde{D}_0}{R}+F'\right)\right)_{nm}Y_{nm}(\theta,\lambda)d\sigma$$
$$= -4\pi GR^2\rho_0^{\mathrm{T}}\sum_{n=0}^{\infty}\left(\frac{R}{r}\right)^{n+1}\sum_{m=-n}^{n}\frac{F_{nm}}{2n+1}Y_{nm}(\theta,\lambda). \tag{6.15}$$

As seen, both sides of Eq. (6.15) are written in terms of spherical harmonic series; therefore, the following spectral relation is achieved by comparing the lhs and the rhs of Eq. (6.15):

$$\left(\Delta\rho^{\mathrm{T}}\left(\frac{\widetilde{D}_0}{R}+F\right)\right)_{nm} = -\rho_0^{\mathrm{T}}F_{nm}. \tag{6.16}$$

As observed, lots of parameters are eliminated and a simple relation between the spherical harmonic coefficients (SHCs) of the lhs and rhs of Eq. (6.15) is established. Now, if these SHCs are inserted into a spherical harmonic series, the following corresponding spatial relation will be achieved:

$$\left(\Delta\rho^{\mathrm{T}}\left(\frac{\widetilde{D}_0}{R}+F\right)\right) = -\rho_0^{\mathrm{T}}F. \tag{6.17}$$

The solution of Eq. (6.17) for $\Delta\rho^{\mathrm{T}}$ is:

$$\Delta\rho^{\mathrm{T}} = -\rho_0^{\mathrm{T}}\frac{RF}{\left(\widetilde{D}_0+RF\right)}. \tag{6.18}$$

Eq. (6.18) is comparable to Eq. (6.5). The main difference is related to the presence of F, which is related to the TB masses. If F is approximated only up to the first term it will be $F = H/R$. By inserting this into Eq. (6.5), Eq. (6.18), is obtained. Therefore, Eq. (6.13) is a general form of the Pratt–Hayford model.

6.3 Airy–Heiskanen model

Airy (1855) assumed that the crust is constructed by discrete columns having the same density. This means that the taller columns thrust the lithosphere more than those which are shorter. In other words, mountains have roots beneath them, meaning that the crust is thicker in mountainous areas than in oceans. In the continents, the pressure of the mountains at the bottom of the crust, which is the boundary between the crust and the upper mantle and known as the Moho surface, is constant; therefore (see Fig. 6.3):

$$\rho^{\mathrm{T}} H^{\mathrm{T}} g + \rho^{\mathrm{T}} \Delta \widetilde{D} g + \rho^{\mathrm{T}} \widetilde{D}_0 g = \rho^{\mathrm{T}} \widetilde{D}_0 g + \rho^{\mathrm{M}} \Delta \widetilde{D} g, \tag{6.19}$$

where $\Delta \widetilde{D}$ is the undulation from the mean compensation $\widetilde{D}_0$ and ρ^{M} stands for upper mantle density. Solution of Eq. (6.19) for $\Delta \widetilde{D}$ is:

$$\Delta \widetilde{D} = \frac{\rho^{\mathrm{T}} H^{\mathrm{T}}}{\Delta \rho}, \tag{6.20}$$

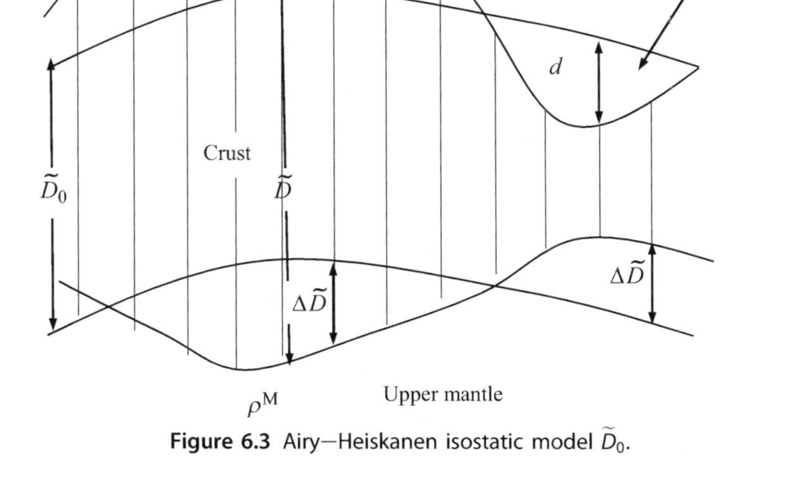

Figure 6.3 Airy–Heiskanen isostatic model $\widetilde{D}_0$.

where $\Delta\rho = \rho^{M} - \rho^{T}$ is the density contrast between the crust and the upper mantle. In oceans, from Fig. 6.3, we have:

$$\rho^{B} dg + \rho^{T}\left(\widetilde{D}_0 - d - \Delta\widetilde{D}\right)g = \rho^{T}\widetilde{D}_0 g - \rho^{M}\Delta\widetilde{D}g. \tag{6.21}$$

Solution of Eq. (6.21) for $\Delta\widetilde{D}$ is:

$$\Delta\widetilde{D} = -\frac{\left(\rho^{B} - \rho_0^{T}\right)d}{\Delta\rho}. \tag{6.22}$$

Comparing Eqs. (6.22) and (6.20) and therefore by considering the solid Earth topography (H), instead of the topographic heights and bathymetric depths, gives:

$$\Delta\widetilde{D} = \frac{\overline{\rho}_0 H}{\Delta\rho}, \tag{6.23}$$

where $\overline{\rho}_0$ was already presented in Eq. (6.5).

Eq. (6.23) shows that the deviation of the compensation depth from its mean value ($\widetilde{D}_0$), i.e. $\Delta\widetilde{D}$, depends on the density contrast $\Delta\rho$ between the crust and upper mantle and $\widetilde{D}_0$, which should be known a priori or obtained from external sources like geological or geophysical measurements. $\widetilde{D}_0$ plays an important role in determining $\Delta\widetilde{D}$. If it is available, $\Delta\widetilde{D}$ can be determined from $\Delta\rho$ and the TB heights and densities; or $\Delta\rho$ can be estimated from $\Delta\widetilde{D}$.

6.3.1 Airy–Heiskanen isostatic model based on gravitational potential in the spherical domain

The Airy–Heiskanen principle can also be presented in the spherical domain. In this case, the gravitational potential of the TB masses should be equal to the potential of the compensating masses; see Fig. 6.3. In the continents, this principle can be represented by the following Newton integrals:

$$G \iint\limits_{\sigma_{\text{continents}}} \rho'^{T} \int\limits_{R-\widetilde{D}}^{R+H'} \frac{r'^2 dr'}{l} d\sigma = G \iint\limits_{\sigma_{\text{continents}}} \rho'^{T} \int\limits_{R-\widetilde{D}_0}^{R} \frac{r'^2 dr'}{l} d\sigma + G \iint\limits_{\sigma_{\text{continents}}} \rho_0'^{M} \int\limits_{R-\widetilde{D}}^{R-\widetilde{D}_0} \frac{r'^2 dr'}{l} d\sigma. \tag{6.24}$$

The integral on the lhs of Eq. (6.24) can be rewritten as:

$$G\iint\limits_{\sigma_{\text{continents}}} \rho'^{\mathrm{T}} \int\limits_{R-\widetilde{D}'}^{R+H'} \frac{r'^2 \mathrm{d}r'}{l} \mathrm{d}\sigma = G\iint\limits_{\sigma_{\text{continents}}} \rho'^{\mathrm{T}} \int\limits_{R-\widetilde{D}'}^{R-\widetilde{D}_0} \frac{r'^2 \mathrm{d}r'}{l} \mathrm{d}\sigma + G\iint\limits_{\sigma_{\text{continents}}} \rho'^{\mathrm{T}} \int\limits_{R-\widetilde{D}_0}^{R} \frac{r'^2 \mathrm{d}r'}{l} \mathrm{d}\sigma + G\iint\limits_{\sigma_{\text{continents}}} \rho'^{\mathrm{T}} \int\limits_{R}^{R+H'} \frac{r'^2 \mathrm{d}r'}{l} \mathrm{d}\sigma. \quad (6.25)$$

Insertion of Eq. (6.25) into Eq. (6.24) and simplification of the results lead to:

$$G\iint\limits_{\sigma_{\text{continents}}} \Delta\rho' \int\limits_{R-\widetilde{D}'}^{R-\widetilde{D}_0} \frac{r'^2 \mathrm{d}r'}{l} \mathrm{d}\sigma = G\iint\limits_{\sigma_{\text{continents}}} \rho'^{\mathrm{T}} \int\limits_{R}^{R+H'} \frac{r'^2 \mathrm{d}r'}{l} \mathrm{d}\sigma. \quad (6.26)$$

In oceans the following integral relation can be written between the compensating and the TB gravitational potential, based on the Newton integral:

$$G\iint\limits_{\sigma_{\text{ocean}}} \rho'^{\mathrm{T}} \int\limits_{R-\widetilde{D}'}^{R-d'} \frac{r'^2 \mathrm{d}r'}{l} \mathrm{d}\sigma + G\iint\limits_{\sigma_{\text{ocean}}} \rho'^{\mathrm{B}} \int\limits_{R-d'}^{R} \frac{r'^2 \mathrm{d}r'}{l} \mathrm{d}\sigma = G\iint\limits_{\sigma_{\text{ocean}}} \rho'^{\mathrm{T}} \int\limits_{R-\widetilde{D}_0}^{R} \frac{r'^2 \mathrm{d}r'}{l} \mathrm{d}\sigma - G\iint\limits_{\sigma_{\text{ocean}}} \rho'^{\mathrm{M}} \int\limits_{R-\widetilde{D}'}^{R-\widetilde{D}_0} \frac{r'^2 \mathrm{d}r'}{l} \mathrm{d}\sigma. \quad (6.27)$$

Now, this equation should be rewritten in such a way that it can be comparable to (6.25):

$$\iint_{\sigma_{\text{ocean}}} \rho'^{\mathrm{T}} \int_{R-\widetilde{D}'}^{R-\widetilde{D}_0} \frac{r'^2 \mathrm{d}r'}{l} \mathrm{d}\sigma + G \iint_{\sigma_{\text{ocean}}} \rho'^{\mathrm{T}} \int_{R-\widetilde{D}_0}^{R} \frac{r'^2 \mathrm{d}r'}{l} \mathrm{d}\sigma$$

$$+ G \iint_{\sigma_{\text{ocean}}} \rho'^{\mathrm{T}} \int_{R}^{R-d'} \frac{r'^2 \mathrm{d}r'}{l} \mathrm{d}\sigma + G \iint_{\sigma_{\text{ocean}}} \rho'^{\mathrm{B}} \int_{R-d'}^{R} \frac{r'^2 \mathrm{d}r'}{l} \mathrm{d}\sigma$$

$$= G \iint_{\sigma_{\text{ocean}}} \rho'^{\mathrm{T}} \int_{R-\widetilde{D}_0}^{R} \frac{r'^2 \mathrm{d}r'}{l} \mathrm{d}\sigma - G \iint_{\sigma_{\text{ocean}}} \rho'^{\mathrm{M}} \int_{R-\widetilde{D}_0}^{R-\widetilde{D}'} \frac{r'^2 \mathrm{d}r'}{l} \mathrm{d}\sigma. \quad (6.28)$$

Simplification of Eq. (6.28) leads to:

$$G \iint_{\sigma_{\text{ocean}}} \Delta\rho' \int_{R-\widetilde{D}'}^{R-\widetilde{D}_0} \frac{r'^2 \mathrm{d}r'}{l} \mathrm{d}\sigma = -G \iint_{\sigma_{\text{ocean}}} \Delta\rho'^{\mathrm{B}} \int_{R}^{R-d'} \frac{r'^2 \mathrm{d}r'}{l} \mathrm{d}\sigma. \quad (6.29)$$

To write an integral formula having the globe as the surface integration domain, Eq. (6.26) is combined with Eq. (6.29):

$$G \iint_{\sigma_{\text{continents}}} \Delta\rho' \int_{R-\widetilde{D}'}^{R-\widetilde{D}_0} \frac{r'^2 \mathrm{d}r'}{l} \mathrm{d}\sigma + G \iint_{\sigma_{\text{ocean}}} \Delta\rho' \int_{R-\widetilde{D}'}^{R-\widetilde{D}_0} \frac{r'^2 \mathrm{d}r'}{l} \mathrm{d}\sigma$$

$$= G \iint_{\sigma_{\text{continents}}} \rho'^{\mathrm{T}} \int_{R}^{R+H'} \frac{r'^2 \mathrm{d}r'}{l} \mathrm{d}\sigma - G \iint_{\sigma_{\text{ocean}}} \Delta\rho'^{\mathrm{B}} \int_{R}^{R-d'} \frac{r'^2 \mathrm{d}r'}{l} \mathrm{d}\sigma, \quad (6.30)$$

when the solid Earth topography (H) as well as the density defined in Eq. (6.5) is applied but with ρ^{T} as the reference density, Eq. (6.30) will be simplified to:

$$V^{\mathrm{CMP}}(r,\theta,\lambda) = G \iint_{\sigma} \Delta\rho' \int_{R-\widetilde{D}'}^{R-\widetilde{D}_0} \frac{r'^2 \mathrm{d}r'}{l} \mathrm{d}\sigma = G \iint_{\sigma} \overline{\rho}'^{\mathrm{T}} \int_{R}^{R+H'} \frac{r'^2 \mathrm{d}r'}{l} \mathrm{d}\sigma$$

$$= V^{\mathrm{TB}}(r,\theta,\lambda). \quad (6.31)$$

In other words, the surface integration is taken all over the globe and but the parameters $\overline{\rho}'^{\mathrm{T}}$ and H' specify where the computations are done over continents or oceans.

Solution of the rhs of Eq. (6.31) in terms of spherical harmonics has already been presented in Chapter 5, Sections 5.1.1.1 and 5.1.1.2, as the TB potential. Now, the lhs is solved for $\widetilde{D}'$. If the Legendre expansion of l^{-1}, Eq. (5.3), is inserted into the lhs of Eq. (6.31) and the radial integral is solved, the following relation is obtained:

$$
\begin{aligned}
V^{\mathrm{CMP}}(r,\theta,\lambda) &= G\iint\limits_{\sigma}\Delta\rho'\int\limits_{R-\widetilde{D}'}^{R-\widetilde{D}_0}\frac{r'^2\mathrm{d}r'}{l}\mathrm{d}\sigma \\
&= G\iint\limits_{\sigma}\Delta\rho'\int\limits_{R-\widetilde{D}'}^{R}\frac{r'^2\mathrm{d}r'}{l}\mathrm{d}\sigma - G\iint\limits_{\sigma}\Delta\rho'\int\limits_{R-\widetilde{D}_0}^{R}\frac{r'^2\mathrm{d}r'}{l}\mathrm{d}\sigma \\
&= V^{\mathrm{CMP}_1}(r,\theta,\lambda) - V^{\mathrm{CMP}_0}(r,\theta,\lambda).
\end{aligned}
\tag{6.32}
$$

The idea of writing the compensation potential in the form presented in Eq. (6.32) comes from Sjöberg (2009, 2013). After solving the radial integrals we obtain:

$$
\begin{aligned}
V^{\mathrm{CMP}_0}(r,\theta,\lambda) &= GR^2\iint\limits_{\sigma}\Delta\rho'\sum_{n=0}^{\infty}\left(\frac{R}{r}\right)^{n+1}P_n(\widetilde{x})\,\frac{1}{n+3}\left[1-\left(1-\frac{\widetilde{D}_0}{R}\right)^{n+3}\right]\mathrm{d}\sigma \\
V^{\mathrm{CMP}_1}(r,\theta,\lambda) &= GR^2\iint\limits_{\sigma}\Delta\rho'\sum_{n=0}^{\infty}\left(\frac{R}{r}\right)^{n+1}P_n(\widetilde{x})\,\frac{1}{n+3}\left[1-\left(1-\frac{\widetilde{D}'}{R}\right)^{n+3}\right]\mathrm{d}\sigma.
\end{aligned}
\tag{6.33}
$$

The term $V^{\mathrm{CMP}_0}(r,\theta,\lambda)$ can be simply written in terms of the spherical harmonics, as $\widetilde{D}_0$ is constant:

$$
\begin{aligned}
V^{\mathrm{CMP}_0}(r,\theta,\lambda) = 4\pi GR^2\sum_{n=0}^{\infty}\frac{1}{(2n+1)(n+3)}\left[1-\left(1-\frac{\widetilde{D}_0}{R}\right)^{n+3}\right]\left(\frac{R}{r}\right)^{n+1} \\
\sum_{n=-m}^{n}\Delta\rho_{nm}Y_{nm}(\theta,\lambda).
\end{aligned}
\tag{6.34}
$$

But the solution of $V^{\mathrm{CMP}_1}(r,\theta,\lambda)$ is not as easy, as $\widetilde{D}'$ is a function of position. Therefore, the binomial term including it should be expanded by the Taylor series, which selected it to the third term here, for determining the potential $V^{\mathrm{CMP}_1}(r,\theta,\lambda)$. In the case of expanding l^{-1} in terms of the Legendre function, Eq. (5.3), the following integral is obtained for deriving $V^{\mathrm{CMP}_1}(r,\theta,\lambda)$:

$$V^{\mathrm{CMP}_1}(r,\theta,\lambda)=GR^2\iint_{\sigma}\Delta\rho'\sum_{n=0}^{\infty}\left(\frac{R}{r}\right)^{n+1}P_n(\widetilde{x})\left(\frac{\widetilde{D}'}{R}-(n+2)\frac{\widetilde{D}'^2}{2R^2}\right)d\sigma. \tag{6.35}$$

This integral can be applied to determine the gravitational potential of the compensating masses around the mean compensation depth, if $\widetilde{D}'$ and $\Delta\rho'$ are known. However, as discussed before as regards isostasy, the goal is to reach an equilibrium for the lithospheric masses floating on the viscous mantle. The main issue is what values $\widetilde{D}'$ and $\Delta\rho'$ should have to reach such an equilibrium. In other words, the unknown parameters which should be estimated to have this property are inherently inside the potential $V^{\mathrm{CMP}_1}(r,\theta,\lambda)$, which is derived simply from the TB potential and mean compensation potential; this means that Eqs. (6.34) and (6.35) are inserted into Eq. (6.33) and the result into Eq. (6.31):

$$V^{\mathrm{CMP}_1}=V^{\mathrm{CMP}_0}+V^{\mathrm{TB}}. \tag{6.36}$$

In fact, Eq. (6.36) is an integral equation, which is non-linear with respect to $\widetilde{D}'$. If the density contrast $\Delta\rho'$ is known from seismic data or external sources in addition to the mean compensation depth $\widetilde{D}_0$, the problem will be the estimation of $\widetilde{D}'$, the compensation depth, or Moho depth based on the Airy–Heiskanen principle. If $\widetilde{D}'$ and $\widetilde{D}_0$ are known parameters from external sources, the problem will be estimation of $\Delta\rho'$ to have the TB masses in isostatic equilibrium. In both cases, a priori knowledge about the value of $\widetilde{D}_0$ plays an important role.

6.3.2 Solutions to the Airy–Heiskanen model

In this section, the goal is to present different ways to estimate the compensation depth $\widetilde{D}'$, with the assumption that $\widetilde{D}_0$ and $\Delta\rho$ are known. However, estimating $\Delta\rho$ from the integral Eq. (6.35) is easier, as this equation is linear with respect to $\Delta\rho$. In the following subsections, some ideas for solving $\widetilde{D}'$ from this integral equation are presented and discussed.

6.3.2.1 Linear approximation of the binomial term

In the case in which the binomial term involving $\widetilde{D}$, in Eq. (6.33), is approximated only to linear terms, the following integral equation is achieved:

$$GR\iint_{\sigma}\Delta\rho'\sum_{n=0}^{\infty}\left(\frac{R}{r}\right)^{n+1}P_n(\widetilde{x})\widetilde{D}'\mathrm{d}\sigma = V^{\mathrm{CMP}_0}(r,\theta,\lambda)+V^{\mathrm{TB}}(r,\theta,\lambda). \tag{6.37}$$

This integral equation is linear with respect to $\widetilde{D}'$ and can be simply solved numerically by one of the methods presented in Chapter 4. However, it should be noted that $\Delta\rho'$ is also inside the integral and should be taken into account for each cell after discretisation of the integral. In the case in which this density contrast is constant, it can be taken outside the integral and the integral equation will change to an integral, which is invertible numerically as long as the computation point is outside the Earth.

The integral Eq. (6.37) can also be solved by spherical harmonics. According to the addition theorem of spherical harmonics (Eq. 1.9):

$$GR\iint_{\sigma}\Delta\rho'\sum_{n=0}^{\infty}\left(\frac{R}{r}\right)^{n+1}P_n(\widetilde{x})\widetilde{D}'\mathrm{d}\sigma = 4\pi GR\sum_{n=0}^{\infty}\frac{1}{2n+1}\left(\frac{R}{r}\right)^{n+1}\sum_{m=-n}^{n}(\Delta\rho D)_{nm}Y_{nm}(\theta,\lambda). \tag{6.38}$$

Using the spherical harmonic expansions of TB potential (see Sections 5.1.1.1 and 5.1.1.2), it will not be difficult to obtain for $r = R$:

$$\left(\Delta\rho\widetilde{D}\right)_{nm} = \frac{R}{(n+3)}\left(\Delta\rho\left[1-\left(1-\frac{\widetilde{D}_0}{R}\right)^{n+3}\right]\right)_{nm} + R(\overline{\rho}F)_{nm}, \tag{6.39}$$

where

$$(\overline{\rho}F)_{nm} = \left(\frac{(\overline{\rho}H)_{nm}}{R}+(n+2)\frac{(\overline{\rho}H^2)_{nm}}{2R^2}+(n+2)(n+1)\frac{(\overline{\rho}H^3)_{nm}}{6R^3}\right). \tag{6.40}$$

Now, let us perform more approximations to see how this equation is related to the Airy–Heiskanen principle, presented in Eq. (6.23). If the binomial term in the first term on the rhs is approximated to the linear term and density contrast generally is considered constant, and the second term is approximated by $(\overline{\rho}F)_{nm} \approx (\overline{\rho}H)_{nm}/R$, Eq. (6.39) will change to:

$$\left(\Delta\widetilde{D}\right)_{nm} = \widetilde{D}_{nm} - \widetilde{D}_0\delta_{n0} = \frac{(\overline{\rho}H)_{nm}}{\Delta\rho}. \tag{6.41}$$

If these SHCs are inserted into a spherical harmonic series, their spatial relation will be obtained, which is the same as Eq. (6.23). Therefore, Eq. (6.38) is a more general form of the Airy–Heiskanen principle.

6.3.2.2 Approximation of the binomial term to second order

Let us rewrite Eq. (6.35) in the following form:

$$GR^2 \iint_\sigma \Delta\rho' \sum_{n=0}^{\infty} \left(\frac{R}{r}\right)^{n+1} P_n(\widetilde{x}) \frac{\widetilde{D}'}{R}\left(1-(n+2)\frac{\widetilde{D}'}{2R^2}\right) d\sigma$$
$$= V^{\mathrm{CMP}_0}(r,\theta,\lambda) + V^{\mathrm{TB}}(r,\theta,\lambda). \tag{6.42}$$

Now, assume that $\widetilde{D}'$ inside the parentheses is approximated by $\widetilde{D}_0$. In this case, Eq. (6.42) changes to:

$$GR \iint_\sigma \Delta\rho' \sum_{n=0}^{\infty} \left(\frac{R}{r}\right)^{n+1} P_n(\widetilde{x})\widetilde{D}'\beta_n d\sigma = V^{\mathrm{CMP}_0}(r,\theta,\lambda) + V^{\mathrm{TB}}(r,\theta,\lambda), \tag{6.43}$$

where

$$\beta_n = 1 - (n+2)\frac{\widetilde{D}_0}{2R}. \tag{6.44}$$

The lhs of Eq. (6.43) has the following spherical harmonic expansion:

$$GR \iint_\sigma \Delta\rho' \sum_{n=0}^{\infty} \left(\frac{R}{r}\right)^{n+1} P_n(\widetilde{x})\widetilde{D}'\beta_n d\sigma = 4\pi GR \sum_{n=0}^{\infty} \frac{1}{2n+1}\left(\frac{R}{r}\right)^{n+1} \beta_n$$
$$\sum_{m=-n}^{n} \left(\Delta\rho\widetilde{D}\right)_{nm} Y_{nm}(\theta,\lambda). \tag{6.45}$$

If the spherical harmonic expansions of the rhs of Eq. (6.43) are inserted into Eq. (6.45), the following relation for the SHC of each expansion will be obtained:

$$\left(\Delta\rho\widetilde{D}\right)_{nm} = R\beta_n^{-1}\frac{1}{n+3}\left[1-\left(1-\frac{\widetilde{D}_0}{R}\right)^{n+3}\right](\Delta\rho)_{nm} + R\beta_n^{-1}(\overline{\rho}F)_{nm}. \quad (6.46)$$

By taking the spherical harmonic synthesis of Eq. (6.46) and assuming a constant $\Delta\rho$ solving the results for $\widetilde{D}'$, we obtain:

$$\widetilde{D} = \frac{R}{\Delta\rho}\sum_{n=0}^{\infty}\beta_n^{-1}\sum_{m=-n}^{n}\left(\frac{1}{n+3}\left[1-\left(1-\frac{\widetilde{D}_0}{R}\right)^{n+3}\right](\Delta\rho)_{nm} + (\overline{\rho}F)_{nm}\right)Y_{nm}(\theta,\lambda). \quad (6.47)$$

In Eq. (6.47), the SHCs of the density contrast $\Delta\rho_{nm}$ are required in addition to $\Delta\rho$. In the case where $\Delta\rho$ is constant, it can be simply taken outside the summation:

$$\widetilde{D} = \frac{R}{3}\left[1-\left(1-\frac{\widetilde{D}_0}{R}\right)^{3}\right]\beta_0^{-1} + \frac{R}{\Delta\rho}\sum_{n=0}^{\infty}\beta_n^{-1}\sum_{m=-n}^{n}(\overline{\rho}F)_{nm}Y_{nm}(\theta,\lambda). \quad (6.48)$$

6.3.2.3 Iterative solution

If we rewrite Eq. (6.36) as:

$$V^{\mathrm{CMP}_1} - V^{\mathrm{CMP}_0} - V^{\mathrm{TB}} = 0, \quad (6.49)$$

we have a non-linear equation with respect to the compensation depth $\widetilde{D}$. Let us linearise the residual compensation potential by the Taylor series:

$$V^{\mathrm{CMP}_1} = \left(V^{\mathrm{CMP}_1}\right)^{(0)} + \frac{\partial\left(V^{\mathrm{CMP}_1}\right)^{(0)}}{\partial\widetilde{D}}\left(\widetilde{D} - \widetilde{D}^{(0)}\right), \quad (6.50)$$

where $\widetilde{D}^{(0)}$ is an approximate value for $\widetilde{D}$, which can be obtained by Eq. (6.41), and $\left(V^{\mathrm{CMP}_1}\right)^{(0)}$ is the residual compensation computed according to $\widetilde{D}^{(0)}$. Insertion of Eq. (6.50) into Eq. (6.49) leads to:

$$\left(V^{\mathrm{CMP}_1}\right)^{(0)} + \frac{\partial\left(V^{\mathrm{CMP}_1}\right)^{(0)}}{\partial\widetilde{D}}\left(\widetilde{D} - \widetilde{D}^{(0)}\right) - V^{\mathrm{CMP}_0} - V^{\mathrm{TB}} = 0 \quad (6.51)$$

In Eq. (6.51) $\widetilde{D}$ is the unknown parameter, which should be estimated. Solving this equation for $\widetilde{D}$ gives a new estimate for $\widetilde{D}$. This principle can be written simply as:

$$\widetilde{D}^{(k+1)} = \widetilde{D}^{(k)} + \left(\frac{\partial (V^{\mathrm{CMP}_1})^{(k)}}{\partial \widetilde{D}} \right)^{-1} \left(V^{\mathrm{CMP}_0} + V^{\mathrm{TB}} - (V^{\mathrm{CMP}_1})^{(k)} \right) \quad (6.52)$$

where $(V^{\mathrm{CMP}_1})^{(k)}$ can be solved simply by inserting $\widetilde{D}^{(k)}$ into Eq. (6.33):

$$\begin{pmatrix} (V^{\mathrm{CMP}_1})^{(k)} \\ \dfrac{\partial (V^{\mathrm{CMP}_1})^{(k)}}{\partial \widetilde{D}} \end{pmatrix} = GR^2 \iint_\sigma \Delta\rho' \sum_{n=0}^{\infty} \frac{1}{n+3} \left(\frac{R}{r} \right)^{n+1} \begin{pmatrix} \left[1 - \left(1 - \dfrac{\widetilde{D}'^{(k)}}{R} \right)^{n+3} \right] \\ \left[\dfrac{(n+3)}{R} \left(1 - \dfrac{\widetilde{D}'^{(k)}}{R} \right)^{n+2} \right] \end{pmatrix} P_n(\widetilde{x}) \mathrm{d}\sigma \quad (6.53)$$

or in spherical harmonic form:

$$\begin{pmatrix} (V^{\mathrm{CMP}_1}(r,\theta,\lambda))^{(k)} \\ \dfrac{\partial (V^{\mathrm{CMP}_1}(r,\theta,\lambda))^{(k)}}{\partial \widetilde{D}} \end{pmatrix} = \sum_{n=0}^{\infty} \left(\frac{R}{r} \right)^{n+1} \sum_{m=-n}^{n} \begin{pmatrix} (v_{nm}^{\mathrm{CMP}_1})^{(k)} \\ \left(\dfrac{\partial v_{nm}^{\mathrm{CMP}_1}}{\partial \widetilde{D}} \right)^{(k)} \end{pmatrix} Y_{nm}(\theta, \lambda) \quad (6.54)$$

where

$$\begin{pmatrix} (v_{nm}^{\mathrm{CMP}_1})^{(k)} \\ \left(\dfrac{\partial v_{nm}^{\mathrm{CMP}_1}}{\partial \widetilde{D}} \right)^{(k)} \end{pmatrix} = \frac{4\pi GR^2}{2n+1} \left(\frac{\Delta\rho}{n+3} \begin{pmatrix} \left[1 - \left(1 - \dfrac{\widetilde{D}^{(k)}}{R} \right)^{n+3} \right] \\ \left[\dfrac{(n+3)}{R} \left(1 - \dfrac{\widetilde{D}^{(k)}}{R} \right)^{n+2} \right] \end{pmatrix} \right)_{nm} \quad (6.55)$$

6.3.2.4 Solving a non-linear integral equation

The integral Eq. (6.36) can also be numerically solved. Let us rewrite it in the following form:

$$GR^2 \iint_\sigma \Delta\rho' K^{AH}\left(r,\psi,\widetilde{D}'\right)d\sigma = V^{CMP_0}(r,\theta,\lambda) + V^{TB}(r,\theta,\lambda) \tag{6.56}$$

where

$$K^{AH}\left(r,\psi,\widetilde{D}'\right) = \sum_{n=0}^{\infty} \frac{1}{n+3}\left(\frac{R}{r}\right)^{n+1}\left[1-\left(1-\frac{\widetilde{D}'}{R}\right)^{n+3}\right]P_n(\widetilde{x}) \tag{6.57}$$

stands for the kernel of the integral Eq. (6.56). As observed, the unknown parameter $\widetilde{D}$ is inside the kernel; in other words, the kernel is non-linear with respect to $\widetilde{D}$. Let us expand this kernel function using the Taylor series to the linear term and insert it into the integral (Eq. 6.56):

$$GR^2 \iint_\sigma \Delta\rho'\left(\left(K^{AH}\left(r,\psi,\widetilde{D}'\right)\right)^{(0)} + \frac{\partial\left(K^{AH}\left(r,\psi,\widetilde{D}'\right)\right)^{(0)}}{\partial \widetilde{D}'}\delta\widetilde{D}'\right)d\sigma$$
$$= V^{CMP_0}(r,\theta,\lambda) + V^{TB}(r,\theta,\lambda) \tag{6.58}$$

To expand the kernel using the Taylor series, an approximate value of $\widetilde{D}'$ is required, which can be derived by Eq. (6.23). Now, instead of estimating $\widetilde{D}'$ directly, the corrections to its approximate value $\delta\widetilde{D}'$ should be estimated by solving the following integral equation:

$$GR^2 \iint_\sigma \Delta\rho' \frac{\partial\left(K^{AH}\left(r,\psi,\widetilde{D}'\right)\right)^{(0)}}{\partial \widetilde{D}'}\delta\widetilde{D}'d\sigma$$
$$= V^{CMP_0}(r,\theta,\lambda) + V^{TB}(r,\theta,\lambda) - GR^2 \iint_\sigma \Delta\rho'\left(K^{AH}\left(r,\psi,\widetilde{D}'\right)\right)^{(0)}d\sigma \tag{6.59}$$

where

$$\frac{\partial\left(K^{AH}\left(r,\psi,\widetilde{D}'\right)\right)^{(0)}}{\partial \widetilde{D}'} = \frac{\widetilde{D}'}{R}\sum_{n=0}^{\infty}\left(\frac{R}{r}\right)^{n+1}P_n(\widetilde{x}). \tag{6.60}$$

Once $\delta\widetilde{D}'$ is estimated, the value of $\widetilde{D}'$ will be updated by adding $\delta\widetilde{D}'$ to it. The updated value of $\widetilde{D}'$ will be used for computing the kernel value of the integral on the rhs and rhs of Eq. (6.59) and the inversion process will be repeated until $\delta\widetilde{D}'$ becomes very small and by iteration no improvement is observed.

6.3.3 Gravimetric isostasy

So far, our goal has been to present the isostatic principles of Pratt–Hayford and Airy–Heiskanen based on the gravitational potential of the loads and compensating masses. It was shown that when the gravitational potential of the loads, TB masses, is equal to the gravitational potential of the compensating masses, the loads are in isostatic equilibrium according to these theories. However, no gravimetric data have been applied in the formulation of these theories. In this section, the goal is to involve the gravimetric data for expressing the isostatic equilibrium. Here, two approaches are discussed; the first is the one presented by Moritz (1990, p. 255), which was developed further by Sjöberg (2009) and named Vening Meinesz–Moritz isostasy. Eshagh (2016a) presented a different solution for this isostatic model, which will be discussed in this section as approach 1. The other approach comes from the idea of Jeffrey (1976), which was developed further by Eshagh (2016a). This method is named approach 2 in this discussion.

6.3.3.1 Approach 1

The principle of the gravimetric isostasy is that the isostatic gravity anomaly should be zero if the TB masses are in isostatic equilibrium. According to Sjöberg (2009) this anomaly is defined by:

$$\Delta g^{\text{Iso}} = \Delta g - \Delta g^{\text{TB}} + \Delta g^{\text{CMP}} \tag{6.61}$$

where Δg is the gravity anomaly, Δg^{TB} the effect of the TB masses and Δg^{CMP} the effect of the compensating masses on the anomaly. It is said that the TB masses are in isostatic equilibrium when $\Delta g^{\text{Iso}} = 0$. Eq. (6.61) contains Δg, which is in fact a gravimetric measurement. Similar to isostatic gravity anomaly, different isostatic anomalies can be defined, for example, isostatic potential anomaly and isostatic gravity disturbance and so on. However, let us continue the discussion based on the isostatic potential so that the differences between the gravimetric approach to isostasy and the classical model of Airy–Heiskanen become more visible. In this case the isostatic potential is (see also Sjöberg, 2013):

$$V^{\text{Iso}} = T - V^{\text{TB}} + V^{\text{CMP}} \tag{6.62}$$

where V^{Iso} stands for the isostatic potential anomaly, T the disturbing potential, V^{TB} the potential of the TB masses and finally V^{CMP} the potential of the compensating masses.

By assuming that $V^{\text{Iso}} = 0$, then Eq. (6.62) can be simply solved for V^{CMP}. The gravitational potential of the TB masses has already been presented in Chapter 5 and V^{CMP} in Eq. (6.32); therefore:

$$V^{\text{CMP}}(r,\theta,\lambda) = G\iint_{\sigma} \Delta\rho' \int_{R-\widetilde{D}'}^{R-\widetilde{D}_0} \frac{r'^2 \mathrm{d}r'}{l} \mathrm{d}\sigma \approx V^{\text{TB}}(r,\theta,\lambda) - T(r,\theta,\lambda). \tag{6.63}$$

The integral on the lhs of Eq. (6.63) has already been written in terms of Legendre polynomials and separated into mean compensation potential and residual compensation potential according to Eq. (6.33). The final formula expressing the relation between the compensation depth $\widetilde{D}$ and the gravitational potential of the TB masses and disturbing potential T is:

$$\begin{aligned} GR^2 \iint_{\sigma} \Delta\rho' \sum_{n=0}^{\infty} \frac{1}{n+3}\left(\frac{R}{r}\right)^{n+1}\left[1-\left(1-\frac{\widetilde{D}'}{R}\right)^{n+3}\right] P_n(\widetilde{x})\mathrm{d}\sigma \\ = V^{\text{CMP}_0}(r,\theta,\lambda) + V^{\text{TB}}(r,\theta,\lambda) - T(r,\theta,\lambda) \end{aligned} \tag{6.64}$$

$\widetilde{D}'$ can be derived similar to the processes presented in Sections 6.3.2.1–6.3.2.4.

Eq. (6.64) is an equation of isostatic equilibrium considering the gravimetric quantity T, which is determined e.g. from a global gravity model. The main difference between this equation and that of Airy–Heiskanen is related to the presence of T on the rhs. This equation shows the relation between $\widetilde{D}'$, $\widetilde{D}_0$ and $\Delta\rho'$. Again $\widetilde{D}_0$ should be known from external sources of information, e.g. seismic data for determining the mean compensation potential. $\Delta\rho'$ can be determined if $\widetilde{D}'$ is known and vice versa. The problem of determining $\Delta\rho'$ is easier because this integral equation is linear with respect to it. However, determination of $\widetilde{D}'$ is more complicated and needs more algebraic manipulations. In the following, we will present different methods for solving Eq. (6.64) for $\widetilde{D}'$.

6.3.4 Linear approximation

The main difficulty in solving Eq. (6.64) is that $\widetilde{D}'$ is inside a binomial term which has the power $n+3$. The simplest way is then to linearise this term using the Taylor series. In this case, after some simplification Eq. (6.64) changes to:

$$GR\iint_{\sigma}\Delta\rho'\widetilde{D}'\sum_{n=0}^{\infty}\left(\frac{R}{r}\right)^{n+1}P_n(\widetilde{x})\mathrm{d}\sigma = V^{\mathrm{CMP}_0}(r,\theta,\lambda) + V^{\mathrm{TB}}(r,\theta,\lambda) - T(r,\theta,\lambda) \tag{6.65}$$

Now, Eq. (6.65) is an integral equation, which can be solved to estimate the product of density contrast and compensation depth $\Delta\rho\widetilde{D}'$. This means that the integral on the lhs of Eq. (6.65) can be discretised according to the resolution of recovery and solved numerically, by a regularisation method discussed in Chapter 4. However, Eq. (6.65) can also be solved in terms of spherical harmonics. To do so, from Eq. (6.38) and applying the addition theorem of spherical harmonics on the lhs of Eq. (6.65), we obtain:

$$4\pi GR\sum_{n=0}^{\infty}\frac{1}{2n+1}\left(\frac{R}{r}\right)^{n+1}\sum_{m=-n}^{n}\left(\Delta\rho\widetilde{D}\right)_{nm}Y_{nm}(\theta,\lambda) = V^{\mathrm{CMP}_0}(r,\theta,\lambda) + V^{\mathrm{TB}}(r,\theta,\lambda) - T(r,\theta,\lambda) \tag{6.66}$$

If both sides of Eq. (6.66) are written in terms of spherical harmonics and the result is solved for $(\Delta\rho\widetilde{D})_{nm}$, after simplifications, it is shown that $\widetilde{D}$ is simply derived by:

$$\widetilde{D} = \frac{1}{4\pi GR\Delta\rho}\sum_{n=0}^{\infty}(2n+1)\sum_{m=-n}^{n}\left(v_{nm}^{\mathrm{CMP}_0} + v_{nm}^{\mathrm{TB}} - t_{nm}\right)Y_{nm}(\theta,\lambda) \tag{6.67}$$

If the mathematical models of $v_{nm}^{\mathrm{CMP}_0}$ and v_{nm}^{TB} are inserted into Eq. (6.67), after necessary simplifications, the results will be:

$$\widetilde{D} = \frac{R}{\Delta\rho}\sum_{n=0}^{\infty}\sum_{m=-n}^{n}\left(\frac{\delta_{n0}}{n+3}\Delta\rho\left[1-\left(1-\frac{\widetilde{D}_0}{R}\right)^{n+3}\right] + (\overline{\rho}F)_{nm} - \frac{t_{nm}}{4\pi G}\right)Y_{nm}(\theta,\lambda) \tag{6.68}$$

Eq. (6.67) can also be written based on the gravity anomaly or gravity disturbance, based on their spectral relations. In this case,

$$\widetilde{D} = \frac{1}{4\pi G\Delta\rho}\sum_{n=0}^{\infty}\frac{2n+1}{n\mp 1}\sum_{m=-n}^{n}\begin{pmatrix}\Delta g_{nm}^{\mathrm{CMP_0}} + \Delta g_{nm}^{\mathrm{TB}} - \Delta g_{nm}\\ \delta g_{nm}^{\mathrm{CMP_0}} + \delta g_{nm}^{\mathrm{TB}} - \delta g_{nm}\end{pmatrix} Y_{nm}(\theta,\lambda). \tag{6.69}$$

.

To use Eqs. (6.67) or (6.69), a global gravity model can simply be applied to compute t_{nm}, Δg_{nm} or δg_{nm}, and a global TB model for v_{nm}^{TB}, $\Delta g_{nm}^{\mathrm{TB}}$ or $\delta g_{nm}^{\mathrm{TB}}$. For computing $v_{nm}^{\mathrm{CMP_0}}$, $\Delta g_{nm}^{\mathrm{CMP_0}}$ or $\delta g_{nm}^{\mathrm{CMP_0}}$ density contrast $\Delta\rho$ with a global coverage is required. Today, different global gravity and TB height models are available in addition to the CRUST1.0 model, which contains the density of the upper mantle and different crustal layers. From Eq. (6.69), $\Delta\rho$ can also be derived if $\widetilde{D}$ is known from external sources.

6.3.5 Second-order approximation

The main difference between the process done for the Airy–Heiskanen model, in Section 6.3.2.2 can be simply applied for the gravimetric approach to isostasy. In this case, Eq. (6.45) will change to:

$$GR\iint_{\sigma}\Delta\rho'\widetilde{D}'\sum_{n=0}^{\infty}\left(\frac{R}{r}\right)^{n+1}\beta_n P_n(\widetilde{x})\mathrm{d}\sigma = V^{\mathrm{CMP_0}}(r,\theta,\lambda) + V^{\mathrm{TB}}(r,\theta,\lambda) - T(r,\theta,\lambda) \tag{6.70}$$

The only difference with Eq. (6.45) is related to the presence of $T(r,\theta,\lambda)$ on the rhs of the equation. Eq. (6.70) is also an integral equation, which can be solved for $\widetilde{D}'$ numerically. However, the spherical harmonic solution of $\widetilde{D}'$ will be (see Eshagh, 2017):

$$\widetilde{D} = \frac{1}{4\pi GR\Delta\rho}\sum_{n=0}^{\infty}\frac{2n+1}{\beta_n}\sum_{m=-n}^{n}\left(v_{nm}^{\mathrm{CMP_0}} + v_{nm}^{\mathrm{TB}} - t_{nm}\right)Y_{nm}(\theta,\lambda) \tag{6.71}$$

and in the case of using gravity anomaly or disturbance:

$$\widetilde{D} = \frac{1}{4\pi G\Delta\rho}\sum_{n=0}^{\infty}\frac{2n+1}{(n\mp1)\beta_n}\sum_{m=-n}^{n}\begin{pmatrix}\Delta g_{nm}^{\mathrm{CMP_0}} + \Delta g_{nm}^{\mathrm{TB}} - \Delta g_{nm}\\ \delta g_{nm}^{\mathrm{CMP_0}} + \delta g_{nm}^{\mathrm{TB}} - \delta g_{nm}\end{pmatrix}Y_{nm}(\theta,\lambda) \tag{6.72}$$

The left panel of Fig. 6.4 shows the values of the binomial term for $\widetilde{D}_0 = 30$ km and its first- and second-order approximation to degree 180. As observed, the linear approximation is far from the function, especially at high degrees, and it leads to a loss of signal power, whilst the second-order approximation is closer to the function but does not coincide at high degrees. The approximation error will be even larger for $\widetilde{D}_0 > 30$ km. What is concluded here is that the approximation, at least up to second order, should be considered for the binomial term up to degree and order 180.

Fig. 6.4, right, shows the signal spectra of the Moho models computed based on Eqs. (6.67) and (6.72) as well as the CRUST1.0 (Laske et al., 2013) Moho model. A density contrast of 600 km/m^3 and $D_0 = 30$ km, EGM08 (Pavlis et al., 2012) and SRTM30 (Far et al., 2007) were considered for these computations. The signal of the Moho depths computed by Eq. (6.67) is weaker than that of CRUST1.0 at high frequencies; as discussed, this is due to linearisation of the binomial term. By using the second-order approximation, Eq. (6.72), the signal will be closer

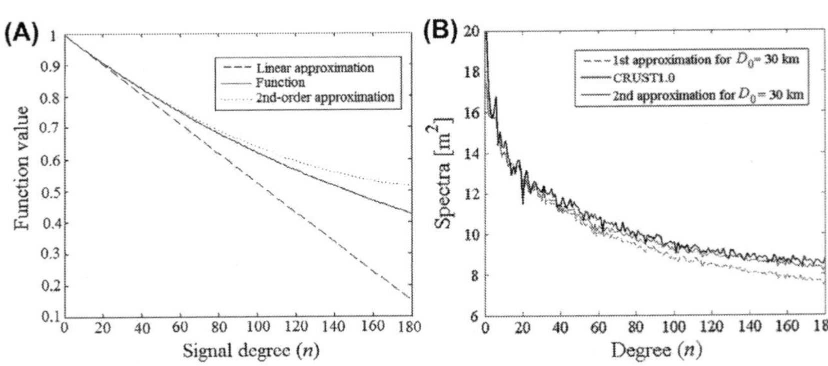

Figure 6.4 (Left) Values of the binomial term for $\widetilde{D}_0 = 30$ and its first- and second-order approximations. (Right) the Moho depth signal based on the first- and second-order approximations done for the binomial term. *(From Eshagh, M., 2016a. A theoretical discussion on Vening Meinesz-Moritz inverse problem of isostasy. Geophys. J. Int. 207, 1420–1431.)*

to the CRUST1.0 signal but still weaker. The signal is closer to that of CRUST1.0 for higher degrees than for lower.

6.3.5.1 Iterative solution

Similar to the iterative solution presented for solving the isostasy problem based on the Airy–Heiskanen assumption, an iterative solution can also be presented for determining the compensation depth $\widetilde{D}$ or $\Delta\rho$. Here, determination of $\widetilde{D}$ is considered and it can be shown that the only difference between this iterative solution and Eq. (6.73) is related to the presence of the gravimetric data on its rhs. The general iterative formula for determining $\widetilde{D}$ from disturbing potential, gravity anomaly and gravity disturbance will be:

$$\widetilde{D}^{(k+1)} = \widetilde{D}^{(k)} + \begin{pmatrix} \left(\dfrac{\partial (V^{\mathrm{CMP}_1})^{(k)}}{\partial \widetilde{D}}\right)^{-1} \\ \left(\dfrac{\partial (\Delta g^{\mathrm{CMP}_1})^{(k)}}{\partial \widetilde{D}}\right)^{-1} \\ \left(\dfrac{\partial (\delta g^{\mathrm{CMP}_1})^{(k)}}{\partial \widetilde{D}}\right)^{-1} \end{pmatrix} \begin{pmatrix} V^{\mathrm{CMP}_0} + V^{\mathrm{TB}} - T - (V^{\mathrm{CMP}_1})^{(k)} \\ \Delta g^{\mathrm{CMP}_0} + \Delta g^{\mathrm{TB}} - \Delta g - (\Delta g^{\mathrm{CMP}_1})^{(k)} \\ \delta g^{\mathrm{CMP}_0} + \delta g^{\mathrm{TB}} - \delta g - (\delta g^{\mathrm{CMP}_1})^{(k)} \end{pmatrix} \tag{6.73}$$

where Δg^{TB} and δg^{TB} are the effects of the TB masses on the gravity anomaly and disturbance, respectively, and $\Delta g^{\mathrm{CMP}_0}$ and $\delta g^{\mathrm{CMP}_0}$ are the effects of the mean compensating masses on them. In addition:

$$\begin{pmatrix} (\Delta g^{\mathrm{CMP}_1}(r,\theta,\lambda))^{(k)} \\ \dfrac{\partial (\Delta g^{\mathrm{CMP}_1}(r,\theta,\lambda))^{(k)}}{\partial \widetilde{D}} \end{pmatrix} = \sum_{n=0}^{\infty} (n-1)\left(\frac{R}{r}\right)^{n+2} \sum_{m=-n}^{n} \begin{pmatrix} (V^{\mathrm{CMP}_1})^{(k)} \\ \left(\dfrac{\partial (V^{\mathrm{CMP}_1})^{(k)}}{\partial \widetilde{D}}\right) \end{pmatrix}_{nm} Y_{nm}(\theta,\lambda) \tag{6.74}$$

In the case of using $n+1$ instead of $n-1$ in Eq. (6.74), the corresponding quantities will be derived for gravity disturbance.

An initial value for the compensation depth $\widetilde{D}$ is inserted into the rhs of Eq. (6.73) to obtain a new update for $\widetilde{D}$; thereafter, this new value is inserted into the rhs of the equations and again another new update is achieved. This process is repeated until the solution converges according to a predefined tolerance.

6.3.5.2 Non-linear inversion

Here, three integral equations are presented in Eq. (6.75), and the main difference between them is related to the gravimetric data used for forming the gravimetric isostasy. This integral can be simply discretised and solved numerically. Since this equation is non-linear, the kernel of the integral should be linearised and the integral equation should be solved iteratively. This linearised form is written in the following:

$$GR^2 \iint_\sigma \Delta\rho' \begin{pmatrix} \dfrac{\partial K_0^{\mathrm{V}}(R,\psi,\widetilde{D}')}{\partial \widetilde{D}'} \\ \dfrac{\partial K_0^{\Delta g}(R,\psi,\widetilde{D}')}{\partial \widetilde{D}'} \\ \dfrac{\partial K_0^{\delta g}(R,\psi,\widetilde{D}')}{\partial \widetilde{D}'} \end{pmatrix} \delta\widetilde{D}' \mathrm{d}\sigma = \begin{pmatrix} V^{\mathrm{CMP}_0} + V^{\mathrm{TB}} - T - GR^2 \iint_\sigma \Delta\rho' K_0^{\mathrm{V}}(R,\psi,\widetilde{D}')\mathrm{d}\sigma \\ \Delta g^{\mathrm{CMP}_0} + \Delta g^{\mathrm{TB}} - \Delta g - GR^2 \iint_\sigma \Delta\rho' K_0^{\Delta g}(R,\psi,\widetilde{D}')\mathrm{d}\sigma \\ \delta g^{\mathrm{CMP}_0} + \delta g^{\mathrm{TB}} - \delta g - GR^2 \iint_\sigma \Delta\rho' K_0^{\delta g}(R,\psi,\widetilde{D}')\mathrm{d}\sigma \end{pmatrix} \tag{6.75}$$

where

$$\begin{pmatrix} K^{\mathrm{V}}(R,\psi,\widetilde{D}') \\ K^{\Delta g}(R,\psi,\widetilde{D}') \\ K^{\delta g}(R,\psi,\widetilde{D}') \end{pmatrix} = \sum_{n=0}^{\infty} \frac{1}{n+3} \begin{pmatrix} 1 \\ (n-1)R^{-1} \\ (n+1)R^{-1} \end{pmatrix} \left[1 - \left(1 - \frac{\widetilde{D}'}{R}\right)^{n+3}\right] P_n(\widetilde{x}) \tag{6.76}$$

are the kernels of the integral, which are generated based on the initial value of $\widetilde{D}$ for each integral equation. The derivatives of these kernel functions with respect to $\widetilde{D}$, which are used for the linearisation of the integral equations are:

$$\frac{\partial}{\partial \widetilde{D}'}\begin{pmatrix} K^{V}\left(r,\psi,\widetilde{D}'\right) \\ K^{\Delta g}\left(r,\psi,\widetilde{D}'\right) \\ K^{\delta g}\left(r,\psi,\widetilde{D}'\right) \end{pmatrix} = \frac{\widetilde{D}'}{R}\sum_{n=0}^{\infty}\frac{1}{n+3}\begin{pmatrix} 1 \\ (n-1)R^{-1} \\ (n+1)R^{-1} \end{pmatrix} P_n(\widetilde{x}) \tag{6.77}$$

Note that the integral Eq. (6.64) is linear with respect to $\Delta\rho'$ and solvable numerically for $\Delta\rho'$. Therefore, no iterative approach is required for this purpose.

6.3.6 Approach 2

The gravimetric isostasy can be presented in a different way. Let us write the compensation potential in the following integral form:

$$\begin{aligned} V^{CMP} &= G\iint_{\sigma} \Delta\rho' \int_{R-\widetilde{D}_0-\Delta\widetilde{D}'}^{R-\widetilde{D}_0} \frac{r'^2}{l}\,\mathrm{d}r'\,\mathrm{d}\sigma \\ &= G\iint_{\sigma} \Delta\rho' \sum_{n=0}^{\infty}\frac{1}{r^{n+1}} \int_{R-\widetilde{D}_0-\Delta\widetilde{D}'}^{R-\widetilde{D}_0} r'^{n+2}\,\mathrm{d}r' P_n(\widetilde{x})\,\mathrm{d}\sigma \end{aligned} \tag{6.78}$$

where $\Delta\widetilde{D}'$ is the variation of Moho around $\widetilde{D}_0$.

As observed $\widetilde{D}_0$ is in the upper and lower limits of the integral and there is no need to consider a mean compensation potential. In other words, the potential of the variations of the masses around $\widetilde{D}_0$ is considered directly in the integral, and not as a residual compensation potential, which was considered in approach 1. Therefore, the main goal will be to determine the variations of the compensation depth $\Delta\widetilde{D}'$ with respect to $\widetilde{D}_0$. To do so, the radial integral is solved in Eq. (6.78) and the result will be:

$$V^{CMP} = G\iint_{\sigma} \Delta\rho' \sum_{n=0}^{\infty}\frac{1}{r^{n+1}}\frac{1}{n+3}\left[\left(R-\widetilde{D}_0\right)^{n+3} - \left(R-\widetilde{D}_0-\Delta\widetilde{D}'\right)^{n+3}\right] P_n(\widetilde{x})\,\mathrm{d}\sigma \tag{6.79}$$

After factorising $\left(R-\widetilde{D}_0\right)^{n+3}$ inside the brackets, the equation will change to:

$$V^{\mathrm{CMP}}=G\iint\limits_{\sigma}\Delta\rho'\sum_{n=0}^{\infty}\frac{1}{r^{n+1}}\frac{\left(R-\widetilde{D}_0\right)^{n+3}}{n+3}\left[1-\left(1-\frac{\Delta\widetilde{D}'}{R-\widetilde{D}_0}\right)^{n+3}\right]P_n(\widetilde{x})\mathrm{d}\sigma \tag{6.80}$$

This compensation potential should be equal to $V^{\mathrm{TB}}(r,\theta,\lambda)-T(r,\theta,\lambda)$ so that the TB masses remain in gravimetric isostatic equilibrium. Mathematically, this will be:

$$G\left(R-\widetilde{D}_0\right)^2\iint\limits_{\sigma}\Delta\rho'\sum_{n=0}^{\infty}\left(\frac{R-\widetilde{D}_0}{r}\right)^{n+1}\frac{1}{n+3}\left[1-\left(1-\frac{\Delta\widetilde{D}'}{R-\widetilde{D}_0}\right)^{n+3}\right]$$
$$P_n(\widetilde{x})\mathrm{d}\sigma=V^{\mathrm{TB}}(r,\theta,\lambda)-T(r,\theta,\lambda). \tag{6.81}$$

Eq. (6.81) is another expression for the gravimetric isostatic equilibrium. The attenuation term $\left(\frac{R-\widetilde{D}_0}{r}\right)^{n+1}$ diminishes the power of the compensation signal from a sphere with the radius $R-\widetilde{D}_0$. Therefore, this factor depends highly on the choice of $\widetilde{D}_0$. The unknown parameter will be either $\Delta\rho'$ or $\Delta\widetilde{D}'$. If the compensation depth $\widetilde{D}'$ is given from external sources, $\widetilde{D}_0$ and $\Delta\widetilde{D}$ can be obtained from it and the kernel function will be evaluable. The integral can be discretised and inverted for obtaining $\Delta\rho'$ from $V^{\mathrm{TB}}(r,\theta,\lambda)-T(r,\theta,\lambda)$. However, when $\Delta\widetilde{D}'$ is desired, the mathematical model (Eq. 6.81) needs to be simplified as it is non-linear with respect to it. In the following, different methods for such simplifications and solutions for $\Delta\widetilde{D}'$ are presented.

6.3.6.1 Linear approximation

If the binomial term in Eq. (6.81) is expanded up to the linear term, the equation will change to:

$$G\left(R-\widetilde{D}_0\right)\iint\limits_{\sigma}\Delta\rho'\Delta\widetilde{D}'\sum_{n=0}^{\infty}\left(\frac{R-\widetilde{D}_0}{r}\right)^{n+1}$$
$$P_n(\widetilde{x})\mathrm{d}\sigma=V^{\mathrm{TB}}(r,\theta,\lambda)-T(r,\theta,\lambda) \tag{6.82}$$

Note that the mean compensation potential does not exist on the rhs of Eq. (6.82), as the residual compensation depth $\Delta\widetilde{D}'$ is derived by solving this equation and not the compensation depth $\widetilde{D}$. After applying the addition theorem of spherical harmonics, the lhs of Eq. (6.82) can be written in terms of spherical harmonics:

$$4\pi G\left(R-\widetilde{D}_0\right)\sum_{n=0}^{\infty}\frac{1}{2n+1}\left(\frac{R-\widetilde{D}_0}{r}\right)^{n+1}\sum_{m=-n}^{n}\left(\Delta\rho\Delta\widetilde{D}\right)_{nm}Y_{nm}(\theta,\lambda). \\ = V^{\mathrm{TB}}(r,\theta,\lambda)-T(r,\theta,\lambda). \tag{6.83}$$

If $V^{\mathrm{TB}}(r,\ \theta,\ \lambda)-T(r,\theta,\lambda)$ is written in terms of spherical harmonics, $(\Delta\rho\Delta\widetilde{D})_{nm}$ can be solved from them. $\Delta\rho\Delta\widetilde{D}$ is solved using spherical harmonic synthesis and after that by assuming that $\Delta\rho$ is constant, and both sides of the results can be divided by $\Delta\rho$ to obtain $\Delta\widetilde{D}$. A general solution for $\Delta\widetilde{D}$ from gravity anomaly and gravity disturbances can be written according to the spectral relations with disturbing potential:

$$\Delta\widetilde{D}=\frac{1}{4\pi GR\Delta\rho}\sum_{n=0}^{\infty}(2n+1)\left(\frac{R}{R-\widetilde{D}_0}\right)^{n+2}\begin{pmatrix}1\\(n-1)^{-1}R\\(n+1)^{-1}R\end{pmatrix} \\ \sum_{m=-n}^{n}\begin{pmatrix}v_{nm}^{\mathrm{TB}}-t_{nm}\\ \Delta g_{nm}^{\mathrm{TB}}-\Delta g_{nm}\\ \delta g_{nm}^{\mathrm{TB}}-\delta g_{nm}\end{pmatrix}Y_{nm}(\theta,\lambda). \tag{6.84}$$

In this solution the main assumption is that $\Delta\widetilde{D} << R-\widetilde{D}_0$, which is a rather good approximation. For example, the average global mean compensation depth $\widetilde{D}_0$ is about 30 km over continents. The variation from this average value reaches about 40 km over special places, like the Himalaya and Andes mountain ranges. Therefore, 40 km << 6348 km, is about 0.6%.

6.3.6.2 Second-order approximation

The binomial term in Eq. (6.80) can be expanded to the third term by the Taylor series. However, if a second-order approximation is considered, the binomial term in Eq. (6.80) is approximated to the third term and Eq. (6.82) changes to:

$$G\left(R-\widetilde{D}_0\right)\iint_{\sigma}\Delta\rho'\Delta\widetilde{D}'\sum_{n=0}^{\infty}\left(\frac{R-\widetilde{D}_0}{r}\right)^{n+1}\beta_n^{*}P_n(\widetilde{x})\mathrm{d}\sigma \\ = V^{\mathrm{TB}}(r,\theta,\lambda)-T(r,\theta,\lambda) \tag{6.85}$$

where

$$\beta_n^* = 1 - (n+2)\frac{\overline{\Delta\widetilde{D}}}{2(R-\widetilde{D}_0)} \tag{6.86}$$

and $\overline{\Delta\widetilde{D}}$ is the average value of the undulations or variations of $\widetilde{D}$ from $\widetilde{D}_0$, which can be obtained from $\Delta\widetilde{D}$ values and Eq. (6.84). Therefore, a spherical harmonic expansion similar to Eq. (6.84) can be derived based on this approximation:

$$\Delta\widetilde{D} = \frac{1}{4\pi GR\Delta\rho}\sum_{n=0}^{\infty}\frac{2n+1}{\beta_n^*}\left(\frac{R}{R-\widetilde{D}_0}\right)^{n+2}\begin{pmatrix}1\\(n-1)^{-1}R\\(n+1)^{-1}R\end{pmatrix}\sum_{m=-n}^{n}\begin{pmatrix}v_{nm}^{\mathrm{TB}}-t_{nm}\\\Delta g_{nm}^{\mathrm{TB}}-\Delta g_{nm}\\\delta g_{nm}^{\mathrm{TB}}-\delta g_{nm}\end{pmatrix}Y_{nm}(\theta,\lambda). \tag{6.87}$$

The appearance of β_n^* in the denominator causes a signal amplification, which depends on $\widetilde{D}_0$. The absence of this factor in Eq. (6.83) means that the solution for $\widetilde{D}$ will be smoother than the case in which Eq. (6.87) is applied for solving $\widetilde{D}$. However, β_n^* amplifies the errors of the data as well and may lead to divergence of the solution when n goes to infinity. Therefore, limiting the spherical harmonic series (Eq. 6.87) to a maximum degree, say to 180, can be an idea to avoid these problems. Definitely, for larger values of $\widetilde{D}_0$ this amplification starts in lower degrees. Nevertheless, on the basis of assuming $\Delta\widetilde{D} << R - \widetilde{D}_0$, β_n^* will be closer to 1 and the signal amplification will not be very strong.

6.3.6.3 Iterative solution

An iterative solution can be obtained based on the principle explained in approach 2. The general form of this method is:

$$\Delta\widetilde{D}^{(k+1)} = \Delta\widetilde{D}^{(k)} + \begin{pmatrix}\left(\dfrac{\partial(V^{\mathrm{CMP}})^{(k)}}{\partial\Delta\widetilde{D}}\right)^{-1}\\\left(\dfrac{\partial(\Delta g^{\mathrm{CMP}})^{(k)}}{\partial\Delta\widetilde{D}}\right)^{-1}\\\left(\dfrac{\partial(\delta g^{\mathrm{CMP}})}{\partial\Delta\widetilde{D}}\right)^{-1}\end{pmatrix}\begin{pmatrix}-V^{\mathrm{TB}}+T-(V^{\mathrm{CMP}})^{(k)}\\-g^{\mathrm{TB}}+\Delta g-(\Delta g^{\mathrm{CMP}})^{(k)}\\-\delta g^{\mathrm{TB}}+\delta g-(\delta g^{\mathrm{CMP}})\end{pmatrix} \tag{6.88}$$

where

$$\left.\begin{array}{l}(V^{\mathrm{CMP}})^{(k)}\\ (\delta g^{\mathrm{CMP}})^{(k)}\\ (\Delta g^{\mathrm{CMP}})^{(k)}\end{array}\right\} = G(R-\widetilde{D}_0)\iint_\sigma \Delta\rho' \sum_{n=0}^{\infty}\frac{1}{n+3}\left\{\begin{array}{c}R^{-1}\\ n+1\\ n-1\end{array}\right\}\left(\frac{R-\widetilde{D}_0}{R}\right)^{n+2}$$
$$\left[1-\left(1-\frac{\Delta\widetilde{D}'^{(k)}}{R-\widetilde{D}_0}\right)^{n+3}\right]P_n(\widetilde{x})\mathrm{d}\sigma \tag{6.89}$$

and

$$\left.\begin{array}{l}\dfrac{\partial(V^{\mathrm{CMP}})^{(k)}}{\partial\Delta\widetilde{D}^{(k)}}\\ \dfrac{\partial(\delta g^{\mathrm{CMP}})^{(k)}}{\partial\Delta\widetilde{D}^{(k)}}\\ \dfrac{\partial(\Delta g^{\mathrm{CMP}})^{(k)}}{\partial\Delta\widetilde{D}^{(k)}}\end{array}\right\} = G(R-\widetilde{D}_0)\iint_\sigma \Delta\rho' \sum_{n=0}^{\infty}\left\{\begin{array}{c}R^{-1}\\ n+1\\ n-1\end{array}\right\} \tag{6.90}$$
$$\left(\frac{R-\widetilde{D}_0}{R}\right)^{n+2}\Delta\widetilde{D}'^{(k)}P_n(\widetilde{x})\mathrm{d}\sigma.$$

The residual compensation potential, $(V^{\mathrm{CMP}})^{(k)}$, or the effect of compensating residual masses on $(\Delta g^{\mathrm{CMP}})^{(k)}$ and $(\delta g^{\mathrm{CMP}})^{(k)}$ is inserted on the rhs of Eq. (6.87) and a new value for $\Delta\widetilde{D}$ is obtained. Based on that, new updates for $(V^{\mathrm{CMP}})^{(k)}$, $(\Delta g^{\mathrm{CMP}})^{(k)}$ or $(\delta g^{\mathrm{CMP}})^{(k)}$ and their derivatives with respect to $\Delta\widetilde{D}$ are derived, and according to Eq. (6.87) another new updated value will be achieved for $\Delta\widetilde{D}$. This process is repeated until the solution converges or the difference between that last to the successive iterative values becomes smaller than a predefined threshold ε, e.g. $|\Delta\widetilde{D}^{(k+1)}-\Delta\widetilde{D}^{(k)}|\leq\varepsilon$.

6.3.6.4 Non-linear inversion

Here, three integral equations are presented in Eq. (6.75), and the main difference between them is which type of gravimetric measurement is used for solving the gravimetric problem of isostasy. This integral can be discretised simply and solved numerically. Since this equation is non-linear the kernel of the integral should be linearised and solved iteratively:

$$+ GR^2 \iint_\sigma \Delta\rho' \begin{pmatrix} \dfrac{\partial K^{\mathrm{V}}\left(r, \psi, \Delta\widetilde{D}'^{(k)}\right)}{\partial \Delta\widetilde{D}'} \\ \dfrac{\partial K^{\Delta g}\left(r, \psi, \Delta\widetilde{D}'^{(k)}\right)}{\partial \Delta\widetilde{D}'} \\ \dfrac{\partial K^{\delta g}\left(r, \psi, \Delta\widetilde{D}'^{(k)}\right)}{\partial \Delta\widetilde{D}'} \end{pmatrix} \Delta\widetilde{D}'^{(k+1)} \mathrm{d}\sigma$$

$$= \begin{pmatrix} +V^{\mathrm{TB}} - T - GR^2 \iint_\sigma \Delta\rho' K^{\mathrm{V}}\left(r, \psi, \Delta\widetilde{D}'^{(k)}\right) \mathrm{d}\sigma \\ +\Delta g^{\mathrm{TB}} - \Delta g - GR^2 \iint_\sigma \Delta\rho' K^{\Delta g}\left(r, \psi, \Delta\widetilde{D}'^{(k)}\right) \mathrm{d}\sigma \\ +\delta g^{\mathrm{TB}} - \delta g - GR^2 \iint_\sigma \Delta\rho' K^{\delta g}\left(r, \psi, \Delta\widetilde{D}'^{(k)}\right) \mathrm{d}\sigma \end{pmatrix} \tag{6.91}$$

where the subscript 0 for the kernels means that they are computed from the initial compensation depths.

$$\begin{pmatrix} K^{\mathrm{V}}\left(R, \psi, \Delta\widetilde{D}'^{(k)}\right) \\ K^{\Delta g}\left(R, \psi, \Delta\widetilde{D}'^{(k)}\right) \\ K^{\delta g}\left(R, \psi, \Delta\widetilde{D}'^{(k)}\right) \end{pmatrix} = \sum_{n=0}^{\infty} \frac{1}{n+3} \begin{pmatrix} 1 \\ (n-1)R^{-1} \\ (n+1)R^{-1} \end{pmatrix} \left(\frac{R - \widetilde{D}_0}{R}\right)^{n+3} \left[1 - \left(1 - \frac{\Delta\widetilde{D}'^{(k)}}{R - \widetilde{D}_0}\right)^{n+3}\right] P_n(\widetilde{x}) \tag{6.92}$$

or

$$\begin{pmatrix} K^{\mathrm{V}}\left(R, \psi, \Delta\widetilde{D}'^{(k)}\right) \\ K^{\Delta g}\left(R, \psi, \Delta\widetilde{D}'^{(k)}\right) \\ K^{\delta g}\left(R, \psi, \Delta\widetilde{D}'^{(k)}\right) \end{pmatrix} = \sum_{n=0}^{\infty} \frac{1}{n+3} \begin{pmatrix} 1 \\ (n-1)R^{-1} \\ (n+1)R^{-1} \end{pmatrix} P_n(\widetilde{x})$$

$$\left(\frac{R - \widetilde{D}_0}{R}\right)^{n+3} - \sum_{n=0}^{\infty} \frac{1}{n+3} \begin{pmatrix} 1 \\ (n-1)R^{-1} \\ (n+1)R^{-1} \end{pmatrix} \left(1 - \frac{\Delta\widetilde{D}'^{(k)}}{R - \widetilde{D}_0}\right)^{n+3} P_n(\widetilde{x}) \tag{6.93}$$

$$\frac{\partial}{\partial \Delta \widetilde{D}'}\begin{pmatrix} K^{V}\left(r,\psi,\Delta \widetilde{D}'^{(k)}\right) \\ K^{\Delta g}\left(r,\psi,\Delta \widetilde{D}'^{(k)}\right) \\ K^{\delta g}\left(r,\psi,\Delta \widetilde{D}'^{(k)}\right) \end{pmatrix} = \frac{\Delta \widetilde{D}'^{(k)}}{R}\sum_{n=0}^{\infty}\frac{1}{n+3}\begin{pmatrix} 1 \\ (n-1)R^{-1} \\ (n+1)R^{-1} \end{pmatrix} P_n(\widetilde{x}). \tag{6.94}$$

Closed-form expressions can also be derived for these kernel functions. However, it should be noted that in isostasy high frequencies of TB load are not compensated. This means that the spectral kernel up to maximum degree ∞ is not required practically.

6.3.7 A numerical example: Moho model based on approach 1 over the Tibet Plateau

In this section, a part of the numerical studies of Eshagh (2016b) over the Tibet Plateau is presented.

The study area is limited between the latitudes 20°N and 50°N and the longitudes 60°E and 110°E. The EGM08 (Pavlis et al., 2012) gravity model and the SRTM (Farr et al., 2007) TB model are used in our computations. The SRTM model contains SHCs of the TB heights to degree and order 2160 and EGM08 those of the Earth's gravitational potential and partially to higher. However, according to Turcotte and Schubert (2014, p. 252) the loads having shorter wavelengths than 100 km are not compensated for and therefore we use the SRTM and EGM08 models to degree and order 180. Fig. 6.5A represents the smooth TB heights over the study area generated from the SRTM model to degree and order 180 with the maximum, mean, minimum and standard deviation 5.5, 1.5, −3.7 and 1.7 km, respectively. Truncating the spherical harmonic series causes the positive and high values of the TB heights to be seen in the central part of the area, which can somehow give an impression of the Moho deepening regime of the region.

Fig. 6.5B is the map of gravity anomalies generated from EGM08 (Δg^{EGM08}) to the same degree. Negative values are clear over the orogenic belt of the Himalayan mountain chain and are immediately followed by high positive values along the mountains and in the northern part of the area in addition to the south of Tarim Basin. Negative values have surrounded this basin except for the eastern part.

The densities and thicknesses of the crustal crystalline layers and density of the upper mantle are used to compute a laterally variable density contrast model over the study area. Furthermore, the density of the TB masses is

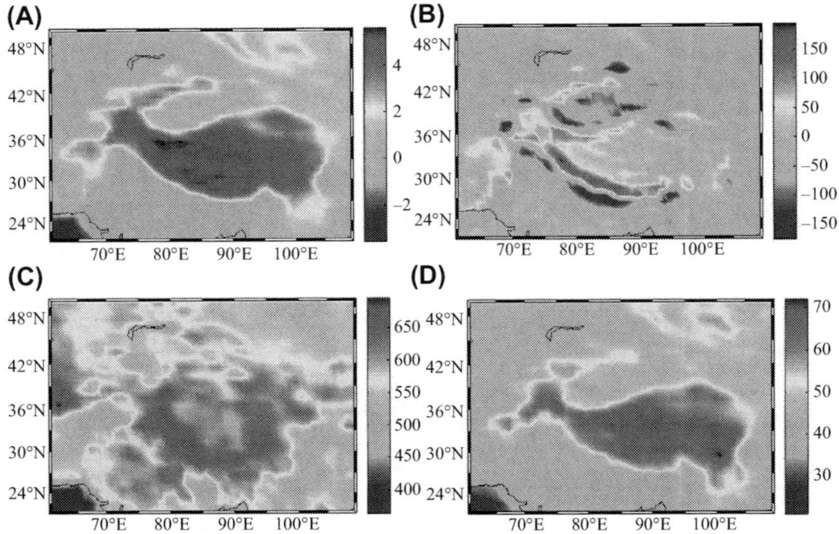

Figure 6.5 (A) Topographic/bathymetric heights (km) from the ETOPO model. (B) Gravity anomalies computed from EGM08 to degree and order 180 (mGal). (C) Density contrast, computed by subtracting a reference density of 2670 kg/m^3 from the upper mantle density given in the CRUST1.0 model (gr/cm^3). (D) The Moho depth computed based on approach 1 and the given information (km). *From Eshagh, M., 2016b. On the Vening Meinesz-Moritz and flexural theories of isostasy. J. Geod. Sci. 6, 139–151.*

considered constant and equal to 2670 kg/m^3 and the density of water 1000 kg/m^3. Fig. 6.5C illustrates the map of the laterally variable density contrast $\Delta\rho$ of the CRUST1.0 model. The large values are seen around the Tibet area and the lowest ones in the southwestern part, over the Oman Sea and Makran subduction zone. $\Delta\rho$ ranges from 343.9 to 694.9 kg/m^3 over the area with the mean and standard deviation of 566.9 and 51.4 kg/m^3. However, Rabbel et al. (2013) investigated some petrological and physical methods for estimating $\Delta\rho$ and concluded that $\Delta\rho$ hardly reaches 600 kg/m^3 over the Earth. The results of Eshagh et al. (2016) confirm their results.

For Moho modelling according to approach 1 and Eq. (6.71), the choice of $\widetilde{D}_0$ plays a significant role, especially for the case in which $\widetilde{D}_0$ contributes to all frequencies of the solution. Here, the idea is to estimate the best value for $\widetilde{D}_0$ so that the Moho model has the least root mean squared error with respect to the seismic Moho model of CRUST1.0. By trial and error a value of 33 km was estimated for $\widetilde{D}_0$. Fig. 6.5D shows the map of the Moho model over the area based on the given data and approach 1.

6.4 Flexure isostasy and the Vening Meinesz principle

Vening Meinesz (1931) presented another vision for isostasy and used the loading and flexure theories to present a compensation mechanism for the TB masses which play the role of the loads on the lithosphere. In his derivations, the mechanical properties of the lithosphere like rigidity, the Poisson ratio, elastic thickness and the Young modulus are seen instead of gravimetric data. In fact, Vening Meinesz (1931) believed that the lithosphere is an elastic shell, with some physical properties, which is deformed downward by the loads. These properties determine how much the lithosphere should be bent under these loads. The more rigid lithosphere flexes less under the loads. Moritz (1990) has mentioned that the gravimetric approach to isostasy is a kind of Vening Meinesz model; but instead of the mechanical properties of the lithosphere he used the gravity data for expressing the compensation mechanism. Sjöberg (2009) called such a solution Vening Meinesz—Moritz, which means that it was the Vening Meinesz problem defined by Moritz (1990). However, the original article of Vening Meinesz (1931) represents the loading theory and its application for explaining the isostatic compensation of TB loads. Abd-Elmotaal (1993) has applied this original idea for determining the Moho depth in Egypt. In this section, the goal is to present the mathematical model of this type of isostasy and show that an isostatic mechanism can be expressed by them. Two types of models are considered for expressing flexure of the lithosphere. The first that will be discussed is called here and hereafter the simple flexure model. It is called simple because the membrane stress inside the lithospheric shell is not considered in it. The second type of model is called the flexure model considering the membrane stresses. In the following, these models are presented and discussed.

6.4.1 Simple flexure model

When the lithosphere is assumed as a thin spherical elastic shell, it will deform under the pressure of loads and be flexed. The following partial differential equation (see Watts, 2001, Stuewe, 2007, p. 176, Artemieva, 2011, p. 554) expresses flexure of a spherical shell under topographic/bathymetric loads:

$$\frac{D^{\mathrm{Rig}}}{R^4}\nabla^4\Delta\widetilde{D}+\Delta\rho\Delta\widetilde{D}g=\overline{\rho}Hg \tag{6.95}$$

where ∇^2 is the Laplacian operator, g stands for gravity, $\overline{\rho}$ has already been introduced in Eq. (6.5), H stands for the TB heights, $\Delta\rho$ is the density contrast between the crust and the upper mantle, and D^{Rig} is the rigidity of the lithosphere, with the following formula:

$$D^{\mathrm{Rig}} = \frac{ET_e^3}{12(1 - \nu^2)} \tag{6.96}$$

and E stands for the Young modulus, ν the Poisson ratio and T_e the elastic thickness of the lithosphere. In most of the literature, Eq. (6.95) is solved in the Fourier domain, but here its solution in terms of spherical harmonics is presented. First let us assume that the density contrast $\Delta\rho$ is constant; in this case the following spherical harmonic expansions can be considered for $\Delta\widetilde{D}$ and $(\overline{\rho}H)$:

$$\Delta\widetilde{D} = \sum_{n=0}^{\infty}\sum_{m=-n}^{n} \Delta\widetilde{D}_{nm} Y_{nm}(\theta, \lambda) \tag{6.97}$$

$$\overline{\rho}H = \sum_{n=0}^{\infty}\sum_{m=-n}^{n} (\overline{\rho}H)_{nm} Y_{nm}(\theta, \lambda) \tag{6.98}$$

where $Y_{nm}(\theta,\lambda)$ is the fully normalised spherical harmonic functions of degree n and order m. θ and λ are, respectively, the co-latitude and longitude, and $\Delta\widetilde{D}_{nm}$ and $(\overline{\rho}H)_{nm}$ are the SHCs of $\Delta\widetilde{D}$ and $\overline{\rho}H$.

Substituting Eqs. (6.97) and (6.98) into Eq. (6.95) and writing the results in terms of the spherical harmonics yield:

$$\frac{D^{\mathrm{Rig}}}{R^4}\Delta\rho\nabla^4\sum_{n=0}^{\infty}\sum_{m=-n}^{n}\left(\Delta\widetilde{D}\right)_{nm} Y_{nm}(\theta, \lambda) + g\Delta\rho\sum_{n=0}^{\infty}\sum_{m=-n}^{n}\left(\Delta\widetilde{D}\right)_{nm} Y_{nm}(\theta, \lambda) = g\sum_{n=0}^{\infty}\sum_{m=-n}^{n}(\overline{\rho}H)_{nm} Y_{nm}(\theta, \lambda) \tag{6.99}$$

It is well known that the Laplacian of the spherical harmonics can be written as (e.g. Turcotte et al., 1981):

$$\nabla^2 Y_{nm}(\theta, \lambda) = -n(n+1)Y_{nm}(\theta, \lambda) = -\kappa_n Y_{nm}(\theta, \lambda) \tag{6.100}$$

According to Eqs. (6.100) and (6.99), it will not be difficult to show that the solution of the partial differential Eq. (6.95) in the spectral domain is:

$$\frac{D^{\mathrm{Rig}}}{R^4}\kappa_n^2\left(\Delta\widetilde{D}\right)_{nm} + g\Delta\rho\left(\Delta\widetilde{D}\right)_{nm} = g(\overline{\rho}H)_{nm} \tag{6.101}$$

Solving Eq. (6.101) for $\Delta\widetilde{D}_{nm}$ and performing the spherical harmonic synthesis from both sides of the result yield:

$$\Delta\widetilde{D} = \sum_{n=0}^{\infty} C_n \sum_{m=-n}^{n} (\overline{\rho}H)_{nm} Y_{nm}(\theta, \lambda) \tag{6.102}$$

where

$$C_n = \left(\kappa_n^2 \frac{D^{\mathrm{Rig}}}{R^4 g} + \Delta\rho\right)^{-1} \text{ and } \kappa_n^2 = n^2(n+1)^2. \tag{6.103}$$

In Eq. (6.102) or (6.103), when $D^{\mathrm{Rig}} = 0$, then the mathematical model will be the same as the isostatic model according to the Airy–Heiskanen principle. For $D^{\mathrm{Rig}} > 0$, it is observed that the coefficients κ_n^2 increase as n increases. The presence of $\kappa_n^2 D^{\mathrm{Rid}}/R^4$ in the denominator of the Moho flexure formula plays a role of regularisation factor depending on the frequency n. When D^{Rig} is small the ratio D^{Rig}/R^4 becomes small and reduces the value of κ_n^2 and since it appears in the denominator of the solution, the solution will be closer to the Airy–Heiskanen one. However, for a large value of D^{Rig}, the nominator will be larger and make the solution smoother.

6.4.2 Flexural model considering membrane stress

Another partial differential equation of flexure for a thin elastic spherical shell was given by Kraus (1967), which considers the membrane stress inside the shell as well. Turcotte et al. (1981) has used this equation for studying the membrane stress in Mars and the Moon. Here, we consider a simplified version of their equation. They mentioned that for the loads having much lower wavelength than the radius of the Earth, the geoid displacement due to the load is negligible. In this case, their partial differential equation changes to:

$$\begin{aligned}&\left[\frac{D^{\mathrm{Rig}}}{R^4}(\nabla^6 + 4\nabla^4) + \frac{ET_e}{R^2}(\nabla^2 + 2)\right]\Delta\widetilde{D} + (\nabla^2 + 1 - \nu)\Delta\widetilde{D}\Delta\rho g \\ &= (\nabla^2 + 1 - \nu)(\overline{\rho}H)g\end{aligned} \tag{6.104}$$

The solution of this differential equation using spherical harmonics, according to Eqs. (6.97) and (6.98), considering the property (Eq. 6.100), is:

$$\begin{aligned}&\left[\frac{D^{\mathrm{Rig}}}{R^4}\left(\kappa_n^3 + 4\kappa_n^2\right) + \frac{ET_e}{R^2}\left(\kappa_n^2 + 2\right)\right]\left(\Delta\widetilde{D}\right)_{nm} + \Delta\rho(\kappa_n + 1 - \nu)\left(\Delta\widetilde{D}\right)_{nm} g \\ &= (\kappa_n + 1 - \nu)(\overline{\rho}H)_{nm} g\end{aligned} \tag{6.105}$$

Solving Eq. (6.105) for $(\Delta\widetilde{D})_{nm}$ and performing the spherical harmonic synthesis from both sides of the result yield a mathematical model similar to the one presented in Eq. (6.102), but with the following compensation degree:

$$C_n = \frac{\kappa_n - (1-\nu)}{\left(\kappa_n^3 - 4\kappa_n^2\right)\frac{D^{\mathrm{Rig}}}{R^4} + \frac{ET_e}{R^2}(\kappa_n - 2) + (\kappa_n - (1-\nu))\Delta\rho} \tag{6.106}$$

The membrane stress is due to the fact that the curvature of any plate drifting north–south must change to conform to the curvature of the geoid (Engelder, 1993, p.383). Turcotte et al. (1981) mentioned that for the loads whose wavelengths are on the order of the radius of the Earth the upward displacement of the geoid due to the load is important. For smaller loads the contribution of the geoid and its curvature can be negligible. Fig. 6.6 shows the plot of the compensation degree C_n computed based on Eqs. (6.103) and (6.106). Here, an elastic thickness of 28 km and a density contrast of $\Delta\rho = 500\ \mathrm{kg/m^3}$ are considered for the computations. The plot shows that under such conditions, the compensation degrees differ in low degrees, say up to degree 60, and they coincide after. This means that the role of the curvatures is important only for low degrees. Later, our numerical studies will show how much these differences influence the Moho flexure and gravity anomaly generated from these isostasy theories.

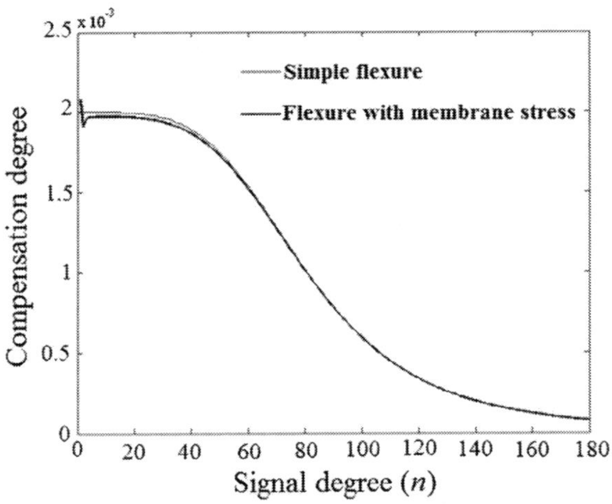

Figure 6.6 Compensation degree C_n according to the simple flexure model, Eq. (6.103), and the flexure model considering membrane stress, Eq. (6.106); $T_e = 28$ km. *(From Eshagh, M., 2016b. On the Vening Meinesz-Moritz and flexural theories of isostasy. J. Geod. Sci. 6, 139–151.)*

6.4.3 A numerical example: Moho model based on flexure isostasy over Tibet Plateau

Now, the same TB heights and density contrast as those presented in Section 6.3.7 are applied for computing two Moho models based on the simple flexure model and the flexure model considering membrane stress. Constant values of $E = 10^{11}$ and $\nu = 0.25$ have been selected for the flexure models with an elastic thickness between 28 and 38, which has been determined in such a way that the resulting Moho model is as close as possible to the gravimetric Moho model, presented in Section 6.3.7. Fig. 6.7A and B shows the Moho model computed over the Tibet Plateau by Eshagh (2016b), based on the simple flexure model and the one considering the membrane stress inside the lithosphere with the same mean compensation depth $\widetilde{D}_0$ and elastics thickness T_e in both cases. The two models have very similar patterns. Recognising the difference between the computed Moho models is not straightforward from the maps. For this reason the map of the differences is plotted in Fig. 6.7C and, as observed in this figure, the differences are smaller than 1 km over the area. However, again in the eastern part of the Tibet Plateau the differences are larger. No gravity anomaly was used to smooth the Moho models. However, the same TB height and density contrast models were considered as the input data in the mathematical models. Therefore, the differences are solely due to the approximations and formulation of the problem.

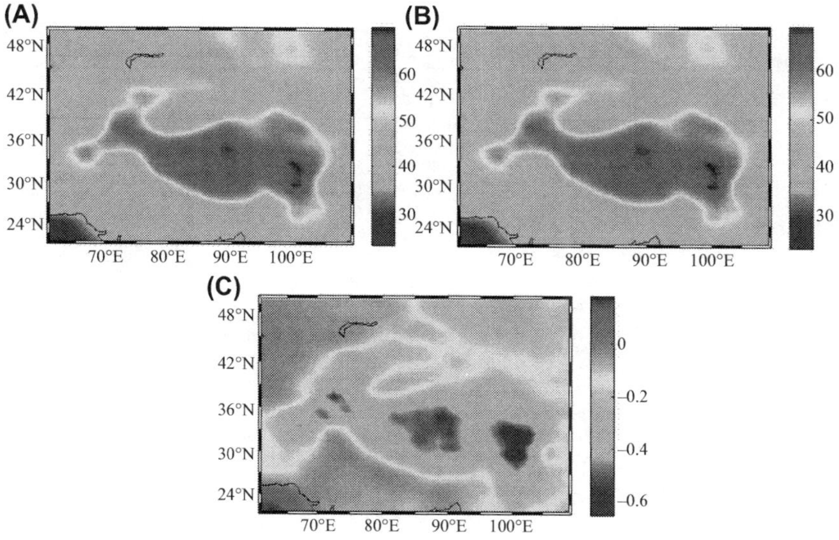

Figure 6.7 Moho model determined by (A) the simple flexure model and (B) the flexure model with membrane stress. (C) The differences between the Moho models (km). *From Eshagh, M., 2016b. On the Vening Meinesz-Moritz and flexural theories of isostasy. J. Geod. Sci. 6, 139–151.*

6.4.4 Combination of gravimetric and flexural isostasy

Two similar compensation depths can be determined from gravimetric and flexural isostatic theories. However, our Moho maps in Figs 6.5A and 6.7A show that the Moho model derived from the flexure theory is smoother than that of the gravimetric method. The reason is that a constant elastic thickness has been considered in the flexural theory, which gives the same compensation to all loads, whilst in the gravimetric isostasy the gravity data vary from one point to another and cause different isostatic compensations for the loads. By looking at the mathematical models of the compensation mechanism of gravimetric and flexural isostasy, we realise that they are some flavour of the Airy–Heiskanen model. This means that the Airy–Heiskanen compensation is smoothed by the gravity data to make the compensation regional in the gravimetric isostasy, but in the flexure theory, these are the mechanical properties of the lithosphere, which make the compensation regional. Fig. 6.8 shows the map of differences between the Moho models, determined by the gravimetric and flexure theories considering constant values of elastic thickness, Young's modulus and Poisson's ratio. These differences reach to about 7 km based on these assumptions. These differences will be even smaller when laterally variable models for these parameters are applied. Now, let us assume that the compensation depths obtained based on these methods are equal. In this case, new mathematical models can be derived connecting the gravity data to mechanical properties of the lithosphere.

In the case of considering the effects of crustal crystalline rocks, sediments and laterally variable density contrast, the mantle anomalies can be computed by subtracting the computed gravity anomalies from the

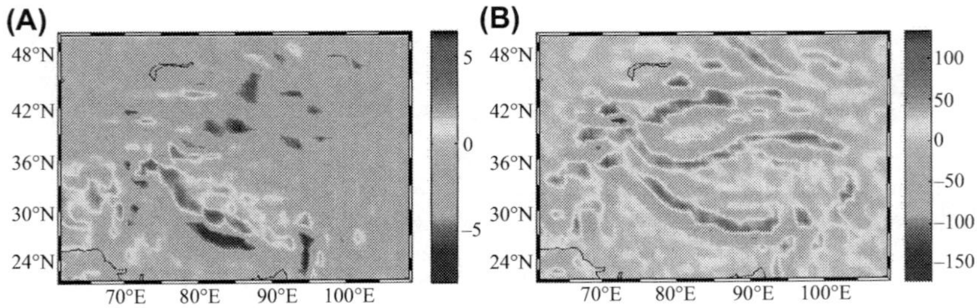

Figure 6.8 (A) Differences between the Moho model computed by the gravimetric method (approach 1) and flexural isostasy with membrane stress (km). (B) The differences between the gravity anomalies generated from EGM08 and combination of gravimetric and flexural isostatic models (mGal). *(From Eshagh, M., 2016b. On the Vening Meinesz-Moritz and flexural theories of isostasy. J. Geod. Sci. 6, 139–151.)*

crustal gravity anomalies. In this case, the density and structural inhomogeneities in the mantle can be studied (Artemieva, 2011, p.554). Such a study requires a good crustal model so that the differences between the observed and the computed anomalies are due purely to the mantle.

Here, gravimetric approaches 1 and 2 are combined with the flexural method. Approach 1 contains a contribution to mean compensation depth. If the density contrast is assumed as constant, the contribution of this mean compensation will contribute only to the zero-degree harmonic of the compensation potential. If the mean compensation is ignored, it will not be difficult to show that the variation of compensation depth around this mean value has the following mathematical form based on approach 1:

$$\Delta\widetilde{D}_{nm} = \widetilde{D}_{nm} - \frac{1}{4\pi G\Delta\rho}\begin{pmatrix} v_0^{\mathrm{CMP}_0} \\ \Delta g_0^{\mathrm{CMP}_0} \\ \delta g_0^{\mathrm{CMP}_0} \end{pmatrix}\delta_{n0}$$
$$= \frac{\beta_n^{-1}}{4\pi G\Delta\rho}(2n+1)\begin{pmatrix} 1 \\ (n-1)^{-1}R \\ (n+1)^{-1}R \end{pmatrix}\begin{pmatrix} v_{nm}^{\mathrm{TB}} - t_{nm} \\ \Delta g_{nm}^{\mathrm{TB}} - \Delta g_{nm} \\ \delta g_{nm}^{\mathrm{TB}} - \delta g_{nm} \end{pmatrix} \tag{6.107}$$

and based on approach 2:

$$\Delta\widetilde{D}_{nm} = \frac{\left(\beta_n^*\right)^{-1}}{4\pi GR\Delta\rho}(2n+1)\left(\frac{R}{R-\widetilde{D}_0}\right)^{n+2}\begin{pmatrix} 1 \\ (n-1)^{-1}R \\ (n+1)^{-1}R \end{pmatrix}\begin{pmatrix} v_{nm}^{\mathrm{TB}} - t_{nm} \\ \Delta g_{nm}^{\mathrm{TB}} - \Delta g_{nm} \\ \delta g_{nm}^{\mathrm{TB}} - \delta g_{nm} \end{pmatrix} \tag{6.108}$$

Solving Eqs. (6.107) and (6.108) for t_{nm}, Δg_{nm} and δg_{nm} and after performing spherical harmonic synthesis, and considering $\Delta D_{nm} = C_n(\overline{\rho}H)_{nm}$ in flexural isostasy, we obtain:

$$\begin{pmatrix} T \\ \Delta g \\ \delta g \end{pmatrix} = \begin{pmatrix} V^{\mathrm{TB}} \\ \Delta g^{\mathrm{TB}} \\ \delta g^{\mathrm{TB}} \end{pmatrix} - 4\pi G\Delta\rho\sum_{n=0}^{\infty}\frac{\beta_n C_n}{(2n+1)}\begin{pmatrix} 1 \\ (n-1)/R \\ (n+1)/R \end{pmatrix} \sum_{m=-n}^{n}(\overline{\rho}H)_{nm}Y_{nm}(\theta,\lambda) \tag{6.109}$$

and

$$\begin{pmatrix} T \\ \Delta g \\ \delta g \end{pmatrix} = \begin{pmatrix} V^{\mathrm{TB}} \\ \Delta g^{\mathrm{TB}} \\ \delta g^{\mathrm{TB}} \end{pmatrix} - 4\pi G R \Delta\rho \sum_{n=0}^{\infty} \frac{\beta_n^* C_n}{(2n+1)} \left(\frac{R}{R-\widetilde{D}_0}\right)^{-n-2} \begin{pmatrix} 1 \\ (n-1)/R \\ (n+1)/R \end{pmatrix} \sum_{m=-n}^{n} (\overline{\rho}H)_{nm} Y_{nm}(\theta,\lambda) \tag{6.110}$$

These equations can be used to generate T, Δg and δg from the TB masses and combination of gravimetric and flexure isostasy theories. If they are subtracted from the real values of the data, the contribution of the signals from the sub-lithosphere will be obtained. Fig. 6.8. shows the difference between the gravity anomalies computed from EGM08 and the crustal structure from Eq. (6.109).

6.4.5 Flexural convolution in the spherical domain

The combination of the gravimetric and flexure isostatic theories can be done in the spatial domain using integral formulae. Braitenberg et al. (2006) presented the flexure convolution analysis method for estimating flexural rigidity and elastic thickness. Their idea is to perform the computations in the spatial domain rather than the spectral. By that, they mean that for each gravity anomaly value several convolutions should be generated with different elastic thicknesses, and the one which generates the closest gravity anomaly to the real one is selected as the best value for the thickness. However, they performed the computations based on the Moho flexure model that they had already computed from the gravimetric approach. In this section, we generalise their method to the spherical domain using spherical harmonics and present spherical convolution for the same purpose.

According to Heiskanen and Moritz (1967, p. 30) the Laplace coefficients $(\overline{\rho}H)_n$ can be written in terms of the Legendre polynomials and integral formula:

$$\sum_{m=-n}^{n} (\overline{\rho}H)_{nm} Y_{nm}(\theta,\lambda) = (\overline{\rho}H)_n = \frac{2n+1}{4\pi} \iint_{\sigma} (\overline{\rho}H) P_n(\widetilde{x}) \mathrm{d}\sigma \tag{6.111}$$

where σ is the unit sphere and $\mathrm{d}\sigma$ the surface integration element. $P_n(\widetilde{x})$ stands for the Legendre polynomial of degree n.

Substituting Eq. (6.111) into Eqs. (6.109) and (6.110) and further simplifications yield:

$$\begin{pmatrix} T \\ \Delta g \\ \delta g \end{pmatrix} = \begin{pmatrix} V^{\mathrm{TB}} \\ \Delta g^{\mathrm{TB}} \\ \delta g^{\mathrm{TB}} \end{pmatrix} - G\Delta\rho \iint_{\sigma} \sum_{n=0}^{\infty} \beta_n C_n \begin{pmatrix} 1 \\ (n-1)/R \\ (n+1)/R \end{pmatrix} P_n(\widetilde{x})(\overline{\rho}' H')\mathrm{d}\sigma \tag{6.112}$$

$$\begin{pmatrix} T \\ \Delta g \\ \delta g \end{pmatrix} = \begin{pmatrix} V^{\mathrm{TB}} \\ \Delta g^{\mathrm{TB}} \\ \delta g^{\mathrm{TB}} \end{pmatrix} - GR\Delta\rho \iint_{\sigma} \sum_{n=0}^{\infty} \beta_n^* C_n \left(\frac{R}{R-\widetilde{D}_0}\right)^{-n-2} \begin{pmatrix} 1 \\ (n-1)/R \\ (n+1)/R \end{pmatrix} P_n(\widetilde{x})(\overline{\rho}' H')\mathrm{d}\sigma. \tag{6.113}$$

These two formulae can be applied for generating T, Δg and δg from the crustal structures locally. No closed-form formula can be written for the kernels of the integrals due to the involvement of C_n, β_n or β_n^*. Note that C_n is a function of the elastic thickness T_e and Young's modulus and Poisson's ratio, or in other words, the mechanical parameters of the lithosphere. Turcotte and Schubert (2014, p.252) mentioned that the loads with wavelengths smaller than 100 km are not compensated for by the lithospheric thickness. In the spherical harmonic domain such a wavelength is equivalent to degree and order 180. Therefore, there will be no need to generate the kernels to degrees higher than 180.

Fig. 6.9 is an example plot of the kernel of the integral involving gravity anomaly in Eq. (6.113) showing the behaviour of the kernel to a geocentric angle of 10° for the three values of 10, 30 and 50 km for elastic thickness (T_e) and the mean compensation depth $\widetilde{D}_0 = 33$ km. The kernels are generated to degree 180 and they are all well behaving and decay to zero after a geocentric angle of about 5°. As observed, when $T_e = 10$ km the kernel has its largest value in the vicinity of the computation point, $\psi = 0°$, and it is larger than the cases of $T_e = 30$ km and $T_e = 50$ km. This means that the integral involving gravity anomaly in Eq. (6.112) delivers smoother solutions for the larger values of T_e than the smaller T_e. This is quite logical and meaningful as a thicker elastic shell resists more against loads and filters out higher frequencies of solution.

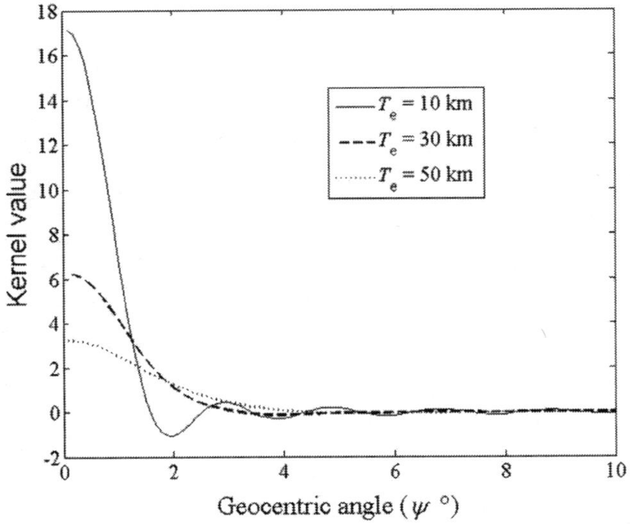

Figure 6.9 The kernel function related to gravity anomaly in Eq. (6.112) generated from the spectral of the kernel to degree 180 for three different values of elastic thickness: 10, 30 and 50 km. The mean compensation depth $\tilde{D}_0$ was chosen as 33 km. *(From Eshagh, M., 2016b. On the Vening Meinesz-Moritz and flexural theories of isostasy. J. Geod. Sci. 6, 139–151.)*

6.5 The effect of sediment, ice and crustal masses in isostasy

In Chapter 5, detailed discussions about the gravitational potential of the sediments and crustal masses have been presented. Sediments are parts of the upper crust and TB masses, which have different density comparing to the upper crust density. Therefore, the effect of sediments is equally important as that of the TB masses. Therefore, the gravitational effects of the sediment should be considered on the gravimetric data or even as loads in the flexural isostasy. One issue, which should be stated again, is that the TB heights, which are used, should not include the sediments, otherwise, the effect of sediment masses will be counted twice. The TB height model presented in CRUST1.0 model does not include the sediment thicknesses and is applicable straightforwardly. However, the sediments, the sediment thickness should be subtracted from the TB heights and the effect of TB masses should be counted from bottom of the sediments.

As presented in Chapter 5, the crust is partitioned to three layers of upper, middle and lower crust having their own thickness and densities. Their effects on the gravimetric quantities should be considered separately as well in isostasy. These crustal density and thickness variations are larger inside the crust and below the spherical Earth, their effect can be considered as the effect of subsurface masses. These effects are large and significant as the thicknesses of the crustal layers are larger and they are denser than TB masses.

Another way for considering the sediments and crustal mass variation effects is to compute the density contrast of these masses from a reference value say 2670 kg m^{-3}. In this case, the isostatic equilibrium can be considered based on a constant density for the TB masses, but the corrections due to this assumptions are computed from the sediments and crustal masses and added to the TB effect of the gravimetric quantities or the loads. This is the approach presented by Tenzer et al. (2015), and applied by Eshagh et al. (2016), Eshagh and Hussain (2016) and Eshagh et al. (2017).

Note that in a correct isostatic equilibrium model, the effect of the sediments and crustal masses need to be considered. This means that the sediment and crustal corrections should be considered to TB effects/loads on all mathematical equations or the error due to considering a constant density for these masses should be subtracted from them.

For example, Eq. (6.87) can be rewritten considering the SHCs of sediment, crustal crystalline and ice potential as:

$$\Delta\widetilde{D} = \frac{1}{4\pi G R \Delta\rho} \sum_{n=0}^{\infty} \frac{2n+1}{\beta_n^*} \left(\frac{R}{R - \widetilde{D}_0} \right)^{n+2} \sum_{m=-n}^{n} \left(v_{nm}^{\mathrm{TBSCI}} - t_{nm} \right) Y_{nm}(\theta, \lambda), \tag{6.114}$$

where $v_{nm}^{\mathrm{TBSCI}} = v_{nm}^{\mathrm{TB}} + v_{nm}^{\mathrm{S}} + v_{nm}^{\mathrm{C}} + v_{nm}^{\mathrm{I}}$ and v_{nm}^{TB}, v_{nm}^{S}, v_{nm}^{C} and v_{nm}^{I} are, respectively, the SHCs of the TB masses, sediments, crustal crystalline and ice. The mathematical model of these SHCs have been presented in Chapter 5.

6.6 Non-isostatic equilibriums

As explained so far, the loads of TB, sediments, ice and crystalline crustal masses push the lithosphere downward and flex it. This flexure is considered as an isostatic compensation and it is assumed in isostasy that the Earth's crust is in isostatic equilibrium. This assumption is partially correct as there are other compensation mechanisms than the isostatic one inside the Earth. Generally, we can say that in isostasy it is assumed that the mountains or loads have roots beneath, this means that the crust is thicker over the mountainous regions than the oceans. However, there are examples showing that the crust is not that thick as isostasy presents under mountains and, in fact, there are other compensation mechanisms which keep them in equilibrium.

Orogenic belts are some of these examples. Two plates moves towards themselves and cause that they move upward and build a mountain, which is kept by the stress due to the plates' convergent movements. Here, it is the stress and pressure that keeps the mountains in equilibrium to some extent. In the orogenic belts, mountains' roots are underestimated than those determined by isostatic equilibrium. Another example is high lands due to mantle plumes, which in fact the mantle plumes and thermal isostasy keep the mountains in equilibrium partly. Over the high lands having a mantle plume below no roots or small roots for the mountains or high lands exists. In fact, these areas are in non-isostatic equilibrium. Subduction zones are those places where an oceanic plate thrust the continental one.

Two areas in the world are famous for post glacial rebound, Fennoscandia and Hudson Bay in Canada. Many years ago, there have been large ice caps over these areas, which cause that the lithosphere thrusts into the mantle, but when the ices were melted the lithosphere started coming back to the isostatic equilibrium, Glacial Isostatic Adjustment. This means that the mountains and loads are not in isostatic equilibrium yet. The land over these areas are continuously and gently moving upwards and this phenomenon is called land uplift. Today, studying such land uplifts are important due to the sea level rise and global warming. We should monitor them to check whether the land uplift rate is larger or smaller than sea level rise rate. Therefore, in areas experiencing post-glacial rebound, isostatic equilibrium is not complete yet. Therefore, if a deep Moho is observed there, it could be due to the pressure of the ice which has been there for long time pressing the lithosphere downward into to the mantle and loads that we observe today.

The non-isostatic equilibriums studied by comparing the compensation depth based on the isostatic equilibrium, and the observed seismic Moho depths. If the loads are in isostatic equilibrium therefore, the differences between the seismic and gravimetric Moho surfaces would be of random nature due to uncertainty of the measurements. However, this is not the case in reality. Combination of the seismic and gravimetric Moho models is meaningful only if the area is indeed in isostatic equilibrium. This could be simply tested statistically by performing a goodness of fit test for the difference between the Moho models. The non-isostatic correction is not a meaningful action in Moho determination as it twists the gravimetric Moho model towards the seismic one, by indirectly adding the differences between them. Non-isostatic equilibriums can be studied for other geophysical and geological purpose for understanding physical and mechanical properties of the lithosphere rather than Moho determining.

Here, we present mathematically how this effect is computed. Assume that we use Eq. (6.115) as the equation for computing the gravimetric Moho model. However, any other formula presented here can also be applied for derivation of non-isostatic effects as the principle is the same. The Moho depth according to Eq. (6.114) will be:

$$\widetilde{D} = \widetilde{D}_0 + \frac{1}{4\pi G R \Delta\rho} \sum_{n=0}^{\infty} \frac{2n+1}{\beta_n^*} \left(\frac{R}{R - \widetilde{D}_0} \right)^{n+2} \sum_{m=-n}^{n} \left(v_{nm}^{\mathrm{TBSCI}} - t_{nm} \right) Y_{nm}(\theta, \lambda) \tag{6.115}$$

Assume that the SHCs of potential of the non-isostatic effects is v_{nm}^{N} and by adding it to SHCs of v_{nm}^{TBSCI}, our model will give the seismic Moho model $\widetilde{D}^{\mathrm{SEI}}$. Therefore, we can write:

$$\widetilde{D}^{\mathrm{SEI}} = \widetilde{D}_0 + \frac{1}{4\pi G R \Delta\rho} \sum_{n=0}^{\infty} \frac{2n+1}{\beta_n^*} \left(\frac{R}{R - \widetilde{D}_0} \right)^{n+2} \sum_{m=-n}^{n} \left(v_{nm}^{\mathrm{TBSCI}} - t_{nm} + v_{nm}^{\mathrm{N}} \right) Y_{nm}(\theta, \lambda). \tag{6.116}$$

If Eq. (6.115) is subtracted from Eq. (6.116) we get:

$$\widetilde{D}^{\mathrm{SEI}} - \widetilde{D} = \frac{1}{4\pi G R \Delta\rho} \sum_{n=0}^{\infty} \frac{2n+1}{\beta_n^*} \left(\frac{R}{R - \widetilde{D}_0} \right)^{n+2} \sum_{m=-n}^{n} v_{nm}^{\mathrm{N}} Y_{nm}(\theta, \lambda). \tag{6.117}$$

By writing Eq. (6.117) in spectral form and solve the result for v_{nm}^{N} we obtain:

$$v_{nm}^{\mathrm{N}} = 4\pi GR\Delta\rho\beta_n^* \left(\frac{R - \widetilde{D}_0}{R}\right)^{n+2} \frac{\widetilde{D}_{nm}^{\mathrm{SEI}} - \widetilde{D}_{nm}}{2n+1} \tag{6.118}$$

where

$$\widetilde{D}_{nm}^{\mathrm{SEI}} - \widetilde{D}_{nm} = \frac{1}{4\pi}\iint_{\sigma} (\widetilde{D}'^{\mathrm{SEI}} - \widetilde{D}')Y_{nm}(\theta', \lambda')\mathrm{d}\sigma. \tag{6.119}$$

Note that considering v_{nm}^{N} as an extra correction to disturbing potential is not correct as by inserting it inside the equation, we will obtain the seismic Moho model. This means that the gravimetric Moho model has no role in computations. Mathematically, if Eq. (6.118) is inserted in Eq. (6.118), we will obtain $\widetilde{D} = \widetilde{D}^{\mathrm{SEI}}$.

Fig. 6.10 shows the map of non-isostatic contributions to gravity anomaly all over the world, computed by Abrehdary (2017). The differences between a global Moho model and seismic Moho model of CRUST1.0 are converted to the gravity anomalies. However, in his computation the main assumption is that the CRUST1.0 Moho model is realistic and the differences are only due to the non-isostatic equilibriums. The effect of non-isostatic equilibrium varies from −500 to 250 mGal.

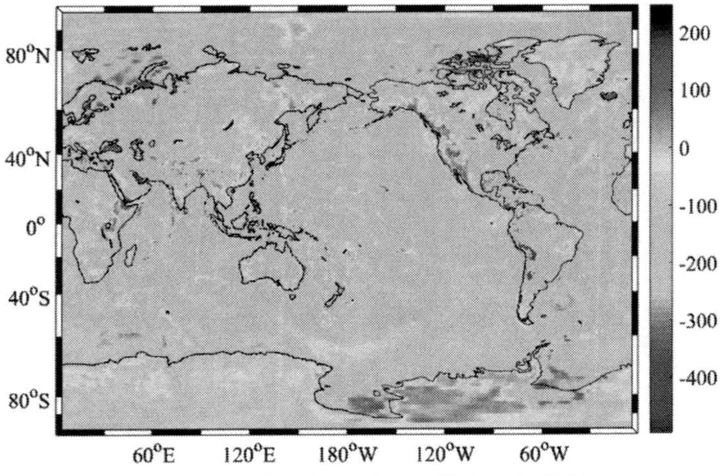

Figure 6.10 Non-isostatic equilibrium [mGal].

References

Abd-Elmotaal, H., 1993. Vening Meinesz Moho depths: traditional, exact and approximated. Manuscripta Geod. 18, 171–181.

Abrehdary, M., Sjöberg, L.E., Bagherbandi, M., Sampietro, D., 2017. Towards the Moho depth and Moho density contrast along with their uncertainties from seismic and satellite gravity observations, 11 (4), 231–247.

Airy, G.B., 1855. On the computations of the effect of the attraction of the mountain masses as disturbing the apparent astronomical latitude of stations in geodetic surveys. Philos. Trans. R. Soc. Series B 145, 101–104.

Artemieva, I., 2011. The Lithosphere, an Interdisciplinary Approach. Cambridge University Press.

Braitenberg, C., Wienecke, S., Wang, Y., 2006. Basement structures from satellite-derived gravity field: south China Sea ridge. J. Geophys. Res. 111, B05407. https://doi.org/10.1029/2005JB003938.

Engelder, T., 1993. Stress Regimes in the Lithosphere. Princeton University Press.

Eshagh, M., 2016a. A theoretical discussion on Vening Meinesz-Moritz inverse problem of isostasy. Geophys. J. Int. 207, 1420–1431.

Eshagh, M., 2016b. On the vening meinesz-moritz and flexural theories of isostasy. J. Geod. Sci. 6, 139–151.

Eshagh, M., 2017. On the approximations in formulation of the Vening Meinesz-Moritz inverse problem of isostasy. Geophys. J. Int. 210, 500–508.

Eshagh, M., Hussain, M., 2016. An approach to Moho discontinuity recovery from on-orbit GOCE data with application over Indo-Pak region. Tectonophysics 690 (B), 253–262.

Eshagh, M., Hussain, M., Tenzer, R., Romeshkani, M., 2016. Moho density contrast in central Eurasia from GOCE gravity gradients. Remote Sens. 418 (8), 1–18.

Eshagh, M., Ebadi, S., Tenzer, R., 2017. Isostatic GOCE Moho model for Iran. J. Asian Earth Sci. 138, 12–24.

Farr, T.G., Paul, A.R., Edward, C., Riley, D., Scott, H., Michael, K., Mimi, P., Rodriguez, E., Ladislav, R., David, S., Shaffer, S., Joanne, S., 2007. The shuttle radar topography mission. Rev. Geophys. 45, RG2004. https://doi.org/10.1029/2005RG000183.

Heiskanen, W., Moritz, H., 1967. Physical Geodesy. W.H. Freeman and company, San Francisco and London.

Jeffrey, H., 1976. The Earth: Its Origin, History and Physical Constitution, sixth ed. Cambridge University Press.

Kraus, H., 1967. Thin Elastic Shells. John Wiley, New York.

Laske, G., Masters, G., Ma, Z., Pasyanos, M., 2013. Update on CRUST1.0 - a 1-degree global model of Earth's crust. In: EGU General Assembly Conference Abstracts, vol. 15, p. 2658.

Moritz, H., 1990. The Figure of the Earth. H Wichmann, Karlsruhe.

Pavlis, N.K., Holmes, S.A., Kenyon, S.C., Factor, J.K., 2012. The development and evaluation of the Earth Gravitational model 2008 (EGM2008). J. Geophys. Res. 117, B04406. https://doi:10.1029/2011JB008916.

Pratt, J.H., 1854. On the attraction of Himalayan mountain and of the elevated regions beyond them, upon the plumb line in India. Philos. Trans. R. Soc. Lond. 146, 53–100.

Rabbel, W., Kaban, M., Tesauro, M., 2013. Contrasts of seismic velocity, density and strength across the Moho. Tectonophysics 609, 437–455.

Sjöberg, L.E., 2009. Solving vening meinesz-moritz inverse problem in isostasy. Geophys. J. Int. 179, 1527–1536.

Sjöberg, L.E., 2013. On the isostatic gravity anomaly and disturbance and their applications to Vening Meinesz-Moritz inverse problem of isostasy. Geophys. J. Int. 193, 1277–1282.

Stuewe, K., 2007. Geodynamics of the Lithosphere, second ed. Springer Verlag.

Turcotte, D., Schubert, G., 2014. Geodynamics, third ed. Cambridge University Press.

Tenzer, R., Chen, W., Tsoulis, D., et al., 2015. Analysis of the Refined CRUST1.0 Crustal Model and its Gravity Field. Surv Geophys 36, 139–165. https://doi.org/10.1007/s10712-014-9299-6.

Turcotte, D., Willemann, W.F., Haxby, F., Norberry, J., 1981. Role of membrane stresses in the support of planetary topography. J. Geophys. Res. 86 (B5), 3951–3959.

Vening Meinesz, F.A., 1931. Une nouvelle method pour la reduction isostatique regionale de l'intensite de la pesanteur. Bull. Geod. 29, 33–51.

Watts, A.B., 2001. Isostasy and Flexure of Lithosphere. Cambridge University Press.

CHAPTER 7

Satellite gravimetry and isostasy

7.1 Smoothing satellite gravimetry data

Satellite gravimetry data are collected along a satellite's orbit and far from the Earth's surface. The satellite's altitude causes these data to become smooth and contain lower frequencies of the Earth's gravity field. Generally, integral inversion of such smooth data is simpler than that of the terrestrial data due to higher frequencies overwhelmed by measurement noise. Different types of satellite gravimetry data have been presented in Chapter 2, where we could observe that such data are, in one way or another, related to satellite acceleration, velocity or their combination. This means that they are not direct measurements of the Earth's gravitational potential, but some other quantities having some relation with it, which are not as smooth as the gravitational potential. Therefore, it is necessary to smooth these data further for simplifying their downward continuation process to sea level.

The main condition for solving the Laplace equation of the gravitational potential, Eq. (1.1), is that no mass exists outside the spherical Earth. The spherical harmonic expansion of the Earth's gravitational potential is valid when the Earth is a sphere fitted to the mean sea level, with no excess mass outside. Such an assumption is not realistic due to the existence of topographic and atmospheric masses. Therefore, their gravitational effects should be removed from the gravimetry data to make the computational space harmonic and suitable for solving the Laplace equation. However, after performing the downward continuation process these effects should be resorted to the result. This process is called remove—compute—restore and was discussed in Chapter 5. Its problem is that we always work with the anomalous quantities, derived after subtracting the effect of the normal gravity field from the measurements, causing a huge reduction of them. The effect of topographic masses is significantly larger than such anomalous quantities, because no compensation mechanism has been considered for reduction of the topographic effect.

Satellite Gravimetry and the Solid Earth
ISBN 978-0-12-816936-0
https://doi.org/10.1016/B978-0-12-816936-0.00007-4

Isostatic reduction is one of the different types of reductions, and it has a rather meaningful geophysical interpretation. Therefore, by considering it, the topographic effect will be reduced to the level of anomalous quantities and by removing them from these quantities, we can simplify the inversion of such anomalies. The satellite gravimetry data are a class of anomalous quantities if they are reduced for a normal gravitational field. Such data can be further reduced by removing the isostatically reduced topographic effect and become smoother because high frequencies of the Earth's gravitational field come mainly from the topographic masses; see e.g. Eshagh and Bagherbandi (2011).

Another issue, which can be discussed as regards to the importance of smoothing the satellite gravimetry data by isostatic reduction, is to remove the known part of the Earth's gravitational signal from the measurements and highlight the gravitational frequencies from unknown sources. For example, by removing the effects of topographic/bathymetric (TB), sediment and crustal crystalline masses, with a suitable compensation mechanism, from the satellite gravimetry data, those parts of the gravitational signal, which are due to sub-lithosphere, can be studied; see e.g. Eshagh et al. (2018).

7.1.1 Reductions based on combination of flexure and gravimetric isostatic theories

In Chapter 6, two ways of formulating the gravimetric—isostatic equilibrium were presented as approaches 1 and 2. The principles behind these methods are the same, and their difference is mainly related to approximations in formulating them. Here, we assume that the TB, sediment, crustal crystalline and ice masses are disturbances which should be reduced from the satellite gravimetry data. Approaches 1 and 2 in combination with flexural isostasy can be applied to isostatically reduce the effects of these masses. In the following, mathematical formulations of such reductions are presented based on the mentioned approaches.

Let us consider approach 1 of gravimetric isostasy considering the gravitational potentials of sediment and crustal crystalline masses and ice in addition to the TB masses, like Eq. (6.71) but for approach 1. The spectral form of the approach solved for the spherical harmonic coefficients (SHCs) of isostatically reduced gravitational potential of these masses is:

$$t_{nm}^{\mathrm{TBSCI/Iso}} = -4\pi GR\frac{\beta_n}{2n+1}\left(\frac{R}{r}\right)^{n+1}\left(\Delta\rho\widetilde{D}\right)_{nm} + v_{nm}^{\mathrm{CMP_0}} + v_{nm}^{\mathrm{TB}} + v_{nm}^{\mathrm{S}} + v_{nm}^{\mathrm{C}} + v_{nm}^{\mathrm{I}}, \tag{7.1}$$

where $t_{nm}^{\mathrm{TBSCI/Iso}}$ is the SHC of the isostatically reduced disturbing potential estimated from the SHCs of TB masses v_{nm}^{TB}, sediments v_{nm}^{S}, crustal crystalline v_{nm}^{C} and ice v_{nm}^{I}. $v_{nm}^{\mathrm{CMP_0}}$ is the SHC of the mean compensation potential.

However, estimation of $t_{nm}^{\mathrm{TBSCI/Iso}}$ is not possible without knowledge about the compensation depth $\widetilde{D}$, density contrast $\Delta\rho$ which is considered constant here and mean compensation depth $\widetilde{D}_0$. Our idea here is to use the flexural isostasy to determine $\widetilde{D}$, and the compensation variation $\Delta\widetilde{D}$ around $\widetilde{D}_0$. Therefore, we write $\widetilde{D}_{nm} = \widetilde{D}_0\delta_{n0} + \Delta\widetilde{D}_{nm}$. Accordingly, Eq. (7.1) will change to:

$$t_{nm}^{\mathrm{TBSC/Iso}} = -4\pi GR\frac{\beta_n}{2n+1}\left(\frac{R}{r}\right)^{n+1}\Delta\rho\left(\widetilde{D}_0\delta_{n0}\right) + v_{nm}^{\mathrm{CMP_0}} - 4\pi GR\frac{\beta_n}{2n+1}\left(\frac{R}{r}\right)^{n+1}\Delta\rho\Delta\widetilde{D}_{nm} + v_{nm}^{\mathrm{TB}} + v_{nm}^{\mathrm{S}} + v_{nm}^{\mathrm{C}} + v_{nm}^{\mathrm{I}}. \tag{7.2}$$

$v_{nm}^{\mathrm{CMP_0}}$ is a function of laterally variable $\Delta\rho$, but when $\Delta\rho$ is constant, the mean compensation potential according to Eq. (6.34), $v_{nm}^{\mathrm{CMP_0}}$, will change to:

$$v_0^{\mathrm{CMP_0}} = \frac{4\pi GR^2\Delta\rho}{3}\left(1-\left(1-\frac{\widetilde{D}_0}{R}\right)^3\right)\left(\frac{R}{r}\right)\delta_{n0} \approx 4\pi GR\Delta\rho\widetilde{D}_0\frac{R}{r} \tag{7.3}$$

and

$$4\pi GR\frac{\beta_n}{2n+1}\left(\frac{R}{r}\right)^{n+1}\left(\Delta\rho\left(\widetilde{D}_0\delta_{n0}\right)\right) = 4\pi G\left(R-\widetilde{D}_0\right)\Delta\rho\widetilde{D}_0\left(\frac{R}{r}\right). \tag{7.4}$$

Note that the SHC variation in the compensation depth, $\Delta\widetilde{D}$, according to flexure theory considering the TB, sediment, crustal crystalline and ice loads is:

$$\Delta\widetilde{D}_{nm} = C_n\overline{K}_{nm}, \tag{7.5}$$

where C_n is the compensation degree based on the flexural isostasy, Eq. (6.103) or (6.106), and

$$\overline{K}_{nm} = (\overline{\rho}H)_{nm} + (\rho^{\mathrm{S}}d_{\mathrm{S}})_{nm} + (\rho^{\mathrm{C}}d_{\mathrm{C}})_{nm} + (\rho^{\mathrm{I}}d_{\mathrm{I}})_{nm}, \tag{7.6}$$

where $(\overline{\rho}H)_{nm}$, $(\rho^{\mathrm{S}}d_{\mathrm{S}})_{nm}$, $(\rho^{\mathrm{C}}d_{\mathrm{C}})_{nm}$ and $(\rho^{\mathrm{I}}d_{\mathrm{I}})_{nm}$ are the SHCs of the products $\overline{\rho}H$, $\rho^{\mathrm{S}}d_{\mathrm{S}}$, $\rho^{\mathrm{C}}d_{\mathrm{C}}$ and $\rho^{\mathrm{I}}d_{\mathrm{I}}$, respectively, and ρ^{S}, ρ^{C} and ρ^{I} are the density of the sediments, crustal crystalline and ice, and d_S, d_C and d_I their corresponding thicknesses.

By inserting Eq. (7.5) into Eq. (7.2), and performing spherical harmonic synthesis from both sides of it,. the spherical harmonic series of disturbing potential generated from the crustal structure based on the combination of the gravimetric and flexural isostasy theories will be obtained:

$$T^{\mathrm{TBSCI/Iso}}(r,\theta,\lambda)=\sum_{n=0}^{\infty}\left(\frac{R}{r}\right)^{n+1}\sum_{m=-n}^{n}t_{nm}^{\mathrm{TBSCI/Iso}}Y_{nm}(\theta,\lambda), \tag{7.7}$$

where

$$t_{nm}^{\mathrm{TBSC/Iso}}=-\frac{4\pi GR\beta_n}{2n+1}C_n\overline{K}_{nm}+v_{nm}^{\mathrm{TB}}+v_{nm}^{\mathrm{S}}+v_{nm}^{\mathrm{C}}+v_{nm}^{\mathrm{I}}. \tag{7.8}$$

The SHCs of Eq. (7.8) can be derived in another way, based on approach 2 (discussed in previous chapter). In this case, after following the same procedure as we did for obtaining Eq. (7.8), we obtain:

$$t_{nm}^{\mathrm{TBSCI/Iso}}=-4\pi G\left(R-\widetilde{D}_0\right)\Delta\rho\left(\frac{R-\widetilde{D}_0}{R}\right)^{n+1}\frac{\beta_n^*}{2n+1}C_n\overline{K}_{nm}+v_{nm}^{\mathrm{TB}}+v_{nm}^{\mathrm{S}}+v_{nm}^{\mathrm{C}}+v_{nm}^{\mathrm{I}}. \tag{7.9}$$

Therefore, to compute the isostatically reduced gravitational potential, it is enough to compute either Eq. (7.8) or Eq. (7.9) and insert the result into Eq. (7.7). Since the high frequencies of the Earth's gravitational field come from the crustal mass and structure variation, by removing their gravitational effects from the satellite gravimetry data, we can smooth them and simplify their inversion process.

By inserting the SHCs presented in Eq. (7.8) or Eq. (7.9) into the spherical harmonic expansions of satellite gravimetry data, presented in Chapter 2 (replacing v_{nm} with $t_{nm}^{\mathrm{TBSC/Iso}}$), the isostatically-reduced effects on the satellite gravimetry data are obtained. The satellite gravimetry data will be smoothed by subtracting the effects from them.

For applying Eq. (7.8) or Eq. (7.9) the global values for the elastic thickness and mean compensation depth are required. Practically, by trial and error, we can find some optimal values for these parameters so that the effect comes as close as possible to the data, and thereafter, we need to check if by removing the effects the satellite gravimetry data become smooth or not. This process is rather complicated for global smoothing of such data, but for local gravity field determination or integral inversion of satellite gravimetry data, it has been rather successful. In the following, we present two numerical examples, performed by Eshagh and Bagherbandi (2011) and Eshagh et al. (2018), of these subjects.

7.1.2 Example: smoothing the gravitational potential tensor

Here, an example is presented, which was done by Eshagh and Bagherbandi (2011), who studied the smoothing effect on the quality of downward continuation of satellite gradiometry data. They simulated all elements of the gravitational tensor at an elevation of 250 km with a resolution of $0.5^\circ \times 0.5^\circ$ and their goal was to smooth these data by applying the Vening Meinesz–Moritz isostasy (Sjöberg, 2009) for estimating compensation depth, which was discussed in Chapter 6 as approach 1 to gravimetry isostasy. They considered solely the TB effects on the generated gradients with an isostatic reduction according to the Vening Meinesz–Moritz method, instead of flexural isostasy. Here, we present one part of their study, which is related to the inversion of T_{zz} down to gravity anomaly at an elevation of 10 km above the sea level.

Fig. 7.1A shows the simulated gravity gradient, T_{zz}, 250 km above sea level from the EGM08 (Pavlis et al., 2012) gravity model to degree and order 360. Fig. 7.1B is the TB effect on this gradient, computed from the SRTM (Shuttle Radar Topography Mission) model (Farr et al., 2007), considering constant values of 2670 and 1000 kg/m^3 as the density of topographic and bathymetric masses, respectively. By comparing Fig. 7.1A,B, we observe that the value of the TB effect is considerably larger than the simulated data. Fig. 7.1C shows the reduced TB effect by gravimetric isostasy and the value of this effect is almost at the level of the simulated data. By subtracting it from the data a smoother map of the gradients should achieved, which is presented in Fig. 7.1D. Eshagh and Bagherbandi (2011) showed that the quality of downward continuation of the gravitational tensor from 250 to 10 km will be twice as good when the satellite gravity gradients are smoothed.

7.1.3 Example: removing the effects of mass density and structure heterogeneities from second-order radial derivatives measured by the GOCE mission

Here, another example is presented from the study of Eshagh et al. (2018) for modelling sub-lithospheric stress due to mantle convection. They used the real satellite gradiometry data, T_{zz}, of the GOCE (Gravity Field and Steady-State Ocean Circulation Explorer) mission from January 2013 with a 10-s data-sampling interval to compute the second-order radial derivative of the gravitational potential (in total 21,920 values) corrected for the

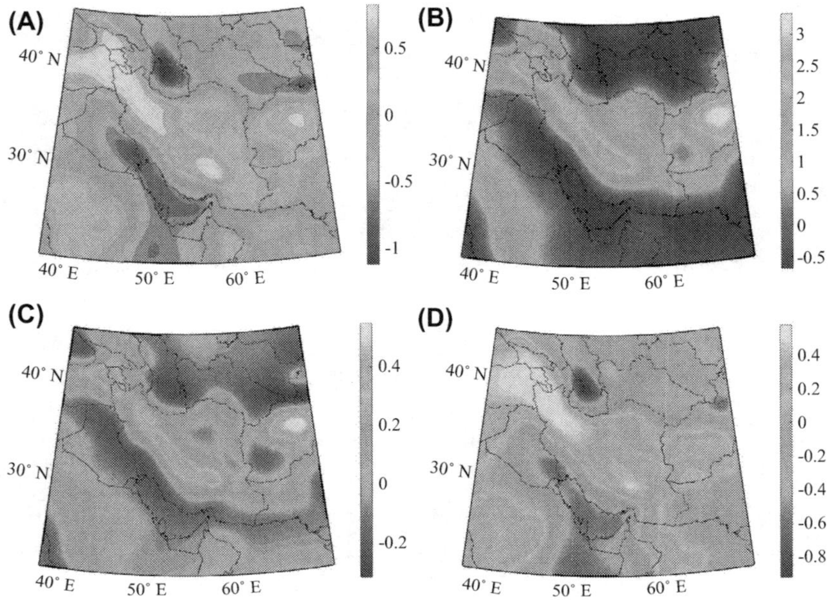

Figure 7.1 (A) Simulated gravity gradient at 250 km level, (B) topographic/bathymetric (TB) effect of the gradients, (C) reduced TB effect by gravimetric isostasy and (D) smoothed gradients. Unit: E.

corresponding one induced by normal gravitational field GRS80 (Moritz, 2000). To recover the gravitational signal induced by sub-lithospheric masses, the gravitational effects of all masses above the sub-lithosphere were removed from T_{zz}. The effects of the TB, sediment and crystalline masses should be computed, according to the methods presented in Chapter 5, and removed from on-orbit T_{zz}. The compensation potential has been computed by the flexural isostasy method (see Chapter 6), with the elastic thickness model presented by Tassara et al. (2007). Fig. 7.2A shows the TB effect on T_{zz} reaching 7 E, which is rather larger than the values of T_{zz}. A constant value of 2670 kg/m^3 has been considered as a reference value for the density of sediments and crystalline masses, and later the deviation of the real densities from these reference values is computed as a required correction to the values based on the constant density value. Fig. 7.2B,C are, respectively, the effects of the density contrast variations of sediments and crustal crystalline masses from the reference density. This is the reason that the effect values are smaller than those in Fig. 7.2A. Fig. 7.2D is the total effect after considering the isostatic reduction of the effects of the TB, sediment and crustal crystalline masses. This isostatically reduced effect should be reduced from the data for smoothing them and recovering sub-lithospheric signals.

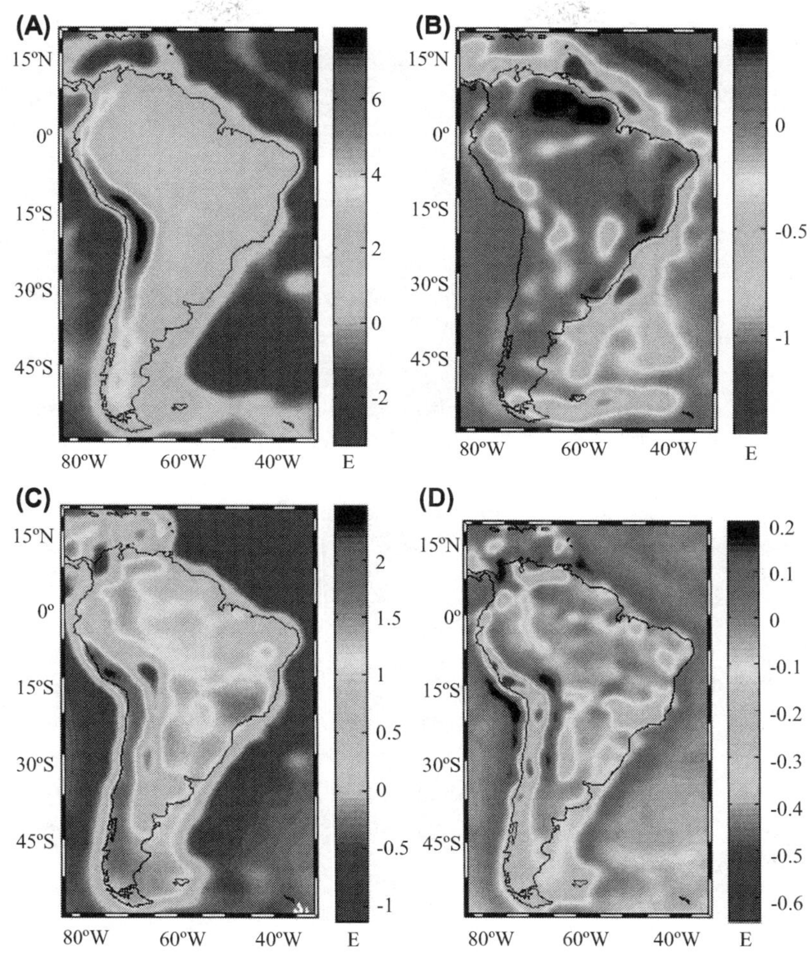

Figure 7.2 The effects of (A) topographic/bathymetric (TB) masses, (B) density contrast of sediment, (C) density contrast of crustal crystalline masses from a reference density of 2670 kg/m^3 and (D) isostatic effect of TB, sediment and crustal crystalline contrast effects by considering the compensation.

7.2 Determination of the product of Moho depth and density contrast

Here and hereafter, we use approach 2 of gravimetric isostasy, which was presented in Chapter 6, but the principles and mathematical foundations of their developments for determining the product of Moho depth variation and density contrast of both approaches are very similar.

In addition, the second-order solution of approach 2, presented in Section 6.3.6.2, is applied. We also select satellite altimetry data, due to their simplicity, to present the recovery of the product of Moho depth variations and density contrast. After that similar processes are performed for the use of satellite gradiometry data and inter-satellite measurements. We emphasise that this principle can be similarly applied for all types of gravimetry data.

7.2.1 Product of Moho depth variation and density contrast from satellite altimetry data

Satellite altimetry data are collected over oceans; therefore, we can simply use them to compute oceanic Moho depths. These data are the measured distances from the satellite to the sea surface. After the altitude of the satellite is removed from these measurements, the sea surface variations or sea surface topography departing from the geoid by about 2 m is obtained. This means that the satellite altimetry data are almost equal to geoid heights over oceans. According to the Bruns formula, Eq. (3.5), we have $T = \gamma N$, where γ is the normal gravity; therefore, the equation of equilibrium presented in Eq. (6.85) can be directly used when T is replaced by γN. One problem in applying this equation is that the satellite altimetry data cannot be measured over land and continents; even the quality of these data is not very good over shallow water and coastal zones. Therefore, the integration domain of the integral will be limited to oceans, where the spherical harmonics are not orthogonal. Eq. (6.85) can be written in the following integral form:

$$G\left(R-\widetilde{D}_0\right)\iint_{\sigma_{\text{ocean}}}\Delta\rho'\Delta\widetilde{D}'\sum_{n=0}^{\infty}\left(\frac{R-\widetilde{D}_0}{R}\right)^{n+1}\beta_n^*P_n(\widetilde{x})\mathrm{d}\sigma = V^{\text{BSC}}(R,\theta,\lambda)-\gamma N(R,\theta,\lambda), \tag{7.10}$$

where $V^{\text{BSC}}(R,\theta,\lambda)$ is the joint effect of bathymetric, sediment and crystalline masses and $N(R,\theta,\lambda)$ is the satellite altimetry–geoid height over ocean. β_n^* has been already defined in Eq. (6.86). Eq. (7.10) is very similar to Eq. (6.85), but the computation domain is at the surface of the spherical Earth, i.e. $r = R$, and instead of the TB effect, we consider the effect of bathymetric, sediment and crustal crystalline masses in Eq. (7.10), and instead of the disturbing potential we use $\gamma N(R,\theta,\lambda)$.

Solving the integral Eq. (7.10) is possible numerically. The surface of oceans should be gridded with a desired resolution, and the integral is discretised based on it and finally inverted numerically to estimate $\Delta\rho'\Delta\widetilde{D}'$ by one of the regularisation methods discussed in Chapter 4. It should be stated that we cannot derive a closed-form formula for the kernel of this internal due to the presence of β_n^* in its spectral form. In the case of assuming $\beta_n^* = 1$, a closed form is achievable having a $\widetilde{D}_0$-dependent behaviour.

In a similar manner, another integral formula can be developed for inverting the satellite altimetry data. To do so, the integral equation presented in Eq. (7.10) is written in spectral form:

$$4\pi G\left(R - \widetilde{D}_0\right)\frac{1}{2n+1}\left(\Delta\rho\Delta\widetilde{D}\right)_{nm}\left(\frac{R - \widetilde{D}_0}{R}\right)^{n+1}\beta_n^* = v_{nm}^{\mathrm{BSC}} - \gamma N_{nm}, \quad (7.11)$$

where $\left(\Delta\rho\Delta\widetilde{D}\right)_{nm}$, v_{nm}^{BSC} and N_{nm} are the SHCs of $\Delta\rho\Delta\widetilde{D}$, $V^{\mathrm{BSC}}(R,\theta,\lambda)$ and $N(R,\theta,\lambda)$, respectively. After solving Eq. (7.11) for $\left(\Delta\rho\Delta\widetilde{D}\right)_{nm}$ and performing spherical harmonic synthesis from both sides of the result, we obtain:

$$\Delta\rho\Delta\widetilde{D} = \frac{1}{4\pi G\left(R - \widetilde{D}_0\right)}\sum_{n=0}^{\infty}\frac{2n+1}{\beta_n^*}\left(\frac{R}{R - \widetilde{D}_0}\right)^{n+1}\sum_{m=-n}^{n} v_{nm}^{\mathrm{BSC}} Y_{nm}(\theta,\lambda) - W_{\Delta\rho\Delta\widetilde{D}}. \quad (7.12)$$

The first term on the right-hand side (rhs) of Eq. (7.12) is the contribution of the bathymetric, sediment and crustal crystalline masses, and the second is the contribution from the satellite altimetry data to $\Delta\rho\Delta\widetilde{D}$, which can be derived by inverting satellite altimetry-based geoid heights over ocean by the following integral equations:

$$G\left(R - \widetilde{D}_0\right)\iint_{\sigma_{\mathrm{ocean}}} W'_{\Delta\rho\Delta\widetilde{D}}\sum_{n=0}^{\infty}\left(\frac{R - \widetilde{D}_0}{R}\right)^{n+1}\beta_n^* P_n(\overline{x})\mathrm{d}\sigma = \gamma N(R,\theta,\lambda). \quad (7.13)$$

Once $W'_{\Delta\rho\Delta\widetilde{D}}$ is estimated it can be inserted into Eq. (7.12) to obtain $\Delta\rho\Delta\widetilde{D}$.

One significant difference between the satellite altimetry data and other types of satellite gravimetry data is that they are already collected at sea level, but others are collected on satellite orbits. Therefore, no downward continuation is needed when altimetry data are used.

The term $\overline{\Delta\widetilde{D}}/(R-\widetilde{D}_0)$ in Eq. (6.86) is rather small; for example, if $R = 6378$ km, $\widetilde{D}_0 = 10$ km over oceans, which is the mean value of the Moho depth. $\overline{\Delta\widetilde{D}}$ is the average of the residual Moho depth from the mean value and it is rather clear that it should be very close to zero and when it is divided by $\left(R-\widetilde{D}_0\right)$ it will be even smaller. If we consider $\overline{\Delta\widetilde{D}} = 0$, then $\beta_n^* = 1$ and a closed-form formula can be obtained for the kernel function of the integral (Eq. 7.13):

$$\sum_{n=0}^{\infty}\left(\frac{R-\widetilde{D}_0}{R}\right)^{n+1} P_n(\widetilde{x}) = \frac{s}{D}, \quad \text{where } s = \frac{R-\widetilde{D}_0}{R},$$

$$D = \sqrt{1+s^2-2s\widetilde{x}}, \quad \widetilde{x} = \cos\psi, \tag{7.14}$$

where ψ is the geocentric angle between the computation and the integration points.

7.2.2 Product of Moho depth variation and density contrast from satellite gravity gradiometry data

Satellite gradiometry data also have simple mathematical formulae compared with the rest of satellite gravimetry data; see Chapter 2. In this section, two methods for recovering the product of Moho depth and density contrast from such data are presented. In the first, the TB, sediment and crystalline effects are continued upward to the satellite level; in other words, their gravitational effects are computed on the satellite gradiometry data. This process is performed by replacing the SHCs of the gravitational potential with those of these effects in the spherical harmonic expansion of the satellite gradiometry data; see Chapter 2. After subtracting the effects from the gradiometry data, the results will be continued downward while computing the product of Moho depth and density contrast. In the second approach, proposed by Eshagh (2014), only the satellite gradiometry data are continued downward, and their contribution to the product is estimated. However, the principle of both methods is the same, but the results will be practically different, and care should be taken when applying them. The difference is that in the first case, both the effects and the satellite gradiometry data are regularised together and simultaneously in one procedure. The long-wavelength portion of the effects should be removed and restored if the inversion is performed locally, as it is difficult to recover them, having a global nature, using data with limited coverage. If this remove–compute–restore scheme is not applied the product will be underestimated, meaning that the maps of the product will be over-smoothed. In the second approach, the effects are computed separately down at the sea level and the satellite gradiometry data will be continued downward to a contribution to the product. In this case, there is no regularisation in the effects except the truncation of their spherical harmonic series. However, the frequency contents of the estimated contribution from the satellite gradiometry data will be different. In this section, we present the mathematical foundations of both methods and leave the choice to the reader.

7.2.2.1 Product of Moho depth variation and density contrast from inversion satellite gradiometry data and effects of topographic/bathymetric, sediment, crustal crystalline and ice masses

Here, Eq. (6.85) is used again which is the equation of the equilibrium based on approach 2 of our gravimetry isostasy, and all effects of TB, sediment, crustal crystalline and ice masses are merged into one parameter, V^{TBSCI}. In this case the equation of equilibrium of the gravitational potential for a point with the spherical coordinates (r, θ, λ) outside the Earth is (see Eq. 6.85):

$$
\begin{aligned}
&G\left(R-\widetilde{D}_0\right) \iint\limits_{\sigma} \Delta\rho' \Delta\widetilde{D}' \sum_{n=0}^{\infty} \left(\frac{R-\widetilde{D}_0}{r}\right)^{n+1} \beta_n^* P_n(\widetilde{x}) \mathrm{d}\sigma \\
&= V^{\mathrm{TBSCI}}(r, \theta, \lambda) - T(r, \theta, \lambda).
\end{aligned} \tag{7.15}
$$

Now, our goal is to relate Eq. (7.15) to the satellite gradiometry data, which can be simply done by taking second-order partial derivatives of both sides of the equations with respect to the x-, y- and z-axes of the local north-oriented frame (LNOF); see e.g. Eqs. (3.80)–(3.92) in Chapter 3. Therefore, six integral equations can be established for inversion of the satellite gradiometry data and the effects of TB, sediment, crustal crystalline and ice masses simultaneously:

$$
\begin{aligned}
&G\left(R-\widetilde{D}_0\right) \iint\limits_{\sigma} \Delta\rho' \Delta\widetilde{D}' M_{ij}\left(r, \widetilde{D}_0, \psi\right) \mathrm{d}\sigma \\
&= V_{ij}^{\mathrm{TBSCI}}(r, \theta, \lambda) - T_{ij}(r, \theta, \lambda), \quad i = x, y, z \text{ and } j = x, y, z,
\end{aligned} \tag{7.16}
$$

where

$$\begin{bmatrix} M_{xx}(r, \widetilde{D}_0, \psi) \\ M_{yy}(r, \widetilde{D}_0, \psi) \end{bmatrix} = \frac{M_r(r, \widetilde{D}_0, \psi)}{r} + \frac{1}{2r^2}\left\{\left[M_{\psi\psi}(r, \widetilde{D}_0, \psi) + \cot\psi M_\psi(r, \widetilde{D}_0, \psi)\right]\right.$$
$$\left. \pm \left[M_{\psi\psi}(r, \widetilde{D}_0, \psi) - \cot\psi M_\psi(r, \widetilde{D}_0, \psi)\right]\cos 2\alpha\right\}, \tag{7.17}$$

$$\begin{bmatrix} M_{xz}(r, \widetilde{D}_0, \psi) \\ M_{yz}(r, \widetilde{D}_0, \psi) \end{bmatrix} = \frac{1}{r}\left[\frac{1}{r}M_\psi(r, \widetilde{D}_0, \psi) - M_{r\psi}(r, \widetilde{D}_0, \psi)\right]\begin{pmatrix} \cos\alpha \\ -\sin\alpha \end{pmatrix}, \tag{7.18}$$

$$M_{xy}(r, \widetilde{D}_0, \psi) = -\frac{1}{2r^2}\left[M_{\psi\psi}(r, \widetilde{D}_0, \psi) - \cot\psi M_\psi(r, \widetilde{D}_0, \psi)\right]\sin 2\alpha, \tag{7.19}$$

and

$$M(r, \widetilde{D}_0, \psi) = \sum_{n=0}^{\infty}\left(\frac{R - \widetilde{D}_0}{r}\right)^{n+1}\beta_n^* P_n(\widetilde{x}), \tag{7.20}$$

$$M_r(r, \widetilde{D}_0, \psi) = -\sum_{n=0}^{\infty}\frac{(n+1)}{r}\left(\frac{R - \widetilde{D}_0}{r}\right)^{n+1}\beta_n^* P_n(\widetilde{x}), \tag{7.21}$$

$$M_{zz}(r, \widetilde{D}_0, \psi) = \sum_{n=0}^{\infty}\frac{(n+1)(n+2)}{r^2}\left(\frac{R - \widetilde{D}_0}{r}\right)^{n+1}\beta_n^* P_n(\widetilde{x}), \tag{7.22}$$

$$M_{r\psi}(r, \widetilde{D}_0, \psi) = -\sum_{n=0}^{\infty}\frac{(n+1)}{r}\left(\frac{R - \widetilde{D}_0}{r}\right)^{n+1}\beta_n^* \frac{dP_n(\widetilde{x})}{d\psi}, \tag{7.23}$$

$$M_{\psi\psi}(r, \widetilde{D}_0, \psi) = \sum_{n=0}^{\infty}\left(\frac{R - \widetilde{D}_0}{r}\right)^{n+1}\beta_n^* \frac{d^2P_n(\widetilde{x})}{d\psi^2}. \tag{7.24}$$

By inverting the integral Eq. (7.16), the product of the density contrast and Moho variation will be estimated. Therefore, if one of them is known, the other one can be derived from this product. As observed, the mean Moho depth $\widetilde{D}_0$ plays a very important role in all steps of the computations. The spectral forms of the kernels can be directly used for the integral inversion purpose, but if we assume that $\beta_n^* = 1$, closed-form formulae for the kernels are achievable. As we discussed about the kernel functions for gravity field recovery, it can be said that the kernels (Eqs. 7.20 and 7.21) are well-behaving and those which involve the derivative of the Legendre polynomial will be bell shaped. The closed-form formulae after approximating $\beta_n^* = 1$ will be:

$$M_{rr}\left(r, \widetilde{D}_0, \psi\right) = \frac{2t}{r^2 D^{1/2}} - \frac{2t^2(t-\widetilde{x})}{r^2 D^{3/2}} - \frac{t^2(3t-2\widetilde{x})}{r^2 D^{3/2}} + \frac{3t^2(t-\widetilde{x})^2}{r^2 D^{5/2}}, \quad (7.25)$$

$$M_r\left(r, \widetilde{D}_0, \psi\right) = -\frac{t}{rD^{1/2}} + \frac{t^2(t-\widetilde{x})}{rD^{3/2}}, \quad (7.26)$$

$$M_\psi\left(r, \widetilde{D}_0, \psi\right) = \frac{-t^2\sqrt{1-\widetilde{x}^2}}{D^{3/2}}, \quad (7.27)$$

$$M_{r\psi}\left(r, \widetilde{D}_0, \psi\right) = \left(\frac{-t\sqrt{1-\widetilde{x}^2}}{D^{3/2}} + \frac{3t(t-x)\sqrt{1-\widetilde{x}^2}}{D^{5/2}}\right)\left(-\frac{t}{r}\right), \quad (7.28)$$

$$M_{\psi\psi}\left(r, \widetilde{D}_0, \psi\right) = -\frac{t\widetilde{x}}{D^{3/2}} + \frac{3t^2(1-\widetilde{x}^2)}{D^{5/2}}, \quad (7.29)$$

where $D = \sqrt{1+t^2-2t\widetilde{x}}$, $t = \left(R - \widetilde{D}_0\right)/r$ and $\widetilde{x} = \cos\psi$.

7.2.2.2 Contribution of satellite gradiometry data to the product of Moho depth variation and density contrast

The TB, sediment and crustal crystalline masses make strong contributions to the product of the Moho variation and density contrast, and Eshagh (2016) mentioned that the gravimetric data play the role of smoother. Therefore, the satellite gradiometry data can be continued downward to the contribution to this product and the gravitational effect of the mentioned masses at sea level can be computed directly. Here, we explain how this process is done for T_{zz}; the method can be simply and similarly developed for the rest of the gradiometry data. To do so, let us write Eq. (7.15) in spectral form and take its second-order radial derivative; in this case, we obtain:

$$\begin{aligned} &4\pi G\left(R-\widetilde{D}_0\right)\left(\Delta\rho\Delta\widetilde{D}\right)_n \frac{(n+1)(n+2)}{r^2}\left(\frac{R-\widetilde{D}_0}{r}\right)^{n+1}\frac{\beta_n^*}{2n+1} \\ &= v_n^{\mathrm{TBSCI}}\frac{(n+1)(n+2)}{r^2}\left(\frac{R}{r}\right)^{n+1} - (T_{rr})_n, \end{aligned} \quad (7.30)$$

where $\left(\Delta\rho\Delta\widetilde{D}\right)_n$, $(T_{rr})_n$ and v_n^{TBSCI} are, respectively, the Laplace coefficients of $\Delta\rho\Delta\widetilde{D}$, T_{rr} and V^{TBSCI}. Here, we write the spectral form in terms of Laplace coefficients, to keep the formulae shorter. However, writing it in terms of SHCs is also possible, and theoretically both approaches are the same.

Now, Eq. (7.30) is solved for $\left(\Delta\rho\Delta\widetilde{D}\right)_n$:

$$\left(\Delta\rho\Delta\widetilde{D}\right)_n = \frac{1}{4\pi G\left(R-\widetilde{D}_0\right)}\left(\frac{R}{R-\widetilde{D}_0}\right)^{n+1}\frac{2n+1}{\beta_n^*}v_n^{\mathrm{TBSCI}} - W_{\Delta\rho\Delta\widetilde{D},n}, \tag{7.31}$$

where

$$W_{\Delta\rho\Delta\widetilde{D},n} = \frac{1}{4\pi G\left(R-\widetilde{D}_0\right)}\left(\frac{r}{R-\widetilde{D}_0}\right)^{n+1}\frac{r^2}{(n+1)(n+2)}\frac{2n+1}{\beta_n^*}\left(T_{rr}\right)_n. \tag{7.32}$$

If we take summation over n from both sides of Eq. (7.31), and write the first term of the result in terms of SHCs, we obtain:

$$\Delta\rho\Delta\widetilde{D} = \frac{1}{4\pi G\left(R-\widetilde{D}_0\right)}\sum_{n=0}^{\infty}\left(\frac{R}{R-\widetilde{D}_0}\right)^{n+1}\frac{2n+1}{\beta_n^*}\sum_{m=-n}^{n}v_{nm}^{\mathrm{TBSCI}}Y_{nm}(\theta,\lambda) - W_{\Delta\rho\Delta\widetilde{D}}, \tag{7.33}$$

where $W_{\Delta\rho\Delta\widetilde{D}}$ is the contribution of T_{rr} to $\Delta\rho\Delta\widetilde{D}$. To estimate it, we solve Eq. (7.32) for $\left(T_{rr}\right)_n$:

$$4\pi G\left(\frac{R-\widetilde{D}_0}{r}\right)^{n+2}\beta_n^*\frac{(n+1)(n+2)}{(2n+1)r}W_{\Delta\rho\Delta\widetilde{D},n} = \left(T_{rr}\right)_n. \tag{7.34}$$

Inserting the integral expansion of $W_{\Delta\rho\Delta\widetilde{D},n}$ (see Eq. 1.11) into Eq. (7.34) and taking summation on n from both sides lead to an integral equation with the unknowns $W_{\Delta\rho\Delta\widetilde{D}}$. Similar integral equations can be obtained for the rest of the satellite gradiometry data. Such integral equations have the following general form by applying the operators presented in Eqs. (1.95)–(1.100) to the integral Eq. (7.15) considering only the disturbing potential T on the rhs.

$$G\left(R-\widetilde{D}_0\right)\iint_\sigma W'_{\Delta\rho\Delta\widetilde{D}}M_{ij}\left(r,\widetilde{D}_0,\psi\right)\mathrm{d}\sigma = T_{ij}, \tag{7.35}$$

$$\text{where } i = x,\ y,\ z \text{ and } j = x,\ y,\ z.$$

Note that the kernel functions are the same as those presented in Eqs. (7.17)–(7.24).

7.2.3 Product of Moho depth variation and density contrast from inter-satellite range rates

Now, we write two equations of isostatic equilibrium, Eq. (7.15), for two satellites moving in the same orbit at two different positions, $(r_1, \theta_1, \lambda_1)$ and $(r_2, \theta_2, \lambda_2)$. After subtracting them from each other, a new equation of equilibrium for these two satellites is obtained. The product of the Moho variation and the density contrast is a common term in this new equation, but the interesting issue is that on one side of this new equation the difference between the gravitational potential at satellites 1 and 2 will appear, which relates to the inter-satellite range rates. In this section, like the previous one, we present two approaches for recovering this product from this type of data. We will present the mathematical foundation of each one of them next.

7.2.3.1 Simultaneous inversion of inter-satellite range rates and effects of topographic/bathymetric, sediment, crustal crystalline and ice masses

As mentioned, if Eq. (7.15) is written for satellites 1 and 2 separately and the two equations are subtracted from each other, we get the following integral formula:

$$G\left(R-\widetilde{D}_0\right)\iint\limits_{\sigma}\Delta\rho'\Delta\widetilde{D}'\left[M\left(r_2,\widetilde{D}_0,\psi_2\right)-M\left(r_1,\widetilde{D}_0,\psi_1\right)\right]\mathrm{d}\sigma$$
$$=V^{\mathrm{TBSCI}}(r_2,\theta_2,\lambda_2)-V^{\mathrm{TBSCI}}(r_1,\theta_1,\lambda_1)-T(r_2,\theta_2,\lambda_2)+T(r_1,\theta_1,\lambda_1), \tag{7.36}$$

where ψ_1 and ψ_2 are the spherical geometric angles between the common integration points, and computation points at satellites 1 and 2, respectively; in addition:

$$V^{\mathrm{TBSCI}}(r_2,\theta_2,\lambda_2)-V^{\mathrm{TBSCI}}(r_1,\theta_1,\lambda_1)$$
$$=\sum_{n=0}^{\infty}\left(v_n^{\mathrm{TBSCI}}(r_2,\theta_2,\lambda_2)-v_n^{\mathrm{TBSCI}}(r_1,\theta_1,\lambda_1)\right), \tag{7.37}$$

where

$$v_n^{\mathrm{TBSCI}}(r_i,\theta_i,\lambda_i)=\sum_{m=-n}^{n}v_{nm}^{\mathrm{TBSCI}}\left(\frac{R}{r_i}\right)^{n+1}Y_{nm}(\theta_i,\lambda_i)\quad \text{with } i=1 \text{ or } 2. \tag{7.38}$$

Then the difference between the disturbing potential measured at satellites 2 and 1 can be written in the following spherical harmonic form:

$$T(r_2,\theta_2,\lambda_2)-T(r_1,\theta_1,\lambda_1)=\sum_{n=0}^{\infty}\sum_{m=-n}^{n}t_{nm}\left[\left(\frac{R}{r_2}\right)^{n+1}Y_{nm}(\theta_2,\lambda_2)-\left(\frac{R}{r_1}\right)^{n+1}Y_{nm}(\theta_1,\lambda_1)\right]. \tag{7.39}$$

However, as was shown in Chapters 2 and 4, Eq. (7.39) has a direct relation with the range rate between the satellites, $\dot{\tilde{\rho}}$, and their mean velocity v_0; see Eq. (2.114 or 3.75). In this case, Eq. (7.36) can be written in the following form:

$$G\left(R-\widetilde{D}_0\right)\iint_{\sigma}\Delta\rho'\Delta\widetilde{D}'\left[M\left(r_2,\widetilde{D}_0,\psi_2\right)-M\left(r_1,\widetilde{D}_0,\psi_1\right)\right]\mathrm{d}\sigma=V^{\mathrm{TBSCI}}(r_2,\theta_2,\lambda_2)-V^{\mathrm{TBSCI}}(r_1,\theta_1,\lambda_1)-v_0\dot{\tilde{\rho}}+\Upsilon, \tag{7.40}$$

where Υ is the difference between the normal potential at satellites 2 and 1; see Eq. (3.75).

The kernel function of the integral Eq. (7.40) is of bipolar type, meaning that it is functions of positions of both satellites. To apply Eq. (7.40) the effects of mass heterogeneities should be computed at both satellite positions and subtracted from each other. This means the upward continuation of the gravitational potential of these masses from sea level to satellite levels, which can be done by inserting the SHCs of the gravitational potential of the masses into the spherical harmonic expansion of the inter-satellite range rates; see Eq. (7.38). Eq. (7.40) shows that they should be considered with the satellites' average velocity and range rates and the results are considered as known parameters on the rhs of the equation, which should be inverted to estimate $\Delta\rho\Delta\widetilde{D}$.

7.2.3.2 Contribution of inter-satellite range rates to the product of Moho depth variation and density contrast

Upward, and after that downward, continuation of satellite data leads to some loss of frequency and power of signal, especially when the downward continuation and inversion is done locally. In this case, we can simply continue the inter-satellite range rates solely downward to contribution to $\Delta\rho\Delta\widetilde{D}$. To do so, Eq. (7.15) is rewritten in the following spectral form for satellite 1:

$$4\pi G\left(R-\widetilde{D}_0\right)\left(\Delta\rho\Delta\widetilde{D}\right)_{nm}\left(\frac{R-\widetilde{D}_0}{r_1}\right)^{n+1}\frac{\beta_n^*}{2n+1}=v_{nm}^{\mathrm{TBSCI}}\left(\frac{R}{r_1}\right)^{n+1} - t_{nm}\left(\frac{R}{r_1}\right)^{n+1}. \tag{7.41}$$

After solving this equation for $\left(\Delta\rho\Delta\widetilde{D}\right)_{nm}$ and performing spherical harmonic synthesis from both sides of it, we obtain:

$$\Delta\rho\Delta\widetilde{D}=\frac{1}{4\pi G\left(R-\widetilde{D}_0\right)}\sum_{n=0}^{\infty}\frac{2n+1}{\beta_n^*}\left(\frac{R}{R-\widetilde{D}_0}\right)^{n+1}\sum_{m=-n}^{n}v_{nm}^{\mathrm{TBSCI}}Y_{nm}(\theta,\lambda)-W_{\Delta\rho\Delta\widetilde{D}}, \tag{7.42}$$

where the SHC of $W_{\Delta\rho\Delta\widetilde{D}}$ is:

$$\left(W_{\Delta\rho\Delta\widetilde{D}}\right)_{nm}=\frac{1}{4\pi G\left(R-\widetilde{D}_0\right)}\left(\frac{r_1}{R-\widetilde{D}_0}\right)^{n+1}\frac{2n+1}{\beta_n^*}t_{nm}\left(\frac{R}{r_1}\right)^{n+1}. \tag{7.43}$$

Again, $\left(W_{\Delta\rho\Delta\widetilde{D}}\right)_{nm}$ is the contribution of the disturbing potential at satellite 1. To relate $W_{\Delta\rho\Delta\widetilde{D}}$ to this observable, let us first solve Eq. (7.43) for $t_{nm}(R/r_1)^{n+1}$, then:

$$4\pi G\left(R-\widetilde{D}_0\right)\left(\frac{R-\widetilde{D}_0}{r_1}\right)^{n+1}\frac{\beta_n^*}{2n+1}\left(W_{\Delta\rho\Delta\widetilde{D}}\right)_{nm}=t_{nm}\left(\frac{R}{r_1}\right)^{n+1}. \tag{7.44}$$

A similar equation can be written for satellite 2:

$$4\pi G\left(R-\widetilde{D}_0\right)\left(\frac{R-\widetilde{D}_0}{r_2}\right)^{n+1}\frac{\beta_n^*}{2n+1}\left(W_{\Delta\rho\Delta\widetilde{D}}\right)_{nm}=t_{nm}\left(\frac{R}{r_2}\right)^{n+1}. \tag{7.45}$$

Now, Eqs. (7.44) and (7.45) are combined in such a way that the inter-satellite range rates appear. When $\left(W_{\Delta\rho\Delta\widetilde{D}}\right)_{nm}$ is written in the spatial form according to Eq. (1.8), and inserted into Eqs. (7.45) and (7.44), and by multiplying both sides of Eq. (7.43) by $Y_{nm}(\theta_1, \lambda_1)$ and both sides of Eq. (7.44) by $Y_{nm}(\theta_2, \lambda_2)$, the results will be:

$$G\left(R-\widetilde{D}_0\right)\frac{\beta_n^*}{2n+1}\iint_\sigma W'_{\Delta\rho\Delta\widetilde{D}}Y_{nm}(\theta', \lambda')\mathrm{d}\sigma\left(\frac{R-\widetilde{D}_0}{r_1}\right)^{n+1}Y_{nm}(\theta_1, \lambda_1)$$

$$=t_{nm}\left(\frac{R}{r_1}\right)^{n+1}Y_{nm}(\theta_1, \lambda_1), \tag{7.46}$$

$$G\left(R-\widetilde{D}_0\right)\frac{\beta_n^*}{2n+1}\iint_\sigma W'_{\Delta\rho\Delta\widetilde{D}}Y_{nm}(\theta', \lambda')\mathrm{d}\sigma\left(\frac{R-\widetilde{D}_0}{r_2}\right)^{n+1}Y_{nm}(\theta_2, \lambda_2)$$

$$=t_{nm}\left(\frac{R}{r_2}\right)^{n+1}Y_{nm}(\theta_2, \lambda_2). \tag{7.47}$$

When Eq. (7.46) is subtracted from Eq. (7.47), a summation over m is taken from $-n$ to n, by taking advantage of the addition theorem of spherical harmonics (Eq. 1.9), and taking another summation over n from 0 to infinity, the following integral relation will be obtained:

$$G\left(R-\widetilde{D}_0\right)^2\iint_\sigma W'_{\Delta\rho\Delta\widetilde{D}}\left[M\left(r_2, \widetilde{D}_0, \psi_2\right)-M\left(r_1, \widetilde{D}_0, \psi_1\right)\right]\mathrm{d}\sigma=v_0\dot{\widetilde{\rho}}-\Upsilon. \tag{7.48}$$

Note that the kernel function of the integral Eqs. (7.48) and (7.40) is the same.

Therefore, the value of $W_{\Delta\rho\Delta\widetilde{D}}$ will be estimated from the inter-satellite range rates and inserted into Eq. (7.42).

7.2.4 Simultaneous inversion of inter-satellite line-of-sight and effects of topographic/bathymetric, sediment, crustal crystalline and ice masses

The principle of connecting the product of Moho variation and density contrast to the inter-satellite line-of-sight (LOS) measurement is rather like the procedures presented for the satellite gravimetry data discussed in the previous sections. The main difference is that the mathematical derivations are longer and slightly complicated, as the gradients of the gravitational potentials of two satellites are used. In other words, two vectors of gravitation at two satellites with different positions are applied for modelling the gravimetric isostatic equilibrium. Similar to the discussions made so far, the mathematical foundations for recovering the product of the Moho variation and density contrast from the LOS measurements are presented in the two scenarios of inversion of the TB, sediment, crustal crystalline and ice masses to and from satellites and direct inversion of the inter-satellite data to the contribution to the product.

7.2.4.1 Inversion of the effects of the crustal mass and structure heterogeneities and line-of-sight measurements

In Section 2.7, we showed how such type of satellite gravimetry data and the gravitational potential of the Earth are connected; see Eq. (2.120). Here, we will find a mathematical relation for the equation of equilibrium and the inter-satellite LOS measurements.

If both sides of Eq. (7.41) are multiplied by $Y_{nm}(\theta_1, \lambda_1)$ and the gradient operator, Eq. (1.45), is applied to the results we obtain:

$$4\pi G\left(R-\widetilde{D}_0\right)\frac{\beta_n^*}{2n+1}\left(\Delta\rho\Delta\widetilde{D}\right)_{nm}\nabla\left[\left(\frac{R-\widetilde{D}_0}{r_1}\right)^{n+1}Y_{nm}(\theta_1,\lambda_1)\right] = \left(v_{nm}^{\mathrm{TBSCI}}-t_{nm}\right)\nabla\left[\left(\frac{R}{r_1}\right)^{n+1}Y_{nm}(\theta_1,\lambda_1)\right]. \tag{7.49}$$

Now, we write $\left(\Delta\rho\Delta\widetilde{D}\right)_{nm}$ in integral form, according to Eq. (1.8), and after performing the gradient operation, the equation of equilibrium (Eq. 7.49) is written in terms of the vector spherical harmonics of poloidal type at satellite 1:

$$G\left(R-\widetilde{D}_0\right)\frac{\beta_n^*}{2n+1}\iint_\sigma \Delta\rho'\Delta\widetilde{D}'Y_{nm}(\theta',\lambda')\mathrm{d}\sigma \left[\frac{\partial}{\partial r_1}\left(\frac{R-\widetilde{D}_0}{r_1}\right)^{n+1}\mathbf{X}_{nm}^{(1)}(\theta_1,\lambda_1)+\frac{1}{r_1}\left(\frac{R-\widetilde{D}_0}{r_1}\right)^{n+1}\mathbf{X}_{nm}^{(2)}(\theta_1,\lambda_1)\right] = \left(v_{nm}^{\mathrm{TBSCI}}-t_{nm}\right)\left[\frac{\partial}{\partial r_1}\left(\frac{R}{r_1}\right)^{n+1}\mathbf{X}_{nm}^{(1)}(\theta_1,\lambda_1)+\frac{1}{r_1}\left(\frac{R}{r_1}\right)^{n+1}\mathbf{X}_{nm}^{(2)}(\theta_1,\lambda_1)\right]. \tag{7.50}$$

According to the addition theorem of vector spherical harmonics, Eqs. (1.61 and 1.62), and further simplifications, we arrive at:

$$G\left(R-\widetilde{D}_0\right)\iint_\sigma \Delta\rho'\Delta\widetilde{D}'\left[M_{r_1}\left(r_1,\widetilde{D}_0,\psi_1\right)\mathbf{e}_{z_1}+\frac{1}{r_1}M_{\theta_1}\left(r_1,\widetilde{D}_0,\psi_1\right)\mathbf{e}_{x_1}\right.$$
$$\left.+\frac{1}{r_1}M_{\lambda_1}\left(r_1,\widetilde{D}_0,\psi_1\right)\mathbf{e}_{y_1}\right]d\sigma=\left(v_{nm}^{\mathrm{TBSCI}}-t_{nm}\right)\left[\frac{\partial}{\partial r_1}\left(\frac{R}{r_1}\right)^{n+1}\mathbf{X}_{nm}^{(1)}(\theta_1,\lambda_1)\right.$$
$$\left.+\frac{1}{r_1}\left(\frac{R}{r_1}\right)^{n+1}\mathbf{X}_{nm}^{(2)}(\theta_1,\lambda_1)\right]. \tag{7.51}$$

By multiplying Eq. (7.41) by $Y_{nm}(\theta_2,\lambda_2)$ applying the gradient operator to the result and performing algebraic simplifications similar to what we did to obtain Eq. (7.51), we obtain another similar formula, but for satellite 2. If we subtract Eq. (7.51) from this equation, we obtain an integral equation connecting the LOS data to the product of the Moho variation and density contrast:

$$G\left(R-\widetilde{D}_0\right)\iint_\sigma \Delta\rho'\Delta\widetilde{D}'M^{\mathrm{B}}\left(r_1,\psi_1,\alpha_1,r_2,\psi_2,\alpha_2,\widetilde{D}_0\right)d\sigma=\Sigma^{\mathrm{TBSCI}}-\ddot{\widetilde{\rho}}$$
$$+\frac{\left(\delta\dot{\mathbf{r}}_{12}\right)^2-\dot{\widetilde{\rho}}^2}{\widetilde{\rho}}+\Sigma, \tag{7.52}$$

where Σ has already been presented in Eq. (3.78), and

$$M^{\mathrm{B}}\left(r_1,\psi_1,\alpha_1,r_2,\psi_2,\alpha_2,\widetilde{D}_0\right)=\left(M_{r_2}\left(r_2,\widetilde{D}_0,\psi_2\right)\mathbf{e}_{z_2}\right.$$
$$\left.+\frac{1}{r_2}M_{\psi_2}\left(r_2,\widetilde{D}_0,\psi_2\right)\left(\cos\alpha_2\mathbf{e}_{x_2}-\sin\alpha_2\mathbf{e}_{y_2}\right)\right)\cdot\mathbf{e}_{\mathrm{LOS}}$$
$$-\left(M_{z_1}\left(r_1,\widetilde{D}_0,\psi_1\right)\mathbf{e}_{r_1}+\frac{1}{r_1}M_{\psi_1}\left(r_1,\widetilde{D}_0,\psi_1\right)\left(\cos\alpha_1\mathbf{e}_{x_1}-\sin\alpha_1\mathbf{e}_{y_1}\right)\right)\cdot\mathbf{e}_{\mathrm{LOS}}. \tag{7.53}$$

Note that the kernel (Eq. 7.53) is bipolar and a function of mean Moho depth $\widetilde{D}_0$.

To compute the effects of the crustal masses on this type of data, it is enough to replace the SHCs of the Earth's gravitational potential by the SHCs of these masses (Eq. 2.126), namely:

$$\Sigma^{\mathrm{TBSCI}} = \sum_{n=0}^{\infty}\sum_{m=-n}^{n} v_{nm}^{\mathrm{TBSCI}}\left[-\frac{n+1}{R}\left(\left(\frac{R}{r_2}\right)^{n+1}\mathbf{X}_{nm}^{(1)}(\theta_2,\lambda_2) - \left(\frac{R}{r_1}\right)^{n+1}\mathbf{X}_{nm}^{(1)}(\theta_1,\lambda_1)\right) + \frac{1}{r_2}\left(\frac{R}{r_2}\right)^{n+1}\mathbf{X}_{nm}^{(2)}(\theta_2,\lambda_2) - \frac{1}{r_1}\left(\frac{R}{r_1}\right)^{n+1}\mathbf{X}_{nm}^{(2)}(\theta_1,\lambda_1)\right]\cdot \mathbf{e}_{\mathrm{LOS}}. \tag{7.54}$$

7.2.4.2 Inversion of inter-satellite line-of-sight measurements to the product of Moho depth variation and density contrast

The process of finding an integral equation for the contribution of the inter-satellite LOS measurements is very similar to what we have done for the satellite gravimetry data, which was discussed so far. According to what we have learnt so far, it will not be difficult to show that the formula for determining the product of density contrast and Moho variations is the same as what we have obtained in Eq. (7.42), but $W'_{\Delta\rho\Delta\widetilde{D}}$ will be derived by solving the following integral equation:

$$G\left(R-\widetilde{D}_0\right)\iint_{\sigma} W'_{\Delta\rho\Delta\widetilde{D}} M^B\left(r_1,\psi_1,\alpha_1,r_2,\psi_2,\alpha_2,\widetilde{D}_0\right)\mathrm{d}\sigma = \ddot{\widetilde{\rho}} - \frac{\left(\delta\dot{\mathbf{r}}_{12}\right)^2 - \dot{\widetilde{\rho}}^2}{\widetilde{\rho}} - \Sigma. \tag{7.55}$$

Therefore, it is enough to solve this integral equation by one of the regularisation methods presented in Chapter 4 and after that insert the estimated $W'_{\Delta\rho\Delta\widetilde{D}}$ into Eq. (7.42) to obtain the product.

7.2.5 Example: oceanic Moho model computed based on gravity model derived from satellite altimetry data

This example has been taken from the study done by Abrehdary et al. (2018) for determining Moho depth beneath oceans. They used 42,965 free-air gravity disturbances from the DNSC08GRA global marine gravity field model (Andersen et al., 2010) in a set of 1° × 1° blocks over the oceans. Full retracking of the 20-Hz waveform data for the entire ERS-1 geodetic mission data and a secondary waveform retracking on all the ocean Brown model waveforms, where information from the first fitting is used to constrain parameters for the secondary retracking. In addition to the reanalysed data from the ERS-1 and GEOSAT geodetic missions, altimetry data from other available satellite missions were included in the derivation of the DNSC08GRA global gravity field. This model includes mean tracks from the exact repeat missions Topex/Poseidon, GFO, ERS-2, and Envisat. Furthermore, high-resolution laser altimetry from the first three epochs of ICESat was included to fill the polar gap. For all altimetry data sets, the best available range and geophysical corrections have been applied.

The considered free-air gravity disturbances of DNSC08GRA were used to obtain simple Bouguer gravity disturbance and corrected for the gravitational contributions of TB and density contrasts of the oceans, ice sheets and sediment basins from the ESCM180 Earth spectral crustal model (Chen and Tenzer, 2015). Another model of the Bouguer gravity disturbance was generated from the SHCs of a satellite-only model, GOGRA04S (Yi et al., 2013), completed to degree and order 180. The SHCs of the normal gravity field were computed according to the parameters of the reference system GRS80 (Moritz, 2000). Fig. 7.3 shows the maps of the

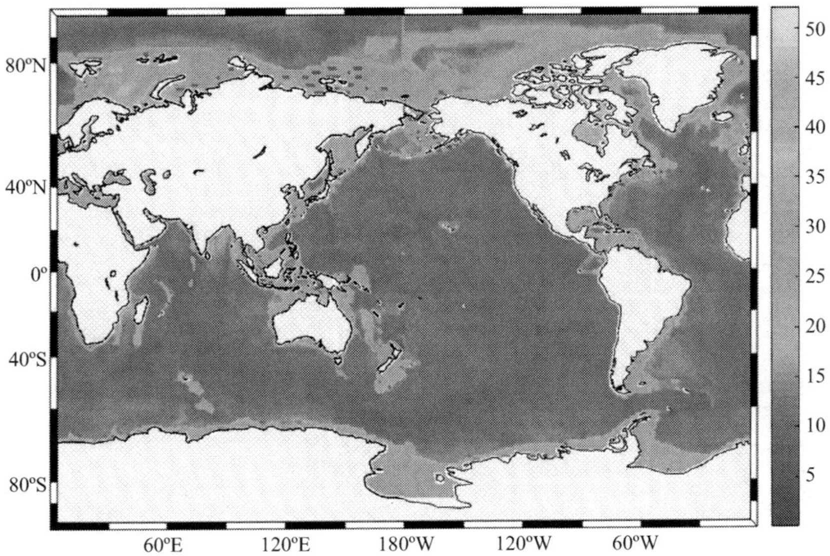

Figure 7.3 Oceanic Moho computed based on gravimetric isostasy and satellite altimetry data of ERS-1 and -2, GEOSAT, ICESat, GFO, Topex/Poseidon and Envisat. Unit: km (Abrehdary et al., 2018).

recovered oceanic Moho by Abrehdary et al. (2018). As seen, the deeper Moho is seen in the Northern Hemisphere and north of Europe. Also, large values are seen in the Gulf of Bohemia and the Baltic Sea in Fennoscandia and Hudson Bay in Canada, where we already know that there were large ice caps during the ice age, which pressed the lithosphere downward. Now, even after the ice has melted, the Moho is still deep there, but it is moving upward due to the viscous mantle towards isostatic equilibrium.

7.2.6 Example: Moho determination over the Indo-Pak region from GOCE data

Here, another example is presented from the studies of Eshagh and Hussain (2016), who determined a Moho model based on the inversion of GOCE gradiometry data. Their study area was bounded by 10°N to 50°N latitude and 50°E to 90°E longitude, mainly comprising the Indo-Pak subcontinent and adjoining areas. The effects of TB, sediment and crustal crystalline masses were determined from the CRUST1.0 model, but we do not present them in this section. Eshagh and Hussain (2016) used the approach of downward continuation of GOCE data, T_{rr} to the contribution to $W_{\Delta\rho\Delta\widetilde{D}}$. The satellite data of GOCE, consisting of second-order radial derivatives of geopotential (Bouman et al. 2011), measured in November 2013 were used. They removed the effects of the normal gravity field from these radial derivatives to change them to derivatives of the disturbing potential (T_{rr}). They used 163,084 data points with a sampling rate of 50 s to reduce the size of the system of equations to 3260. The data are presented in Fig. 7.4A. The density contrast at the Moho interface is presented in Fig. 7.4B. They continued with the GOCE data, T_{rr}, down to $W_{\Delta\rho\Delta\widetilde{D}}$ directly and added the results to the corresponding signal coming from the crustal masses. Fig. 7.4C shows the map of $W_{\Delta\rho\Delta\widetilde{D}}$ over the study area, and finally, Fig. 7.4D is the map of the recovered Moho model over the Indo-Pak region.

7.2.7 Example: Moho model of Iran from GOCE gradiometry data

The GOCE gradiometry data have also been used for Moho recovery over the territory of Iran, limited between the latitudes 25°N and 45°N and the longitudes 40°E and 65°E, by Eshagh et al. (2017). They selected a larger area by 5° from each side of the study area and selected the GOCE data over this area for their inversion purpose. In this case, the spatial

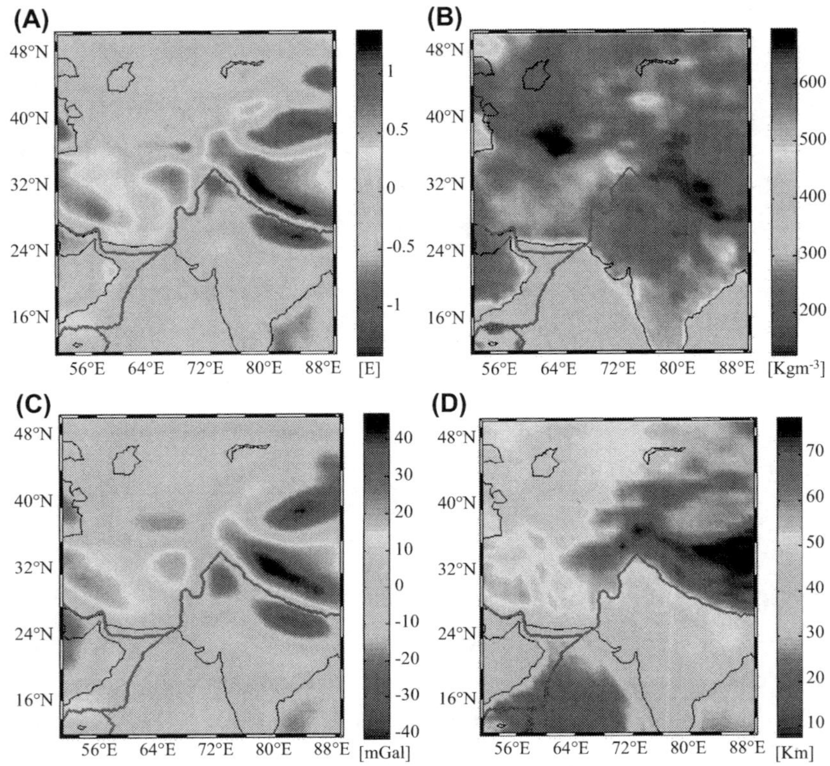

Figure 7.4 (A) GOCE gradiometry data (T_{rr}), (B) density contrast from CRUST1.0 model, (C) $W_{\Delta\rho\Delta\tilde{D}}$ recovered from the GOCE data and (D) the recovered Moho model over the Indo-Pak region. *(From Eshagh, M., Hussain, M., 2016. An approach to Moho discontinuity recovery from on-orbit GOCE data with application over Indo-Pak region. Tectonophysics 690 (B), 253–262.)*

truncation error; see Section 4.2, of the integral formula was reduced when the inversion was performed over this large area, but they selected those values, which are in the central part. They also used the second-order radial derivative of the geopotential, as this gradient amongst the others has the strongest signal and its mathematical model is simpler than the rest of the gradients, due to the involvement of only the spherical harmonics and not their derivatives after removing the effect of the normal gravity field from these data based on the GRS80 (Moritz, 1990) normal gravity field model.

Fig. 7.5A shows the GOCE gradiometry data over the study area with 5° extension from each side for the reduction of the spatial truncation

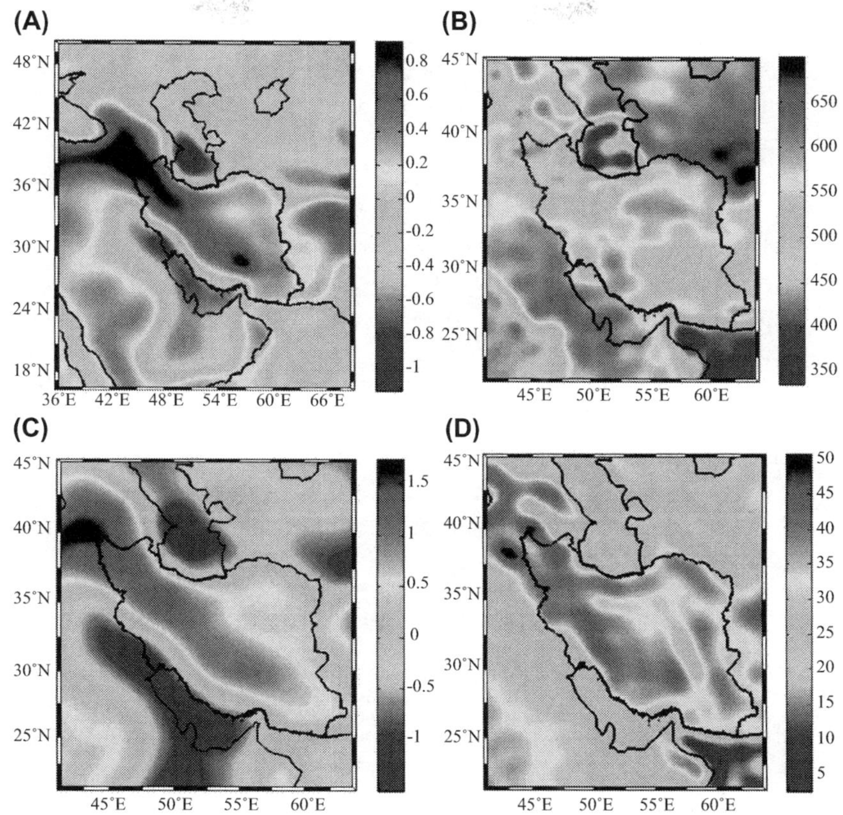

Figure 7.5 (A) GOCE gradiometry data, T_{rr} (unit: E), (B) density contrast from CRUST1.0 (unit: kg/m^3), (C) $W_{\Delta\rho\Delta\widetilde{D}}$ (unit: mGal) and (D) Moho depths (unit: km). *(From Eshagh, M., Ebadi, S., Tenzer, R., 2017. Isostatic GOCE Moho model for Iran. J. Asian Earth Sci. 138, 12–24.)*

error. Fig. 7.5B is the map of the density contrast of crust and upper mantle and Fig. 7.5C is the recovered $W_{\Delta\rho\Delta\widetilde{D}}$ from the data presented in Fig. 7.5A. After the contribution of the TB, sediment and crustal crystalline masses to the Moho is computed and $W_{\Delta\rho\Delta\widetilde{D}}$ is subtracted from them, the product of the Moho variation and density contrast will be obtained. Dividing the result by the density contrasts presented in Fig. 7.5B and adding it to the mean Moho depth derives the Moho model presented in Fig. 7.5D. It shows that the Moho depth reaches about 50 km according to the gravimetric isostasy approach 2, from the GOCE data.

7.3 Determination of density contrast

The problem of estimating density contrast is rather simpler than the Moho depth determination, because the integral equation of isostatic equilibrium is linear with respect to it. In the previous sections, we developed mathematical models for the estimation of the product of the density contrast and Moho variation. Once this product is estimated, we can simply estimate the Moho depth if the density contrast is already known, or vice versa, we can determine the density contrast if the Moho depth is given. However, we should mention that this product is the result of assuming some simplification and linearisation of mathematical models. In this section, we present the mathematical models for estimating the density contrast when the Moho variation is known; in this case, fewer approximations will be required for determining the density contrast compared with the method presented in Section 7.2. Here, we try to keep the section short and do not repeat the mathematical developments as they are very similar to what we discussed in the previous section.

7.3.1 Determination of density contrast from satellite altimetry data

Our assumption here is that the satellite altimetry data are the geoid heights over the ocean. In analogy to Eq. (7.10), the equation of equilibrium at the surface of the sea is:

$$G\left(R-\widetilde{D}_0\right)\iint\limits_{\sigma_{\text{ocean}}}\Delta\rho'\sum_{n=0}^{\infty}\left(\frac{R-\widetilde{D}_0}{R}\right)^{n+1}\kappa_n\left(\Delta\widetilde{D}\right)P_n(\widetilde{x})d\sigma = V^{\text{BSC}}(r,\theta,\lambda)-\gamma N(r,\theta,\lambda), \tag{7.56}$$

where

$$\kappa_n\left(\Delta\widetilde{D}\right)=\frac{1}{n+3}\left[1-\left(1-\frac{\Delta\widetilde{D}}{R-\widetilde{D}_0}\right)^{n+3}\right]. \tag{7.57}$$

Note that $\kappa_n\left(\Delta\widetilde{D}\right)$ is considered as a known parameter, as $\Delta\widetilde{D}$ and $\widetilde{D}_0$ are already given from independent sources like seismic data. If Eq. (7.56) is written in the spectral form, based on the addition theorem of the spherical harmonics, Eq. (1.9), we get:

$$G\left(R-\widetilde{D}_0\right)\frac{4\pi}{2n+1}\Delta\rho_{nm}\left(\frac{R-\widetilde{D}_0}{R}\right)^{n+1}\kappa_n\left(\Delta\widetilde{D}\right)=v_{nm}^{\text{BSC}}-\gamma N_{nm}. \tag{7.58}$$

After solving Eq. (7.58) for $\Delta\rho_{nm}$ and performing spherical harmonic synthesis from both sides of the result, we obtain:

$$\Delta\rho = \frac{1}{4\pi G\left(R-\tilde{D}_0\right)} \sum_{n=0}^{\infty} \frac{2n+1}{\kappa_n\left(\Delta\tilde{D}\right)} \left(\frac{R}{R-\tilde{D}_0}\right)^{n+1} \sum_{m=-n}^{n} v_{nm}^{\mathrm{BSC}} Y_{nm}(\theta,\lambda) - W_{\Delta\rho}. \tag{7.59}$$

The first term on the rhs of Eq. (7.59) is the contribution of the bathymetric, sediment and crustal crystalline masses to the density contrast and the second one the contribution from the satellite altimetry data $W_{\Delta\rho}$, which can be derived by inverting satellite altimetry-based geoid heights over the ocean by the following integral equations:

$$G\left(R-\tilde{D}_0\right) \iint_{\sigma_{\mathrm{ocean}}} W'_{\Delta\rho} \sum_{n=0}^{\infty} \left(\frac{R-\tilde{D}_0}{R}\right)^{n+1} \kappa_n\left(\Delta\tilde{D}\right) P_n(\tilde{x})\mathrm{d}\sigma = \gamma N(R,\theta,\lambda). \tag{7.60}$$

Once $W_{\Delta\rho}$ is estimated, it can be inserted back into Eq. (7.59) to derive the density contrast.

Finding a closed-form formula for the kernel function of the integral Eq. (7.60) is not possible, due to the involvement of $\kappa_n\left(\Delta\tilde{D}\right)$. Therefore, its spectral form can be applied instead, but it is needed to limit the kernel series to a maximum degree. For the purpose of Moho or density contrast determination, a maximum value of 180 should be enough. Since $\left(R-\tilde{D}_0\right) < R$ we can expect that the kernel is divergent, but $\tilde{D}_0$ is, e.g., about 30 km over continents; therefore the ratio $\left(R-\tilde{D}_0\right)/R$ will be very close to 1. Therefore, $\kappa_n\left(\Delta\tilde{D}\right)$ will play an important role in the behaviour of the kernel function.

7.3.2 Determination of density contrast from satellite gradiometry data

Similarly, inversion of satellite gradiometry data to density contrast can be performed in two ways: (1) inversion of crustal mass heterogeneities and satellite gradiometry data or (2) inversion of these data to contribution to density contrast ($W_{\Delta\rho}$), which is contribution of the gradiometry data to density contrast. The following integral equation can be applied for the first approach:

$$G\left(R-\widetilde{D}_0\right)\iint\limits_{\sigma}\Delta\rho' M_{ij}^{*}\left(r,\widetilde{D}_0,\Delta\widetilde{D},\psi\right)\mathrm{d}\sigma = V_{ij}^{\mathrm{TBSCI}}(r,\theta,\lambda)-T_{ij}(r,\theta,\lambda),\quad i=x,y,z \text{ and } j=x,y,z, \tag{7.61}$$

where again $V_{ij}^{\mathrm{TBSCI}}(r,\theta,\lambda)$ is the upward continued crustal mass heterogeneities effect on the gradiometry data $T_{ij}(r,\theta,\lambda)$, and i and j are indexes showing partial derivatives in the LNOF. The kernel of this integral equation is a function of Moho depth variation $\Delta\widetilde{D}$ in addition to the geocentric distance of the satellite, mean Moho depth $\widetilde{D}_0$, and geocentric angle between the computation and the integration points. The mathematical formulae of these kernels are:

$$\begin{bmatrix} M_{xx}^{*}\left(r,\widetilde{D}_0,\Delta\widetilde{D},\psi\right) \\ M_{yy}^{*}\left(r,\widetilde{D}_0,\Delta\widetilde{D},\psi\right)\end{bmatrix}=\frac{M_r^{*}\left(r,\widetilde{D}_0,\Delta\widetilde{D},\psi\right)}{r}+\frac{1}{2r^2}\left\{\left[M_{\psi\psi}^{*}\left(r,\widetilde{D}_0,\Delta\widetilde{D},\psi\right)+\cot\psi M_{\psi}^{*}\left(r,\widetilde{D}_0,\Delta\widetilde{D},\psi\right)\right]\pm\left[M_{\psi\psi}^{*}\left(r,\widetilde{D}_0,\Delta\widetilde{D},\psi\right)-\cot\psi M_{\psi}^{*}\left(r,\widetilde{D}_0,\Delta\widetilde{D},\psi\right)\right]\cos 2\alpha\right\}, \tag{7.62}$$

$$\begin{bmatrix} M_{xz}^{*}\left(r,\widetilde{D}_0,\Delta\widetilde{D},\psi\right) \\ M_{yz}^{*}\left(r,\widetilde{D}_0,\Delta\widetilde{D},\psi\right)\end{bmatrix}=\frac{1}{r}\left[\frac{1}{r}M_{\psi}^{*}\left(r,\widetilde{D}_0,\Delta\widetilde{D},\psi\right)-M_{r\psi}^{*}\left(r,\widetilde{D}_0,\Delta\widetilde{D},\psi\right)\right]\begin{pmatrix}\cos\alpha \\ -\sin\alpha\end{pmatrix}, \tag{7.63}$$

$$M_{xy}^{*}\left(r,\widetilde{D}_0,\Delta\widetilde{D},\psi\right)=-\frac{1}{2r^2}\left[M_{\psi\psi}^{*}\left(r,\widetilde{D}_0,\Delta\widetilde{D},\psi\right)+\cot\psi M_{\psi}^{*}\left(r,\widetilde{D}_0,\Delta\widetilde{D},\psi\right)\right]\sin 2\alpha, \tag{7.64}$$

and

$$M^{*}\left(r,\widetilde{D}_0,\Delta\widetilde{D},\psi\right)=\sum_{n=0}^{\infty}\left(\frac{R-\widetilde{D}_0}{r}\right)^{n+1}\kappa_n\left(\Delta\widetilde{D}\right)P_n(\widetilde{x}), \tag{7.65}$$

$$M_r^{*}\left(r,\widetilde{D}_0,\Delta\widetilde{D},\psi\right)=-\sum_{n=0}^{\infty}\frac{(n+1)}{r}\left(\frac{R-\widetilde{D}_0}{r}\right)^{n+1}\kappa_n\left(\Delta\widetilde{D}\right)P_n(\widetilde{x}), \tag{7.66}$$

$$M^*_{zz=rr}\left(r,\widetilde{D}_0,\Delta\widetilde{D},\psi\right)=\sum_{n=0}^{\infty}\frac{(n+1)(n+2)}{r^2}\left(\frac{R-\widetilde{D}_0}{r}\right)^{n+1}\kappa_n\left(\Delta\widetilde{D}\right)P_n(\widetilde{x}),\tag{7.67}$$

$$M^*_{r\psi}\left(r,\widetilde{D}_0,\Delta\widetilde{D},\psi\right)=-\sum_{n=0}^{\infty}\frac{(n+1)}{r}\left(\frac{R-\widetilde{D}_0}{r}\right)^{n+1}\kappa_n\left(\Delta\widetilde{D}\right)\frac{dP_n(\widetilde{x})}{d\psi},\tag{7.68}$$

$$M^*_{\psi\psi}\left(r,\widetilde{D}_0,\Delta\widetilde{D},\psi\right)=\sum_{n=0}^{\infty}\left(\frac{R-\widetilde{D}_0}{r}\right)^{n+1}\kappa_n\left(\Delta\widetilde{D}\right)\frac{d^2P_n(\widetilde{x})}{d\psi^2}.\tag{7.69}$$

In the case where our goal is not to perform upward and downward continuation of the crustal mass heterogeneities effects, it will not be difficult to show that the density contrast is derived by:

$$\Delta\rho=\frac{1}{4\pi G\left(R-\widetilde{D}_0\right)}\sum_{n=0}^{\infty}\left(\frac{R}{R-\widetilde{D}_0}\right)^{n+1}\frac{2n+1}{\kappa_n\left(\Delta\widetilde{D}\right)}\sum_{m=-n}^{n}v_{nm}^{\mathrm{TBSCI}}Y_{nm}(\theta,\lambda)-W_{\Delta\rho},\tag{7.70}$$

where the first term on the rhs of Eq. (7.70) is the contribution of crustal masses on the Moho and the second term is the contribution from satellite gradiometry data, which is estimated by solving the following integral equation:

$$G\left(R-\widetilde{D}_0\right)\iint_{\sigma}W'_{\Delta\rho}M^*_{ij}\left(r,\widetilde{D}_0,\Delta\widetilde{D},\psi\right)d\sigma=T_{ij}(r,\theta,\lambda).\tag{7.71}$$

It is emphasised that $W'_{\Delta\rho}$ is the contribution to density contrast.

7.3.3 Determination of density contrast from inter-satellite range rates

The effect of crustal masses can be computed for both satellites 1 and 2 according to their positions. The difference between the disturbing potentials at both satellites will be replaced by $v_0\,\dot{\widetilde{\rho}}-\Upsilon$; therefore, the following integral equation can be constructed for determining the density contrast:

$$G\left(R-\widetilde{D}_0\right)\iint\limits_{\sigma}\Delta\rho'\left(M^*\left(r_2,\widetilde{D}_0,\Delta\widetilde{D},\psi_2\right)-M^*\left(r_1,\widetilde{D}_0,\Delta\widetilde{D},\psi_1\right)\right)\mathrm{d}\sigma$$
$$=V^{\mathrm{TBSCI}}(r_2,\theta_2,\lambda_2)-V^{\mathrm{TBSCI}}(r_1,\theta_1,\lambda_1)-v_0\dot{\widetilde{\rho}}+\Upsilon. \tag{7.72}$$

In the case of only downward continuation of range rates being desired, the mathematical formula will be the same as Eq. (7.70), but $W_{\Delta\rho}$ is estimated by solving the following integral equation:

$$G\left(R-\widetilde{D}_0\right)\iint\limits_{\sigma}W'_{\Delta\rho}\left(M^*\left(r_2,\widetilde{D}_0,\Delta\widetilde{D},\psi_2\right)-M^*\left(r_1,\widetilde{D}_0,\Delta\widetilde{D},\psi_1\right)\right)\mathrm{d}\sigma=v_0\dot{\widetilde{\rho}}-\Upsilon. \tag{7.73}$$

The kernel function of Eq. (7.73) is bipolar, meaning that it is a function of the satellites' positions in addition to $\widetilde{D}_0$ and $\Delta\widetilde{D}$. The estimated $W'_{\Delta\rho}$ can be inserted into Eq. (7.59) for determining the density contrast, as the first term of this equation is independent of any satellite gravimetry data.

7.3.4 Determination of density contrast from the combination of inter-satellite line-of-sight measurements

The integral equation for simultaneous downward continuation of the crustal mass variation effects on the inter-satellite LOS data is:

$$G\left(R-\widetilde{D}_0\right)\iint\limits_{\sigma}\Delta\rho' M^{*\mathrm{B}}\left(r_1,\psi_1,\alpha_1,r_2,\psi_2,\alpha_2,\widetilde{D}_0,\Delta\widetilde{D}\right)\mathrm{d}\sigma$$
$$=\Sigma^{\mathrm{TBSCI}}-\ddot{\widetilde{\rho}}+\frac{\left(\delta\dot{\mathbf{r}}_{12}\right)^2-\dot{\widetilde{\rho}}^2}{\widetilde{\rho}}+\Sigma, \tag{7.74}$$

and if only downward continuation of the inter-satellite LOS data is desired, Eq. (7.70) should be applied, but $W_{\Delta\rho}$, or the contribution of the satellite data to density contrast, is estimated from these data through the following integral equation.

$$G\left(R-\widetilde{D}_0\right)\iint\limits_{\sigma} W'_{\Delta\rho}M^{*\mathrm{B}}\left(r_1,\psi_1,\alpha_1,r_2,\psi_2,\alpha_2,\widetilde{D}_0,\Delta\widetilde{D}\right)\mathrm{d}\sigma$$
$$=\ddot{\widetilde{\rho}}-\frac{\left(\delta\dot{\mathbf{r}}_{12}\right)^2-\dot{\widetilde{\rho}}^2}{\widetilde{\rho}}-\Sigma. \tag{7.75}$$

The kernel function of both integral Eqs. (7.74) and (7.75) is:

$$M^{*\mathrm{B}}\left(r_1,\psi_1,\alpha_1,r_2,\psi_2,\alpha_2,\widetilde{D}_0,\Delta\widetilde{D}\right)=\Big(M^{*}_{r_2}\left(r_2,\widetilde{D}_0,\Delta\widetilde{D},\psi_2\right)\mathbf{e}_{z_2}$$
$$+\frac{1}{r_2}M^{*}_{\psi_2}\left(r_2,\widetilde{D}_0,\Delta\widetilde{D},\psi_2\right)\left(\cos\alpha_2\mathbf{e}_{x_2}-\sin\alpha_2\mathbf{e}_{y_2}\right)\Big)\cdot\mathbf{e}_{\mathrm{LOS}}$$
$$-\left(M^{*}_{r_1}\left(r_1,\widetilde{D}_0,\Delta\widetilde{D},\psi_1\right)\mathbf{e}_{z_1}+\frac{1}{r_1}M^{*}_{\psi_1}\left(r_1,\widetilde{D}_0,\Delta\widetilde{D},\psi_1\right)\right.$$
$$\left.\left(\cos\alpha_1\mathbf{e}_{x_1}-\sin\alpha_1\mathbf{e}_{y_1}\right)\right)\cdot\mathbf{e}_{\mathrm{LOS}}. \tag{7.76}$$

The only difference between the kernel Eq. (7.76) and kernel Eq. (7.53) is related to the presence of $\Delta\widetilde{D}$ inside the kernel Eq. (7.76); in other words, this kernel is a function of $\Delta\widetilde{D}$ in addition to other parameters.

7.3.5 Example: Moho density contrast from CryoSat-2 and Jason-1 marine gravity model

Abrehdary et al. (2018) made use of the GMG2014 global marine gravity model (Sandwell et al., 2014), which was provided by new radar altimeter measurements from the satellites CryoSat-2 and Jason-1. This model is two times more accurate than previous models. The GMG2014 model has an accuracy of about 2 mGal and covers all oceans between −80.738 and 80.738 latitude, and it has a very dense and homogeneous coverage (Abrehdary et al., 2018). The Vening Meinesz—Moritz (Sjöberg, 2009), approach 1, isostasy method has been applied in addition to the crustal information of the CRUST1.0 model (Laske et al., 2013) for determining crustal effects, in addition to Moho depths. The map of the density contrast over ocean is shown in Fig. 7.6 and it shows that the density contrast is smaller along the mid-oceanic ridges, as expected. The large values of the density contrast are seen in the Northern Hemisphere, in the north of Europe and Hudson Bay in Canada.

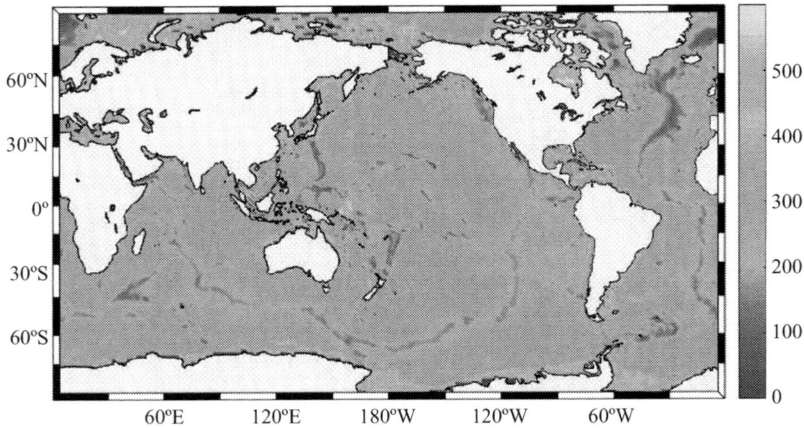

Figure 7.6 Density contrast determined over oceans from the GMG2014 model derived from radar measurements of CryoSat-2 and Jason-1. Unit: kg/m^3. *(From Abrehdary, M., Sjöberg, L.E., Sampietro, D., 2018. Contribution of satellite altimetry in modelling Moho density contrast in oceanic areas. J. Appl. Geod. 13, 33–40. DOI: 10.1515/jag-2018-0034.)*

7.3.6 Example: density contrast determination over central Eurasia from GOCE data

Here we present a part of the study done by Eshagh et al. (2016), which was related to density contrast determination from the second-order radial derivative of the real GOCE data. The regional study area of central Eurasia is bounded by the northern parallels of 0°N and 70°N and the meridians of 40°E and 120°E. The tectonic configuration of this study area comprises parts of the Indian, Eurasian, Arabian and African plates, including the Tibetan and Iranian tectonic blocks. The most prominent tectonic feature is the active continent-to-continent collision of the Indian plate with the Tibetan block, responsible for the uplift of Himalaya. Tomographic evidence also suggests the southward subduction of the Eurasian lithosphere underneath the Tibetan block. The subduction of the Indian and Eurasian lithosphere underneath the Tibetan block resulted in a significant compressional tectonism, followed by the uplift of the whole Tibetan block. The study area is characterised by zones of compressional, extensional and strike–slip tectonism. Compressional tectonism is evident from the existence of the Himalayas and its subordinate ranges and the Makran subduction zone (e.g., Eshagh et al., 2016). Extensional tectonism appears along the Mid-Indian oceanic rift zone.

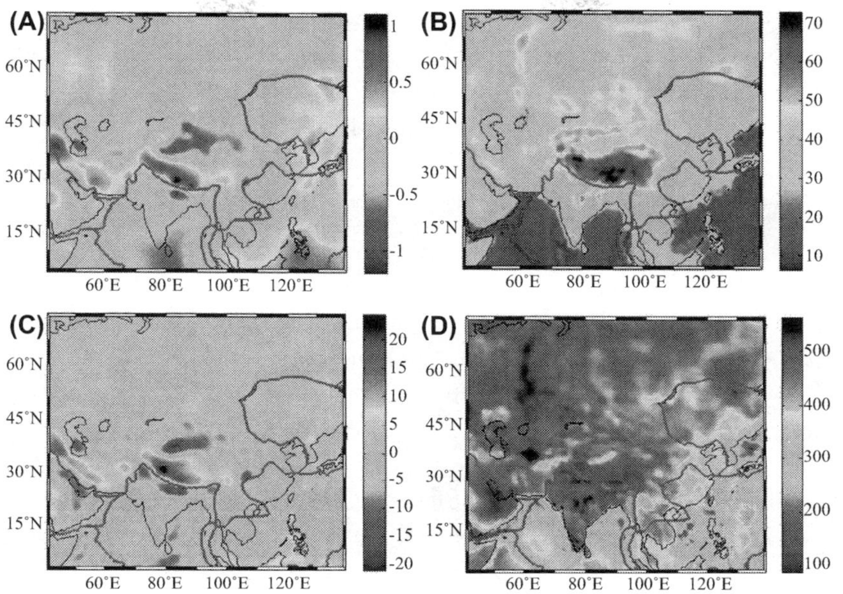

Figure 7.7 (A) GOCE data, T_{rr} (unit: E), (B) CRUST1.0 Moho depth model (unit: km), (C) $W_{\Delta\rho\Delta\widetilde{D}}$ (unit: mGal) and (D) the recovered density contrast after constraining the solution to the seismic data (unit: kg/m^3). *(From Eshagh, M., Hussain, M., Tenzer, R., Romeshkani, M., 2016. Moho density contrast in central Eurasia from GOCE gravity gradients. Remote Sens. 8 (418), 1–18.)*

The topographic heights on land and the bathymetric depths offshore from the SRTM30 database were used to compute the TB effect with a spectral resolution complete to a spherical harmonic degree 180. The average density of the upper continental crust of 2670 kg/m^3, the reference crustal density, was adopted as the topographic density.

T_{rr} was used to determine the Moho density contrast over the study area with a resolution of $1° \times 1°$. These data reach positive values up to 1.3 E as well as negative values to -1.4 E, with the largest horizontal spatial changes across profiles intersecting the continental basins and orogens (see Fig. 7.7A). The GOCE data are inverted to $W_{\Delta\rho\Delta\widetilde{D}}$. For this purpose, the integral equation was discretised and $W_{\Delta\rho\Delta\widetilde{D}}$ were then estimated by applying a least-squares method. The system of observation equations and the regularisation were solved by applying the conjugate-gradient technique (see Chapter 4). This method is very fast compared with a direct regularisation method, which uses the matrix inversion. The L-curve method (Hansen, 1998) was used to select the regularisation parameter. The estimated $W_{\Delta\rho\Delta\widetilde{D}}$ varies between −21 and 24 mGal (see Fig. 7.7B). Since the values T_{rr} and $W_{\Delta\rho\Delta\widetilde{D}}$ are highly correlated, their spatial patterns are very similar. By dividing the derived $W_{\Delta\rho\Delta\widetilde{D}}$ by the Moho depth variations obtained from the Moho model of CRUST1.0, presented in Fig. 7.7B, the density contrast model of the area is obtained. However, it should be mentioned that Eshagh et al. (2016) had constrained the density contrast solution to the seismic data through a condition adjustment model. They considered a mathematical model which relates $W_{\Delta\rho\Delta\widetilde{D}}$ to the density contrast and Moho depth and some a priori values of the Moho depth and the density contrast; see Fig. 7.7C. Later, they performed a minimum norm least-squares solution to find the smallest corrections to these initial values. It should be mentioned that in density contrast determination, there is a risk of obtaining very large and unrealistic values or even negative values, meaning that the density of the crust is larger than the density of the upper mantle, which is not physically meaningful. In addition, the problem of non-isostatic effect always exists, as the isostatic equilibrium is not valid in all parts of the Earth. Constraining the solution for the density contrast seems to be very important to obtain meaningful values; see Fig. 7.7D.

7.4 Determination of lithospheric elastic thickness and rigidity

In Chapter 6, we provided the mathematical models for connecting the lithospheric mechanical parameters to the gravity data, by combining the flexural and gravimetric isostasy theories. The main assumption is that the Moho flexure derived from the gravimetric isostasy and the flexural isostasy are equal. This assumption was applied by different geoscientists for modelling the elastic thickness T_e or the rigidity, e.g. Stewart and Watts (1997) or Braitenberg et al. (2006). Here, we apply only the gravimetric isostatic approach 2 and the simple flexural model to explain how to mathematically determine the elastic thickness or rigidity. The solution is of the forward computation type, meaning that different values of T_e are selected from 0 to a large value, say 150 km, by a certain interval, e.g. 1 km. For each elastic thickness a flexure Moho is computed and compared with the corresponding one obtained by the gravimetric method. The value which gives the closest flexure Moho to that of the gravimetric one is chosen as the elastic thickness value. This method has been presented by Eshagh (2018) and successfully applied over South America, by Eshagh et al. (2018) over central Eurasia, and by Eshagh and Pitonak (2018) over Africa. In the following, we explain how to develop this method for the estimation of flexural rigidity and elastic thickness from satellite altimetry, satellite gradiometry, and inter-satellite measurement.

7.4.1 The mathematical foundation

The combined solution of approach 2 of gravimetric isostasy and the flexure model can be solved for determining the SHCs of the disturbing potential (see Eq. 7.9):

$$t_{nm}^{\mathrm{TBSCI}} = -4\pi G\left(R-\widetilde{D}_0\right)\Delta\rho\left(\frac{R-\widetilde{D}_0}{R}\right)^{n+1}\frac{\beta_n^*}{2n+1}C_n\overline{K}_{nm} + \left(v_{nm}^{\mathrm{TB}}+v_{nm}^{\mathrm{S}}+v_{nm}^{\mathrm{C}}+v_{nm}^{\mathrm{I}}\right). \tag{7.77}$$

Let us perform a spherical harmonic synthesis from both sides of Eq. (7.77); then the relation between the disturbing potential generated from the crustal masses and structure and the flexural rigidity becomes:

$$\begin{aligned} T^{\mathrm{TBSCI}}(R,\theta,\lambda) &= V^{\mathrm{TBSCI}}(R,\theta,\lambda) - 4\pi G\left(R-\widetilde{D}_0\right)\Delta\rho\sum_{n=0}^{\infty}\left(\frac{R-\widetilde{D}_0}{R}\right)^{n+1} \\ &\quad \frac{\beta_n^*}{2n+1}\left(\frac{n^2(n+1)^2}{R^4 g}\widetilde{\Theta}+\Delta\rho\right)^{-1}\sum_{m=-n}^{n}\overline{K}_{nm}Y_{nm}(\theta,\lambda) \\ &= V^{\mathrm{TBSCI/Iso}}(R,\theta,\lambda), \end{aligned} \tag{7.78}$$

where

$$\widetilde{\Theta} = \begin{cases} D^{\mathrm{Rig}} & \text{if flexural rigidity is desired} \\ \dfrac{ET_e^3}{12(1-v^2)} & \text{if elastic thickness is desired} \end{cases}. \tag{7.79}$$

Eq. (7.79) specifies which quantity is desired; the flexural rigidity can be estimated directly, or the elastic thickness when (T_e), the Young modulus (E) and the Poisson ratio (v) are given.

Eq. (7.78) is the core equation connecting the gravitational potential of the crustal masses to the mechanical properties of the lithosphere. The parameter $\widetilde{\Theta}$ can be simply considered as the lithospheric rigidity D^{Rig}, which can be obtained directly from Eq. (7.79). Also, $\widetilde{\Theta}$ can be considered as a function of E, T_e and v, and T_e can be estimated from the equation when E and v are given. As Eq. (7.78) shows, $\widetilde{\Theta}$ does not depend on the degree and order of the spherical harmonic series. This means that $\widetilde{\Theta}$ is estimable from each degree and order, but there is no guarantee that the estimated values for $\widetilde{\Theta}$ will be the same. Therefore, different values of $\widetilde{\Theta}$ will be obtained at each point with the spherical coordinates (R, θ, λ), which cannot be realistic. Therefore, the best way to determine it is to use the whole spherical harmonic expansion, or in other words, the spatial form, rather than using the SHCs or spectral form.

Eq. (7.78) gives the isostatically reduced disturbing potential. However, the real disturbing potential contains all frequencies of the gravity field, including those coming from the crustal masses. Therefore, that portion of the disturbing potential due to the sub-lithosphere should be removed if the goal is to estimate the mechanical properties of the lithosphere; otherwise, the estimates will contain biases from sub-lithospheric masses. Stewart and Watts (1997) mentioned that the SHCs of the gravity field limited to degree and order 15 come mainly from the sub-lithosphere. This means that this portion of signal should be removed from the real gravitational potential to remove the potential of the crustal masses. In this case, a gravity model is required for removing these low-degree SHCs. It should be noted that degree and order 15 is for the case in which the Earth is assumed as a two-body layer. In other words, the lithosphere is assumed as a spherical shell. In other words, the lithospheric depth is constant. However, we should note that lithospheric depth is larger or smaller in some places, meaning that the sub-lithospheric contributions to the gravity field is in even lower or higher degrees and orders than 15. And vice versa, over some areas the lithosphere is thin and higher degrees than 15 should be removed from the real data. Therefore, it is better to rewrite Eq. (7.78) in the following form:

$$T(R, \theta, \lambda) - \sum_{n=0}^{N_{\text{Litho}}} \sum_{m=-n}^{n} t_{nm} Y_{nm}(\theta, \lambda) = V^{\text{TBSCI/Iso}}(R, \theta, \lambda). \tag{7.80}$$

N_{Litho} is the degree to which the gravitational signal from the sub-lithosphere comes, e.g. 15, according to Stewart and Watts (1997).

A forward computation process should be done to find an optimum value for $\widetilde{\Theta}$. This means that an interval from 0 to a maximum number for the rigidity or elastic thickness should be considered and divided to smaller specified intervals. Thereafter, all values are inserted into Eq. (7.80) to see which one of them makes the rhs closer to the left-hand side (lhs). That value will be chosen as the optimal value for the elastic thickness or flexural rigidity.

Next, we will explain how Eq. (7.80) and the principle for the chosen optimal value for elastic thickness or rigidity are developed for some selected satellite gravimetry data.

7.4.2 Determination of elastic thickness from satellite altimetry, gradiometry and inter-satellite data

According to Eshagh (2018), the absolute value of the disturbing potential generated from compensated crustal masses and the real disturbing potential excluding the low degrees coming from the sub-lithosphere should be minimised. This principle can be shown mathematically by:

$$\min_{\underset{\Theta}{\sim}}\left|V^{\mathrm{TBSCI/Iso}}(R,\theta,\lambda)-\left(\gamma N(R,\theta,\lambda)-\sum_{n=2}^{N_{\mathrm{Litho}}}\sum_{m=-n}^{n}N_{nm}Y_{nm}(\theta,\lambda)\right)\right|. \tag{7.81}$$

The term N_{nm} is the SHC of geoid height, which should be computed from an Earth gravity model. Therefore, the results will not be solely from satellite altimetry data but from their combination with an independent Earth gravity model from these data.

The isostatic disturbing potential, which is derived from the crustal masses and structures, and gravimetric and flexural isostasy theories, at a point with a geocentric distance r outside the Earth is:

$$t_{nm}^{\mathrm{TBSCI}}\left(\frac{R}{r}\right)^{n+1}=-4\pi Gr\Delta\rho\left(\frac{R-\widetilde{D}_0}{r}\right)^{n+2}\frac{\beta_n^*C_n}{2n+1}\overline{K}_{nm}+\left(\frac{R}{r}\right)^{n+1}\left(v_{nm}^{\mathrm{TB}}+v_{nm}^{\mathrm{S}}+v_{nm}^{\mathrm{C}}+v_{nm}^{\mathrm{I}}\right). \tag{7.82}$$

After performing the spherical harmonic synthesis from both sides of Eq. (7.82), we obtain:

$$T^{\mathrm{TBSCI}}(r,\theta,\lambda) = V^{\mathrm{TBSCI}}(r,\theta,\lambda) - 4\pi G r \Delta\rho \sum_{n=0}^{\infty} \left(\frac{R-\widetilde{D}_0}{r}\right)^{n+2} \frac{\beta_n^*}{2n+1}\left(\frac{n^2(n+1)^2}{R^4 g}\widetilde{\Theta} + \Delta\rho\right)^{-1} \left(\frac{R}{r}\right)^{n+1} \sum_{m=-n}^{n} \overline{K}_{nm} Y_{nm}(\theta,\lambda) = V^{\mathrm{TBSCI/Iso}}(r,\theta,\lambda). \tag{7.83}$$

By applying the operators defined in Eqs. (2.129)–(2.134) for defining the spherical harmonic expansions of the satellite gradiometry data in the LNOF, the crustal/isostatic effect on the gradiometry data can be determined. The elastic thickness or flexural rigidity can be estimated by minimising the following objective function:

$$\min_{\widetilde{\Theta}} \left| V_{ij}^{\mathrm{TBSCI/Iso}}(r,\theta,\lambda) - \left(T_{ij}(r,\theta,\lambda) - \sum_{n=2}^{N_{\mathrm{Litho}}} \sum_{m=-n}^{n} (T_{ij})_{nm} Y_{nm}(\theta,\lambda) \right) \right|. \tag{7.84}$$

The long-wavelength portion of the satellite gradiometry data, which mainly comes from the sub-lithospheric masses, should be removed to enrich the gravitational signal induced by the lithosphere.

To derive an objective function for finding the elastic thickness/rigidity from inter-satellite range rates, the isostatically compensated crustal effects should be computed for each satellite separately and subtracted from each other. In addition, the differences between their gravitational potential can be written as the product of the mean velocity of them and the range rates. It should be noted that the contribution to the sub-lithosphere should be subtracted from the satellite data as well. Mathematically, the objective function for finding the elastic thickness/rigidity is:

$$\min_{\widetilde{\Theta}} \left| V^{\mathrm{TBSCI/Iso}}(r_2,\theta_2,\lambda_2) - V^{\mathrm{TBSCI/Iso}}(r_2,\theta_2,\lambda_2) - v_0\dot{\tilde{\rho}} + \Upsilon + \sum_{n=2}^{N_{\mathrm{Litho}}} \sum_{m=-n}^{n} t_{nm} \left(\left(\frac{R}{r_2}\right)^{n+1} Y_{nm}(\theta_2,\lambda_2) - \left(\frac{R}{r_1}\right)^{n+1} Y_{nm}(\theta_1,\lambda_1) \right) \right|. \tag{7.85}$$

Here, different values of elastic thickness/rigidity are inserted into the objective function to see which one gives the minimum absolute value of the function. Then that value is chosen as the elastic thickness or flexural rigidity.

In determining the elastic thickness or flexural rigidity from the inter-satellite LOS measurements the mathematical form of the objective function is more complicated due to its vector form, which is:

$$\min_{\widetilde{\Theta}} \left| \Sigma^{\mathrm{TBSCI/Iso}} - \ddot{\widetilde{\rho}} + \frac{(\delta\dot{\mathbf{r}}_{12})^2 - \dot{\widetilde{\rho}}^2}{\widetilde{\rho}} + \Sigma + \Xi \right|, \tag{7.86}$$

where

$$\begin{pmatrix} \Sigma^{\mathrm{TBSCI/Iso}} \\ \Xi \end{pmatrix} = \sum_{n=0}^{N_{\max}} \sum_{m=-n}^{n} \begin{pmatrix} t_{nm}^{\mathrm{TBSCI/Iso}} \\ t_{nm} \end{pmatrix} \left[-\frac{n+1}{R} \left(\left(\frac{R}{r_2}\right)^{n+1} \mathbf{X}_{nm}^{(1)}(\theta_2, \lambda_2) - \left(\frac{R}{r_1}\right)^{n+1} \mathbf{X}_{nm}^{(1)}(\theta_1, \lambda_1) \right) + \frac{1}{r_2} \left(\frac{R}{r_2}\right)^{n+1} \mathbf{X}_{nm}^{(2)}(\theta_2, \lambda_2) - \frac{1}{r_1} \left(\frac{R}{r_1}\right)^{n+1} \mathbf{X}_{nm}^{(2)}(\theta_1, \lambda_1) \right] \cdot \mathbf{e}_{\mathrm{LOS}}, \tag{7.87}$$

and $N_{\max}$ will be infinity when the crustal/isostatic effect $\Sigma^{\mathrm{TBSCI/Iso}}$ is computed and N_{Litho} for computing Ξ.

7.4.3 Example: determination of effective elastic thickness from GOCE gradiometry data over Africa

Here, an example of elastic thickness determination from on-orbit GOCE data, which was done by Eshagh and Pitonak (2019), is presented. They used the GOCE level 2 EGG_TRF_2 data product (Gruber et al., 2010) from 4 August 2013 to 30 September 2013 containing satellite gravitational gradients in the LNOF (Gruber et al., 2010). The EGG_TRF_2 data with flags higher than 2 were removed from the next processing. Since the full coverage from 2 months contains 403,301 points, before the outliers were removed, they decided to apply a 10-s data-sampling interval. So, they considered 38,959 measurements of the second-order radial derivative. The long-wavelength portion, from degree 0 to 16 (see Stewart and Watts, 1997), of the gradiometric data, which come mainly from deep mantle, was estimated from TIM-r5 (Brockmann et al., 2014) and subtracted from the GOCE data. The maps of on-orbit GOCE

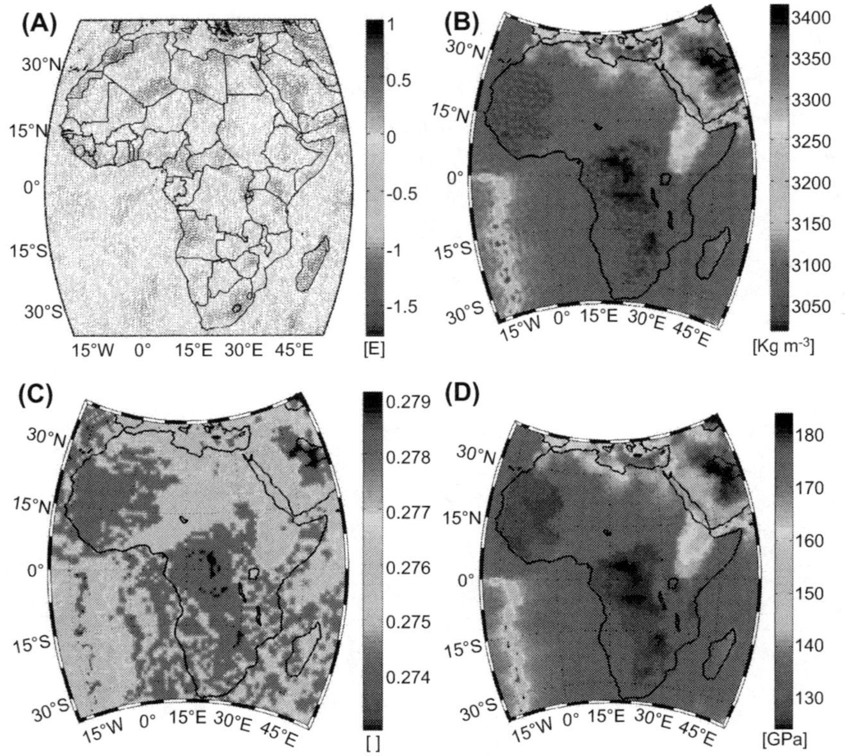

Figure 7.8 (A) On-orbit GOCE gradiometry data (second-order radial derivative of disturbing potential) (unit: E), (B) density of the upper mantle (unit: kg/m^3), (C) Poisson ratio and (D) Young's modulus (unit: GPa) from CRUST1.0. *(From Eshagh, M., Pitonak, M., 2019. Elastic thickness determination from on-orbit GOCE data and CRUST1.0. Pure Appl. Geophys. 176, 685–696.)*

gradiometry data are presented in Fig. 7.8A over the study area, showing that they vary from about −1.5 to 1 E.

They used the CRUST1.0 model for computing the effects of TB, sediment and crustal masses over Africa, and used the Vening Meinesz–Moritz method, approach 1 of gravimetric isostasy, presented in Chapter 6. They applied the flexure model considering the membrane stress, which was discussed in Section 6.4.2. The density of the upper mantle, which is needed for computing the density contrast between the crust and the upper mantle, was taken from the CRUST1.0 model as well; see Fig. 7.8B.

For elastic thickness determination and flexural isostasy, the Poisson ratio and Young's modulus are required. Eshagh and Pitonak (2019) used the code provided by Michael Bevis from Ohio State University, which is available at the CRUST1.0 homepage, to compute these parameters from the S and P waves in addition to the density of the upper mantle. The maps of the Poisson ratio and Young's modulus are illustrated in Fig. 7.8C, D, respectively.

The methodology of estimating the effective T_e from the GOCE satellite measurements is rather simple. Different values of T_e, say from 0 to 110 km in this example, are inserted into the equations to find which one gives the closer gradient to that of GOCE. This process should be repeated for each one of 38,959 GOCE gradient values. An ideal case of finding the local minima of such function for the GOCE data is plotted in Fig. 7.9A.

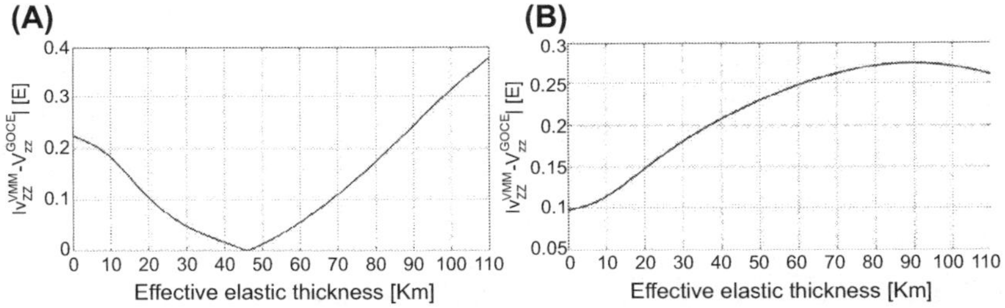

Figure 7.9 (A) Ideal and (B) non-ideal examples of finding the effective elastic thickness as a local minimum of absolute differences between the generated and the measured by the GOCE data. *(From Eshagh, M., Pitonak, M., 2019. Elastic thickness determination from on-orbit GOCE data and CRUST1.0. Pure Appl. Geophys. 176, 685–696.)*

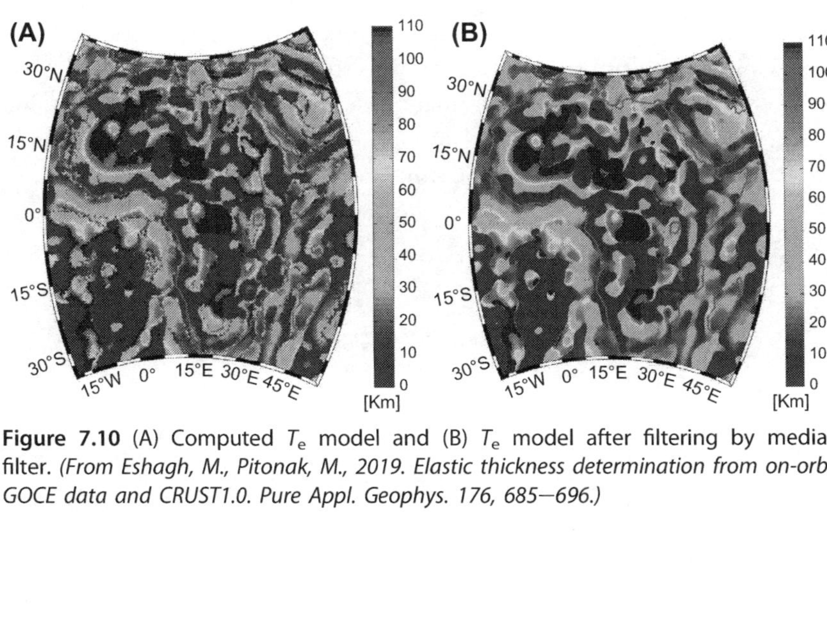

Figure 7.10 (A) Computed T_e model and (B) T_e model after filtering by median filter. *(From Eshagh, M., Pitonak, M., 2019. Elastic thickness determination from on-orbit GOCE data and CRUST1.0. Pure Appl. Geophys. 176, 685–696.)*

However, in some cases this local minimum does not exist; see Fig. 7.9B for where it is not clear whether the real value of T_e is zero or not at this point. In some cases, this absolute difference decays monotonically, meaning that by increasing the value of T_e the difference becomes smaller. Therefore, some a priori information about the range of this thickness is required to limit its maximum value in the search technique.

The map of estimated T_e is presented in Fig. 7.10A (Eshagh and Pitonak, 2019). We can see those artefacts mainly around the coast line in the west and the south. A median filter was applied to decrease the value of noise in the results. The problem was how to set the length of the filter. We decided to apply the median filter with length of 2 arc-deg, which corresponds to approximately 180 km. The results after filtering are plotted in Fig. 7.9B.

Fig. 7.10A, B shows large values along the mid-oceanic ridge in the western part of the area, which cannot be correct. However, over the continents the presented model is in agreement with the models presented so far for Africa.

7.5 Determination of oceanic bathymetry

Satellite altimetry or gravimetry data provide a good coverage of data all over the oceans, where it is not easy to go and perform a sounding process for measuring depth. Satellite data are very useful in this respect, but they have their own challenges and limitations. In this section, the goal is to show that bathymetry depths can be approximated from satellite gravimetry data and isostatic theories. Again, the gravimetric isostatic equilibrium with the simple flexural isostasy will be considered.

Let us use Eq. (7.15), but considering that over ocean, the topographic effect vanishes, as there are no topographic masses at a point with a geocentric distance r outside the Earth:

$$t_{nm}^{\mathrm{BSC}}\left(\frac{R}{r}\right)^{n+1} = v_{nm}^{\mathrm{BSC}}\left(\frac{R}{r}\right)^{n+1} - 4\pi G r \Delta\rho\left(\frac{R-\widetilde{D}_0}{r}\right)^{n+2}\frac{\beta_n^*}{2n+1}C_n\overline{K}_{nm}, \quad (7.88)$$

where

$$v_{nm}^{\mathrm{BSC}} = v_{nm}^{\mathrm{B}} + v_{nm}^{\mathrm{S}} + v_{nm}^{\mathrm{C}}, \quad (7.89)$$

$$\overline{K}_{nm} = (\overline{\rho}d)_{nm} + (\rho^{S}d_{S})_{nm} + (\rho^{C}d_{C})_{nm}. \tag{7.90}$$

Therefore, Eq. (7.88) can be written in the following expanded form:

$$\begin{aligned} t_{nm}^{\mathrm{BSCI}}\left(\frac{R}{r}\right)^{n+1} &= v_{nm}^{\mathrm{B}}\left(\frac{R}{r}\right)^{n+1} + \left(v_{nm}^{\mathrm{S}} + v_{nm}^{\mathrm{C}}\right)\left(\frac{R}{r}\right)^{n+1} \\ &\quad - 4\pi Gr\Delta\rho\left(\frac{R-\widetilde{D}_0}{r}\right)^{n+2}\frac{\beta_n^{*}}{2n+1}C_n \\ &\quad \left((\overline{\rho}d)_{nm} + (\rho^{S}d_{S})_{nm} + (\rho^{C}d_{C})_{nm}\right). \end{aligned} \tag{7.91}$$

Our goal for rewriting Eq. (7.88) is to show the known and unknown parameters and separate them from one another. Assume that the lhs of Eq. (7.91) is estimable from satellite altimetry data, and all SHCs presented in Eq. (7.91) are known except for v_{nm}^{B}, and all other parameters are known except d. After rearranging Eq. (7.91) and taking all known parameters to the lhs and all unknown to the rhs, we obtain:

$$\begin{aligned} & t_{nm}^{\mathrm{BSC}}\left(\frac{R}{r}\right)^{n+1} - \left(\left(v_{nm}^{\mathrm{S}} + v_{nm}^{\mathrm{C}}\right) + 4\pi Gr\Delta\rho\left(\frac{R-\widetilde{D}_0}{r}\right)^{n+2}\right. \\ & \qquad \left.\frac{\beta_n^{*}}{2n+1}C_n\left((\rho^{S}d_{S})_{nm} + (\rho^{C}d_{C})_{nm}\right)\right)\left(\frac{R}{r}\right)^{n+1} \\ & = v_{nm}^{\mathrm{B}}\left(\frac{R}{r}\right)^{n+1} - 4\pi Gr\Delta\rho\left(\frac{R-\widetilde{D}_0}{r}\right)^{n+2}\frac{\beta_n^{*}}{2n+1}C_n(\overline{\rho}d)_{nm}. \end{aligned} \tag{7.92}$$

Eq. (7.92) is the core equation of gravimetric bathymetry.

7.5.1 Mathematical foundations

Here, two approaches are presented for determining bathymetry depth. The first deals with determination of the depth directly and in the second a constrained solution to the mean depth is provided if it is available. However, the mathematical basis of both cases is Eq. (7.92). In this section, the idea of bathymetry from some selected satellite gravimetry data is theoretically presented and discussed.

7.5.1.1 Direct linear estimation of bathymetry depth

The lhs of Eq. (7.92) contains the gravitational potential of sediments and crustal crystalline with their isostatic compensations. The rhs is the SHC bathymetry potential and its isostatic compensation. To shorten the mathematical derivations let us write Eq. (7.92) in the following short form:

$$
\begin{aligned}
& t_{nm}^{\mathrm{BSC}}\left(\frac{R}{r}\right)^{n+1}-\left(v_{nm}^{\mathrm{S/Iso}}+v_{nm}^{\mathrm{C/Iso}}\right)\left(\frac{R}{r}\right)^{n+1} \\
& \quad = v_{nm}^{\mathrm{B}}\left(\frac{R}{r}\right)^{n+1}-4\pi Gr\Delta\rho\left(\frac{R-\widetilde{D}_0}{r}\right)^{n+2}\frac{\beta_n^*}{2n+1}C_n(\overline{\rho}d)_{nm},
\end{aligned} \tag{7.93}
$$

where $v_{nm}^{\mathrm{S/Iso}}$ and $v_{nm}^{\mathrm{C/Iso}}$ are, respectively, the SHCs of the isostatically reduced gravitational potential of sediment and crustal crystalline masses.

The gravitational potential of the bathymetric masses has been modelled and presented in Chapter 5. Eq. (5.31) shows the SHCs of the bathymetric masses. If it is truncated to the first term, the SHCs of the bathymetric potential will be:

$$
v_{nm}^{\mathrm{B}}=\frac{4\pi GR\overline{\rho}}{2n+1}d_{nm}. \tag{7.94}
$$

Once the density contrast of water and the basement is considered constant, by inserting Eq. (7.94) into Eq. (7.93) and performing further simplifications, we obtain:

$$
\begin{aligned}
& t_{nm}^{\mathrm{BSC}}\left(\frac{R}{r}\right)^{n+1}-\left(v_{nm}^{\mathrm{S/Iso}}+v_{nm}^{\mathrm{C/Iso}}\right)\left(\frac{R}{r}\right)^{n+1}=\frac{4\pi G\overline{\rho}R}{2n+1}\left(\frac{R}{r}\right)^{n+1} \\
& \quad \left(1-R\Delta\rho\left(\frac{R-\widetilde{D}_0}{R}\right)^{n+2}\beta_n^*C_n\right)d_{nm}.
\end{aligned} \tag{7.95}
$$

Now, we solve Eq. (7.95) for d_{nm}:

$$
\begin{aligned}
& d_{nm}=\frac{2n+1}{4\pi G\overline{\rho}R}\Gamma_n^{-1}\left(t_{nm}^{\mathrm{BSC}}-v_{nm}^{\mathrm{S/Iso}}-v_{nm}^{\mathrm{C/Iso}}\right)\quad \text{where} \\
& \Gamma_n=1-R\Delta\rho\left(\frac{R-\widetilde{D}_0}{R}\right)^{n+2}\beta_n^*C_n.
\end{aligned} \tag{7.96}
$$

The spherical harmonic synthesis of Eq. (7.96) leads to:

$$
d=\frac{-1}{4\pi G\overline{\rho}R}\sum_{n=0}^{\infty}(2n+1)\Gamma_n^{-1}\sum_{m=-n}^{n}\left(v_{nm}^{\mathrm{S/Iso}}+v_{nm}^{\mathrm{C/Iso}}\right)Y_{nm}(\theta,\lambda)+W_d, \tag{7.97}
$$

where the Laplace coefficient of W_d multiplied by $\left(\frac{R}{r}\right)^{n+1}$ is:

$$\left(\frac{R}{r}\right)^{n+1}(W_d)_n=\frac{2n+1}{4\pi G\overline{\rho}R}\left(\frac{R}{r}\right)^{n+1}\Gamma_n^{-1}t_n^{\mathrm{BSC}}. \tag{7.98}$$

Now, we solve Eq. (7.98) for t_n^{BSC}:

$$t_n^{\mathrm{BSC}}\left(\frac{R}{r}\right)^{n+1}=\frac{4\pi G\overline{\rho}R}{(2n+1)\Gamma_n^{-1}}\left(\frac{R}{r}\right)^{n+1}(W_d)_n. \tag{7.99}$$

According to Eq. (1.11), $(W_d)_n$ can be written in integral form and the lhs as the disturbing potential excluding the contribution from the sub-lithosphere:

$$\begin{aligned}T(r,\theta,\lambda)-\sum_{n=0}^{N_{\mathrm{Litho}}}\left(\frac{R}{r}\right)^{n+1}t_n&=T^{\mathrm{BSC}}(r,\theta,\lambda)\\&=G\overline{\rho}R\iint_{\sigma}W_d'\sum_{n=0}^{\infty}\left(\frac{R}{r}\right)^{n+1}\Gamma_nP_n(\widetilde{x})\mathrm{d}\sigma.\end{aligned} \tag{7.100}$$

Eq. (7.100) is an integral equation which should be solved numerically by one of the methods presented in Chapter 4. Once W_d is estimated, it can be inserted into Eq. (7.97) for computing the ocean depth.

One important parameter for bathymetry based on this approach is the elastic thickness over oceans. The oceanic lithosphere is rather weak and very much influenced by the thermal state of the lithosphere/asthenosphere. Therefore, determination of the elastic thickness over oceans is not an easy task and there are mechanisms other than isostasy for keeping the lithosphere in equilibrium. Ramillien and Cazenave (1997) applied the mathematical model presented for elastic thickness based on the classical expression for the half-space cooling model of Carslaw and Jaeger (1959) for determining the global bathymetry from the satellite altimetry data of ERS-1. This model has the following mathematical form:

$$T_e=2\sqrt{kt}\,\mathrm{erf}^{-1}\left(\frac{\widetilde{T}}{\widetilde{T}_m}\right), \tag{7.101}$$

where k is the thermal diffusivity and t is the plate age, and $\widetilde{T}$ and $\widetilde{T}_m$ are, respectively, the lithospheric isotherm ranging from 450 to 600°C (Watts,

1978; Calmant et al., 1990) and the mantle temperature, approximately 1250°C (Ramillien and Cazenave, 1997). erf stands for the inverse error function.

7.5.1.2 Constrained solution to mean depth

If the mean depth of the ocean is available from an external source, the mathematical formulae can be constrained to it and the approximation error in formulation will be reduced. Let us start with the gravitational potential of the isostatically reduced bathymetry masses as it is used in isostasy. This can be written directly from Eq. (7.93), but we write the rhs of this equation in the spatial form rather than spectral. In this case, the SHCs of the bathymetric masses can be replaced by a Newton integral, and the second term, coming from the compensation of the bathymetric masses as in terms of Laplace coefficients of the depth:

$$V^{\mathrm{B/Iso}}(r,\theta,\lambda) = G\overline{\rho}\iint\limits_{\sigma}\int\limits_{R-d_0-\Delta d'}^{R}\frac{r'^2\mathrm{d}r'\mathrm{d}\sigma}{l} - 4\pi G r\overline{\rho}\Delta\rho \sum_{n=0}^{\infty}\left(\frac{R-\widetilde{D}_0}{r}\right)^{n+2}\frac{\overset{*}{\beta}_n C_n}{2n+1}\left(d_0\delta_{n0}-\Delta d_n\right). \tag{7.102}$$

By looking at the radial integral limit in Eq. (7.102), we can see that the lower limit contains d_0, which stands for mean depth, and Δd, the variation of depth from d_0. The Laplace coefficient $d_n = d_0\delta_{n0} + \Delta d_n$ in the second term, meaning that it is written in terms of the mean depth and departure from it. Now, the question is how to solve Eq. (7.102) for Δd_n, assuming the lhs of the equation is estimable from satellite gravimetry data, and the effects of sediments and crustal crystalline masses on them.

To begin with, let us name the first term on the rhs of Eq. (7.102) I_1; in this case we can write this integral as a summation of two integrals:

$$I_1 = G\overline{\rho}\iint\limits_{\sigma}\int\limits_{R-d_0}^{R}\frac{r'^2\mathrm{d}r'\mathrm{d}\sigma}{l} + G\overline{\rho}\iint\limits_{\sigma}\int\limits_{R-d_0-\Delta d'}^{R-d_0}\frac{r'^2\mathrm{d}r'\mathrm{d}\sigma}{l} \tag{7.103}$$

where l stands for the distance between computation and integration points. The first integral gives the gravitational potential of the bathymetric masses having a spherical shell form with thickness d_0, and all parameters of this term are known. The second term contains $\Delta d'$, which is the unknown and should be determined. After solving the first integral, and the radial one in the second, and after using the Legendre expansion of l^{-1}, Eq. (7.103) changes to:

$$I_1 = 4\pi G\overline{\rho}\frac{R^2}{3r}\left(1-\left(1-\frac{d_0}{R}\right)^3\right) + 4\pi G r^2\overline{\rho}\iint_{\sigma}\sum_{n=0}^{\infty}\left(\frac{R-d_0}{r}\right)^{n+3}\frac{1}{n+3}\left(1-\left(1-\frac{\Delta d'}{R-d_0}\right)^{n+3}\right)P_n(\widetilde{x})\mathrm{d}\sigma. \tag{7.104}$$

After expansion of the binomial term in the second term of Eq. (7.104) up to linear term by the Taylor series, and application of the Legendre integral in Eq. (1.11), this term can be written in terms of Δd_n:

$$I_1 = 4\pi G\overline{\rho}\frac{R^2}{3r}\left(1-\left(1-\frac{d_0}{R}\right)^3\right) + 4\pi G r\overline{\rho}\sum_{n=0}^{\infty}\left(\frac{R-d_0}{r}\right)^{n+2}\frac{\Delta d_n}{2n+1}. \tag{7.105}$$

Now, we simplify the second term of Eq. (7.102) and separate the terms containing d_0 and Δd. Let us name this term I_2; then:

$$I_2 = -4\pi G\overline{\rho}\Delta\rho\frac{R^2}{r}\left(\frac{R-\widetilde{D}_0}{R}\right)^2\beta_0^* C_0 d_0 + 4\pi G r\overline{\rho}\Delta\rho\sum_{n=0}^{\infty}\left(\frac{R-\widetilde{D}_0}{r}\right)^{n+2}\frac{\beta_n^* C_n}{2n+1}\Delta d_n. \tag{7.106}$$

Let us write Eq. (7.105) and Eq. (7.106) in spectral form and add them together; then the results will be:

$$v_n^{\mathrm{B/Iso}} = (I_1)_n + (I_2)_n = 4\pi G\overline{\rho}R^2 A(r)\delta_{n0} + 4\pi G\overline{\rho}B_n(r)\frac{\Delta d_n}{2n+1}, \tag{7.107}$$

where

$$A(r) = \frac{1}{3r}\left(1-\left(1-\frac{d_0}{R}\right)^3\right) - \Delta\rho\frac{1}{r}\left(\frac{R-\widetilde{D}_0}{R}\right)^2\beta_0^* C_0 d_0, \tag{7.108}$$

$$B_n(r) = r\left(\frac{R-d_0}{r}\right)^{n+2} + r\Delta\rho\left(\frac{R-\widetilde{D}_0}{r}\right)^{n+2}\beta_n^* C_n. \tag{7.109}$$

The solution of Eq. (7.107) for Δd_n will be:

$$\Delta d_n = (2n+1)B_n^{-1}(r)\left(\frac{v_n^{\mathrm{B/Iso}}}{4\pi G\overline{\rho}} - R^2A(r)\delta_{n0}\right). \tag{7.110}$$

After taking summation on n we obtain:

$$\Delta d = -B_0^{-1}(r)A(r)R^2 - \frac{1}{4\pi G\overline{\rho}}\sum_{n=0}^{\infty}(2n+1)B_n^{-1}(r)\left(v_n^{\mathrm{S/Iso}} + v_n^{\mathrm{C/Iso}}\right) + W_{\Delta d}. \tag{7.111}$$

The Laplace coefficient of $W_{\Delta d}$ and its solution for t_n^{BSC} are:

$$(W_{\Delta d})_n = \frac{(2n+1)B_n^{-1}(r)t_n^{\mathrm{BSCI}}}{4\pi G\overline{\rho}} \quad \text{or} \quad t_n^{\mathrm{BSC}} = \frac{4\pi G\overline{\rho}(W_{\Delta d})_n}{(2n+1)B_n^{-1}(r)} \tag{7.112}$$

By writing the second equation of Eq. (7.112) in integral form we get:

$$\begin{aligned} T^{\mathrm{BSC}}(r,\theta,\lambda) &= T(r,\theta,\lambda) - \sum_{n=0}^{N_{\mathrm{Litho}}}\left(\frac{R}{r}\right)^{n+1} t_n \\ &= -G\overline{\rho}\iint_{\sigma} W'_{\Delta d}\sum_{n=0}^{\infty} B_n(r)P_n(\widetilde{x})\mathrm{d}\sigma. \end{aligned} \tag{7.113}$$

The integral in Eq. (7.113) should be solved numerically for $W_{\Delta d}$, and the result is inserted into Eq. (7.111) to determine the depth variation from d_0.

7.5.2 Bathymetry from satellite altimetry, gradiometry and inter-satellite data

Eq. (7.113) is an integral equation, which can be inverted numerically for estimating $W'_{\Delta d}$ from the disturbing potential at a point outside the Earth induced by the bathymetric, sediment and crustal crystalline masses and their corresponding isostatic reduction. In the case of using the satellite altimetry-derived geoid heights, we can simply replace r with R, and therefore, Eq. (7.111) will change to:

$$\begin{aligned} \Delta d = &-B_0^{-1}(R)A(R)R^2 - \frac{1}{4\pi G\overline{\rho}}\sum_{n=0}^{\infty}(2n+1)B_n^{-1}(R) \\ &\sum_{m=-n}^{n} v_{nm}^{\mathrm{SC/Iso}}Y_{nm}(\theta,\lambda) + W_{\Delta d}, \end{aligned} \tag{7.114}$$

where $\nu_{nm}^{\mathrm{SC/Iso}} = \nu_{nm}^{\mathrm{S/Iso}} + \nu_{nm}^{\mathrm{C/Iso}}$ and $W_{\Delta d}$ can be estimated by solving the following integral equation:

$$\gamma N - \gamma \sum_{n=0}^{N_{\mathrm{Litho}}} N_n = T^{\mathrm{BSC/Iso}} = G\overline{\rho} \iint_{\sigma} W'_{\Delta d} \sum_{n=0}^{\infty} B_n(R) P_n(\widetilde{x}) \mathrm{d}\sigma. \quad (7.115)$$

In Eq. (7.115), $T^{\mathrm{BSC/Iso}}$ is replaced by $\gamma N - \gamma \sum_{n=0}^{N_{\mathrm{Litho}}} N_n$; this means that the satellite altimetry-geoid height without low frequencies coming from sub-lithosphere.

The background theory for bathymetry has been presented before; here, we present just the final formulae for recovering $W_{\Delta d}$. Here, Eq. (7.114) is our main mathematical formula for bathymetry and we try to write the integral equations for inverting the satellite gradiometry data for determining $W_{\Delta d}$. Since such data are measured at satellite altitude, we must use Eq. (7.115), which is the relation between the disturbing potential induced by the sediment and crustal crystalline mass and their isostatic compensation. If we apply the partial derivative operators that we applied for expressing the gradiometry data, in Section 1.4.2, Eqs. (1.95)–(1.100), we reach:

$$G\overline{\rho} \iint_{\sigma} W'_{\Delta d} B^*_{ij}\left(r, d_0, \widetilde{D}_0, \psi\right) \mathrm{d}\sigma \approx T_{ij}(r, \theta, \lambda) - \sum_{n=0}^{N_{\mathrm{Litho}}} T_{ij,n}(r, \theta, \lambda), \quad (7.116)$$

where

$$\begin{bmatrix} B^*_{xx}\left(r, d_0, \widetilde{D}_0, \psi\right) \\ B^*_{yy}\left(r, d_0, \widetilde{D}_0, \psi\right) \end{bmatrix} = \frac{B^*_r\left(r, d_0, \widetilde{D}_0, \psi\right)}{r} + \frac{1}{2r^2}\left\{\left[B^*_{\psi\psi}\left(r, d_0, \widetilde{D}_0, \psi\right) + \cot\psi B^*_{\psi}\left(r, d_0, \widetilde{D}_0, \psi\right)\right] \pm \left[B^*_{\psi\psi}\left(r, d_0, \widetilde{D}_0, \psi\right) - \cot\psi B^*_{\psi}\left(r, d_0, \widetilde{D}_0, \psi\right)\right]\cos 2\alpha\right\}, \quad (7.117)$$

$$\begin{bmatrix} B^*_{xz}\left(r, d_0, \widetilde{D}_0, \psi\right) \\ B^*_{yz}\left(r, d_0, \widetilde{D}_0, \psi\right) \end{bmatrix} = \frac{1}{r}\left[\frac{1}{r} B^*_{\psi}\left(r, d_0, \widetilde{D}_0, \psi\right) - B^*_{r\psi}\left(r, d_0, \widetilde{D}_0, \psi\right)\right]\begin{pmatrix} \cos\alpha \\ -\sin\alpha \end{pmatrix}, \quad (7.118)$$

$$B^*_{xy}\left(r, d_0, \widetilde{D}_0, \psi\right) = -\frac{1}{2r^2}\left[B^*_{\psi\psi}\left(r, d_0, \widetilde{D}_0, \psi\right) + \cot\psi B^*_{\psi}\left(r, d_0, \widetilde{D}_0, \psi\right)\right]\sin 2\alpha, \quad (7.119)$$

and

$$B^*\left(r,\widetilde{D}_0,\Delta\widetilde{D},\psi\right)=\sum_{n=0}^{\infty}B_n(r)P_n(\widetilde{x}), \tag{7.120}$$

$$B_r^*\left(r,\widetilde{D}_0,\Delta\widetilde{D},\psi\right)=\sum_{n=0}^{\infty}\frac{\partial B_n(r)}{\partial r}P_n(\widetilde{x}), \tag{7.121}$$

$$B_{zz=rr}^*\left(r,\widetilde{D}_0,\Delta\widetilde{D},\psi\right)=\sum_{n=0}^{\infty}\frac{\partial^2 B_n(r)}{\partial r^2}P_n(\widetilde{x}), \tag{7.122}$$

$$B_{r\psi}^*\left(r,\widetilde{D}_0,\Delta\widetilde{D},\psi\right)=\sum_{n=0}^{\infty}\frac{\partial B_n(r)}{\partial r}\frac{\partial P_n(\widetilde{x})}{\partial\psi}, \tag{7.123}$$

$$B_{\psi\psi}^*\left(r,\widetilde{D}_0,\Delta\widetilde{D},\psi\right)=\sum_{n=0}^{\infty}B_n(r)\frac{\mathrm{d}^2 P_n(\widetilde{x})}{\mathrm{d}\psi^2}. \tag{7.124}$$

The effect of crustal masses can be computed for both satellites 1 and 2 according to their positions. The difference between the disturbing potentials at both satellites will be replaced by $v_0\dot{\widetilde{\rho}}-\Upsilon$; therefore, the following integral equation can be constructed for determining the density contrast:

$$\begin{aligned}&G\overline{\rho}\iint_{\sigma}W'_{\Delta d}\left(B^*\left(r_2,d_0,\widetilde{D}_0,\psi_2\right)-B^*\left(r_1,d_0,\widetilde{D}_0,\psi_1\right)\right)\mathrm{d}\sigma\\&=v_0\dot{\widetilde{\rho}}-\Upsilon-\sum_{n=2}^{N_{\mathrm{Litho}}}\left(t_n(r_2,\theta_2,\lambda_2)-t_n(r_1,\theta_1,\lambda_1)\right).\end{aligned} \tag{7.125}$$

The integral equation for simultaneous inversion of the crustal mass variation effects on the inter-satellite LOS measurements is:

$$G\overline{\rho}\iint_{\sigma}W'_{\Delta d}B^{*B}\left(r_1,\psi_1,\alpha_1,r_2,\psi_2,\alpha_2,d_0,\widetilde{D}_0\right)\mathrm{d}\sigma=\ddot{\widetilde{\rho}}-\frac{\left(\delta\dot{\mathbf{r}}_{12}\right)^2-\dot{\widetilde{\rho}}^2}{\widetilde{\rho}}-\Xi, \tag{7.126}$$

and if only inversion of the inter-satellite LOS data is desired, Eq. (7.70) should be applied, but $W_{\Delta\rho}$, or the contribution of satellite data to density contrast, is estimated from these data through the following integral equation:

$$B^{*B}\left(r_1, \psi_1, \alpha_1, r_2, \psi_2, \alpha_2, d_0, \widetilde{D}_0\right) = \left(B^*_{r_2}\left(r_2, d_0, \widetilde{D}_0, \psi_2\right)\mathbf{e}_{z_2} + \frac{1}{r_2}B^*_{\psi_2}\left(r_2, d_0, \widetilde{D}_0, \psi_2\right)\right.$$
$$\left.(\cos\alpha_2\mathbf{e}_{x_2} - \sin\alpha_2\mathbf{e}_{y_2})\right)\cdot\mathbf{e}_{\mathrm{LOS}}$$
$$-\left(B^*_{r_1}\left(r_1, d_0, \widetilde{D}_0, \psi_1\right)\mathbf{e}_{z_1} + \frac{1}{r_1}B^*_{\psi_1}\left(r_1, d_0, \widetilde{D}_0, \psi_1\right)\right.$$
$$\left.(\cos\alpha_1\mathbf{e}_{x_1} - \sin\alpha_1\mathbf{e}_{\lambda_1})\right)\cdot\mathbf{e}_{\mathrm{LOS}}. \tag{7.127}$$

The only difference between the kernel Eq. (7.76) and the kernel Eq. (7.53) is related to the presence of $\Delta\widetilde{D}$ inside the kernel Eq. (7.76); in other words, this kernel is a function of $\Delta\widetilde{D}$ in addition to other parameters.

7.6 Continental ice thickness determination

The combination of gravimetric and flexural isostasy can be used for determining the disturbing potential from topography, sediment and crustal consolidated crystalline and the isostatic compensation. In this section, we present the mathematical foundation for the idea behind the approach and after that develop it further based on the selected satellite gravimetry data, except for satellite altimetry, as the goal is the ice thickness determination over continents and not oceans.

7.6.1 Mathematical foundation

The SHCs of the disturbing potential induced by crustal masses at a point with geocentric distance r over continents is:

$$t^{\mathrm{TSCI}}_{nm}\left(\frac{R}{r}\right)^{n+1} = v^{\mathrm{TSCI}}_{nm}\left(\frac{R}{r}\right)^{n+1} - 4\pi G\left(R - \widetilde{D}_0\right)\Delta\rho\left(\frac{R - \widetilde{D}_0}{r}\right)^{n+1}$$
$$\frac{\beta^*_n}{2n+1}C_n\overline{K}_{nm}, \tag{7.128}$$

where

$$v^{\mathrm{TSCI}}_{nm} = v^{\mathrm{T}}_{nm} + v^{\mathrm{S}}_{nm} + v^{\mathrm{C}}_{nm} + v^{\mathrm{I}}_{nm}, \tag{7.129}$$

$$\overline{K}_{nm} = (\overline{\rho}d)_{nm} + (\rho^{\mathrm{S}}d_{\mathrm{S}})_{nm} + (\rho^{\mathrm{C}}d_{\mathrm{C}})_{nm} + (\rho^{\mathrm{I}}d_{\mathrm{I}})_{nm}. \tag{7.130}$$

The unknown parameter in Eq. (7.128) is d_{I}, which should be derived. To do so, we separate the terms including this parameter, namely:

$$
\begin{aligned}
& t_{nm}^{\mathrm{TSC/Iso}}\left(\frac{R}{r}\right)^{n+1} - v_{nm}^{\mathrm{TSC}}\left(\frac{R}{r}\right)^{n+1} + 4\pi Gr\Delta\rho\left(\frac{R-\widetilde{D}_0}{r}\right)^{n+2}\frac{\beta_n^* C_n}{2n+1}\left((\overline{\rho}d)_{nm}\right. \\
& \quad \left. + (\rho^{\mathrm{S}} d_{\mathrm{S}})_{nm} + (\rho^{\mathrm{C}} d_{\mathrm{C}})_{nm}\right) \\
& \quad = v_{nm}^{\mathrm{I}}\left(\frac{R}{r}\right)^{n+1} - 4\pi Gr\Delta\rho\left(\frac{R-\widetilde{D}_0}{r}\right)^{n+2}\frac{\beta_n^*}{2n+1}C_n(\rho^{\mathrm{I}} d_{\mathrm{I}})_{nm}.
\end{aligned}
\tag{7.131}
$$

After performing the spherical harmonic synthesis of Eq. (7.131), and after that writing the series in terms of the Laplace coefficients; see Eq. (1.11), we obtain:

$$
\begin{aligned}
& T^{\mathrm{TSC/Iso}}(r,\theta,\lambda) - V^{\mathrm{TSC}}(r,\theta,\lambda) \\
& \quad + 4\pi Gr\Delta\rho\sum_{n=0}^{\infty}\left(\frac{R-\widetilde{D}_0}{r}\right)^{n+2}\frac{\beta_n^* C_n}{2n+1}\left((\overline{\rho}d)_n + (\rho^{\mathrm{S}} d_{\mathrm{S}})_n + (\rho^{\mathrm{C}} d_{\mathrm{C}})_n\right) \\
& \quad = V^{\mathrm{I}}(r,\theta,\lambda) - 4\pi Gr\Delta\rho\sum_{n=0}^{\infty}\left(\frac{R-\widetilde{D}_0}{r}\right)^{n+2}\frac{\beta_n^* C_n}{2n+1}(\Delta\rho^{\mathrm{I}} d_{\mathrm{I}})_n.
\end{aligned}
\tag{7.132}
$$

The gravitational potential of ice, which is the first term on the rhs of Eq. (7.132), can be written as:

$$
\begin{aligned}
V^{\mathrm{I}}(r,\theta,\lambda) &= G\Delta\rho^{\mathrm{I}}\iint_{\sigma}\int_{R+H'-d_{\mathrm{I}}'}^{R+H'}\frac{r'^2\,\mathrm{d}r'\,\mathrm{d}\sigma}{l} \\
&= G\Delta\rho^{\mathrm{I}}\iint_{\sigma}\sum_{n=0}^{\infty}\frac{P_n(\widetilde{x})(R+H')^{n+3}}{(r^{n+1})(n+3)}\left[1-\left(1-\frac{d_{\mathrm{I}}'}{R+H'}\right)^{n+3}\right]\mathrm{d}\sigma,
\end{aligned}
\tag{7.133}
$$

where $\Delta\rho^{\mathrm{I}}$ is the density contrast between the ice and the upper crust density. Since d_{I}' is practically less than 5 km and significantly smaller than $R+H'$, the binomial term inside the bracket on the rhs of Eq. (7.133) can be simply approximated by the Taylor series up to the linear term. After linearisation and some simplifications, Eq. (7.133) changes to:

$$
V^{\mathrm{I}}(r,\theta,\lambda) \approx G\Delta\rho^{\mathrm{I}} r\iint_{\sigma} d_{\mathrm{I}}'\sum_{n=0}^{\infty}\left(\frac{R+H'}{r}\right)^{n+2}P_n(\widetilde{x})\,\mathrm{d}\sigma. \tag{7.134}
$$

This is an integral equation with respect to d_{I}'. Let us now rewrite Eq. (7.132) according to Eq. (7.134):

$$T^{\mathrm{TSC/Iso}}(r,\theta,\lambda) - V^{\mathrm{TSC/Iso}}(r,\theta,\lambda) = G\Delta\rho^{\mathrm{I}} r \iint_{\sigma} d_{\mathrm{I}}' \sum_{n=0}^{\infty} \left(\frac{R+H}{r}\right)^{n+2} P_n(\widetilde{x})\mathrm{d}\sigma - 4\pi G r \Delta\rho \sum_{n=0}^{\infty} \left(\frac{R-\widetilde{D}_0}{r}\right)^{n+2} \frac{\beta_n^*}{2n+1} C_n\left(\Delta\rho^{\mathrm{I}} d_{\mathrm{I}}\right)_n. \tag{7.135}$$

Now, we write the second term on the rhs in integral form according to Eq. (1.11). In this case, we obtain a new integral equation for solving d_{I}':

$$G\Delta\rho^{\mathrm{I}} \iint_{\sigma} d_{\mathrm{I}}' L\left(r,\widetilde{D}_0,\psi\right)\mathrm{d}\sigma \approx T^{\mathrm{TSC/Iso}}(r,\theta,\lambda) - V^{\mathrm{TSC/Iso}}(r,\theta,\lambda), \tag{7.136}$$

where

$$L\left(r,\widetilde{D}_0,\psi\right) = \sum_{n=0}^{\infty} \left(r\left(\frac{R+H}{r}\right)^{n+2} - r\Delta\rho'\left(\frac{R-\widetilde{D}_0}{r}\right)^{n+2} \beta_n^* C_n \right) P_n(\widetilde{x}). \tag{7.137}$$

The kernel Eq. (7.137) is a function of topographic heights and the density contrast between the crust and the mantle in addition to the elastic thickness. On the other hand,

$$T^{\mathrm{TSC/Iso}}(r,\theta,\lambda) = T(r,\theta,\lambda) - \sum_{n=0}^{N_{\mathrm{Litho}}} t_n(r,\theta,\lambda), \tag{7.138}$$

meaning that the long wavelength of the disturbing potential needs to be removed from the disturbing potential.

7.6.2 Continental ice from satellite gradiometry and inter-satellite data

To obtain the mathematical models for inverting satellite gravimetry data, the contribution of the sub-lithosphere should be removed from the data. Again, a gravity field up to degree and order N_{Litho} can be used for generating them, using the spherical harmonic expansion of the satellite gravimetry data, and removed from the real data. The isostatically reduced effects of sediment, topography and crustal crystalline masses can be reduced from the data. In this case, the satellite data will be even smoother and the isostatically reduced gravitational potential of the ice will be highlighted. The SHCs of the effects can be inserted into the spherical harmonic expansions of the satellite gravimetry data for computing the effects on them.

The integral Eq. (7.136) relates the ice thickness to the gravitational potential at a point outside the Earth. It can be converted to the satellite gravimetry data by applying the required mathematical operator to the integral equation. For example, for satellite gradiometry data, the integral equation will be:

$$G\Delta\rho^{\mathrm{I}} \iint_{\sigma} d_{\mathrm{I}}' L_{ij}\left(r, \widetilde{D}_0, \psi\right) \mathrm{d}\sigma \approx T_{ij}(r,\theta,\lambda) - \sum_{n=0}^{N_{\mathrm{Litho}}} T_{ij,n}(r,\theta,\lambda) - V_{ij,n}^{\mathrm{TBC/Iso}}(r,\theta,\lambda). \tag{7.139}$$

The kernel of these integral equations has very similar structure to those presented in Eqs. (7.117)–(7.124), but the kernel $B^*\left(r, d_0, \widetilde{D}_0, \psi\right)$ should be replaced by $L\left(r, \widetilde{D}_0, \psi\right)$, presented in Eq. (7.137).

Similarly, an integral equation can be derived for recovering the ice thickness from the inter-satellite range rates:

$$\begin{aligned} G\Delta\rho^{\mathrm{I}} \iint_{\sigma} d_{\mathrm{I}}'\left(L\left(r_2, \widetilde{D}_0, \psi_2\right) - L\left(r_1, \widetilde{D}_0, \psi_1\right)\right) \mathrm{d}\sigma &= v_0 \dot{\tilde{\rho}} \\ &- \Upsilon - \sum_{n=0}^{\infty}\left(v_n^{\mathrm{TSC/Iso}}(r_2,\theta_2,\lambda_2) - v_n^{\mathrm{TSC/Iso}}(r_1,\theta_1,\lambda_1)\right) \\ &- \sum_{n=0}^{N_{\mathrm{Litho}}}\left(t_n(r_2,\theta_2,\lambda_2) - t_n(r_1,\theta_1,\lambda_1)\right), \end{aligned} \tag{7.140}$$

and finally, the integral equation for recovering the ice thickness from the inter-satellite LOS measurements:

$$
\begin{aligned}
G\Delta\rho^{\mathrm{I}} \iint_{\sigma} d_{\mathrm{I}}' L^{*B}\left(r_1, \psi_1, \alpha_1, r_2, \psi_2, \alpha_2, \widetilde{D}_0\right) \mathrm{d}\sigma \\
= \ddot{\widetilde{\rho}} - \frac{\left(\delta\dot{\mathbf{r}}_{12}\right)^2 - \dot{\widetilde{\rho}}^2}{\widetilde{\rho}} - \Xi - \Sigma^{\mathrm{TSC/Iso}},
\end{aligned}
\tag{7.141}
$$

where $\Sigma^{\mathrm{TSC/I}}$ is the isostatically reduced effects of topography, sediment and crustal crystalline masses on the inter-satellite LOS data. It can be computed by inserting the SHCs of the potential of these masses into the spherical harmonic expansion of the inter-satellite LOS data, in other words, by replacing $v_{nm}^{\mathrm{TBSCI/Iso}}$ with $v_{nm}^{\mathrm{TSC/Iso}}$ in Eq. (7.87). $L^{*\mathrm{B}}\left(r_1, \psi_1, \alpha_1, r_2, \psi_2, \alpha_2, \widetilde{D}_0\right)$ is a bipolar kernel and has a form similar to that of the kernel presented in Eq. (7.53), but $M\left(r_1, \widetilde{D}_0, \psi_1\right)$, and $M\left(r_2, \widetilde{D}_0, \psi_2\right)$ should be replaced by $L\left(r_1, \widetilde{D}_0, \psi_1\right)$ and $L\left(r_2, \widetilde{D}_0, \psi_2\right)$, respectively.

7.7 Sediment basement determination

The mathematical derivation of an equation based on isostatic equilibrium for sediment basement (thickness) determination is like what we have shown so far, but the difference is seen when the sediment masses are between the ice and the upper crust (topography). In Chapter 5, we presented different models for the radial variation of sediment density and showed that even the forward computation of sediment potentials is rather complicated. Here, we do not use those radial density models to keep the derivations short and simpler. Considering that the density of sediments is radially variable makes the mathematical derivations very complicated and we should perform a lot of approximations to simplify them. Therefore, we decided here to give just an idea, like we did for ice and bathymetry depth determination.

7.7.1 Mathematical foundation

Let us write the gravimetric equation of isostatic equilibrium for a point with a geocentric distance r, in the following form.

$$
t_{nm}^{\mathrm{TBCI}} \left(\frac{R}{r}\right)^{n+1} = v_{nm}^{\mathrm{TBCI/Iso}} \left(\frac{R}{r}\right)^{n+1} + \Delta v_{nm}^{\mathrm{S/Iso}} \left(\frac{R}{r}\right)^{n+1}, \tag{7.142}
$$

where t_{nm}^{TBCI} is the disturbing potential generated from the TB masses and crustal crystalline and ice and $v_{nm}^{\mathrm{TBCI/Iso}}$ is the isostatically reduced gravitational potential of the corresponding masses. $\Delta\rho^{\mathrm{S}}$ is the density contrast between the density of sediments and that of the upper crust. We assume that the sediment density layer has the same density as the upper crust, or topographic masses, and later the defect of this assumption will be eliminated by considering the contrast gravitational potential of the sediment. This is the reason that we write $\Delta v_{nm}^{\mathrm{S/Iso}}$ instead of $v_{nm}^{\mathrm{S/Iso}}$. However, if the gravitational potentials of each layer of TB masses, sediments, crustal crystalline and ice are considered separately, as we saw in the previous sections, the density can be considered directly instead of the density contrast and the potential instead of the potential contrast. In Eq. (7.142), we write the isostatic equilibrium for the gravitational potential contrast.

In Chapter 5, we discussed that the density of sediments increased with depth, or in other words, the density contrast between the sediment density and the upper crust decreases quickly to zero. Different models and methods were developed in that chapter for modelling the gravitational potential of sediment. Once this potential is derived it can be used for determining the sediment thickness assuming that its density contrast is known.

Over the continents and in places where ice exists, the sediment layer will be between the ice and the topographic masses. If no ice exists in the area, the contribution of the ice will be set to zero and it is assumed that the sediments are the upper part of the upper crust. Over oceans, sediments are located between the bathymetric masses (water) and the upper crust. Therefore, there will not be any effect from topographic masses. In the case where we know the mean value of the sediment thickness:

$$\Delta V^{\mathrm{S}}(r,\theta,\lambda) = G\iint_{\sigma}\int_{R+H-d'_{\mathrm{I}}-d_{\mathrm{S}}}^{R+H-d'_{\mathrm{I}}} \Delta\rho'^{\mathrm{S}}\frac{r'^2\mathrm{d}r'\mathrm{d}\sigma}{l}$$
$$\approx Gr\iint_{\sigma}\Delta\rho'^{\mathrm{S}}\sum_{n=0}^{\infty}\left(\frac{R+H'-d'_{\mathrm{I}}}{r}\right)^{n+2}\left(d'_{\mathrm{S}}+\frac{(n+2)d'^2_{\mathrm{S}}}{2(R+H'-d')}\right)$$
$$P_n(\widetilde{x})\mathrm{d}\sigma. \tag{7.143}$$

When the mean value of the sediment thickness d'_{S_0} is given it will be straightforward to simplify Eq. (7.143) by approximating $d'^2_{\mathrm{S}} \approx d'_{\mathrm{S}_0}d'_{\mathrm{S}}$; therefore, Eq. (7.143) will simplify to:

$$\Delta V^{\mathrm{S}}(r,\theta,\lambda) = Gr\iint\limits_{\sigma} \Delta\rho_0'^{\mathrm{S}} d_{\mathrm{S}}' \sum_{n=0}^{\infty}\left(\frac{R+H'-d_{\mathrm{I}}'}{r}\right)^{n+2} \delta_n^* P_n(\widetilde{x})\mathrm{d}\sigma, \quad (7.144)$$

where

$$\delta_n^* = 1 - \frac{(n+2)d_{\mathrm{S}_0}}{2\left(R+H'-d_{\mathrm{I}}'\right)}. \quad (7.145)$$

Eq. (7.144) is a linear integral equation with respect to d_{S}', solvable. The kernel function of this integral is dependent on the topographic height as well. If we consider the compensation of the sediment masses, we obtain:

$$\Delta V^{\mathrm{S/Iso}}(r,\theta,\lambda) = Gr\iint\limits_{\sigma} \Delta\rho'^{\mathrm{S}} d_{\mathrm{S}}' \sum_{n=0}^{\infty}\left(\frac{R+H'-d_{\mathrm{I}}'}{r}\right)^{n+2} \delta_n^* P_n(\widetilde{x})\mathrm{d}\sigma$$
$$- 4\pi Gr\Delta\rho \sum_{n=0}^{\infty}\left(\frac{R-\widetilde{D}_0}{r}\right)^{n+2} \frac{\beta_n^* C_n}{2n+1}\left(\Delta\rho^{\mathrm{S}} d_{\mathrm{S}}\right)_n. \quad (7.146)$$

According to the integral (Eq. 1.11), the Laplace coefficients $\left(\Delta\rho^{\mathrm{S}} d_{\mathrm{S}}\right)_n$ can be written in integral form. After substituting that integral instead of $\left(\Delta\rho^{\mathrm{S}} d_{\mathrm{S}}\right)_n$ in Eq. (7.146):

$$\Delta V^{\mathrm{S/Iso}}(r,\theta,\lambda) = G\iint\limits_{\sigma} \Delta\rho'^{\mathrm{S}} d_{\mathrm{S}}' K_{\mathrm{S}}\left(r,\widetilde{D}_0,d_{\mathrm{S}_0},\psi\right)\mathrm{d}\sigma, \quad (7.147)$$

where

$$K_{\mathrm{S}}\left(r,\widetilde{D}_0,d_{\mathrm{S}_0},\psi\right) = \sum_{n=0}^{\infty}\left(r\left(\frac{R+H'-d_{\mathrm{I}}'}{r}\right)^{n+2}\right.$$
$$\left.\delta_n^* - r\Delta\rho\left(\frac{R-\widetilde{D}_0}{r}\right)^{n+2}\beta_n^* C_n\right)P_n(\widetilde{x}). \quad (7.148)$$

Eq. (7.147) with the kernel (Eq. 7.148) construct an integral equation for determining the product of the sediment density contrast and thickness.

Now, let us start with Eq. (7.142) and perform the spherical harmonic synthesis from both sides of it to obtain the formula in spatial form. In this case, we obtain:

$$
\begin{aligned}
\Delta V^{S/\mathrm{Iso}}(r,\theta,\lambda) &= T^{\mathrm{TBCI/Iso}}(r,\theta,\lambda) - V^{\mathrm{TBCI/Iso}}(r,\theta,\lambda) \\
&= T(r,\theta,\lambda) - V^{\mathrm{TBCI/Iso}}(r,\theta,\lambda) - \sum_{n=0}^{N_{\mathrm{Litho}}} t_n(r,\theta,\lambda).
\end{aligned} \tag{7.149}
$$

This is the equation of isostatic equilibrium for a point outside the Earth. Now, the goal is to find its relationship with different satellite gravimetry data.

7.7.2 Sediment thickness from satellite altimetry, gradiometry and inter-satellite measurements

Satellite altimetry data can be converted to the geoid height over oceans. Therefore, it is rather simple just to consider $r = R$ in Eq. (7.149) and change the disturbing potential to the altimetry-derived geoid height:

$$
G\iint_{\sigma} \Delta\rho'^{S} d'_S K_S\left(R,\widetilde{D}_0,d_{S_0},\psi\right)\mathrm{d}\sigma = \gamma N(R,\theta,\lambda) - V^{\mathrm{BC/Iso}}(R,\theta,\lambda) - \sum_{n=0}^{N_{\mathrm{Litho}}} t_n(R,\theta,\lambda). \tag{7.150}
$$

Eq. (7.150) is a linear integral equation having the altimetry-geoid height as one of the measurements. The gravitational potential of the isostatically reduced bathymetric masses and crustal crystalline is $V^{\mathrm{BC/Iso}}(R,\theta,\lambda)$. The kernel of Eq. (7.150) will be simplified to:

$$
K_S\left(R,\widetilde{D}_0,d_{S_0},\psi\right) = \sum_{n=0}^{\infty}\left(R\left(\frac{R+H'}{R}\right)^{n+2}\delta_n^* - R\Delta\rho'\left(\frac{R-\widetilde{D}_0}{R}\right)^{n+2}\beta_n^* C_n\right) P_n(\widetilde{x}). \tag{7.151}
$$

The last term on the rhs of Eq. (7.150) can be computed from an existing Earth gravity model to degree and order N_{Litho}, to remove the sub-lithospheric gravitational potential of sub-lithospheric masses.

To find integral equations for inverting satellite gradiometry data down to $\Delta\rho'^{S} d'_S$, it is enough to apply the derivative operators defined in Eqs. (1.95)–(1.100) to the integral Eq. (7.150). In this case, the following integral equation will be obtained:

$$G\iint_{\sigma}\Delta\rho'^{S}d'_{S}K_{S,ij}\left(r,\widetilde{D}_0,d_{S_0},\psi\right)d\sigma = T_{ij}(r,\theta,\lambda) - V_{ij}^{TBC/Iso}(r,\theta,\lambda) - \sum_{n=0}^{N_{Litho}} T_{ij,n}(r,\theta,\lambda). \tag{7.152}$$

The kernels of these integral equations have forms similar to those presented in Eqs. (7.117)–(7.124), but $B^{*}_{ij}\left(r,d_0,\widetilde{D}_0,\psi\right)$ should be replaced by $K_{S,ij}\left(r,\widetilde{D}_0,d_{S_0},\psi\right)$.

For the inter-satellite range rates, we can obtain the following integral equation in analogy to Eq. (7.140):

$$\begin{aligned} G\iint_{\sigma}\Delta\rho^{S}d'_{S}\left(K_S\left(r_2,d_{S_0},\widetilde{D}_0,\psi_2\right)-K_S\left(r_1,d_{S_0},\widetilde{D}_0,\psi_1\right)\right)d\sigma \\ = v_0\dot{\widetilde{\rho}} - \Upsilon - \sum_{n=0}^{\infty}\left(v_n^{TBC/Iso}(r_2,\theta_2,\lambda_2) - v_n^{TBC/Iso}(r_1,\theta_1,\lambda_1)\right) \\ - \sum_{n=0}^{N_{Litho}}\left(t_n(r_2,\theta_2,\lambda_2) - t_n(r_1,\theta_1,\lambda_1)\right). \end{aligned} \tag{7.153}$$

Finally, the integral equation for recovering $\Delta\rho'^{S}d'_{S}$ from the inter-satellite LOS measurements is:

$$\begin{aligned} G\iint_{\sigma}\Delta\rho'^{S}d'_{S}K_S^{*B}\left(r_1,\psi_1,\alpha_1,r_2,\psi_2,\alpha_2,\widetilde{D}_0,d_{S_0}\right)d\sigma \\ = \ddot{\widetilde{\rho}} - \frac{\left(\delta\dot{\mathbf{r}}_{12}\right)^2 - \dot{\widetilde{\rho}}^2}{\widetilde{\rho}} - \Xi - \Sigma^{TBC/Iso}, \end{aligned} \tag{7.154}$$

where $\Sigma^{TBC/Iso}$ is the isostatically reduced effects of topography, bathymetry and crustal crystalline masses on the inter-satellite LOS data, computable from the SHCs of the potential of these masses into the spherical harmonic expansion of the inter-satellite LOS data. $K_S^{*B}\left(r_1,\psi_1,\alpha_1,r_2,\psi_2,\alpha_2,\widetilde{D}_0,d_{S_0}\right)$ is a bipolar kernel and has a form similar to that of the kernel presented in Eq. (7.53) but $M\left(r_1,\widetilde{D}_0,\psi_1\right)$ and $M\left(r_2,\widetilde{D}_0,\psi_2\right)$ should be replaced by $K_S\left(r_1,d_{S_0},\widetilde{D}_0,\psi_1\right)$ and $K_S\left(r_2,d_{S_0},\widetilde{D}_0,\psi_2\right)$, respectively.

References

Abrehdary, M., Sjöberg, L.E., Sampietro, D., 2018. Contribution of satellite altimetry in modelling Moho density contrast in oceanic areas. J. Appl. Geod. https://doi.org/10.1515/jag-2018-0034.

Andersen, O.B., Knudsen, P., Berry, P.A.M., 2010. DNSC08GRA global marine gravity field from double retracked satellite altimetry. J. Geod. 84 (3), 191–199.

Braitenberg, C., Wienecke, S., Wang, Y., 2006. Basement structures from satellite-derived gravity field: south China Sea ridge. J. Geophys. Res. 111, B05407. https://doi.org/10.1029/2005JB003938.

Brockmann, J.M., Zehentner, N., Höck, E., Pail, R., Loth, I., Mayer-Guerr, T., Schuh, W.D., 2014. EGM_TIM_RL05: an independent geoid with centimetre accuracy purely based on the GOCE mission. Geophsy. Res. Lett. 41 (22), 8089–8099.

Calmant, S., Francheteau, J., Cazenave, A., 1990. Elastic layer thickening with age of oceanic lithosphere: a toll for predicting the age of volcanoes and oceanic crust. Geophys. J. Int. 90 (100), 59–67.

Carslaw, H.S., Jaeger, J.C., 1959. Conduction of Heat in Solids. Oxford Univ. Press, London.

Chen, W., Tenzer, R., 2015. Harmonic coefficients of the Earth's spectral crustal model 180 – ESCM180. Earth Sci. Inf. 8 (1), 147–159.

Eshagh, M., 2014. Determination of Moho discontinuity from satellite gradiometry data: linear approach. GRIB 1 (2), 1–13.

Eshagh, M., 2016. A theoretical discussion on Vening Meinesz-Moritz inverse problem of isostasy. Geophys. J. Int. 207, 1420–1431.

Eshagh, M., 2018. Elastic thickness determination based on Vening Meinesz-Moritz and flexural theories of isostasy. Geophys. J. Int. 213 (3), 1682–1692, 1.

Eshagh, M., Hussain, M., 2016. An approach to Moho discontinuity recovery from on-orbit GOCE data with application over Indo-Pak region. Tectonophysics 690 (B), 253–262.

Eshagh, M., Pitonak, M., 2019. Elastic thickness determination from on-orbit GOCE data and CRUST1.0. Pure Appl. Geophys. 176, 685–696.

Eshagh, M., Bagherbandi, M., 2011. Smoothing impact of isostatic crustal thickness models on local integral inversion of satellite gravity gradiometry data. Acta Geophys. 59 (5), 891–906.

Eshagh, M., Steinberger, B., Tenzer, R., Tassara, A., 2018. Comparison of gravimetric and mantle flow solutions for lithospheric stress modelling and their combination. Geophys. J. Int. 213 (2), 1013–1028.

Eshagh, M., Hussain, M., Tenzer, R., Romeshkani, M., 2016. Moho density contrast in central Eurasia from GOCE gravity gradients. Remote Sens. 8 (418), 1–18.

Eshagh, M., Ebadi, S., Tenzer, R., 2017. Isostatic GOCE Moho model for Iran. J. Asian Earth Sci. 138, 12–24.

Farr, T.G., Paul, A.R., Edward, C.R.C., Riley, D., Scott, H., Michael, K., Mimi, P., Rodriguez, E., Ladislav, R., David, S., Shaffer, S., Joanne, S., 2007. The shuttle radar topography mission. Rev. Geophys. 45, RG2004. https://doi.org/10.1029/2005RG000183.

Gruber, T., Rummel, R., Abrikosov, O., van Hees, R., 2010. GOCE Level 2 Product Data Handbook, GO-MA-HPF-GS-0110, Issue 4.2.

Laske, G., Masters, G., Ma, Z., Pasyanos, M., 2013. Update on CRUST1.0 - a 1-degree global model of Earth's crust. In: EGU General Assembly Conference Abstracts 15, p. 2658.

Moritz, H., 1990. The Figure of the Earth. H Wichmann, Karlsruhe.

Moritz, H., 2000. Geodetic reference system 1980. J. Geod. 74 (1), 128–612.

Pavlis, N.K., Holmes, S.A., Kenyon, S.C., Factor, J.K., 2012. The development and evaluation of the Earth Gravitational model 2008 (EGM2008). J. Geophys. Res. 117, B04406.

Ramillien, G., Cazenave, A., 1997. Global bathymetry derived from altimeter data of the ERS-1 geodetic mission. J. Geodyn. 23 (2), 129–149.

Sandwell, D.T., Müller, R.D., Smith, W.H.F., Garcia, E., Francis, R., 2014. New global marine gravity model from Cryo-Sat-2 and jason-1 reveals buried tectonic structure. Science 346 (6205), 65–67.

Sjöberg, L.E., 2009. Solving Vening Meinesz-Moritz inverse problem in isostasy. Geophys. J. Int. 179, 1527–1536.

Stewart, J., Watts, A.B., 1997. Gravity anomalies and spatial variations of flexural rigidity at mountain ranges. J. Geophys. Res. 102, B3.

Tassara, A., Swain, C., Hackney, R., Kirby, J., 2007. Elastic thickness structure of South America estimated using wavelets and satellite-derived gravity data. Earth Planet. Sci. Lett. 253, 17–36.

Watts, A.B., 1978. An analysis of isostasy in the World's oceans in Hawaiian-Emperor seamount chain. J. Geophys. Res. 83, 5989–6004.

Yi, W., Rummel, R., Gruber, T., 2013. Gravity field contribution analysis of GOCE gravitational gradient components. Studia Geophys. Geod. 57 (2), 174–202.

CHAPTER 8

Gravity field and lithospheric stress

8.1 Runcorn's theory for sub-lithospheric stress modelling

The sub-lithospheric stress induced by mantle convection can be studied by the Earth's gravity field. Studying this stress at this surface or below is important for geophysicists and geologists to interpret seismicity, volcanicity, kimberlite magmatism, ore concentration and tectonic and magnetic features (Liu, 1977). Kaula (1963), for instance, developed a method based on minimising strain energy and using low-degree gravitational and topographic harmonics to estimate the minimum stresses in an elastic Earth. McKenzie (1967) studied heat flow by gravity data and concluded that the long-wavelength harmonics of the external gravity field cannot be supported by the strength of the lithosphere. Runcorn (1964, 1967) solved the Navier—Stokes equations for sub-lithospheric shear stresses towards north and east. By applying this theory, Liu (1977, 1978) modelled the convection pattern and stress system under the African plate and Asia. Later, Liu (1979) developed a theory for the sub-lithospheric stress concentration and its relation with seismogenic models for the Tangshan earthquake. Furthermore, Liu (1980) studied the convection stress field and intraplate volcanism. McNutt (1980) implemented the regional gravity field to study stress in the crust and upper mantle and stated that stresses implied in the regional compensation scheme are an order of magnitude larger than those corresponding to local ones. Souriau and Souriau (1983) studied the global tectonic using a geoid model derived from gravimetric data. Ricard et al. (1984) investigated the connection between the lithospheric stress and the geoid height. Fu and Huang (1983) extended Runcorn's definition for the full stress tensor by using the shell theory, and used that method to study crustal deformation of the Tibetan Plateau. Ricard et al. (1988) investigated the connection between global plate motion and geoid undulations.

Satellite Gravimetry and the Solid Earth
ISBN 978-0-12-816936-0
https://doi.org/10.1016/B978-0-12-816936-0.00008-6

Runcorn (1964, 1967) discussed and found a direct mathematical model between the spherical harmonic coefficients (SHCs) of the Earth's gravity field and the sub-lithospheric stress caused by mantle convection. However, main objection to his theory is that he uses a constant viscosity and this viscosity structure primarily affects the long-wavelength portion of the gravity field. In this section, the Runcorn approach is presented step-by-step and the approximations that he applied will be discussed. However, Runcorn's solution is the only direct way to connect the mantle convection to the Earth's gravity field, even if it is oversimplified. In the following sections, each step of his theory is presented and discussed.

8.1.1 Poloidal and toroidal flows

Consider a two-layered model for the Earth, core and mantle. This means that the mantle convection occurs inside a spherical shell between the upper boundary of the core, the core–mantle boundary (CMB) and the sub-lithosphere. The mantle flows have two components: poloidal and toroidal. Fig. 8.1A and B presents these components schematically.

Runcorn considered that vector potential **A** was composed of two scalar fields of $\widetilde{V}$ and $\widetilde{W}$ in the form:

$$\mathbf{A} = \mathbf{r}\widetilde{V} + \mathbf{r} \times \nabla\widetilde{W} \tag{8.1}$$

where **r** stands for the vector of position of a point inside the mantle, × is the cross-product operator and **∇** the gradient operator. The goal is to show how these two components of **A** are connected to the poloidal and toroidal

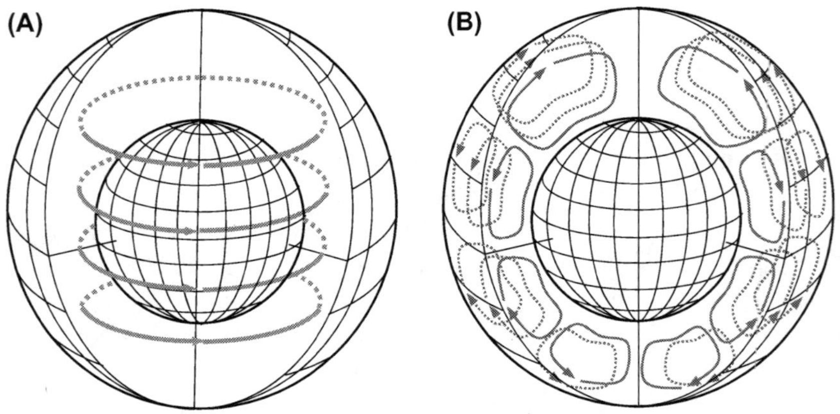

Figure 8.1 Schematic presentation of (A) toroidal and (B) poloidal flow circulation in mantle.

flows. It is known in physics and vector algebra that the curl of a vector represents a rotation. Therefore, the infinitesimal rotation of **A** is obtained by taking its curl:

$$\widetilde{\mathrm{v}}' = \mathrm{curl}\ \mathbf{A} = \nabla \times \mathbf{A} = -\nabla \times \mathbf{r}\widetilde{V} + \nabla \times \mathbf{r} \times \nabla\widetilde{W}. \tag{8.2}$$

In this instance, the toroidal flow is in fact the first term of Eq. (8.2); based on curl identities, one can write:

$$\widetilde{\mathbf{v}}_{\mathrm{tor}} = -\mathbf{r} \times \nabla\widetilde{V}. \tag{8.3}$$

From the second term of Eq. (8.2), the poloidal component is obtained by:

$$\widetilde{\mathbf{v}} = -\frac{\nabla \mathrm{d}\left(r\widetilde{W}\right)}{\mathrm{d}r} + \mathbf{r}\nabla^2\widetilde{W} = \nabla \times (\mathbf{r} \times \nabla\widetilde{W}) = -\nabla \times \nabla \times \mathbf{r}\widetilde{W}. \tag{8.4}$$

The circulation inside the mantle, which has been assumed to be a spherical shell, has a radial component; therefore, that can be described by the poloidal components of the field or the function $\widetilde{W}$. Runcorn (1964) presented the northward and eastward horizontal velocity of the flow using $\widetilde{W}$ by:

$$\overline{v}_x = \frac{1}{r}\frac{\mathrm{d}}{\mathrm{d}\theta}\frac{\mathrm{d}}{\mathrm{d}r}(r\widetilde{W}) \tag{8.5}$$

$$\overline{v}_y = -\frac{1}{r\sin\theta}\frac{\mathrm{d}}{\mathrm{d}\lambda}\frac{\mathrm{d}}{\mathrm{d}r}(r\widetilde{W}). \tag{8.6}$$

As observed, the *toroidal flow is ignored* to compute the velocity of the mantle; only the poloidal one is used. This is the first approximation in the Runcorn theory, whereas the real mantle contains toroidal flows as well.

8.1.2 Navier–Stokes equation

Later, Runcorn used the Navier–Stokes equation of hydrodynamic equations presented by Milne-Thompson (1950):

$$\widetilde{\eta}\nabla \times \nabla \times \widetilde{\mathbf{v}} - \frac{4}{3}\widetilde{\eta}\nabla(\nabla \cdot \widetilde{\mathbf{v}}) - 2(\nabla\widetilde{\eta} \cdot \nabla)\widetilde{\mathbf{v}} - \nabla\widetilde{\eta} \times (\nabla \times \widetilde{\mathbf{v}}) + \frac{2}{3}(\nabla \cdot \widetilde{\mathbf{v}})\nabla\widetilde{\eta} = -\nabla p + g\rho \tag{8.7}$$

where $\widetilde{\eta}$ is the mantle viscosity, p the pressure, g the gravity value and ρ the density of the mantle.

To solve Eq. (8.7), another assumption that Runcorn applied is the incompressibility of the mantle. Mathematically, this means that the divergence of the poloidal velocity field is zero, namely:

$$\nabla \cdot \widetilde{\mathbf{v}} = 0. \tag{8.8}$$

However, the real *mantle is compressible.* By this assumption, the Navier–Stokes Eq. (8.7) will change to:

$$-\widetilde{\eta}\nabla^2\widetilde{\mathbf{v}} - 2\frac{\mathrm{d}\widetilde{\eta}}{\mathrm{d}r}\frac{\mathrm{d}\widetilde{\mathbf{v}}}{\mathrm{d}r} - \frac{1}{r}\frac{\mathrm{d}\widetilde{\eta}}{\mathrm{d}r}\mathbf{r} \times (\nabla \times \widetilde{\mathbf{v}}) = -\nabla p + g\rho. \tag{8.9}$$

Based on the first assumption, the poloidal function $\widetilde{W}$ is inserted into Eq. (8.9) and substitution of Eq. (8.4) into Eq. (8.9) reads:

$$\begin{aligned}&\widetilde{\eta}\nabla\nabla^2\frac{\mathrm{d}(r\widetilde{W})}{\mathrm{d}r} - \widetilde{\eta}\nabla^2(\mathbf{r}\nabla^2\widetilde{W}) + 2\frac{\mathrm{d}\widetilde{\eta}}{\mathrm{d}r}\frac{\mathrm{d}}{\mathrm{d}r}\nabla \times \nabla \times (\mathbf{r}\widetilde{W}) - r\frac{\mathrm{d}\widetilde{\eta}}{\mathrm{d}r}\nabla\nabla^2\widetilde{W}\\ &+ \left(\frac{\mathrm{d}\widetilde{\eta}}{\mathrm{d}r}\frac{\mathrm{d}\nabla^2\widetilde{W}}{\mathrm{d}r}\right)\mathbf{r}\\ &= \nabla p + g\rho.\end{aligned} \tag{8.10}$$

Taking the curl of both sides leads to

$$\begin{aligned}\mathbf{r}\frac{g}{r} \times \nabla\rho = {}& \frac{\mathrm{d}\widetilde{\eta}}{r\mathrm{d}r}\mathbf{r} \times \nabla\nabla^2\frac{\mathrm{d}(r\widetilde{W})}{\mathrm{d}r} - \frac{1}{r}\frac{\mathrm{d}}{\mathrm{d}r}\left(r\frac{\mathrm{d}\widetilde{\eta}}{\mathrm{d}r}\right)\mathbf{r} \times \nabla\nabla^2\widetilde{W} + \mathbf{r}\\ &\times \nabla\left(\frac{\mathrm{d}\widetilde{\eta}}{\mathrm{d}r}\frac{\mathrm{d}\nabla^2\widetilde{W}}{\mathrm{d}r}\right) + \mathbf{r} \times \nabla(\widetilde{\eta}\nabla^4\widetilde{W}) + \left(\frac{2}{r^2}\frac{\mathrm{d}^2\widetilde{\eta}}{\mathrm{d}r^2} - \frac{2}{r^3}\frac{\mathrm{d}\widetilde{\eta}}{\mathrm{d}r}\right)\mathbf{r}\\ &\times \nabla\left(r^2\frac{\mathrm{d}^2\widetilde{W}}{\mathrm{d}r^2} + r\frac{\mathrm{d}\widetilde{W}}{\mathrm{d}r} - \widetilde{W}\right).\end{aligned} \tag{8.11}$$

After further simplifications and rearrangements, Eq. (8.11) changes to

$$\begin{aligned}\mathbf{r} \times \nabla\left(\frac{g\rho}{r}\right) = {}& \mathbf{r} \times \nabla\left(\frac{\mathrm{d}\widetilde{\eta}}{r\mathrm{d}r}\nabla^2\frac{\mathrm{d}(r\widetilde{W})}{\mathrm{d}r}\right) - \mathbf{r} \times \nabla\left(\frac{1}{r}\frac{\mathrm{d}}{\mathrm{d}r}\left(r\frac{\mathrm{d}\widetilde{\eta}}{\mathrm{d}r}\right)\nabla^2\widetilde{W}\right) + \mathbf{r}\\ &\times \nabla\left(\frac{\mathrm{d}\widetilde{\eta}}{\mathrm{d}r}\frac{\mathrm{d}\nabla^2\widetilde{W}}{\mathrm{d}r}\right) + \mathbf{r} \times \nabla(\widetilde{\eta}\nabla^4\widetilde{W}) + \mathbf{r}\\ &\times \nabla\left(\frac{2}{r^2}\frac{\mathrm{d}^2\widetilde{\eta}}{\mathrm{d}r^2} - \frac{2}{r^3}\frac{\mathrm{d}\widetilde{\eta}}{\mathrm{d}r}\right)\left(r^2\frac{\mathrm{d}^2\widetilde{W}}{\mathrm{d}r^2} + r\frac{\mathrm{d}\widetilde{W}}{\mathrm{d}r} - \widetilde{W}\right).\end{aligned} \tag{8.12}$$

In the end, the following equation is obtained:

$$g\rho = \frac{d\widetilde{\eta}}{dr}\nabla^2\frac{rd\widetilde{W}}{dr} + \frac{d\widetilde{\eta}}{dr}r\frac{d\nabla^2\widetilde{W}}{dr} + \widetilde{\eta}r\nabla^4\widetilde{W} - 2\frac{d\widetilde{\eta}}{dr}\left(\frac{d^2\widetilde{W}}{dr^2} + \frac{1}{r}\frac{d\widetilde{W}}{dr} - \frac{\widetilde{W}}{r^2}\right)$$
$$+ \frac{d^2\widetilde{\eta}}{dr^2}\left(-r\nabla^2\widetilde{W} + 2r\frac{d^2\widetilde{W}}{dr^2} + 2\frac{d\widetilde{W}}{dr} - 2\frac{\widetilde{W}}{r}\right). \tag{8.13}$$

This equation is a partial differential equation with respect to poloidal function $\widetilde{W}$ and $\widetilde{\eta}$, which is directly related to gravity and density. The issue is now to solve this differential equation. Runcorn (1967) simplified Eq. (8.13) based on the Elsasser (1963) model:

$$g\rho = \nabla^2(\widetilde{\eta}r\nabla^2\widetilde{W}) - \nabla^2\left(2\widetilde{W}\left(r\frac{d^2\widetilde{\eta}}{dr^2} + \frac{d\widetilde{\eta}}{dr} + \frac{\widetilde{\eta}}{r}\right)\right) - \nabla^2\left(2\widetilde{\eta}\frac{d\widetilde{W}}{dr}\right). \tag{8.14}$$

On the other hand, solution of the Poisson equation for a point inside the Earth is:

$$\nabla^2 V = -4\pi G\rho \tag{8.15}$$

where V stands for the gravitational potential of the Earth, and G the Newtonian gravitational constant.

Comparison of Eqs. (8.14) and (8.15) yields

$$V = -\frac{4\pi G\widetilde{\eta}r}{g}\nabla^2\widetilde{W} + \frac{8\pi G}{g}\widetilde{W}\left(r\frac{d^2\widetilde{\eta}}{dr^2} + \frac{d\widetilde{\eta}}{dr} + \frac{\widetilde{\eta}}{r}\right) - \frac{8\pi G\widetilde{\eta}}{g}\frac{d\widetilde{W}}{dr}. \tag{8.16}$$

As observed, Eq. (8.15) is solved for density ρ and the result is inserted into Eq. (8.14). This means that from ρ the Laplacian of the Earth's gravitational potential is determined. This is the key issue in Runcorn's theory connecting the convection poloidal flow to the gravity field of the Earth. In Eq. (8.16), first- and second-order radial derivatives of the mantle viscosity $\widetilde{\eta}$ are observed in the second-term, which means that $\widetilde{\eta}$ changes only radially. This equation represents the Earth's gravitational potential as the Laplacian of the poloidal function $\widetilde{W}$, its radial derivative and $\widetilde{W}$.

8.1.3 Gravity and sub-lithospheric stress caused by mantle convection

To derive the well-known formula of Runcorn (1967) for the sub-lithospheric stress caused by mantle convection, a constant viscosity is assumed for the mantle. The main objection to this theory is this assumption because the long wavelengths of geoid and gravity strongly depend on the viscosity structure. In addition, at the upper boundary of the mantle, the poloidal function $\widetilde{W}$ and its radial derivative $\frac{\mathrm{d}\widetilde{W}}{\mathrm{d}r}$ will be zero; therefore, the second and third terms of Eq. (8.14) disappear. By such assumptions, the density can be derived from Eq. (8.14):

$$\rho = -\frac{\widetilde{\eta}r}{g}\nabla^4\widetilde{W}. \tag{8.17}$$

Substitution of Eq. (8.17) into the Newton integral (e.g. Eq. 5.1) leads to

$$V(r,\theta,\lambda) = G\iiint_{\Omega}\frac{\rho'\mathrm{d}\Omega}{l} = -\frac{G\widetilde{\eta}r}{g}\iiint_{\Omega}\frac{\nabla^4\widetilde{W}'\mathrm{d}\Omega}{l} \tag{8.18}$$

where Ω is the whole mantle volume, $\mathrm{d}\Omega$ is the volume integration element and l stands for the distance between any point outside the mantle and a point inside.

Eq. (8.18) contains a volume integral. For simplification, the Green theorem is applied to change this volume integral to surface integrals at the surface of a sphere with a radius of r:

$$V(r,\theta,\lambda) = -\frac{G\widetilde{\eta}r^3}{g}\left[\iint_{\sigma}\frac{1}{l}\frac{\mathrm{d}}{\mathrm{d}r}\nabla^2\widetilde{W}'\mathrm{d}\sigma - \iint_{\sigma}\nabla^2\widetilde{W}'\frac{\mathrm{d}}{\mathrm{d}r}\left(\frac{1}{l}\right)\mathrm{d}\sigma\right]. \tag{8.19}$$

Runcorn (1967) mentioned, "*If a fluid is enclosed in a rigid boundary which make distort but which causes the tangential velocity to vanish, there is a layer of fluid above the spherical surface of height p/(ρg)*". This hydrostatic layer provides the necessary pressure distribution to the surface of the sphere. From the equation of creep flow:

$$\begin{aligned}\nabla p &= \widetilde{\eta}\nabla^2\nabla\frac{\mathrm{d}}{\mathrm{d}r}(\widetilde{W}) - \widetilde{\eta}\nabla^2(\mathbf{r}\nabla^2\widetilde{W}) + \mathbf{r}\widetilde{\eta}\nabla^4\widetilde{W} \\ &= \widetilde{\eta}\nabla^2\nabla\frac{\mathrm{d}}{\mathrm{d}r}(r\widetilde{W}) - \widetilde{\eta}\nabla^2\nabla(2\widetilde{W}).\end{aligned} \tag{8.20}$$

Thus, the pressure can be written based on the poloidal function $\widetilde{W}$:

$$p = \widetilde{\eta}\nabla^2\left(\frac{\mathrm{d}}{\mathrm{d}r}(\mathrm{r}\widetilde{W}) - 2\widetilde{W}\right). \tag{8.21}$$

Runcorn (1967) added Eq. (8.21) to the integrant of the first surface integral of Eq. (8.19):

$$V(r,\theta,\lambda) = -\frac{G\tilde{\eta}r}{g}\left[\iint\limits_{\sigma}\frac{1}{l}\left[\frac{\mathrm{d}}{\mathrm{d}r}\nabla^2\widetilde{W}' - \frac{1}{r}\nabla^2\left(\frac{\mathrm{d}}{\mathrm{d}r}\left(r\widetilde{W}'\right) - 2\widetilde{W}'\right)\right]\mathrm{d}\sigma - \iint\limits_{\sigma}\nabla^2\widetilde{W}'\frac{\mathrm{d}}{\mathrm{d}r}\left(\frac{1}{l}\right)\mathrm{d}\sigma\right] \tag{8.22}$$

and after simplifications and algebraic manipulations, he presented:

$$\begin{aligned} V(r,\theta,\lambda) = -\frac{G\tilde{\eta}r}{g}\Bigg\{&\iint\limits_{\sigma}\frac{1}{l}\Bigg[-\frac{1}{r}\frac{\mathrm{d}^2\widetilde{W}'}{\mathrm{d}r^2} - \frac{2}{r^2}\frac{\mathrm{d}\widetilde{W}'}{\mathrm{d}r} \\ &-\frac{1}{r^3}\left(\frac{\partial^2\widetilde{W}'}{\partial\theta^2} + \cot\theta\frac{\partial\widetilde{W}'}{\partial\theta} + \frac{1}{\sin^2\theta}\frac{\partial^2\widetilde{W}'}{\partial\lambda^2}\right)\Bigg]\mathrm{d}\sigma \\ &- \iint\limits_{\sigma}\nabla^2\widetilde{W}'\frac{\mathrm{d}}{\mathrm{d}r}\left(\frac{1}{l}\right)\mathrm{d}\sigma\Bigg\} \end{aligned} \tag{8.23}$$

The Laplacian of $\widetilde{W}$ is (see Eq. 1.2):

$$\nabla^2\widetilde{W} = \frac{\mathrm{d}^2\widetilde{W}}{\mathrm{d}r^2} + \frac{2}{r}\frac{\mathrm{d}\widetilde{W}}{\mathrm{d}r} + \frac{1}{r^2}\left(\frac{\partial^2\widetilde{W}}{\partial\theta^2} + \cot\theta\frac{\partial\widetilde{W}}{\partial\theta} + \frac{1}{\sin^2\theta}\frac{\partial^2\widetilde{W}}{\partial\lambda^2}\right). \tag{8.24}$$

Comparison of Eq. (8.24) and Eq. (8.23) yields:

$$V(r,\theta,\lambda) = -\frac{G\tilde{\eta}r^3}{g}\left[\iint\limits_{\sigma} -\frac{1}{l}\frac{\nabla^2\widetilde{W}'}{r}\mathrm{d}\sigma - \iint\limits_{\sigma}\nabla^2\widetilde{W}'\frac{\mathrm{d}}{\mathrm{d}r}\left(\frac{1}{l}\right)\mathrm{d}\sigma\right]. \tag{8.25}$$

The reciprocal distance l^{-1} can be expanded in terms of Legendre polynomials (Eq. 5.3):

$$\frac{1}{l} = \frac{1}{r}\sum_{n=0}^{\infty}\left(\frac{r}{r'}\right)^{n+1}P_n(\tilde{x}) \tag{8.26}$$

where l is the distance between the computation and integration points and $P_n(\tilde{x})$ is the Legendre polynomial of degree n and argument $\tilde{x} = \cos\psi$, where ψ is the geocentric angle between these points. r and r' are the geocentric distance of the computation and integration points, respectively.

According to Eq. (8.26), Eq. (8.25) will be further simplified to:

$$V(r,\theta,\lambda) = -\frac{G\widetilde{\eta}r^3}{g}\left[\iint_\sigma \sum_{n=0}^{\infty} -\frac{1}{r}\left(\frac{r}{r'}\right)^{n+1} P_n(\widetilde{x})\frac{\nabla^2 \widetilde{W}'}{r}\mathrm{d}\sigma - \iint_\sigma \nabla^2 \widetilde{W}'\frac{\mathrm{d}}{\mathrm{d}r}\left(\sum_{n=0}^{\infty}\frac{1}{r}\left(\frac{r}{r'}\right)^{n+1} P_n(\widetilde{x})\right)\mathrm{d}\sigma\right] \tag{8.27}$$

After simplifications:

$$V(r,\theta,\lambda) = -\frac{G\widetilde{\eta}r^3}{g}\iint_\sigma \sum_{n=0}^{\infty}\left(\frac{r}{r'}\right)^{n+1}\left(-\frac{\nabla^2 \widetilde{W}'(n+1)}{r^2}\right)P_n(\widetilde{x})\mathrm{d}\sigma \tag{8.28}$$

and finally,

$$V(r,\theta,\lambda) = \frac{4\pi G\widetilde{\eta}r}{g}\left[\sum_{n=0}^{\infty}\frac{n+1}{2n+1}\left(\frac{r}{r'}\right)^{n+1}\left(\nabla^2 \widetilde{W}\right)_n\right]. \tag{8.29}$$

The spectral form of Eq. (8.29) will be:

$$v_n(r,\theta,\lambda) = +\frac{4\pi G\widetilde{\eta}r^3}{g}\frac{1}{r^2}\frac{n+1}{2n+1}\left(\frac{r}{r'}\right)^{n+1}\left(\nabla^2 \widetilde{W}\right)_n. \tag{8.30}$$

and its solution for $\left(\nabla^2 \widetilde{W}\right)_n$:

$$\left(\nabla^2 \widetilde{W}\right)_n = \frac{g}{4\pi G\widetilde{\eta}r}\frac{2n+1}{n+1}\left(\frac{r'}{r}\right)^{n+1} v_n(r,\theta,\lambda). \tag{8.31}$$

Therefore, the relation between the harmonics of the Laplacian of the poloidal function $\widetilde{W}$ and those of the Earth's gravity field are obtained.

Now, imagine $\widetilde{W}$ is separable into the radial and horizontal parts:

$$\widetilde{W}(r,\theta,\lambda) = \widetilde{W}_1(r)\widetilde{W}_2(\theta,\lambda) \tag{8.32}$$

The Laplacian of this function assuming that changes only radially will be:

$$\nabla^2 \widetilde{W} = \frac{\mathrm{d}^2 \widetilde{W}}{\mathrm{d}r^2} + \frac{2}{r}\frac{\mathrm{d}\widetilde{W}}{\mathrm{d}r}. \tag{8.33}$$

Inserting Eq. (8.33) into Eq. (8.31) yields:

$$\frac{d^2\widetilde{W}_n}{dr^2}+\frac{2}{r}\frac{d\widetilde{W}_n}{dr}=\frac{g}{4\pi G\widetilde{\eta}r}\frac{2n+1}{n+1}\left(\frac{r'}{r}\right)^{n+1}v_n(r,\theta,\lambda). \tag{8.34}$$

If the upper bound of the mantle has the radius $R - D_{\text{Lith}}$, then

$$\left(\frac{d^2\widetilde{W}_n}{dr^2}+\frac{2}{R-D_{\text{Lith}}}\frac{d\widetilde{W}_n}{dr}\right)\Bigg|_{R-D_{\text{Lith}}}\widetilde{\eta}=\tau_z$$
$$=\frac{g}{4\pi G(R-D_{\text{Lith}})}\frac{2n+1}{n+1}\left(\frac{r'}{R-D_{\text{Lith}}}\right)^{n+1}v_n(r,\theta,\lambda). \tag{8.35}$$

Runcorn (1967) ignored the contribution from the lower boundary, CMB, which is justifiable if the convection extended throughout the mantle to the CMB.

As mentioned, when the lithosphere is thought to be a rigid shell, $\frac{d\widetilde{W}_n}{dr}=0$. The shear stresses at the upper boundary will be:

$$\tau_{zx,n}=\widetilde{\eta}\frac{\partial}{\partial\theta}\frac{\partial^2\widetilde{W}_n}{\partial r^2}=\frac{g}{4\pi G(R-D_{\text{Lith}})}\frac{2n+1}{n+1}\left(\frac{R}{R-D_{\text{Lith}}}\right)^{n+1}\frac{\partial v_n}{\partial\theta}. \tag{8.36}$$

$$\tau_{zy,n}=\widetilde{\eta}\frac{\partial}{\sin\theta\partial\lambda}\frac{\partial^2\widetilde{W}_n}{\partial r^2}=\frac{g}{4\pi G(R-D_{\text{Lith}})}\frac{2n+1}{n+1}\left(\frac{R}{R-D_{\text{Lith}}}\right)^{n+1}\frac{\partial v_n}{\sin\theta\partial\lambda} \tag{8.37}$$

which can also be written in terms of the vector spherical harmonics of poloidal type $\mathbf{X}^{(2)}_{nm}(\theta,\lambda)$(see Eq. 1.49)

$$\tau_{zx,n}\mathbf{e}_x+\tau_{zy,n}\mathbf{e}_y=\frac{g}{4\pi G(R-D_{\text{Lith}})}\sum_{n=2}^{\infty}\frac{2n+1}{n+1}\left(\frac{R}{R-D_{\text{Lith}}}\right)^{n+1}\sum_{m=-n}^{n}v_{nm}\mathbf{X}^{(2)}_{nm}(\theta,\lambda). \tag{8.38}$$

Eqs. (8.36) and (8.37) or Eq. (8.38) are the well-known formulae of Runcorn for the sub-lithospheric stresses caused by mantle convection. In these equations, v_{nm} has the unit of potential (m^2/s^2); however, Runcorn used the unitless coefficients for these harmonics.

Here, the Earth Gravitational Model 2008 (EGM08) from degree and order 2 to 25, reduced for the normal gravity field, is used to compute τ_z, Eq. (8.35) for a lithospheric thickness of 100 km in Fig. 8.2A, and the northward and eastward shear sub-lithospheric stress components τ_{zx} and τ_{zy}, respectively in Figs. 8.2B and C. Finally, the magnitude of sub-

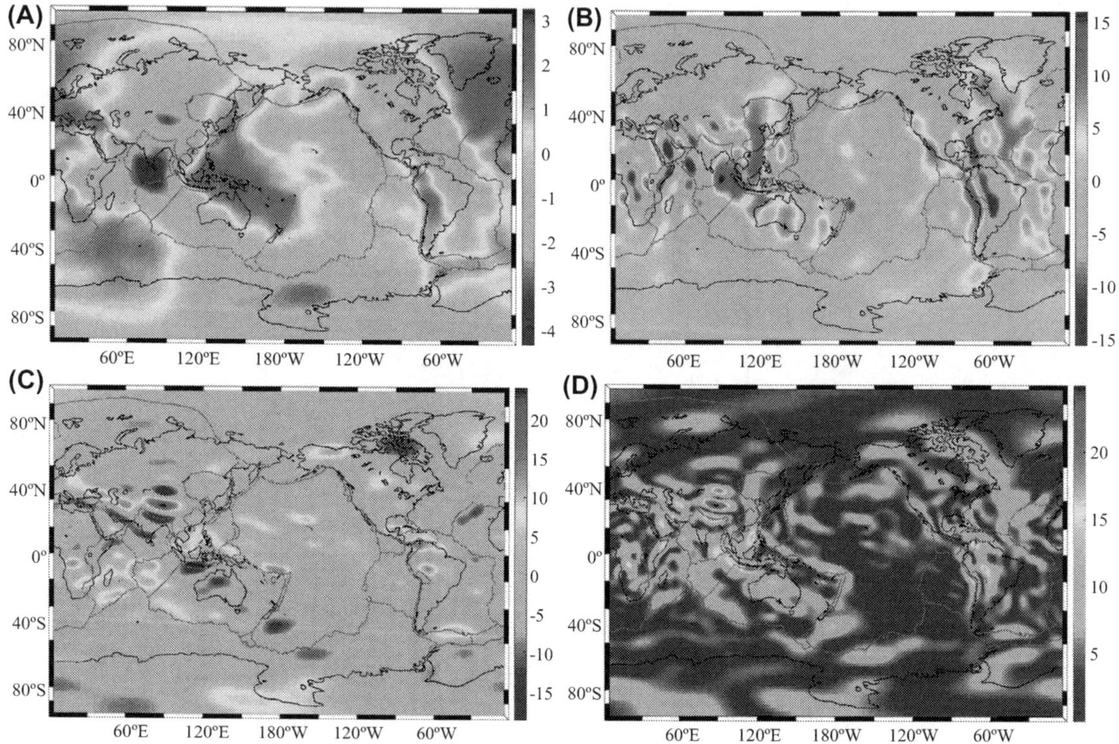

Figure 8.2 (A) τ_z, (B) τ_{zx}, (C) τ_{zy} and (D) $\sqrt{\tau_{zx}^2 + \tau_{zy}^2}$ (MPa).

lithospheric stress $\sqrt{\tau_{zx}^2 + \tau_{zy}^2}$ is shown in Figs. 8.2D. The tectonic boundaries are shown by red lines and relatively good agreement is seen between the stress maps and these tectonic boundaries.

8.2 Hager and O'Connell theory for sub-lithospheric stress modelling

The general equations describing the conservation of mass and momentum, gravitation and the compressible rheology for a fluid are (Corrieu et al., 1995):

$$\rho \nabla \widetilde{\mathbf{v}} + \frac{\partial \rho}{\partial r} \widetilde{v_r} = 0 \tag{8.39}$$

$$\nabla \tau + \rho \nabla V + \delta \rho g = 0 \tag{8.40}$$

$$\nabla^2 V = -4\pi G \delta \rho \tag{8.41}$$

$$\boldsymbol{\tau} = \widetilde{\eta} \left(\nabla \widetilde{\mathbf{v}} + (\nabla \widetilde{\mathbf{v}})^{\mathrm{T}} - \frac{2}{3} (\nabla \cdot \widetilde{\mathbf{v}}) \mathbf{I} \right) - p\mathbf{I} \tag{8.42}$$

where $\widetilde{\mathbf{v}}$ is the velocity vector, $\mathbf{I}$ stands for the identity matrix, $\widetilde{\eta}$ is the mantle viscosity, V the gravitational potential, $\delta\rho$ nonadiabatic density perturbation, δ_{ij} the Kronecker delta, p the nonhydrostatic pressure, g the gravity attraction and $\boldsymbol{\tau}$ the stress tensor.

Solving these equations is complicated, and some assumptions and simplification should be taken to obtain an analytical solution. In Eq. (8.39), which is known as the equation of continuity, the density time derivative is neglected, which means that the fluid is steady. In addition, for the lateral density variation and the third term of Eq. (8.40), the effects of thermal buoyancy, the Reynolds number for mantle are ignored. Hager and O'Connell (1979) consider that the mantle is a Newtonian fluid. Showed that the resulting convections are not significantly different if the mantle is considered Newtonian or non-Newtonian. Incompressibility is another assumption that Hager and O'Connell (1979) applied to model mantle convection. The divergence of the velocity flow vector of an incompressible fluid is zero. Hager and O'Connell (1981) included lateral density variations and the effect of thermal buoyancy. In both studies, the Reynold number was ignored because of its small effect. The complete solution of Eqs. (8.39)–(8.42) considering a compressible mantle was presented by Corrieu et al. (1995).

To solve Eqs. (8.39)−(8.42), Hager and O'Connell (1981) considered the spherical harmonic expansions for the elements of the velocity vector:

$$\widetilde{v}_z(\theta,\lambda) = \sum_{n=0}^{\infty}\sum_{m=-n}^{n} y_{nm}^{(1)} Y_{nm}(\theta,\lambda)\mathbf{e}_z. \tag{8.43}$$

$$\widetilde{v}_x(\theta,\lambda) = \sum_{n=0}^{\infty}\sum_{m=-n}^{n} \left(y_{nm}^{(2)}\mathbf{e}_x + y_{nm}^{(9)}\mathbf{e}_y\right)\cdot\mathbf{X}_{nm}^{(2)}(\theta,\lambda) \tag{8.44}$$

$$\widetilde{v}_y(\theta,\lambda) = -\sum_{n=0}^{\infty}\sum_{m=-n}^{n} \left(y_{nm}^{(2)}\mathbf{e}_x + y_{nm}^{(9)}\mathbf{e}_y\right)\cdot\mathbf{X}_{nm}^{(3)}(\theta,\lambda) \tag{8.45}$$

where $Y_{nm}(\theta,\lambda)$ is the scalar spherical harmonics and $\mathbf{X}_{nm}^{(2)}(\theta,\lambda)$ and $\mathbf{X}_{nm}^{(3)}(\theta,\lambda)$ are poloidal and toroidal harmonics Eqs. (1.49) and (1.50).

The radial and northward and eastward stresses will be:

$$\tau_{rr}(\theta,\lambda) = \sum_{n=0}^{\infty}\sum_{m=-n}^{n} y_{nm}^{(3)} Y_{nm}(\theta,\lambda) \tag{8.46}$$

$$\tau_{zx}(\theta,\lambda) = \sum_{n=0}^{\infty}\sum_{m=-n}^{n} \left(y_{nm}^{(4)}\mathbf{e}_x + y_{nm}^{(10)}\mathbf{e}_y\right)\cdot\mathbf{X}_{nm}^{(2)}(\theta,\lambda) \tag{8.47}$$

$$\tau_{zy}(\theta,\lambda) = -\sum_{n=0}^{\infty}\sum_{m=-n}^{n} \left(y_{nm}^{(4)}\mathbf{e}_x + y_{nm}^{(10)}\mathbf{e}_y\right)\cdot\mathbf{X}_{nm}^{(3)}(\theta,\lambda) \tag{8.48}$$

and the nonequilibrium gravitational potential, its radial derivative, nonhydrostatic pressure and nonadiabatic density perturbation are:

$$V(\theta,\lambda) = \sum_{n=0}^{\infty}\sum_{m=-n}^{n} y_{nm}^{(5)} Y_{nm}(\theta,\lambda) \tag{8.49}$$

$$V_r(\theta,\lambda) = \sum_{n=0}^{\infty}\sum_{m=-n}^{n} y_{nm}^{(6)} Y_{nm}(\theta,\lambda) \tag{8.50}$$

$$\delta p(\theta,\lambda) = \sum_{n=0}^{\infty}\sum_{m=-n}^{n} y_{nm}^{(7)} Y_{nm}(\theta,\lambda) \tag{8.51}$$

$$\Delta\rho(\theta,\lambda) = \sum_{n=0}^{\infty}\sum_{m=-n}^{n} y_{nm}^{(8)} Y_{nm}(\theta,\lambda). \tag{8.52}$$

Hager and O'Connell (1979) converted the partial differential Eqs. (8.39)−(8.42) into a set of first-order ordinary differential equations. For poloidal flow they obtained:

$$\frac{\mathrm{d}}{\mathrm{d}r}y_{nm}^{(1)} = -\frac{2}{r}y_{nm}^{(1)} + \frac{n(n+1)}{r}y_{nm}^{(2)} \tag{8.53}$$

$$\frac{\mathrm{d}}{\mathrm{d}r}y_{nm}^{(2)} = -\frac{1}{r}y_{nm}^{(1)} + \frac{1}{r}y_{nm}^{(2)} + \frac{1}{\widetilde{\eta}}y_{nm}^{(4)} \tag{8.54}$$

$$\frac{\mathrm{d}}{\mathrm{d}r}y_{nm}^{(3)} = \frac{12\widetilde{\eta}}{r^2}y_{nm}^{(1)} - 6\widetilde{\eta}\frac{n(n+1)}{r^2}y_{nm}^{(2)} + \frac{n(n+1)}{r}y_{nm}^{(4)} - \rho y_{nm}^{(6)} + g y_{nm}^{(8)} \tag{8.55}$$

$$\frac{\mathrm{d}}{\mathrm{d}r}y_{nm}^{(4)} = \frac{-6\widetilde{\eta}}{r^2}y_{nm}^{(1)} + 2\widetilde{\eta}\frac{2n(n+1)-1}{r^2}y_{nm}^{(2)} - \frac{1}{r}y_{nm}^{(3)} - \frac{3}{r}y_{nm}^{(4)} - \frac{\rho}{r}y_{nm}^{(5)} \tag{8.56}$$

$$\frac{\mathrm{d}}{\mathrm{d}r}y_{nm}^{(5)} = y_{nm}^{(6)} \tag{8.57}$$

$$\frac{\mathrm{d}}{\mathrm{d}r}y_{nm}^{(6)} = \frac{n(n+1)}{r^2}y_{nm}^{(5)} - \frac{2}{r}y_{nm}^{(6)} - 4\pi G y_{nm}^{(8)} \tag{8.58}$$

where r is the geocentric distance of any point inside the mantle. The perturbed pressure can be eliminated from the system because it has the following relation to $y_{nm}^{(1)}$, $y_{nm}^{(2)}$ and $y_{nm}^{(3)}$. Once these parameters are estimated, $y_{nm}^{(7)}$ can be derived by:

$$y_{nm}^{(7)} = -\frac{4\widetilde{\eta}}{r}y_{nm}^{(1)} + \frac{2\widetilde{\eta}n(n+1)}{r}y_{nm}^{(2)} - y_{nm}^{(3)}. \tag{8.59}$$

The toroidal equations will be:

$$\frac{\mathrm{d}}{\mathrm{d}r}y_{nm}^{(9)} = \frac{1}{r}y_{nm}^{(9)} + \frac{1}{\widetilde{\eta}}y_{nm}^{(10)} \tag{8.60}$$

$$\frac{\mathrm{d}}{\mathrm{d}r}y_{nm}^{(10)} = \frac{\widetilde{\eta}(n(n+1)-2)}{r^2}y_{nm}^{(9)} - \frac{3}{r}y_{nm}^{(10)} \tag{8.61}$$

The poloidal Eqs. (8.53)–(8.58) and the toroidal ones, Eqs. (8.60) and (8.61), can be solved numerically (e.g., using Runge-Kutta technique). To solve these equations analytically, new variables were defined by Hager and O'Connell (1979):

$$\widetilde{\widetilde{u}}_{1,nm} = y_{nm}^{(1)},\ \widetilde{\widetilde{u}}_{2,nm} = y_{nm}^{(2)},\ \widetilde{\widetilde{u}}_{3,nm} = \frac{r y_{nm}^{(3)}}{\widetilde{\eta}_0},\ \widetilde{\widetilde{u}}_{4,nm} = \frac{r y_{nm}^{(4)}}{\widetilde{\eta}_0},\ \widetilde{\widetilde{u}}_{5,nm} = \frac{\rho_0 r y_{nm}^{(5)}}{\widetilde{\eta}_0},\ \widetilde{\widetilde{u}}_{6,nm} = \frac{\rho_0 r^2 y_{nm}^{(6)}}{\widetilde{\eta}_0},\ \widetilde{\widetilde{v}}_{1,nm} = y_{nm}^{(9)},\ \widetilde{\widetilde{v}}_{2,nm} = \frac{r y_{nm}^{(10)}}{\widetilde{\eta}_0} \tag{8.62}$$

where ρ_0 and $\widetilde{\eta}_0$ are the reference density and viscosity. These two sets of equations can be written in matrix forms. For the poloidal equations:

$$\frac{\mathrm{d}\tilde{\tilde{\mathbf{u}}}_{nm}}{\mathrm{d}\nu} = \mathbf{A}_n \tilde{\tilde{\mathbf{u}}}_{nm} + \mathbf{b}_{nm} \tag{8.63}$$

where $\nu = \ln\left(\frac{r}{R}\right)$

$$\mathbf{A}_n = \begin{bmatrix} -2 & n(n+1) & 0 & 0 & 0 & 0 \\ -1 & 1 & 0 & \left(\tilde{\eta}^*\right)^{-1} & 0 & 0 \\ 12\tilde{\eta}^* & 6n(n+1)\tilde{\eta}^* & 1 & n(n+1) & 0 & -\rho^* \\ -6\tilde{\eta}^* & 2\tilde{\eta}^*(2n(n+1)-1) & -1 & -2 & -\rho^* & 0 \\ 0 & 0 & 0 & 0 & 1 & 1 \\ 0 & 0 & 0 & 0 & n(n+1) & 0 \end{bmatrix};$$

$$\mathbf{b}_{nm} = \begin{bmatrix} 0 \\ 0 \\ \dfrac{r^2 g y_{nm}^{(8)}}{\tilde{\eta}_0} \\ 0 \\ 0 \\ -4\pi r^3 G\rho_0 y_{nm}^{(8)}/\tilde{\eta}_0 \end{bmatrix}. \tag{8.64}$$

where $\tilde{\eta}^* = \tilde{\eta}/\tilde{\eta}_0$ and $\rho^* = \rho/\rho_0$ were named normalisation factors by Hager and O'Connell (1981) and are dimensionless.

The solution of Eq. (8.63) is:

$$\begin{aligned} \tilde{\tilde{\mathbf{u}}}_{nm}(\nu) &= e^{\mathbf{A}(\nu-\nu_0)}\tilde{\tilde{\mathbf{u}}}_{nm}(\nu_0) + \int_{\nu_0}^{\nu} e^{\mathbf{A}_n(\nu-\varepsilon)}\mathbf{b}_{nm}(\varepsilon)\mathrm{d}\varepsilon \\ &= \mathbf{P}_{\mathbf{A}_n}(\nu,\nu_0)\tilde{\tilde{\mathbf{u}}}_{nm}(\nu_0) + \int_{\nu_0}^{\nu} \mathbf{P}_{\mathbf{A}}(\nu, t\varepsilon)\mathbf{b}_{nm}(\varepsilon)\mathrm{d}\varepsilon \end{aligned} \tag{8.65}$$

where ν_0 is the reference value of ν and

$$\mathbf{P}_{\mathbf{A}_n}(\nu,\nu_0)=\sum_{i=1}^{2}\frac{e^{\lambda_i(\nu-\nu_0)}}{\phi_i}\left\{\left[\mathbf{I}+\left(\nu-\nu_0-\frac{2}{\lambda_i-\lambda_{3-i}}-\sum_{k=3}^{4}\frac{1}{\lambda_i-\lambda_k}\right)(\mathbf{A}_n-\lambda_i\mathbf{I})\right]\right.$$
$$\left.(\mathbf{A}_n-\lambda_j\mathbf{I})^2\prod_{k=3}^{4}(\mathbf{A}_n-\lambda_k\mathbf{I})\right\}+\sum_{k=3}^{4}\frac{e^{\lambda_k(\nu-\nu_0)}}{\phi_k}\left\{(\mathbf{A}_n-\lambda_{5-k}\mathbf{I})\prod_{i=1}^{2}(\mathbf{A}_n-\lambda_i\mathbf{I})^2\right\}$$
(8.66)

$$\phi_i=(\lambda_i-\lambda_{3-i})^2\prod_{k=3}^{4}(\lambda_i-\lambda_k),\quad i=1,2 \tag{8.67}$$

$$\phi_k=(\lambda_k-\lambda_{5-k})^2\prod_{k=1}^{2}(\lambda_k-\lambda_i)^2,\quad k=3,4 \tag{8.68}$$

$$\lambda_1=n+1,\ \lambda_2=-n,\ \lambda_3=n-1\ \text{and}\ \lambda_4=-n-2.$$

For a toroidal equation, the following matrix form was presented:

$$\frac{\mathrm{d}\widetilde{\widetilde{\mathbf{v}}}_{nm}}{\mathrm{d}\nu}=\mathbf{B}_n\widetilde{\widetilde{\mathbf{v}}}_{nm} \tag{8.69}$$

where

$$\mathbf{B}_n=\begin{bmatrix}1 & \left(\widetilde{\eta}^*\right)^{-1}\\ (n(n+1)-2)\widetilde{\eta}^* & -2\end{bmatrix}. \tag{8.70}$$

with the solution

$$\widetilde{\widetilde{\mathbf{v}}}_{nm}(\nu)=e^{\mathbf{B}_n(\nu-\nu_0)}\widetilde{\widetilde{\mathbf{v}}}_{nm}(\nu_0)=\mathbf{P}_{\mathbf{B}_n}(\nu,\nu_0)\widetilde{\widetilde{\mathbf{v}}}_{nm}(\nu_0) \tag{8.71}$$

where

$$\mathbf{P}_{\mathbf{B}_n}(\nu,\nu_0)=\frac{e^{\frac{1}{2}(\nu-\nu_0)}}{\phi}\begin{bmatrix}3S+\phi C & 2S/\widetilde{\eta}^*\\ 2(n(n+1)-2)\widetilde{\eta}^*S & -3S+\phi C\end{bmatrix} \tag{8.72}$$

$$S=\sinh\left(\frac{\phi}{2}(\nu-\nu_0)\right)\ \text{and}\ C=\cosh\left(\frac{\phi}{2}(\nu-\nu_0)\right) \tag{8.73}$$

$$\phi=\sqrt{1+4n(n+1)}. \tag{8.74}$$

Boundary conditions should be considered while solving the differential equations. At the CMB,

$$\widetilde{\widetilde{\mathbf{u}}}_{nm}\big|_{R-D_{\text{CMB}}}=\begin{bmatrix}0 & \widetilde{\widetilde{u}}_{2,nm} & \widetilde{\widetilde{u}}_{3,nm} & 0 & 0 & 0\end{bmatrix}^{\text{T}}\Big|_{R-D_{\text{CMB}}} \text{ and}$$
$$\widetilde{\widetilde{\mathbf{v}}}_{nm}\big|_{R-D_{\text{CMB}}}=\begin{bmatrix}\widetilde{\widetilde{v}}_{1,nm} & 0\end{bmatrix}^{\text{T}}\Big|_{R-D_{\text{CMB}}}. \tag{8.75}$$

At the surface of the Earth or at the sub-lithosphere, the radial velocity is neglected and only the horizontal components of the velocity are used. Therefore, at the upper boundary of the mantle:

$$\widetilde{\widetilde{\mathbf{u}}}_{nm}\big|_{R-D_{\text{Lith}}}=\begin{bmatrix}0 & \widetilde{\widetilde{u}}_{2,nm} & \widetilde{\widetilde{u}}_{3,nm} & \widetilde{\widetilde{u}}_{4,nm} & 0 & 0\end{bmatrix}^{\text{T}}\Big|_{R-D_{\text{Lith}}} \text{ and}$$

$$\widetilde{\widetilde{\mathbf{v}}}_{nm}\big|_{R-D_{\text{Lith}}}=\begin{bmatrix}\widetilde{\widetilde{v}}_{1,nm} & \widetilde{\widetilde{v}}_{2,nm}\end{bmatrix}^{\text{T}}\Big|_{R-D_{\text{Lith}}} \tag{8.76}$$

where $\widetilde{\widetilde{v}}_{2,nm}\Big|_{R-D_{\text{Lith}}}$ and $\widetilde{\widetilde{v}}_{1,nm}\Big|_{R-D_{\text{Lith}}}$ are the poloidal and toroidal SHCs of the plate velocities derived based on poloidal and toroidal vector spherical harmonic analyses (Eq. 1.53):

$$\widetilde{\widetilde{v}}_{2,nm}\Big|_{R-D_{\text{Lith}}}=\frac{1}{4\pi n(n+1)}\iint_{\sigma}\left(\widetilde{v}_x(\theta',\lambda')\mathbf{e}_x+\widetilde{v}_y(\theta',\lambda')\mathbf{e}_y\right)\mathbf{X}_{nm}^{(2)}(\theta',\lambda')\mathrm{d}\sigma \tag{8.77}$$

$$\widetilde{\widetilde{v}}_{1,nm}\Big|_{R-D_{\text{Lith}}}=\frac{-1}{4\pi n(n+1)}\iint_{\sigma}\left(\widetilde{v}_x(\theta',\lambda')\mathbf{e}_x-\widetilde{v}_y(\theta',\lambda')\mathbf{e}_y\right)\mathbf{X}_{nm}^{(3)}(\theta',\lambda')\mathrm{d}\sigma. \tag{8.78}$$

This means that the observed horizontal velocity vectors (e.g., by global navigation satellite systems) are not directly used in Hager and O'Connell's theory. The SHCs of the poloidal and toroidal components are determined from the velocity vectors by performing vector spherical harmonic analyses (Eqs. 8.77 and 8.78); see also Forte and Peltier (1987) for the importance of poloidal-toroidal coupling of the mantle (Fig. 8.3).

Steinberger et al. (2001) have applied Hager and O'Connell's (1981) method to model a large-scale lithospheric stress field and topography by global mantle convection. The interesting result they found in their study was that stress does not strongly depend on the radial mantle viscosity structure, lithospheric rheology or plate motion model. They also stated that radial and poloidal flows have significant roles, but the toroidal one is less important. Steinberger et al. (2001) concluded that lithospheric rheology cannot strongly affect the result, either. In both methods of Hager

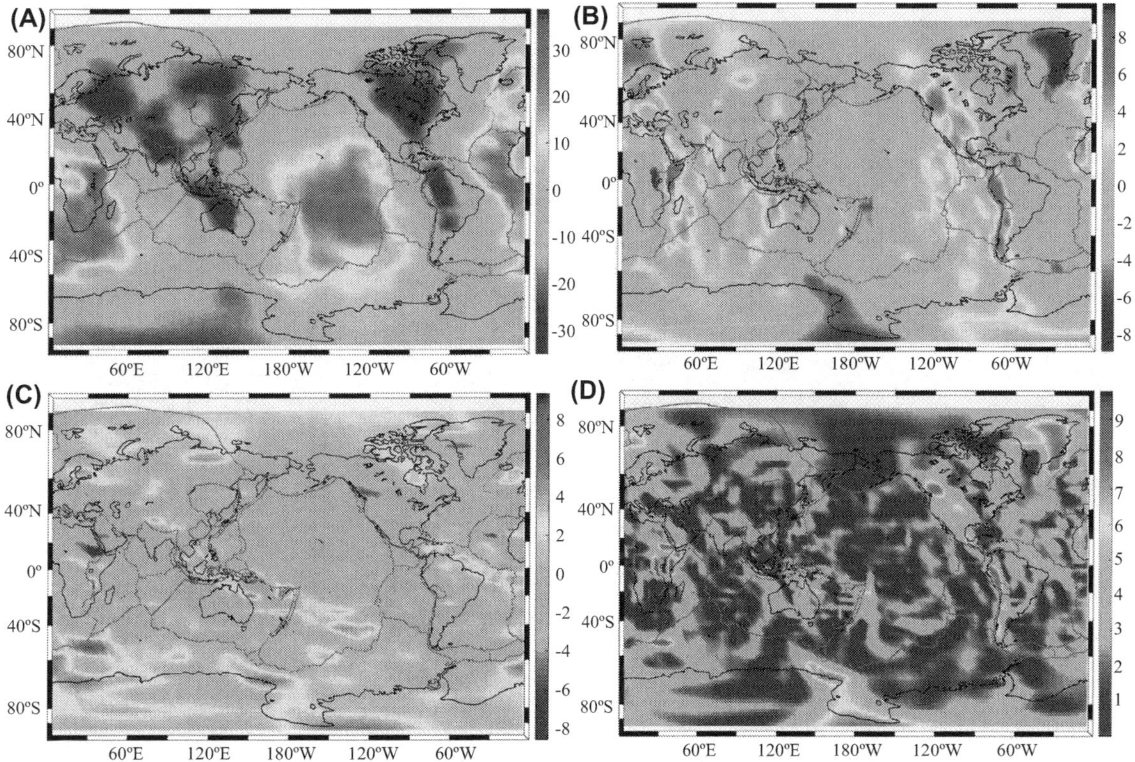

Figure 8.3 (A) Vertical stress τ_{zz}, (B) northward τ_{xz} and (C) eastward τ_{yz} stress components caused by mantle convection, and (D) $\sqrt{\tau_{zx}^2 + \tau_{zy}^2}$ (MPa). *(Computed by Steinberger in a personal communication.)*

and O'Connell and Runcorn, the mantle is assumed to be Newtonian and incompressible. Finally, Runcorn ignored the toroidal flows of the mantle, and Steinberger et al. (2001) supported this assumption by mentioning that such flows have less importance in sub-lithospheric stress determination. However, the Earth's mantle is a complicated medium and modelling its convection flow is not straightforward solely by the gravity field; some other constraints need to be considered for stress modeling using the Runcorn theory.

8.3 Stress propagation from sub-lithosphere to lithosphere

As observed, sub-lithospheric stress caused by mantle convection can be computed using the theory of either Runcorn or Hager and O'Connell. Their results are not equivalent; the reason is that to formulate each theory, different assumptions and data are applied. The gravity field of the Earth does not have a significant role in Hager and O'Connell's theory, whereas the Runcorn solution is solely based on it. On the contrary, the velocity field of tectonic motions is absent in Runcorn's theory, but it is essential to Hager and O'Connell's; however, both theories present sub-lithospheric stress caused by mantle convection. The issue is to see the relation between stress inside the lithosphere and the gravity field of the Earth. Liu (1983) and Fu and Huang (1983) used the solution of the boundary-value problem of elasticity by Love (1944) and applied Runcorn's stresses as the boundary conditions for such a purpose. In this section, the goal is to explain the mathematical foundations of this method.

8.3.1 Partial differential equation of elasticity for a spherical shell

Considering the Earth's lithosphere as a spherical shell is a common assumption in studying the lithosphere. In this instance, this shell is assumed to be elastic and its displacement and deformation behaviour are expressed by the partial differential equation of elasticity for a thin spherical thin shell (Kraus, 1967). Assuming the lithosphere to be a thin shell allows us to use linear elasticity theories, explaining how such a solid shell deforms and is stressed internally owing to prescribed loading conditions. Such an equation has the following general form in spherical coordinates (Fu and Huang, 1983):

$$(\widetilde{\lambda}+\widetilde{\mu})\nabla(\nabla\cdot\mathbf{s})+\widetilde{\mu}\nabla^2\mathbf{s}=0 \tag{8.79}$$

where ∇^2 is the Laplacian and ∇ the gradient operator. $\widetilde{\lambda}$ and $\widetilde{\mu}$ are elasticity parameters, and $\mathbf{s}$ the vector of displacements in the local north-oriented frame. The Laplacian of a vector field such as this displacement vector in a spherical coordinate system is:

$$\begin{aligned}\nabla^2\mathbf{s}=&\left(\nabla^2 s_r-\frac{2}{r^2}s_r-\frac{2}{r^2}\frac{\partial s_\theta}{\partial\theta}-\frac{2}{r^2}s_\theta\cot\theta-\frac{2}{r^2\sin\theta}\frac{\partial s_\lambda}{\partial\lambda}\right)\mathbf{e}_z\\&+\left(\nabla^2 s_\theta+\frac{2}{r^2}\frac{\partial s_r}{\partial\theta}-\frac{s_\theta}{r^2\sin^2\theta}-\frac{2\cos\theta}{r^2\sin^2\theta}\frac{\partial s_\lambda}{\partial\lambda}\right)\mathbf{e}_x\\&+\left(\nabla^2 s_\lambda-\frac{s_\lambda}{r^2\sin^2\theta}+\frac{2}{r^2\sin\theta}\frac{\partial s_r}{\partial\lambda}+\frac{2\cos\theta}{r^2\sin^2\theta}\frac{\partial s_\theta}{\partial\lambda}\right)\mathbf{e}_y.\end{aligned} \tag{8.80}$$

where s_r, s_θ and s_λ are derivatives of s with respect to the radial distance, colatitude and longitude.

In Eq. (8.79) $\nabla\cdot\mathbf{s}$ is called the volumetric strain, which is the divergence of $\mathbf{s}$, with the mathematical formula:

$$\begin{aligned}\nabla\cdot\mathbf{s}&=\frac{1}{r^2}\frac{\partial(r^2 s_r)}{\partial r}+\frac{1}{r\sin\theta}\frac{\partial}{\partial\theta}(s_\theta\sin\theta)+\frac{1}{r\sin\theta}\frac{\partial s_\lambda}{\partial\lambda}\\&=\frac{\partial s_r}{\partial r}+\frac{2}{r}s_r+\frac{1}{r}\frac{\partial s_\theta}{\partial\theta}+\frac{1}{r}s_\theta\cot\theta+\frac{1}{r\sin\theta}\frac{\partial s_\lambda}{\partial\lambda}.\end{aligned} \tag{8.81}$$

and the gradient of Eq. (8.81) is:

$$\begin{aligned}\nabla(\nabla\cdot\mathbf{s})=&\left(-\frac{2}{r^2}s_r+\frac{2}{r}\frac{\partial s_r}{\partial r}+\frac{\partial^2 s_r}{\partial r^2}-\frac{1}{r^2}\left(\frac{\partial s_\theta}{\partial\theta}+s_\theta\cot\theta\right)-\frac{1}{r^2\sin\theta}\frac{\partial s_\lambda}{\partial\lambda}\right)\mathbf{e}_z\\&+\left(\frac{2}{r^2}\frac{\partial}{\partial\theta}s_r+\frac{1}{r}\frac{\partial}{\partial\theta}\frac{\partial s_r}{\partial r}+\frac{1}{r^2}\left(\frac{\partial^2 s_\theta}{\partial\theta^2}+\frac{\partial s_\theta}{\partial\theta}\cot\theta+s_\theta(1+\cot^2\theta)\right)\right.\\&\left.+\frac{\cot\theta}{r^2\sin\theta}\frac{\partial s_\lambda}{\partial\lambda}\right)\mathbf{e}_x\\&+\left(\frac{2}{r^2\sin\theta}\frac{\partial}{\partial\lambda}s_r+\frac{1}{r\sin\theta}\frac{\partial^2 s_r}{\partial\lambda\partial r}+\frac{1}{r^2\sin\theta}\left(\frac{\partial^2 s_\theta}{\partial\lambda\partial\theta}+\frac{\partial s_\theta}{\partial\lambda}\cot\theta\right)\right.\\&\left.+\frac{1}{r^2\sin^2\theta}\frac{\partial^2 s_\lambda}{\partial\lambda^2}\right)\mathbf{e}_y.\end{aligned} \tag{8.82}$$

Solving differential Eq. (8.79) by spherical coordinates is relatively long and complicated. Here, the discussion is continued by the general solution Eq. (8.79) that was presented by Love (1944). The displacement vector **s** based on this solution is (Fu and Huang, 1983):

$$\mathbf{s} = \sum_{n=2}^{\infty}\left[A(r^2\nabla\omega_n + r\alpha_n\omega_n\mathbf{e}_z) + B(r^2\nabla\overline{\omega}_n + r\overline{\alpha}_n\overline{\omega}_n\mathbf{e}_z) + C\nabla\phi_n + D\nabla\overline{\phi}_n\right] \tag{8.83}$$

where ω_n, ϕ_n, $\overline{\omega}_n$ and $\overline{\phi}_n$ are the Laplace coefficients of the scalar functions ω, ϕ, $\overline{\omega}$ and $\overline{\phi}$, respectively. Love (1944) showed that the displacement vector is a linear combination of these functions with coefficients A, B, C and D. Furthermore,

$$\alpha_n = -2\frac{n\widetilde{\lambda} + (3n+1)\widetilde{\mu}}{(n+3)\widetilde{\lambda} + (n+5)\widetilde{\mu}} \tag{8.84}$$

$$\overline{\alpha}_n = 2\frac{(n+1)\widetilde{\lambda} + (3n+2)\widetilde{\mu}}{(2-n)\widetilde{\lambda} + (4-n)\widetilde{\mu}} \tag{8.85}$$

are two parameters carrying the mechanical properties of the lithosphere (1.e. elasticity parameters).

Eq. (8.83) is the displacement vector (**s**) that satisfies partial differential Eq. (8.79). This vector is connected to the mechanical properties of the shell by $\widetilde{\lambda}$ and $\widetilde{\mu}$. In other words, it says how a shell is deformed according to its mechanical properties to satisfy partial differential Eq. (8.79).

8.3.2 Displacement and the gravity field

To relate the solution of the partial differential equation of elasticity for a thin shell, Eq. (8.83), to the Earth's gravity field, the assumption that Runcorn (1967) used is applied; see also Steinberger et al. (2001). Here, **s** is generated from the disturbing potential (T) representing the separation between the reference ellipsoid and geoid. In other words, the geoid height is seen as a displaced physical shape of the Earth compared with its mathematical shape, or the reference ellipsoid. The following generating functions have been considered for the Laplace coefficients ω_n, ϕ_n, $\overline{\omega}_n$ and $\overline{\phi}_n$ (Fu and Huang, 1992):

$$\omega_n = \phi_n = \frac{1}{R}\left(\frac{r}{R}\right)^n t_n \tag{8.86}$$

$$\overline{\omega}_n = \overline{\phi}_n = \frac{1}{R_1}\left(\frac{R_1}{r}\right)^{n+1} t_n \quad \text{with } R_1 = R - D_{\text{Lith}} \tag{8.87}$$

where t_n is the Laplace coefficients of T and r is the geocentric radius of any point inside the lithosphere. R is the radius of the upper bound of the lithospheric shell or the surface of the Earth, and R_1 the lower one. $(R_1/r)^{n+1}$ is the upward continuation factor, which continues t_n upward from surface of a sphere with radius of R_1 to a point with the geocentric angle r inside the lithosphere. By inserting these Laplace coefficients in Eq. (8.83) and further simplifications, the relation between t_n and displacement vector (**s**) is obtained:

$$\begin{aligned}\mathbf{s} = \sum_{n=2}^{\infty}\Bigg[& A\left(\frac{r}{R}\right)^{n+1}\left((n+\alpha_n)t_n\mathbf{e}_z + \frac{\partial t_n}{\partial \theta}\mathbf{e}_x + \frac{1}{\sin\theta}\frac{\partial t_n}{\partial \lambda}\mathbf{e}_y\right) \\ & + B\left(\frac{R_1}{r}\right)^{n}\left((-(n+1)+\overline{\alpha}_n)t_n\mathbf{e}_z + \frac{\partial t_n}{\partial \theta}\mathbf{e}_x + \frac{1}{\sin\theta}\frac{\partial t_n}{\partial \lambda}\mathbf{e}_y\right) \\ & + \frac{C}{R^2}\left(\frac{r}{R}\right)^{n-1}\left(nt_n\mathbf{e}_z + \frac{\partial t_n}{\partial \theta}\mathbf{e}_x + \frac{1}{\sin\theta}\frac{\partial t_n}{\partial \lambda}\mathbf{e}_y\right) + \frac{D}{R_1^2}\left(\frac{R_1}{r}\right)^{n+2} \\ & \left(-(n+1)t_n\mathbf{e}_z + \frac{\partial t_n}{\partial \theta}\mathbf{e}_x + \frac{1}{\sin\theta}\frac{\partial t_n}{\partial \lambda}\mathbf{e}_y\right)\Bigg].\end{aligned} \tag{8.88}$$

A, B, C and D are the coefficients, which should be solved from the boundary values. This subject will be discussed later. Consider **s**, presented in Eq. (8.88), to have the form:

$$\mathbf{s}(r,\theta,\lambda) = s_z\mathbf{e}_z + s_x\mathbf{e}_x + s_y\mathbf{e}_y \tag{8.89}$$

where

$$\begin{aligned}s_z = \sum_{n=0}^{\infty}\Bigg(& A\left(\frac{r}{R}\right)^{n+1}(n+\alpha_n) + B\left(\frac{R_1}{r}\right)^{n}(-(n+1)+\overline{\alpha}_n) + \frac{C}{R^2}\left(\frac{r}{R}\right)^{n-1} n \\ & - \frac{D(n+1)}{R_1^2}\left(\frac{R_1}{r}\right)^{n+2}\Bigg) t_n\end{aligned} \tag{8.90}$$

$$s_x = \sum_{n=0}^{\infty}\left(A\left(\frac{r}{R}\right)^{n+1} + B\left(\frac{R_1}{r}\right)^{n} + \frac{C}{R^2}\left(\frac{r}{R}\right)^{n-1} + \frac{D}{R_1^2}\left(\frac{R_1}{r}\right)^{n+2}\right)\frac{\partial t_n}{\partial \theta}. \tag{8.91}$$

$$s_y = \sum_{n=0}^{\infty}\left(A\left(\frac{r}{R}\right)^{n+1} + B\left(\frac{R_1}{r}\right)^{n} + \frac{C}{R^2}\left(\frac{r}{R}\right)^{n-1} + \frac{D}{R_1^2}\left(\frac{R_1}{r}\right)^{n+2}\right)\frac{1}{\sin\theta}\frac{\partial t_n}{\partial\lambda}. \tag{8.92}$$

By writing **s** and its elements, which are radial, northward and eastward displacements, the displacements are related directly to t_n. Therefore, filtering or reducing such low-frequency signals from the geoid model is necessary to model lithospheric stress.

8.3.3 Displacement, strain and stress

In this section, the relations between the displacement vector (**s**) and strain tensor (**S**), and accordingly **S** and stress tensor ($\boldsymbol{\tau}$), are presented in spherical coordinates. By solving the partial differential equation of elasticity, **s** is obtained, and not $\boldsymbol{\tau}$. Therefore, knowledge about the relations between **s** and **S**, and accordingly, **S** and $\boldsymbol{\tau}$, is of great importance. Generally, a strain tensor in 3D space is defined by a 3 × 3 matrix:

$$\mathbf{S} = \begin{bmatrix} s_{xx} & s_{xy} & s_{xz} \\ s_{xy} & s_{yy} & s_{yz} \\ s_{xz} & s_{yz} & s_{zz} \end{bmatrix} \tag{8.93}$$

where s_{xx}, s_{yy} and s_{zz}, which are the diagonal elements of the tensor, are known as the normal strains and s_{xy}, s_{xz} and s_{yz}, the off-diagonal elements, are called shear strains. As observed, **S** is coordinate system dependent, meaning that its elements will change by changing the coordinate system. However, the trace of this tensor, or the summation of its diagonal elements, is invariant, which means that it will have the same value in any coordinate system. This trace is well-known as dilatation, showing volume expansion and contraction of the solid material.

In a spherical coordinate system, the elements of **s** are related to those of **S** by the well-known equations:

$$s_{zz} = \frac{\partial s_z}{\partial z} \tag{8.94}$$

$$s_{xx} = \frac{1}{r}\left(\frac{\partial s_x}{\partial\theta} + s_z\right) \tag{8.95}$$

$$s_{yy} = \frac{1}{r\sin\theta}\left(\frac{\partial s_y}{\partial\lambda} + s_z\sin\theta + s_x\cos\theta\right) \tag{8.96}$$

$$s_{xz} = \frac{1}{2}\left(\frac{1}{r}\frac{\partial s_z}{\partial \theta} - \frac{\partial s_x}{\partial r} + \frac{s_x}{r}\right) \tag{8.97}$$

$$s_{yz} = \frac{1}{2}\left(\frac{1}{r\sin\theta}\frac{\partial s_z}{\partial \lambda} + \frac{\partial s_y}{\partial r} - \frac{s_y}{r}\right) \tag{8.98}$$

$$s_{xy} = \frac{1}{2r}\left(\frac{1}{\sin\theta}\frac{\partial s_x}{\partial \lambda} + \frac{\partial s_y}{\partial \theta} - s_y\cot\theta\right). \tag{8.99}$$

Practically, the displacement fields are measured in discrete forms in geodesy and geophysics, and to obtain the derivatives of this displacement vector, numerical differentiation should be applied or a mathematical model used to model the displacement field over the study area. After that, the gradient operator will be applied to it to determine the derivatives of the displacements.

The tensor of stress $\boldsymbol{\tau}$ is presented in the matrix form

$$\boldsymbol{\tau} = \begin{bmatrix} \tau_{xx} & \tau_{xy} & \tau_{xz} \\ \tau_{xy} & \tau_{yy} & \tau_{yz} \\ \tau_{xz} & \tau_{yz} & \tau_{zz} \end{bmatrix} \tag{8.100}$$

where τ_{xx}, τ_{yy} and τ_{zz} are named normal stresses and τ_{xy}, τ_{xz} and τ_{yz} the shear stresses.

According to the linear elasticity theory, once $\mathbf{S}$ is determined from $\mathbf{s}$, the elements of $\boldsymbol{\tau}$ can be obtained directly from those of $\mathbf{S}$. The mathematical relation between the strain and displacement is:

$$\tau_{ij} = \tilde{\lambda}\delta_{ij}\nabla\cdot\mathbf{S} + 2\tilde{\mu}s_{ij} \quad \text{where } i,j = x, y \text{ and } z \tag{8.101}$$

and δ_{ij} stands for Kronecker's delta. s_{ij} are the elements of $\mathbf{S}$.

When $i=j$, it means that the normal stresses are connected to the normal strains as well as the volumetric strain. This means that the divergence of the strain must be computed from the mathematical model presented for the displacement field, or it should be computed numerically. This divergence is needed when the normal stresses are desired because the shear stresses are directly related to shear strain without involving divergence of displacement vector $\nabla\cdot\mathbf{s}$.

By substituting the elements of $\mathbf{S}$ into Eq. (8.101), the elements of $\boldsymbol{\tau}$ are derived in terms of strain elements and $\mathbf{s}$:

$$\tau_{zz} = \tilde{\lambda}\nabla\cdot\mathbf{s} + 2\tilde{\mu}\frac{\partial s_z}{\partial r} \tag{8.102}$$

$$\tau_{xx} = \tilde{\lambda}\nabla\cdot\mathbf{s} + \frac{2\tilde{\mu}}{r}\left(\frac{\partial s_x}{\partial\theta} + s_z\right) \tag{8.103}$$

$$\tau_{yy} = \tilde{\lambda}\nabla\cdot\mathbf{s} + \frac{2\tilde{\mu}}{r\sin\theta}\left(\frac{\partial s_y}{\partial\lambda} + s_z\sin\theta + s_x\cos\theta\right) \tag{8.104}$$

$$\tau_{xz} = \tilde{\mu}\left(\frac{1}{r}\frac{\partial s_z}{\partial\theta} + \frac{\partial s_x}{\partial r} - \frac{s_x}{r}\right) \tag{8.105}$$

$$\tau_{yz} = \tilde{\mu}\left(\frac{1}{r\sin\theta}\frac{\partial s_z}{\partial\lambda} + \frac{\partial s_y}{\partial r} - \frac{s_y}{r}\right) \tag{8.106}$$

$$\tau_{xy} = \frac{\tilde{\mu}}{r}\left(\frac{1}{\sin\theta}\frac{\partial s_\theta}{\partial\lambda} + \frac{\partial s_\lambda}{\partial\theta} - s_\lambda\cot\theta\right). \tag{8.107}$$

8.3.4 Stress and gravity field

To obtain the mathematical relations between the disturbing potential (T) of the gravity field and the elements of the stress tensor, the general solution of the partial differential equation of elasticity, Eq. (8.83), should be inserted into Eqs. (8.102)–(8.107). To derive the normal stresses, the volumetric strain $\nabla\cdot\mathbf{s}$ is required. By substituting $\mathbf{s}$, presented in Eq. (8.89) and its elements, Eqs. (8.90)–(8.92), into Eq. (8.81), $\nabla\cdot\mathbf{s}$ is derived (Fu and Huang, 1992):

$$\nabla\cdot\mathbf{s} = \sum_{n=2}^{\infty}\left\{A\left[2n + \alpha_n(3+n)\left(\frac{r}{R}\right)^{n+1}\right]\right.$$
$$\left. + B\left[-2(n+1) + \overline{\alpha}_n(2-n)\left(\frac{R_1}{r}\right)^{n}\right]\right\}\frac{t_n}{r}. \tag{8.108}$$

From Eqs. (8.89)–(8.92) and Eq. (8.102), will be:

$$\tau_{zz} = \frac{1}{r}\sum_{n=2}^{\infty}\left(\tilde{\lambda}K_n^1 + 2\tilde{\mu}K_n^2\right)t_n \tag{8.109}$$

$$\tau_{xx} = \frac{1}{r}\sum_{n=2}^{\infty}\left\{\left(\tilde{\lambda}K_n^1 + 2\tilde{\mu}K_n^3\right)t_n + 2\tilde{\mu}K_n^5\frac{\partial^2 t_n}{\partial\theta^2}\right\}, \tag{8.110}$$

$$\tau_{yy} = \frac{1}{r}\sum_{n=2}^{\infty}\left\{\left(\tilde{\lambda}K_n^1 + 2\tilde{\mu}K_n^3\right)t_n + 2\tilde{\mu}K_n^5\left[\frac{1}{\sin^2\theta}\frac{\partial^2 t_n}{\partial\lambda^2} + \cot\theta\frac{\partial t_n}{\partial\theta}\right]\right\}, \tag{8.111}$$

$$\tau_{xz} = \frac{\widetilde{\mu}}{r} \sum_{n=2}^{\infty} K_n^4 \frac{\partial t_n}{\partial \theta}, \tag{8.112}$$

$$\tau_{yz} = \frac{\widetilde{\mu}}{r \sin \theta} \sum_{n=2}^{\infty} K_n^4 \frac{\partial t_n}{\partial \lambda}, \tag{8.113}$$

$$\tau_{xy} = \frac{\widetilde{\mu}}{r \sin \theta} \sum_{n=2}^{\infty} K_n^5 \left(\frac{\partial^2 t_n}{\partial \theta \partial \lambda} - \cot \theta \frac{\partial t_n}{\partial \lambda} \right) \tag{8.114}$$

where

$$K_n^1 = A[2n + \alpha_n(3+n)] \left(\frac{r}{R}\right)^{n+1} + B[-2(n+1) + \overline{\alpha}_n(2-n)] \left(\frac{R_1}{r}\right)^n \tag{8.115}$$

$$K_n^2 = A(n+\alpha_n)(n+1)\left(\frac{r}{R}\right)^{n+1} - Bn[\overline{\alpha}_n - (n+1)]\left(\frac{R_1}{r}\right)^n + \frac{C}{R^2}(n-1)n\left(\frac{r}{R}\right)^{n-1} + \frac{D}{R_1^2}(n+1)(n+2)\left(\frac{R_1}{r}\right)^{n+2} \tag{8.116}$$

$$K_n^3 = A(n+\alpha_n)\left(\frac{r}{R}\right)^{n+1} + B[\overline{\alpha}_n - (n+1)]\left(\frac{R_1}{r}\right)^n + \frac{C}{R^2} n \left(\frac{r}{R}\right)^{n+1} - \frac{D}{R_1^2}(n+1)\left(\frac{R_1}{r}\right)^{n+2} \tag{8.117}$$

$$K_n^4 = A\left(\frac{r}{R}\right)^{n+1}(2n+\alpha_n) + B\left(\frac{R_1}{r}\right)^n(\overline{\alpha}_n - 2(n+1)) + \frac{2C}{R^2}\left(\frac{r}{R}\right)^{n-1}(n-1) - \frac{2D}{R_1^2}(n+2)\left(\frac{R_1}{r}\right)^{n+2} \tag{8.118}$$

$$K_n^5 = A\left(\frac{r}{R}\right)^{n+1} + B\left(\frac{R_1}{r}\right)^n + \frac{C}{R^2}\left(\frac{r}{R}\right)^{n-1} + \frac{D}{R_1^2}\left(\frac{R_1}{r}\right)^{n+2}. \tag{8.119}$$

8.3.5 Boundary-values and their role

Eqs. (8.109)–(8.114) are the mathematical expressions of the elements of $\boldsymbol{\tau}$ in terms of t_n. However, without the coefficients A, B, C and D, computation of these stresses is impossible. Therefore, additional information about the stress is required to find a special solution for the stress field from its general solution. By assuming the lithosphere is driven only by the shear stresses caused by the mantle convection, the sub-lithospheric shear stresses can be considered boundary-values at the sub-lithosphere, and the radial force can be considered zero. At the upper boundary of the lithosphere, all forces are assumed to be zero. This means that sub-lithospheric shear stresses should have their maximum values at the bottom of the lithosphere and decrease by increasing the distance from the sub-lithosphere. Fu and Huang (1983) considered the following boundary-values:

$$F_r = 0, \quad F_\theta = 0, \quad F_\lambda = 0 \quad \text{for } r = R \tag{8.120}$$

$$F_r = 0, \quad F_\theta = \tau_{xz}, \quad F_\lambda = \tau_{yz} \quad \text{for } r = R - D_{\text{Lith}} \tag{8.121}$$

where F means force. F_r is related to the radial stress and from which the dynamic topography can be obtained, if it is not zero.

The mantle convection produces the shear stresses τ_{xz} and τ_{yz}, and the models presented by Runcorn (1967) in Eq. (8.38) is considered for such stresses. At least four equations must be derived from these boundary-values to solve the unknown coefficients A, B, C and D. Two equations can be derived from the values of the radial stress at the upper and lower boundaries:

$$\tau_{zz}\Big|_R = \frac{1}{R}\sum_{n=2}^{\infty}\left(\tilde{\lambda}K_n^1\big|_R + 2\tilde{\mu}K_n^2\big|_R\right) t_n = 0 \quad \text{and}$$

$$\tau_{zz}\Big|_{R_1} = \frac{1}{R_1}\sum_{n=2}^{\infty}\left(\tilde{\lambda}K_n^1\big|_{R_1} + 2\tilde{\mu}K_n^2\big|_{R_1}\right) t_n = 0. \tag{8.122}$$

By considering $r = R$ for the upper boundary and $R_1 = R - D_{\text{Lith}}$ for the lower Eqs. (8.115) and (8.116) we obtain:

$$\tilde{\lambda}K_n^1\Big|_R + 2\tilde{\mu}K_n^2\Big|_R = AH_{1,n} + BH_{2,n}\left(\frac{R_1}{R}\right)^n + \frac{C}{R^2}H_{3,n} + \frac{D}{R_1^2}H_{4,n}\left(\frac{R_1}{R}\right)^{n+2} = 0 \tag{8.123}$$

$$\tilde{\lambda}K_n^1\Big|_{R_1} + 2\tilde{\mu}K_n^2\Big|_{R_1} = AH_{1,n}\left(\frac{R_1}{R}\right)^{n+1} + BH_{2,n} + \frac{C}{R^2}H_{3,n}\left(\frac{R_1}{R}\right)^{n-1} + \frac{D}{R_1^2}H_{4,n} = 0 \tag{8.124}$$

where

$$H_{1,n} = \widetilde{\mu}n(2n+\alpha_n) + \left\{2n\left(\widetilde{\lambda}+\widetilde{\mu}\right) + \alpha_n\left[(n+3)\widetilde{\lambda} + (n+2)\widetilde{\mu}\right]\right\} \quad (8.125)$$

$$H_{2,n} = \left[\widetilde{\lambda}(-2(n+1)+\overline{\alpha}_n(2-n)) - 2\widetilde{\mu}n[\overline{\alpha}_n - (n+1)]\right] \quad (8.126)$$

$$H_{3,n} = 2\widetilde{\mu}(n-1)n \quad (8.127)$$

$$H_{4,n} = 2\widetilde{\mu}(n+1)(n+2). \quad (8.128)$$

Two other equations can be written based on the values of shear stresses at the upper and lower boundaries:

$$\tau_{xz}\Big|_R = \widetilde{\mu}\sum_{n=2}^{\infty}\left\{K_n^4\Big|_R \frac{\partial t_n}{r\partial\theta}\right\} = 0 \quad (8.129)$$

$$\tau_{xz}\Big|_{R_1} = \widetilde{\mu}\sum_{n=2}^{\infty}K_n^4\Big|_{R_1}\frac{\partial t_n}{r\partial\theta} = \frac{g}{4\pi G(R-D_{\text{Lith}})}\sum_{n=2}^{\infty}\frac{2n+1}{n+1}\left(\frac{R}{R_1}\right)^{-(n+1)}\frac{\partial t_n}{\partial\theta} \quad (8.130)$$

where

$$K_n^4\Big|_R = AG_{1,n} + BG_{2,n}\left(\frac{R_1}{R}\right)^n + \frac{C}{R^2}G_{3,n} + \frac{D}{R_1^2}G_{4,n}\left(\frac{R_1}{R}\right)^{n+1} \quad (8.131)$$

$$K_n^4\Big|_{R_1} = A\left(\frac{R_1}{R}\right)^{n+1}G_{1,n} + BG_{2,n} + \frac{C}{R^2}\left(\frac{R_1}{R}\right)^{n-1}G_{3,n} + \frac{D}{R_1^2}G_{4,n} \quad (8.132)$$

with

$$G_{1,n} = \widetilde{\mu}(2n+\alpha_n) \quad (8.133)$$

$$G_{2,n} = \widetilde{\mu}(-2(n+1)+\overline{\alpha}_n) \quad (8.134)$$

$$G_{3,n} = 2\widetilde{\mu}(n-1) \quad (8.135)$$

$$G_{4,n} = -2\widetilde{\mu}(n+2). \quad (8.136)$$

Now, four equations are derived, Eqs. (8.123), (8.124), (8.131) and (8.132), and a system of equations can be established to determine A, B, C and D. This system can be written in the matrix form:

$$\mathbf{A}_n\mathbf{x}_n = \mathbf{L}_n \quad (8.137)$$

where the coefficients matrix $\mathbf{A}_n$, and the vector of observations $\mathbf{L}_n$, and the vector of unknown parameters $\mathbf{x}_n$ are:

$$\mathbf{A}_n = \begin{bmatrix} H_{1,n} & H_{2,n}\left(\frac{R_1}{R}\right)^{n} & H_{3,n} & H_{4,n}\left(\frac{R_1}{R}\right)^{n+2} \\ H_{1,n}\left(\frac{R_1}{R}\right)^{n+1} & H_{2,n} & H_{3,n}\left(\frac{R_1}{R}\right)^{n-1} & H_{4,n} \\ G_{1,n} & G_{2,n}\left(\frac{R_1}{R}\right)^{n} & G_{3,n} & G_{4,n}\left(\frac{R_1}{R}\right)^{n+2} \\ G_{1,n}\left(\frac{R_1}{R}\right)^{n+1} & G_{2,n} & G_{3,n}\left(\frac{R_1}{R}\right)^{n-1} & G_{4,n} \end{bmatrix},$$

$$\mathbf{L}_n = \begin{bmatrix} 0 \\ 0 \\ 0 \\ F_n \end{bmatrix} \text{ and } \mathbf{x}_n = \begin{bmatrix} A \\ B \\ C' \\ D' \end{bmatrix} \tag{8.138}$$

where

$$F_n = \frac{rg}{4\pi G R_1}\frac{2n+1}{n+1}\left(\frac{R}{R_1}\right)^{n+1}, \quad C' = \frac{C}{R^2} \text{ and } D' = \frac{D}{R_1^2} \tag{8.139}$$

The solution of vector $\mathbf{x}_n$, containing coefficients A, B, C and D, is:

$$\mathbf{x}_n = \mathbf{A}_n^{-1}\mathbf{L}_n \tag{8.140}$$

where $\mathbf{A}_n^{-1}$ is the regular inverse of $\mathbf{A}_n$.

One important point that should be considered to solve this system of equations is that this system is degree-dependent. This means that for each degree of Laplace coefficients, the system should be organised and solved. The determined coefficients A, B, C and D will be different from one degree to another. In addition, writing the matrix $\mathbf{A}_n$ in the presented form, Eq. (8.138), should not have the singularity or ill-conditioning problem, and simply a regular inverse matrix operator can be applied to solve the coefficients.

Once these coefficients are determined, they can simply be inserted in the mathematical models of the elements of $\boldsymbol{\tau}$, Eqs. (8.109)–(8.114). The elements of $\mathbf{S}$ will be:

$$s_{zz} = \frac{1}{r}\sum_{n=2}^{\infty} K_n^2 t_n \tag{8.141}$$

$$s_{xx} = \frac{1}{r}\sum_{n=2}^{\infty}\left(K_n^3 t_n + K_n^5 \frac{\partial^2 t_n}{\partial \theta^2}\right), \tag{8.142}$$

$$s_{yy} = \frac{1}{r}\sum_{n=2}^{\infty}\left\{K_n^3 t_n + K_n^5\left(\frac{1}{\sin^2\theta}\frac{\partial^2 t_n}{\partial \lambda^2} + \cot\theta \frac{\partial t_n}{\partial \theta}\right)\right\}, \tag{8.143}$$

$$s_{xz} = \frac{1}{2r}\sum_{n=2}^{\infty} K_n^4 \frac{\partial t_n}{\partial \theta}, \tag{8.144}$$

$$s_{yz} = \frac{\widetilde{\mu}}{2r\sin\theta}\sum_{n=2}^{\infty} K_n^4 \frac{\partial t_n}{\partial \lambda}, \tag{8.145}$$

$$s_{xy} = \frac{1}{2r\sin\theta}\sum_{n=2}^{\infty}\left\{K_n^5\left(\frac{\partial^2 t_n}{\partial\theta\partial\lambda} - \cot\theta \frac{\partial t_n}{\partial \lambda}\right)\right\}. \tag{8.146}$$

8.3.6 Application: global subcrustal stress

In this section, our goal is to present a simple application of the presented theory to determine the stress at the Moho surface. The necessary data for such computations are the lithospheric depth, crustal depth, elasticity parameters and a global gravity model. The maps of topographic heights and bathymetric depths, the lithospheric thickness computed by Conrad and Lithgow-Bertelloni (2006) and the seismic Moho depth of the CRUST1.0 model (Laske et al., 2013) are presented in Fig. 8.5A–C, respectively.

To compute the sub-lithospheric stress at the base of the lithosphere, the depth of the base of the lithosphere is needed. This depth can be determined by subtracting the topographic heights and bathymetric depths, presented in Fig. 8.4A, from the lithospheric thickness model, Fig. 8.4B. Fig. 8.4C is the map of the Moho depth presented in the CRUST1.0 model from which the stress is computed. The goal is to estimate the stress at the base of the crust using the solution of the boundary-value problem of elasticity with stresses as the boundary-values at the base of the lithosphere.

To determine the stress inside the lithosphere, two elasticity parameters of $\widetilde{\mu}$ and $\widetilde{\lambda}$ are needed. Normally, they are considered constants, but from the seismic velocities and density of the upper mantle, laterally variable models for them can be determined. Such information is available in the CRUST1.0 model. Michal Bevis, of the Ohio State University, provided a

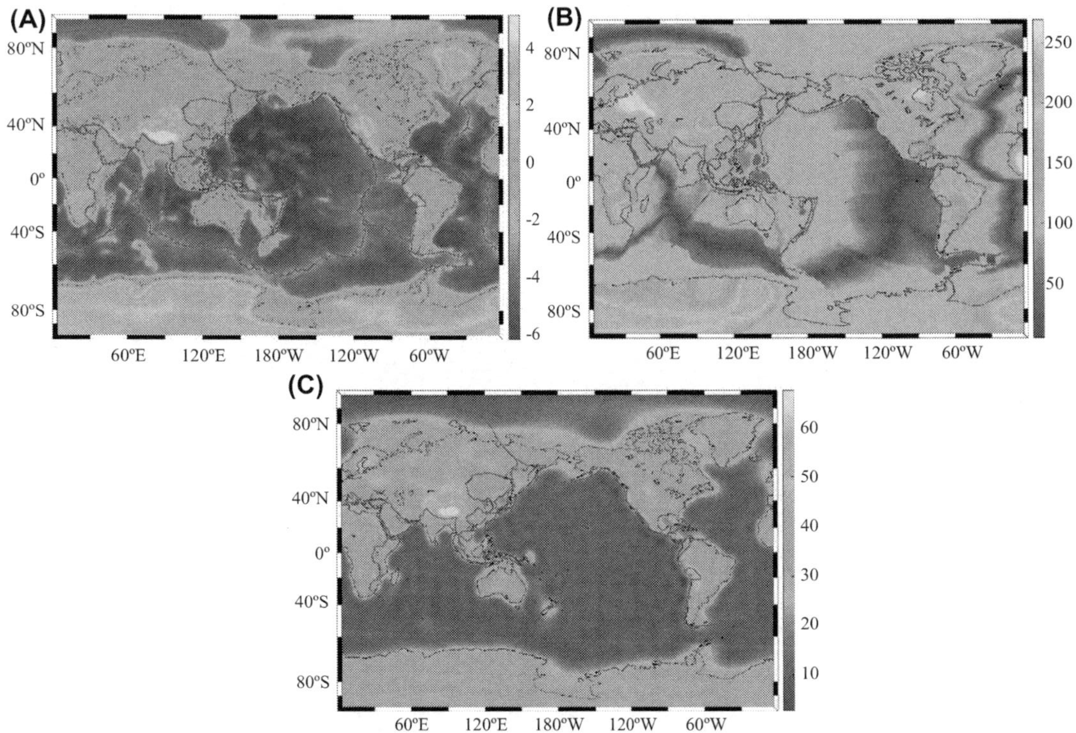

Figure 8.4 (A) Topographic heights and bathymetric depths. (B) Lithospheric thickness of Conrad and Lithgow-Bertelloni (2006). (C) seismic Moho depth of CRUST1.0 (Laske et al., 2013) (km).

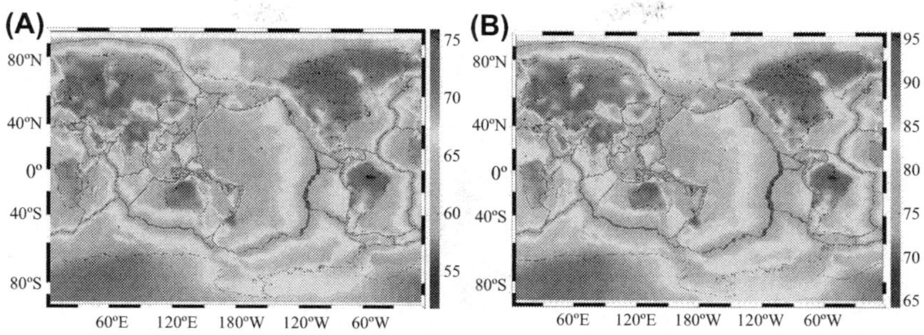

Figure 8.5 (A) $\widetilde{\mu} \times 10^{-10}$. (B) $\widetilde{\lambda} \times 10^{-10}$.

code to determine these parameters from CRUST1.0. Fig. 8.5A and B presents the maps of these parameters.

Fig. 8.5A and B show that the global pattern of $\widetilde{\mu}$ and $\widetilde{\lambda}$ are similar, but the values are different: $\widetilde{\mu}$ ranges from 50×10^{10} to 75×10^{10}, but $\widetilde{\lambda}$ is from 65×10^{10} to 95×10^{10}. Along the tectonic boundaries, both have their smallest values; over continents, they are largest.

The EGM08 gravity model, which is limited between 10 and 30 degrees, is applied to compute the lithospheric stress at the base of the crust. Maps of the elements of $\boldsymbol{\tau}$ computed from the presented method and theory are shown in Fig. 8.6. The dots on the maps shows the distribution of the stress points of the World Stress Map 2016 database (Heidbach et al., 2018). Fig. 8.6A and B are the normal stresses τ_{xx} and τ_{yy} at the subcrust. Fig. 8.6C and D are maps of τ_{zz} and τ_{xy}, respectively. As seen, τ_{zz} has smaller values than τ_{xy}. The reason is that the values of τ_{zz} at the upper and lower boundaries of the lithosphere are zero; therefore, τ_{zz} cannot be large compared with the other stresses. τ_{xy} is about one order of magnitude smaller than the τ_{xx} and τ_{yy}. Fig. 8.6E and F are maps of the shear stresses τ_{xz} and τ_{yz}, which are propagated through the elastic lithosphere from the bottom of the lithosphere to the subcrust.

By applying Eqs. (8.141)−(8.146), the elements of the strain tensor **S** can also be computed. Fig. 8.7A and B are maps of s_{xx} and s_{yy}. These maps resemble maps of the corresponding stresses in Fig. 8.7E and F. Fig. 8.7C and D are maps of s_{zz} and s_{xy}, and finally, s_{xz} and s_{yz} are maps of the shear strains in Fig. 8.7E and F.

Based on the computed coefficients, the elements of the vector of displacements can be computed by using Eqs. (8.89)−(8.92). In Fig. 8.8A, the eastward displacements s_x, reaching about 25 m, are visualised by positive values and the westward ones by the negative. Fig. 8.8B is a similar map showing the northward displacements s_y, and Fig. 8.8C, the upward

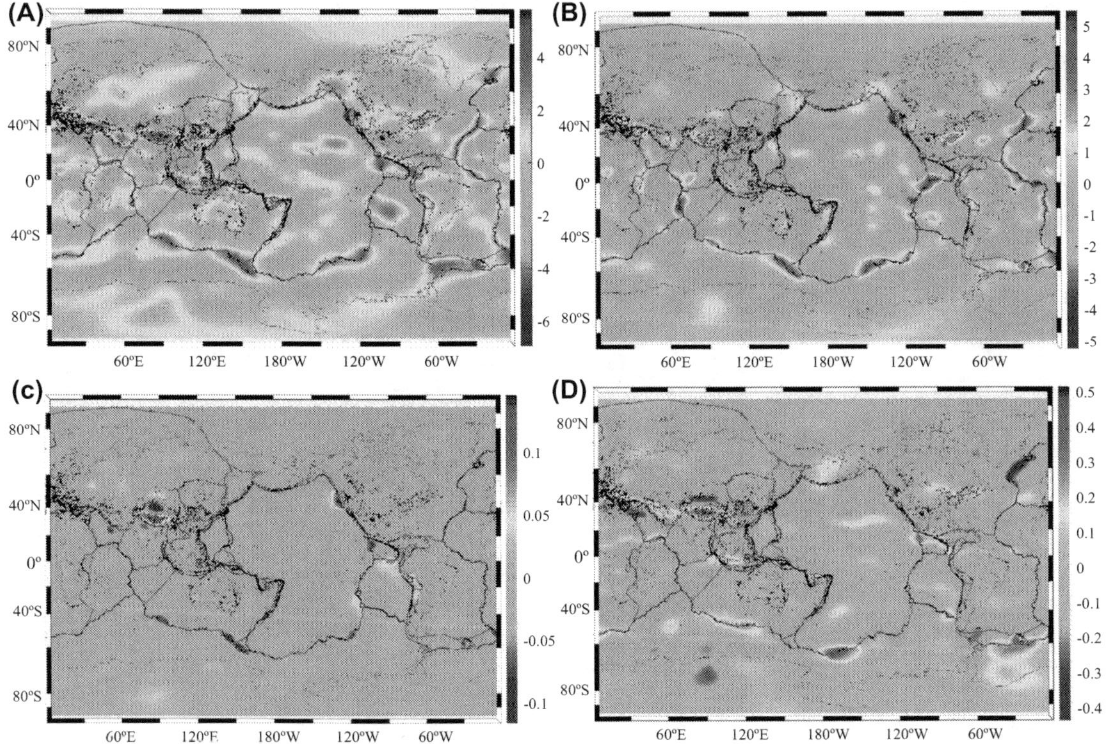

Figure 8.6 (A) τ_{xx}, (B) τ_{yy}, (C) τ_{zz}, (D) τ_{xy}, (E) τ_{xz} and (F) τ_{yz} (MPa). *Dots* represents stress points of the World Stress Map 2016 database (Heidbach et al., 2018).

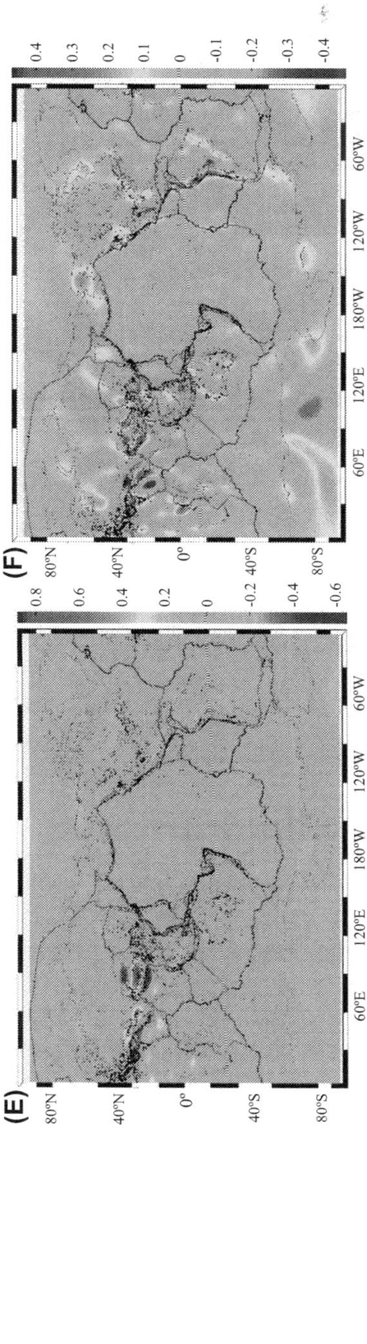

Figure 8.6 cont'd

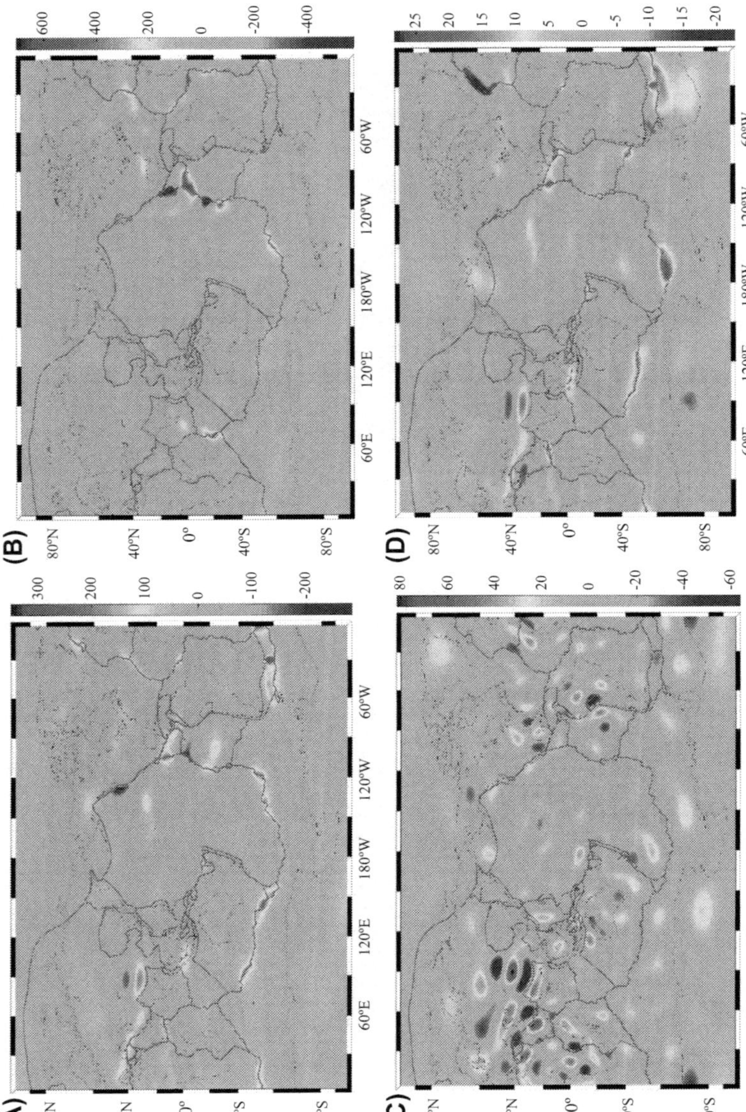

Figure 8.7 (A) s_{xx}, (B) s_{yy}, (C) s_{zz}, (D) s_{xy}, (E) s_{xz} and (F) s_{yz}.

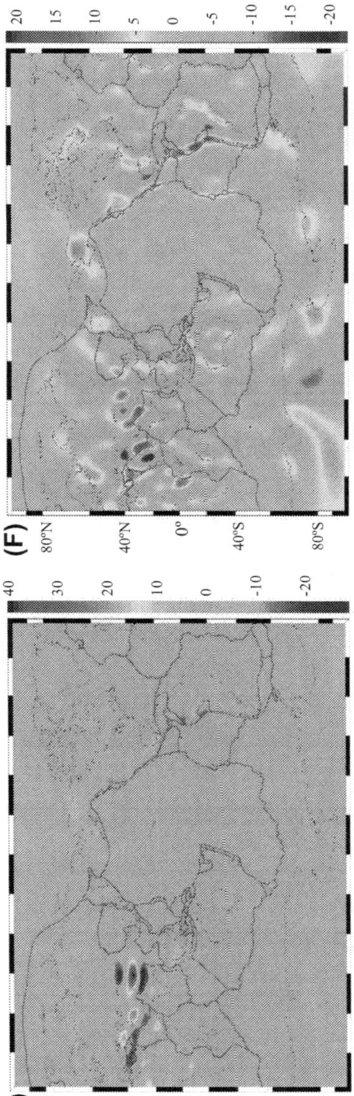

Figure 8.7 cont'd

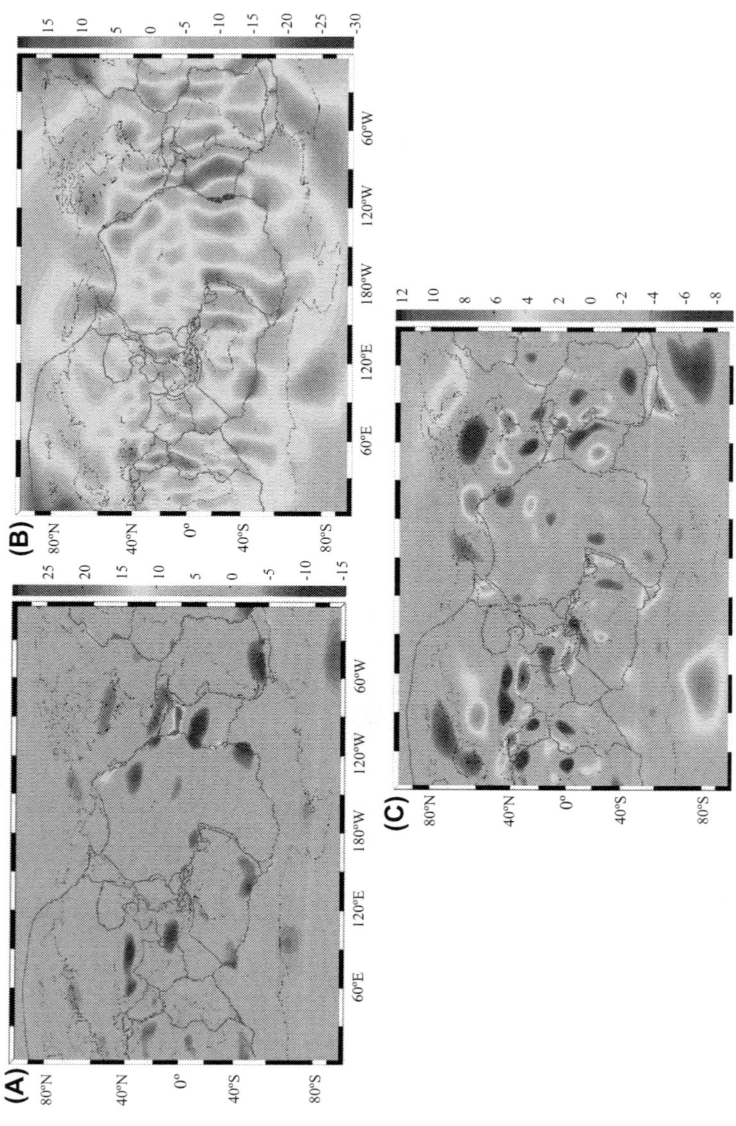

Figure 8.8 Maps of displacements (A) s_x, (B) s_y and (C) s_z at subcrust (m).

and downward s_z. Such displacements come from the geoid height from 10 to 30 degrees and should not be compared with the tectonic displacements. They indicate the contribution of the gravity field between these degrees in the formation of the subcrust.

Acknowledgements

The author is thankful to Professor Bernhard Steinberger for his guidance and advices about the Hagel and O'Connell theory for mantle convection. Andenet Ashagrie Gedamu is appreciated for checking the mathematical formulae. Seyyed Mohammad Ali Noori Rahim Abadi is acknowledged for the discussion about the mantle flow model.

References

Conrad, C.P., Lithgow-Bertelloni, C., 2006. Influence of continental roots and asthenosphere on plate-mantle coupling. Geophys. Res. Lett. 33, L05312. https://doi.org/10.1029/2005GL025621.

Corrieu, V., Thoraval, C., Ricard, Y., 1995. Mantle dynamics and geoid Green functions. Geophys. J. Int. 120, 516–523.

Elsasser, W.M., 1963. Earth Science and Meteoritics. In: Geiss, Goldberg (Eds.). North-Holland Publi. Co, Amsterdam, pp. 1–30.

Forte, A.M., Peltier, W.R., 1987. Plate tectonic and aspherical structure: the importance of poloidal and toroidal coupling. J. Geophys. Res. 92 (B4), 3645–3679.

Fu, R., Huang, P., 1983. The global stress field in the lithosphere obtained from satellite gravitational harmonics. Phys. Earth Planet. Inter. 31, 269–276.

Fu, R., Huang, J., 1992. Deep mantle flow, global tectonic pattern and the background of seismic stress field in China. Acta Seimologia Sinicia 5 (2), 271–281.

Hager, B.H., O'Connell, R.J., 1979. Kinematic model of large scale flow in the Earth's mantle. J. Geophys. Res. 84 (B3), 1031–1048.

Hager, B.H., O'Connell, R.J., 1981. A simple global model of plate dynamics and mantle convection. J. Geophys. Res. 86 (B6), 4843–4867.

Heidbach, O., Rajabi, M., Cui, X., Fuchs, K., Müller, B., Reinecker, J., Reiter, K., Tingay, M., Wenzel, F., Xie, F., Ziegler, M.O., Zoback, M.-L., Zoback, M.D., 2018. The World Stress Map database release 2016: crustal stress pattern across scales. Tectonophysics 744, 484–498.

Kaula, W.M., 1963. Elastic models of the mantle corresponding to variations in the external gravity field. J. Geophys. Res. 68 (17), 4967–4978.

Kraus, H., 1967. Thin Elastic Shells: An Introduction to the Theoretical Foundations and the Analysis of Their Static and Dynamic Behavior. John Wiley, New York, 476 pp.

Laske, G., Masters, G., Ma, Z., Pasyanos, M.E., 2013. Update on CRUST1.0—a 1-degree global model of Earth's crust. Geophys. Res. Abstr. 15 (Abstract EGU2013-2658).

Liu, H.S., 1977. Convection pattern and stress system under the African plate. Phys. Earth Planet. Inter. 15, 60–68.

Liu, H.S., 1978. Mantle convection pattern and subcrustal stress under Asia. Phys. Earth Planet. Inter. 16, 247–256.

Liu, H.S., 1979. Convection-generated stress concentration and seismogenic models of the Tangshan Earthquake. Phys. Earth Planet. Inter. 19, 307–318.

Liu, H.S., 1980. Convection generated stress field and intra-plate volcanism. Tectonophysics 65, 225–244.

Liu, H.S., 1983. A dynamical basis for crustal deformation and seismotectonic block movements in central Europe. Phys. Earth Planet. Inter. 32, 146–159.

Love, A.E.H., 1944. A Treatise on the Mathematical Theory of Elasticity. Dover Publication, New York, p. 249.

McKenzie, D.P., 1967. Some remarks on heat flow and gravity anomalies. J. Geophys. Res. 72 (24), 6261–6273.

McNutt, M., 1980. Implication of regional gravity for state of stress in the Earth's crust and upper mantle. J. Geophys. Res. 85 (B11), 6377–6396.

Milne-Thompson, L.M., 1950. Theoretical Hyrodynamic, second ed. Macmillan, London.

Ricard, Y., Fleitout, L., Froidevaux, C., 1984. Geoid heights and lithospheric stresses for a dynamic earth. Ann. Geophys. 2, 267–286.

Ricard, Y., Froidevaux, C., Fleitout, L., 1988. Global plate motion and the geoid: a physical model. Geophys. J. 93, 477–484.

Runcorn, S.K., 1964. Satellite gravity measurements and laminar viscous flow model of the earth mantle. J. Geophys. Res. 69 (20), 4389–4394.

Runcorn, S.K., 1967. Flow in the mantle inferred from the low degree harmonics of the geopotential. Geophys. J. R. Astron. Soc. 14, 375–384.

Souriau, M., Souriau, A., 1983. Global tectonics and the geoid. Phys. Earth Planet. Inter. 33, 126–136.

Steinberger, B., Schmeling, H., Marquart, G., 2001. Large-scale lithospheric stress field and topography induced by global mantle circulation. Earth Planet. Sci. Lett. 186, 75–91.

CHAPTER 9

Satellite gravimetry and lithospheric stress

9.1 Mathematical foundation based on Runcorn's formula

The determination of a stress-generating function (τ_z) from the disturbing potential (T) using the Runcorn theory (Runcorn, 1964, 1967) is straightforward. To develop this theory further for inverting the satellite gravimetry data, let us write Eq. (8.35) into the form:

$$\tau_z(R - D_{\text{Lith}}, \theta, \lambda) = k \sum_{n=0}^{\infty} \frac{2n+1}{n+1} \left(\frac{R}{R - D_{\text{Lith}}} \right)^{n+1} t_n(r, \theta, \lambda) \tag{9.1}$$

with

$$k = \frac{g}{4\pi G(R - D_{\text{Lith}})} \quad \text{and} \quad t_n(\theta, \lambda) = \sum_{m=-n}^{n} t_{nm} Y_{nm}(\theta, \lambda) \tag{9.2}$$

where g is the gravity attraction, G the Newtonian gravitational constant, R the radius of the spherical Earth, D_{Lith} the lithosphere depth, t_{nm} the disturbing potential spherical harmonic coefficient (SHC) of degree n and order m, and $Y_{nm}(\theta, \lambda)$ the spherical harmonics with arguments of colatitude θ and longitude λ.

This equation is the foundation for relating the SHCs of the Earth's gravitational potential to the stress. We can write Eq. (9.1) in terms of Laplace coefficients as well:

$$\tau_{z,n}(R - D_{\text{Lith}}, \theta, \lambda) = k \frac{2n+1}{n+1} \left(\frac{R}{R - D_{\text{Lith}}} \right)^{n+1} t_n(r, \theta, \lambda) \tag{9.3}$$

where t_n is the Laplace coefficient of T, defined at the surface of the spherical Earth. T is given at the satellite level; we show its Laplace coefficient at satellite level by $t_n(r, \theta, \lambda)$, where r stands for the geocentric distance of the satellite. $t_n(r, \theta, \lambda)$ should be continued downward to the surface of the spherical Earth so that the Runcorn theory can be applied. This means:

Satellite Gravimetry and the Solid Earth
ISBN 978-0-12-816936-0
https://doi.org/10.1016/B978-0-12-816936-0.00009-8

$$\tau_{z,n}(R - D_{\text{Lith}}, \theta, \lambda) = k\frac{2n+1}{n+1}\left(\frac{R}{R - D_{\text{Lith}}}\right)^{n+1}\left(\frac{r}{R}\right)^{n+1} t_n(r, \theta, \lambda) \tag{9.4}$$

where $(r/R)^{n+1}$ is the downward continuation factor.

Solving Eq. (9.4) for $t_n(r, \theta, \lambda)$ reads:

$$\frac{1}{k}\frac{n+1}{2n+1}\left(\frac{R - D_{\text{Lith}}}{r}\right)^{n+1} \tau_{z,n}(R - D_{\text{Lith}}, \theta, \lambda) = t_n(r, \theta, \lambda) \tag{9.5}$$

The spectral relation presented in Eq. (9.5) is between T at satellite level and τ_z at the sub-lithosphere. Eq. (9.5) should be presented in the spatial form to obtain an integral equation for recovering τ_z from T. On the basis of Eq. (1.11), the Laplace coefficient of degree n for τ_z will be:

$$\tau_{z,n}(R - D_{\text{Lith}}, \theta, \lambda) = \frac{2n+1}{4\pi}\iint_\sigma \tau_z(R - D_{\text{Lith}}, \theta', \lambda') P_n(\widetilde{x}) d\sigma \tag{9.6}$$

where $P_n(\widetilde{x})$ is the Legendre polynomial of degree n and $\widetilde{x} = \cos\psi$, and ψ is the geocentric angle between the computation and integration point with spherical coordinates of (θ, λ) and (θ', λ'), respectively.

By inserting Eq. (9.6) into Eq. (9.5), the following integral equation will be obtained:

$$\frac{1}{4\pi k}\iint_\sigma \Xi(r, \psi)\tau_z(R - D_{\text{Lith}}, \theta', \lambda') d\sigma = T(r, \theta, \lambda) \tag{9.7}$$

with the kernel

$$\Xi(r, \psi) = \sum_{n=0}^{\infty}(n+1)s^{n+1}P_n(\widetilde{x}) = \frac{s}{D} - \frac{s(s-\widetilde{x})}{D^3} \quad \text{with } s = \frac{R - D_{\text{Lith}}}{r} \tag{9.8}$$

The integral (9.7) is the mathematical foundation of the connection of the gravitational potential outside the Earth's surface to τ_z. However, this function is not the northward and eastward shear stresses at the sub-lithosphere. To obtain these stresses, the recovered τ_z should be differentiated numerically. This idea was presented by Eshagh (2014) and further implemented by Eshagh and Romeshkani (2015), Eshagh and Tenzer (2015), and Tenzer and Eshagh (2015).

9.2 Sub-lithospheric shear stresses from satellite gradiometry data

The idea of determining sub-lithospheric shear stresses from satellite gradiometry data and the development of the Runcorn theory were presented by Eshagh (2014). He offered different integral approaches to determine such stresses, including integral equations and combined integral estimators by a spectral combination of satellite gradiometry data. In this section, some parts of his mathematical derivations are presented and discussed.

Based on the integral Eq. (9.7) and applying the operators Eqs. (1.95)–(1.100) to this integral, the following integral equations are derived for the inversion of the satellite gravimetry data in the local north-oriented frame (LNOF) to the stress generating function at the base of Lithosphere:

$$\frac{1}{4\pi k}\iint_{\sigma} \Xi_{ij}(r,\psi)\tau_z(\theta',\lambda')\mathrm{d}\sigma = T_{ij}(r,\theta,\lambda) \ \text{ with } \ i,j = x,y,z \tag{9.9}$$

where

$$\begin{bmatrix} \Xi_{xx}(r,D_{\mathrm{Lith}},\psi) \\ \Xi_{yy}(r,D_{\mathrm{Lith}},\psi) \end{bmatrix} = \frac{\Xi_r(r,D_{\mathrm{Lith}},\psi)}{r} + \frac{1}{2r^2}\{[\Xi_{\psi\psi}(r,D_{\mathrm{Lith}},\psi) + \cot\psi\,\Xi_\psi(r,D_{\mathrm{Lith}},\psi)] \pm [\Xi_{\psi\psi}(r,D_{\mathrm{Lith}},\psi) - \cot\psi\,\Xi_\psi(r,D_{\mathrm{Lith}},\psi)]\cos 2\alpha\} \tag{9.10}$$

$$\begin{bmatrix} \Xi_{xz}(r,D_{\mathrm{Lith}},\psi) \\ \Xi_{yz}(r,D_{\mathrm{Lith}},\psi) \end{bmatrix} = \frac{1}{r}\left[\frac{1}{r}\Xi_\psi(r,D_{\mathrm{Lith}},\psi) - \Xi_{r\psi}(r,D_{\mathrm{Lith}},\psi)\right]\begin{pmatrix} \cos\alpha \\ -\sin\alpha \end{pmatrix} \tag{9.11}$$

$$\Xi_{xy}(r,D_{\mathrm{Lith}},\psi) = -\frac{1}{2r^2}[\Xi_{\psi\psi}(r,D_{\mathrm{Lith}},\psi) - \cot\psi\,\Xi_\psi(r,D_{\mathrm{Lith}},\psi)]\sin 2\alpha \tag{9.12}$$

and

$$\Xi_r(r,D_{\mathrm{Lith}},\psi) = \frac{1}{-r}\sum_{n=2}^{\infty}(n+1)^2\left(\frac{R-D_{\mathrm{Lith}}}{r}\right)^{n+1}P_n(\tilde{x}) \tag{9.13}$$

$$\Xi_{zz}(r,D_{\mathrm{Lith}},\psi) = \sum_{n=2}^{\infty}\frac{(n+1)^2(n+2)}{r^2}\left(\frac{R-D_{\mathrm{Lith}}}{r}\right)^{n+1}P_n(\tilde{x}) \tag{9.14}$$

$$\Xi_{r\psi}(r,D_{\mathrm{Lith}},\psi) = \frac{1}{-r}\sum_{n=2}^{\infty}(n+1)^2\left(\frac{R-D_{\mathrm{Lith}}}{r}\right)^{n+1}\frac{\mathrm{d}P_n(\tilde{x})}{\mathrm{d}\psi} \tag{9.15}$$

$$\Xi_{\psi\psi}(r, D_{\text{Lith}}, \psi) = \sum_{n=2}^{\infty}(n+1)\left(\frac{R - D_{\text{Lith}}}{r}\right)^{n+1} \frac{d^2 P_n(\widetilde{x})}{d\psi^2} \tag{9.16}$$

After a relatively long simplification, the closed-form formulae of these kernels are:

$$\begin{aligned}\Xi_{zz}(r, D_{\text{Lith}}, \psi) =& \frac{s(2-s^2)}{r^2 D} + \frac{2s^2(-8s+5\widetilde{x})}{r^2 D^3} + \frac{3s^3(s-\widetilde{x})(10s-7\widetilde{x})}{r^2 D^5} \\ & - \frac{15s^4(s-\widetilde{x})^3}{r^2 D^7} - \frac{1}{r^2} - \frac{\widetilde{x}s}{r^2}\end{aligned} \tag{9.17}$$

$$\Xi_r(r, D_{\text{Lith}}, \psi) = -\frac{s}{rD} + \frac{s^2 + s(s-\widetilde{x})(s+1)}{rD^3} - \frac{3s^2(s-\widetilde{x})^2}{rD^5} + \frac{s+4s^2\widetilde{x}}{-r} \tag{9.18}$$

$$\begin{aligned}\Xi_{r\psi}(r, D_{\text{Lith}}, \psi) =& \left(\frac{s(2s+1)}{rD^3} - \frac{3s^2(s+(s-\widetilde{x})(s+3))}{rD^5}\right. \\ & \left. + \frac{15s^3(s-\widetilde{x})^2}{rD^7} - \frac{4s^2}{r}\right)\sqrt{1-\widetilde{x}^2}\end{aligned} \tag{9.19}$$

$$\begin{aligned}\Xi_{\psi\psi}(r, D_{\text{Lith}}, \psi) =& -\frac{s\widetilde{x}(s+1)}{D^3} + \frac{3(1-\widetilde{x}^2)s^2(s+2) + 3s^2\widetilde{x}(s-\widetilde{x})}{D^5} \\ & - \frac{15s^3(1-\widetilde{x}^2)(s-\widetilde{x})}{D^7} + 2s^2\widetilde{x}\end{aligned} \tag{9.20}$$

where $\widetilde{x} = \cos\psi$ and $D = \sqrt{1+s^2-2s\widetilde{x}}$.

9.3 Sub-lithospheric stress from vertical-horizontal satellite gravity gradients

The northward and eastward shear stresses can be determined directly by inverting the vertical-horizontal (VH) derivatives of T measured by a satellite. Unlike the previous section, no τ_z needs to be estimated by inverting these gradients. According to Eqs. (8.36) and (8.37) we have:

$$\left.\begin{aligned}\tau_{zx}(R - D_{\text{Lith}}, \theta, \lambda) \\ \tau_{zy}(R - D_{\text{Lith}}, \theta, \lambda)\end{aligned}\right\} = k\sum_{n=2}^{\infty}\frac{2n+1}{n+1}\left(\frac{R}{R - D_{\text{Lith}}}\right)^{n+1}\left\{\begin{aligned}&\frac{\partial t_n(R,\theta,\lambda)}{\partial\theta} \\ &\frac{\partial t_n(R,\theta,\lambda)}{\sin\theta\partial\lambda}\end{aligned}\right. . \tag{9.21}$$

If Eq. (9.21) is written in terms of the Laplace coefficients, we obtain:

$$\left.\begin{matrix}\tau_{zx,n}(R-D_{\text{Lith}},\theta,\lambda)\\ \tau_{zy,n}(R-D_{\text{Lith}},\theta,\lambda)\end{matrix}\right\}=k\frac{2n+1}{n+1}\left(\frac{R}{R-D_{\text{Lith}}}\right)^{n+1}\left\{\begin{matrix}\dfrac{\partial t_n(R,\theta,\lambda)}{\partial\theta}\\ \dfrac{\partial t_n(R,\theta,\lambda)}{\sin\theta\partial\lambda}\end{matrix}\right.. \tag{9.22}$$

From Eqs. (2.133) and (2.134), similar SHCs for the VH gradients will be:

$$\left.\begin{matrix}T_{zx,n}(r,\theta,\lambda)\\ T_{zy,n}(r,\theta,\lambda)\end{matrix}\right\}=\frac{1}{r^2}(n+2)\left(\frac{R}{r}\right)^{n+1}\left\{\begin{matrix}\dfrac{\partial t_n(R,\theta,\lambda)}{\partial\theta}\\ \dfrac{\partial t_n(R,\theta,\lambda)}{\sin\theta\partial\lambda}\end{matrix}\right.. \tag{9.23}$$

Comparison of Eqs. (9.22) and (9.23) leads to

$$\left.\begin{matrix}\tau_{zx,n}(R-D_{\text{Lith}},\theta,\lambda)\\ \tau_{zy,n}(R-D_{\text{Lith}},\theta,\lambda)\end{matrix}\right\}\frac{1}{k}\frac{n+1}{2n+1}\left(\frac{R-D_{\text{Lith}}}{R}\right)^{n+1}=\frac{r^2}{n+2}\left(\frac{r}{R}\right)^{n+1}\left\{\begin{matrix}T_{zx,n}(r,\theta,\lambda)\\ T_{zy,n}(r,\theta,\lambda)\end{matrix}\right.. \tag{9.24}$$

Solving Eq. (9.24) for the VH gradients yields:

$$\left.\begin{matrix}\tau_{zx,n}(R-D_{\text{Lith}},\theta,\lambda)\\ \tau_{zy,n}(R-D_{\text{Lith}},\theta,\lambda)\end{matrix}\right\}\frac{1}{k}\frac{(n+1)(n+2)}{(2n+1)r^2}\left(\frac{R-D_{\text{Lith}}}{r}\right)^{n+1}=\left\{\begin{matrix}T_{zx,n}(r,\theta,\lambda)\\ T_{zy,n}(r,\theta,\lambda)\end{matrix}\right.. \tag{9.25}$$

Taking summation over n from both sides of the result reads:

$$\frac{1}{kr^2}\sum_{n=2}^{\infty}\frac{(n+1)(n+2)}{(2n+1)}\left(\frac{R-D_{\text{Lith}}}{r}\right)^{n+1}\left\{\begin{matrix}\tau_{zx,n}(R-D_{\text{Lith}},\theta,\lambda)\\ \tau_{zy,n}(R-D_{\text{Lith}},\theta,\lambda)\end{matrix}\right.=\left\{\begin{matrix}T_{zx}(r,\theta,\lambda)\\ T_{zy}(r,\theta,\lambda)\end{matrix}\right.. \tag{9.26}$$

The Laplace coefficients of the shear stresses are obtained by Eq. (1.11):

$$\left.\begin{matrix}\tau_{zx,n}(R-D_{\text{Lith}},\theta,\lambda)\\ \tau_{zy,n}(R-D_{\text{Lith}},\theta,\lambda)\end{matrix}\right\}=\frac{2n+1}{4\pi}\iint_{\sigma}\left\{\begin{matrix}\tau_{zx}(R-D_{\text{Lith}},\theta',\lambda')\\ \tau_{zy}(R-D_{\text{Lith}},\theta',\lambda')\end{matrix}\right\}P_n(\widetilde{x})d\sigma. \tag{9.27}$$

Inserting Eq. (9.27) into Eq. (9.26), Eq. (1.9), leads to these integral equations for the direct determination of sub-lithospheric shear stresses:

$$\frac{1}{4\pi kr^2}\iint_\sigma \Xi_{\mathrm{VH}}(r, D_{\mathrm{Lith}}, \psi)\begin{Bmatrix} \tau_{zx}(R-D_{\mathrm{Lith}}, \theta', \lambda') \\ \tau_{zy}(R-D_{\mathrm{Lith}}, \theta', \lambda') \end{Bmatrix} d\sigma = \begin{Bmatrix} T_{zx}(r, \theta, \lambda) \\ T_{zy}(r, \theta, \lambda) \end{Bmatrix} \tag{9.28}$$

with

$$\Xi_{\mathrm{VH}}(r, D_{\mathrm{Lith}}, \psi) = \sum_{n=2}^{\infty}(n+1)(n+2)\left(\frac{R-D_{\mathrm{Lith}}}{r}\right)^{n+1} P_n(\widetilde{x}) \tag{9.29}$$

which has the closed-form formula:

$$\Xi_{\mathrm{VH}}(r, D_{\mathrm{Lith}}, \psi) = \frac{2s}{D} + \frac{s(1-(s-\widetilde{x})(1-2s))}{D^3} - \frac{3s(s-\widetilde{x})^2}{D^5} - s - 6s^2\widetilde{x}. \tag{9.30}$$

9.4 Example: application of Gravity Field and Ocean Circulation Explorer data for determining sub-lithospheric shear stresses in Iran

Here, the numerical study performed by Eshagh and Romeshkani (2015) is presented. Their goal was to determine northward and eastward sub-lithospheric shear stresses resulting from mantle convection for a lithospheric depth of 100 km from gradiometry data of the gravity field and steady-state ocean circulation explorer (GOCE, ESA, 1999). Their study area was limited to latitudes 10°N and 60°N and longitudes 25°E and 80°E containing Iran and its neighboring countries. Fig. 9.1A shows the maps of the second-order radial derivatives of T determined from Gravity Field and Ocean Circulation Explorer (GOCE) over the study area. The maximum, mean, minimum, and standard deviation of these data at a rate of 10 s were 1.41, 0.00, −1.41, and 0.33E, respectively.

The spatial truncation error (STE) of the integral formula is a limiting factor in the local inversion of satellite gradiometry. To reduce it, an area larger than the desired one is selected and the result in the central area is selected as the final result after the inversion process. Because the second-order radial derivatives of disturbing potential ($T_{rr} = T_{zz}$) is used and derivatives of spherical harmonics or Legendre polynomials are not involved with the mathematical modelling, the kernel $\Xi_{zz}(r, \widetilde{D}_{\mathrm{Lith}}, \psi)$ will be well-behaved (Eshagh, 2011) and the effect of STE is reducible. Fig. 9.1B illustrates that this kernel has values close to zero after a

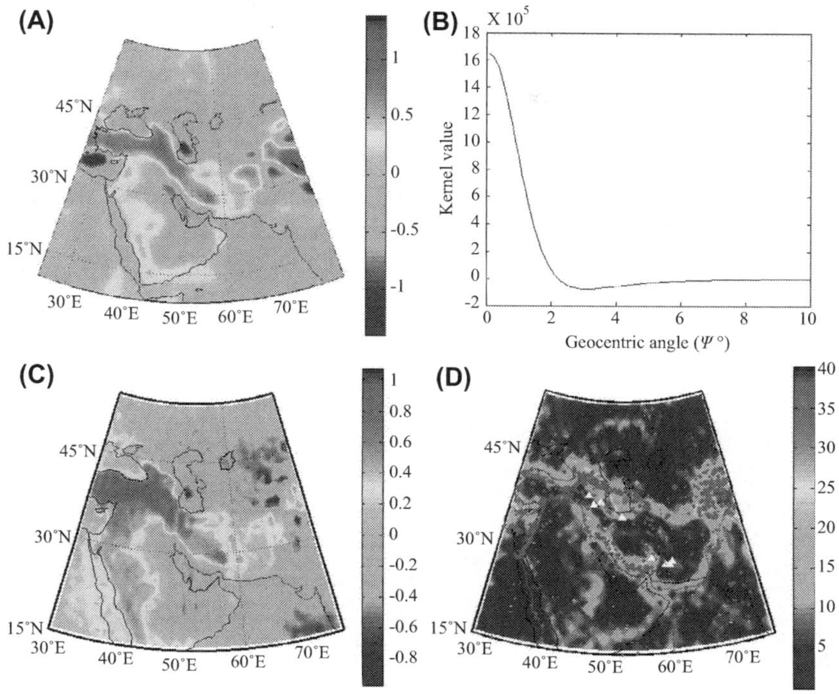

Figure 9.1 (A) Real T_{rr} measured by Gravity Field and Ocean Circulation Explorer [E], (B) Behaviour of the kernel $\Xi_{zz}(r, D_{\text{Lith}}, \psi)$ up to a geometric angle of 10 degrees, (C) recovered τ_z(MPa), and (D) $\sqrt{\tau_{zx}^2 + \tau_{zy}^2}$ and distribution of seismic points (MPa). *(From Eshagh M, Romeshkani M. Determination of sub-lithospheric stress due to mantle convection using GOCE gradiometric data over Iran. J. Appl. Geophys 2015;122:11–17.)*

geocentric angle of about 5 degrees. This means that the central area should be smaller than the whole area by 5 degrees from each side to reduce the effect of the STE sufficiently.

Fig. 9.1C shows the maps of the recovered τ_z with a resolution of 1×1 degrees from T_{rr} measured by GOCE. The shear stresses τ_{zx} and τ_{zy} should be computed numerically from this recovered τ_z. Fig. 9.1D is the map of the stress magnitude computed by $\sqrt{\tau_{zx}^2 + \tau_{zy}^2}$, as a background of seismic points of the World Stress Map (Heidbach et al. 2010). It displays good agreement between the stress magnitude and the distribution of seismic points over the area.

9.5 Example: application of Gravity Field and Ocean Circulation Explorer and seismic data for sub-lithospheric stress modelling over Indo-Pak region

This section summarises GOCE and seismic data for sub-lithospheric stress modelling that were applied by Eshagh et al. (2016). Here, the gravimetric isostasy theory was applied for the seismic data of a CRUST1.0 model and two τ_z were obtained considering the effect of sediments, crustal crystalline stress and density contrast between the crust and upper mantle. This means that this approach provides more information about the properties of the lithosphere that are considered for sub-lithospheric stress modelling of the Indo-Pak plate.

The method presented by Eshagh (2014), presenting the relation between the Moho depth and sub-lithospheric stress is applied to determine the contribution of the crustal heterogeneities. Topographic-bathymetric data, the EGM08 gravity model (Pavlis et al., 2012), to degree and order 180, and the effect of sediments and consolidated crust are computed according to seismic data presented in the CRUST1.0 model (Laske et al., 2013).

The study area was limited to latitudes 0°N to 60°N and longitudes 40°E to 120°E covering Pakistan, Afghanistan, India, Iran, and some parts of surrounding countries. The SRTM30 (Farr et al., 2007) digital elevation model is used to compute the effect of topographic-bathymetric masses on the gravity field. Fig. 9.2A shows the topographic-bathymetric heights generated by this model to degree and order 2160 with $5' \times 5'$ resolution over the study area. Fig. 9.2B is the map of the seismic Moho of CRUST1.0. The maximum, mean, minimum, and standard deviation of the topographic-bathymetric heights are respectively 6.6, −0.1, −5.6, and 2.2 km; the corresponding statistics for the seismic Moho depths of CRUST1.0 are 72.9, 35.3, 5.3, and 14.1 km.

To compute the gravimetric Moho model of the area, the continent and ocean are considered separately, and the mean Moho depth over these areas is taken from the seismic Moho model of CRUST1.0. Density contrasts of 600 and 480 kg/m^3 are considered over the land and ocean areas, respectively. EGM08 and SRTM30 are used to compute the gravimetric model based on approach 1 (see Section 6.3.3.1). The map of this Moho model is illustrated in Fig. 9.2C.

In addition, the on-orbit T_{rr} measured by GOCE is shown in Fig. 9.2D. These data were collected for less than 5 months, from the Jan 1 to the end of May 2012, with an interval of 100 s containing 6475 gradients. The maximum, mean, minimum, and standard deviation of these data are 1.32, −0.11, −1.38, and 0.36, respectively, in the Eötvös unit.

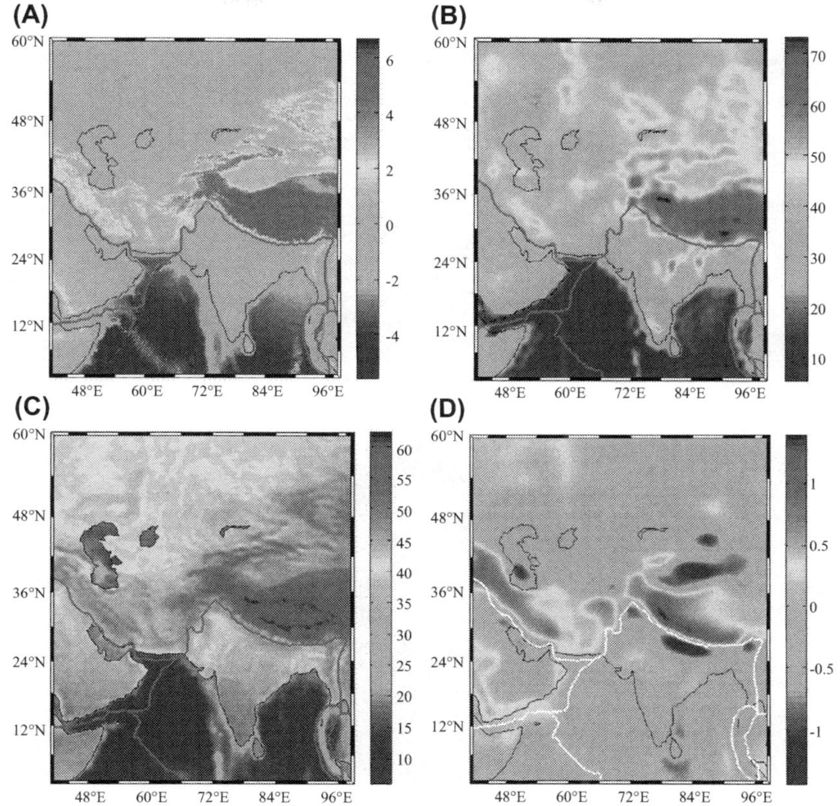

Figure 9.2 (A) Topographic-bathymetric heights (km), (B) seismic Moho model of CRUST1.0 (km), (C) gravimetric Moho model (km), and (D) Gravity Field and Ocean Circulation Explorer data T_{rr}[E] over the study area. *(From Eshagh M., Hussain M. and Tiampo K.F., Towards sub-lithospheric stress determination from seismic Moho, topographic heights and GOCE data, J. Asian Earth Sci. 129, 2016, 1–12.)*

Fig. 9.3A–C are maps of τ_z computed from the gravimetric Moho model, the seismic Moho model of CRUST1.0, and an inversion of the on-orbit GOCE data.

Fig. 9.4A shows the result of the joint inversion of the gravimetric, seismic, and GOCE data to τ_z over the area. Fig. 9.4B shows the distribution of active points of the WSM08 database as well as the tectonic boundaries on the background of the magnitude of stress recovered from the combined τ_z. It shows that the seismic points have a limited distribution in the northern part of the study area. In the northwestern part, we observe one seismic point and some close to the middle part of the northern area,

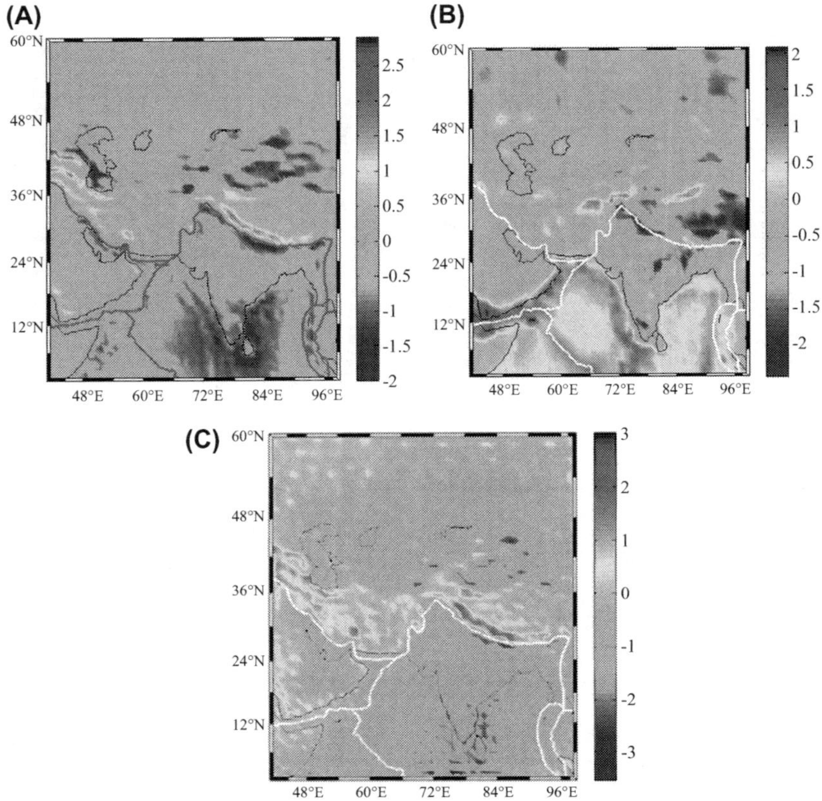

Figure 9.3 τ_z computed from (A) gravimetric Moho model, (B) seismic Moho model, and (C) Gravity Field and Ocean Circulation Explorer data (MPa). *(From Eshagh M., Hussain M. and Tiampo K.F., Towards sub-lithospheric stress determination from seismic Moho, topographic heights and GOCE data, J. Asian Earth Sci. 129, 2016, 1–12.)*

which are extended south. The map has good agreement with WSM08. For more details and geophysical interpretations, see Eshagh et al. (2016).

9.6 Example: considering lithospheric mass and structure heterogeneities to determine sub-lithospheric shear stress

In the first example, Section 9.4, performed by Eshagh and Romeshkani (2015) the GOCE data were inverted to τ_z according to the Runcorn theory. No additional information has been incorporated to generate sub-lithospheric stress. In Section 9.5, more data have been used to

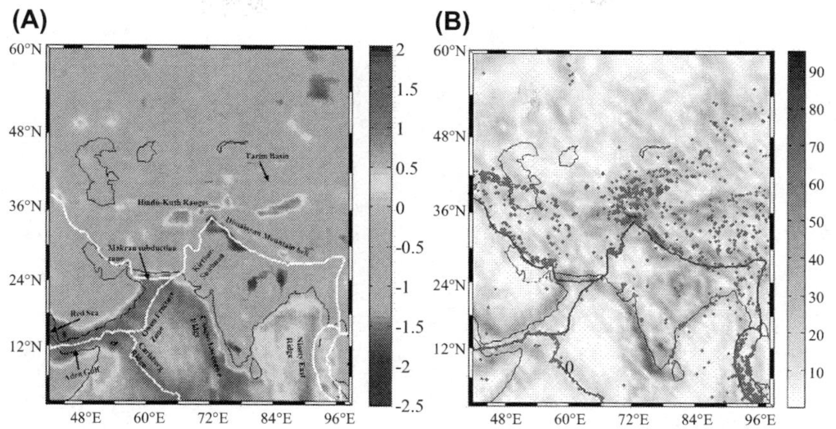

Figure 9.4 (A) τ_z computed by combining topographic-bathymetric data of SRTM30, seismic Moho model of CRUST1.0 and Gravity Field and Ocean Circulation Explorer data (MPa), and (B) magnitude of the shear stress and the distribution of the seismic points (MPa). *(From Eshagh M., Hussain M. and Tiampo K.F., Towards sub-lithospheric stress determination from seismic Moho, topographic heights and GOCE data, J. Asian Earth Sci. 129, 2016, 1–12.)*

compute τ_z (Eshagh et al. 2016). Both studies have a common issue: the contribution of high frequencies of the gravity field on the generated stress field. As shown in Chapter 8, the high frequencies of the gravity field should not be considered to determine sub-lithospheric stress because they come mainly from the lithosphere. This means that to apply real satellite gravimetry data, the gravitational effects of the lithospheric mass and structure heterogeneities (see Chapter 5) should be removed from them. In this case, those frequencies related to sub-lithospheric masses can be enhanced.

Here, a combination of the gravimetric and flexural isostasy is applied to reduce the real on-orbit data of GOCE (T_{rr}), for the lithospheric mass and structure heterogeneities; thereafter, the reduced GOCE data are inverted to τ_z.

9.6.1 Data and area

Here, South America is selected as the study area owing to its special geological features. The GOCE T_{rr} during Jan. 2013 with a 10-s data-sampling interval (21,920 values in total) is shown in Fig. 9.5A. Large positive values are seen along the Andes with maxima (up to 1.5E) in its central part. The largest negative values mark locations of the oceanic subductions along the Puerto Rico and Peru–Chile trenches. Elsewhere, the values are typically within the interval of ±0.5E.

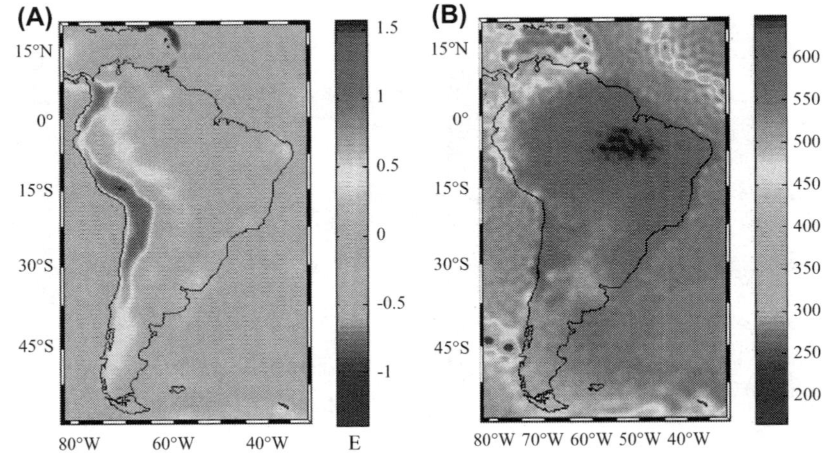

Figure 9.5 (A) T_{rr} measured by Gravity Field and Ocean Circulation Explorer [E], and (B) density contrast model from CRUST1.0 (kg/m^3). *(From Eshagh M, Steinberger B. Tenzer R, Tassara A. Comparison of gravimetric and mantle flow solutions for lithospheric stress modelling and their combination. Geophys. J. Int. 2018;213(2):1013–1028.)*

The 1 × 1 arc-deg data from CRUST1.0 are used to compute the variable Moho density contrast of the area (Fig. 9.5B), with maxima exceeding 600 kg/m^3, in agreement with locations of the Brazilian Shield and São Francisco Craton, and minima are typically along the midoceanic ridges. A relatively small Moho density contrast is also detected along the oceanic subduction under the Puerto Rico Trench, whereas such a feature is absent along the Peru–Chile trench.

9.6.2 The effect of topographic-bathymetric, sediment and consolidated crustal masses on the on-orbit Gravity Field and Ocean Circulation Explorer data

The effects of the topography, bathymetry, sediments and consolidated crustal masses on T_{rr} of GOCE are presented in Fig. 9.6A–C, respectively. The combined topographic-bathymetric correction was computed using SRTM30 data for the average density of the upper continent crust of 2670 kg/m^3 (Hinze, 2003). The same density value was adopted to define the density contrasts for the corrections. The seawater density of 1027 kg/m^3 was adopted for the oceanic density contrast. A comparison of the values of these corrections and T_{rr} shows that the corrections are considerably larger than T_{rr}. This is normal because no compensation has been considered for these corrections or effects. Therefore, it is necessary to consider a

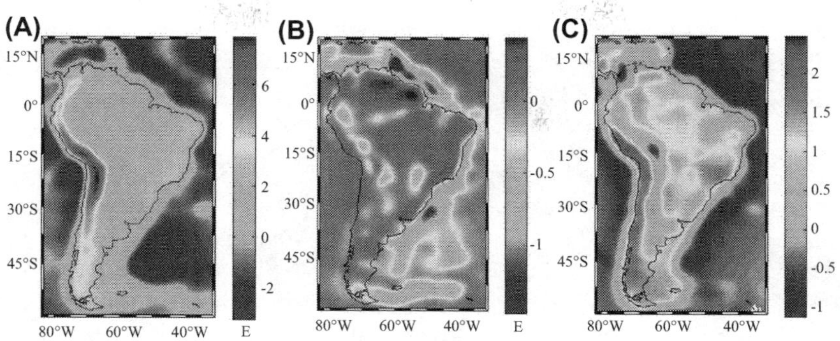

Figure 9.6 Effects of (A) topography and bathymetry, (B) sediments and (C) consolidated crust on T_{rr}. *(From Eshagh M, Steinberger B. Tenzer R, Tassara A. Comparison of gravimetric and mantle flow solutions for lithospheric stress modelling and their combination. Geophys. J. Int. 2018;213(2):1013–1028.)*

compensation mechanism to reduce their values. Here, the flexural isostasy is applied to create this compensation mechanism. This theory requires elastic thickness of the area; here, the one presented by Tassara et al. (2007) has been adopted for such a purpose. The effects of the sediment, topography-bathymetry and consolidated crustal loads have been considered in the flexural isostasy to generate the required compensation. Fig. 9.7A shows the elastic thickness model determined by Tassara et al. (2007), and Fig. 9.7B, the determined Moho model based on this elastic thickness model and the mentioned loads considering the density contrast model, presented in

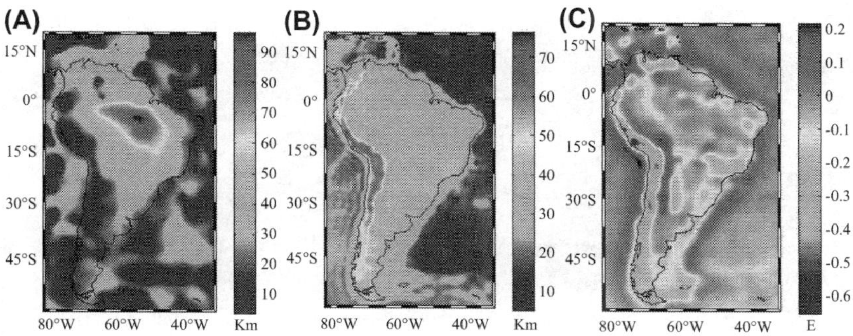

Figure 9.7 Regional maps of (A) the elastic thickness of the lithosphere according to Tassara et al. (2007), and (B) the Moho depth and (C) combined contribution of crustal density heterogeneities and the Moho depth on the gravitational gradient. *(From Eshagh M, Steinberger B. Tenzer R, Tassara A. Comparison of gravimetric and mantle flow solutions for lithospheric stress modelling and their combination. Geophys. J. Int. 2018;213(2):1013–1028.)*

Fig. 9.5B. The combined contribution from the topography, bathymetry, sediments and density heterogeneities within the remaining crust (down to and including the Moho interface) is shown in Fig. 9.7C, with values mostly between −0.7E and 0.2E. The isostatic signature of the Andes marked by the largest negative values is flanked on both sides of the central Andes by large positive values along the subduction zone of the Nazca plate and the zone of convergence between the Altiplano and South American plates.

9.6.3 Sub-lithospheric shear stresses from Gravity Field and Ocean Circulation Explorer data with and without considering lithospheric mass and structure heterogeneities

A lithospheric depth of 100 km is assumed to estimate the sub-lithospheric shear stresses resulting from mantle convection from the GOCE data. The integral Eq. (9.9) with the kernel (9.14) is discretised based on the resolution $1° \times 1°$ corresponding to 4290 values over the study area. The computation was realised with the spectral kernel limited to 31 degrees. τ_z is determined first without considering the lithospheric heterogeneities; afterwards, the northward and eastward shear stresses are determined numerically from it. Fig. 9.8A and B shows these stresses, and Fig. 9.8C, the magnitude of the shear stress and the stress pattern, which agrees closely with a tectonic configuration of South America. A prevailing convergent stress-vector orientation along the central Andes agrees with the compressional tectonism of orogenic formations. This stress pattern indicates that the estimation of the stress field from the uncorrected GOCE data reflects more shallow stresses within the crust rather than stresses deeper in the mantle (Eshagh et al., 2018). Elsewhere, the stress intensity is relatively small. Overall, the maximum stress intensity occurs in tectonically active regions, whereas old, stable cratons are characterised by the lower stress magnitude.

By considering the contributions of topography and lithospheric density heterogeneities, the northward and eastward shear stresses are computed. Figs. 9.8D and E show these respective stresses after the reductions. Similarly, the spectral resolution of the integral kernel is 31 for determining τ_z at a depth of 100 km. The GOCE data comprise the gravitational signal of crustal density heterogeneities that should be subtracted to reveal the mantle signature. The spatial distribution of the recovered τ_z becomes smoother, reducing the intensity of the resulting stress field (compared with the result from the uncorrected GOCE data). However, the isostatic

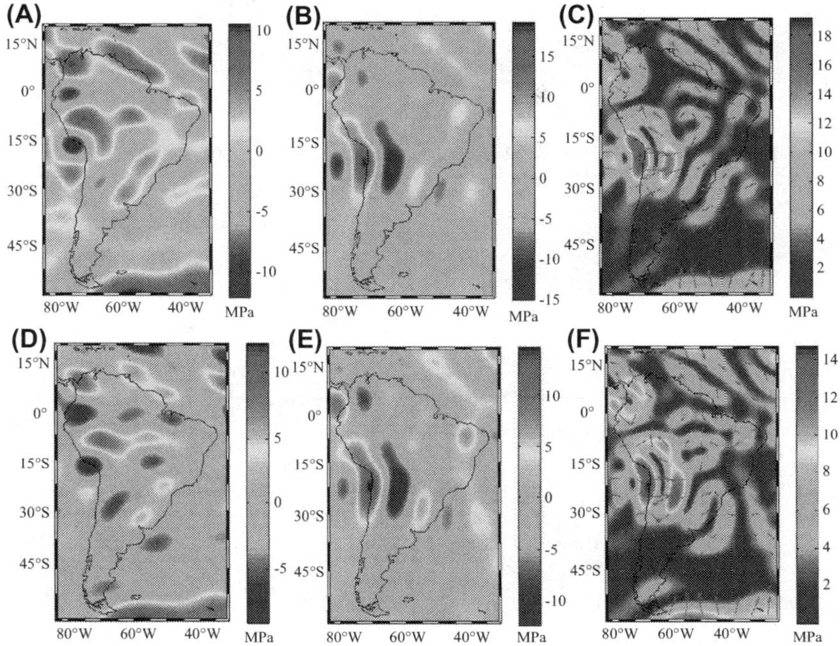

Figure 9.8 Regional map of stress field (MPa) in South America derived based on Runcorn's theory from the Gravity Field and Ocean Circulation Explorer data uncorrected for the crust density heterogeneities: (A) the northward stress component, (B) the eastward stress component and (C) the stress-vector orientation and intensity, and corrected for the contributions of topography and crustal density heterogeneities: (D) the northward stress component, (E) the eastward stress component and (F) the stress vectors orientation and intensity. *(From Eshagh M, Steinberger B. Tenzer R, Tassara A. Comparison of gravimetric and mantle flow solutions for lithospheric stress modelling and their combination. Geophys. J. Int. 2018;213(2):1013–1028.)*

equilibrium does not hold everywhere, so that some large stress anomalies remain, especially in regions of subduction or continental convergence. This is demonstrated in Fig. 9.8F, where the stress field obtained from the GOCE data corrected for the contributions of topography and crustal density heterogeneities is plotted. By comparing Figs. 9.8C and F, the overall spatial pattern of the stress field does not change considerably, but the stress intensity decreases (by up to about 4 MPa). The maximum stress intensity computed according to Runcorn's theory reaches about 18 MPa; after reducing the GOCE data for the contributions of topography and crustal density heterogeneities, it is reduced to about 14 MPa.

9.6.4 Sub-lithospheric shear stresses from Gravity Field and Ocean Circulation Explorer data corrected for crust density heterogeneities constrained by mantle flow model

In contrast to the recovered stress magnitude based on the reduced GOCE data for the crustal mass heterogeneities, the maximum stress intensity inferred from the mantle flow solution is only about 8 MPa (see Chapter 8). Because of the presence of the factor $(R/(R-D_{\mathrm{Lith}}))^{n+1}$, the higher frequencies of the solution recovered τ_z are amplified considerably. By discretising and solving integral equations, the effect of this amplification and ill-conditioning of the system are mixed, and by the regularisation both are controlled simultaneously. Some other constraining parameters are needed when Runcorn's theory is applied. Phillips and Ivins (1979) mentioned combining Runcorn's solutions with other geological parameters requires compensation for its oversimplifications. In this case, the northward and eastward shear stresses determined by Hager and O'Connell's theory and the velocity vectors of the tectonic plates are considered constraints here. However, inversion of the reduced GOCE data for the crustal mass heterogeneities leads to τ_z and not the shear stresses directly. The system of equations should be solved so that the derivatives of τ_z towards north and east become equivalent to those determined from Hager and O'Connell's theory and plate velocity model. For details about the solution of the system and constraining it to these shear stresses, see Eshagh et al. (2018). τ_z determined in this way is differentiated numerically towards north and east and is presented in Fig. 9.9A and B, respectively. The stress magnitude and pattern are seen in Fig. 9.9C.

9.7 Satellite gradiometry data and lithospheric stress tensor

In Chapter 8, propagation of sub-lithospheric shear stresses into the stress tensor inside the lithosphere was discussed. The spherical boundary-value problem of elasticity should be solved considering these sub-lithospheric shear stresses as boundary values at the sub-lithosphere and assuming they vanish at the surface of the lithosphere. In addition, the vertical force acting at the sub-lithosphere and its surface is assumed to be zero. This means that the plate motions are solely the result of shear stresses induced by poloidal flow inside the mantle. The main goal of this section is to develop mathematical models for the stress tensor inside the lithosphere based on

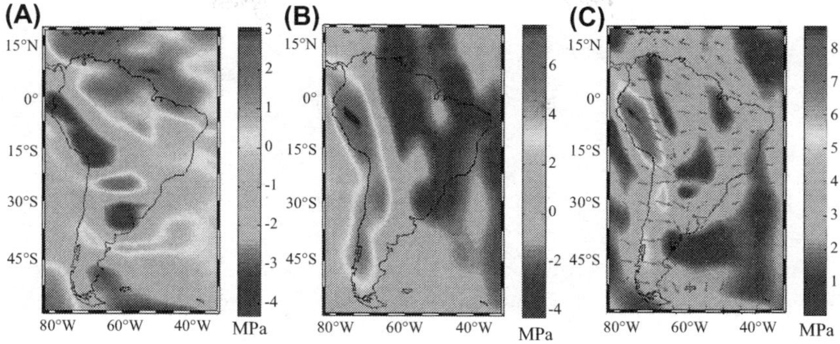

Figure 9.9 Regional map of stress field (MPa) in South America derived from the Gravity Field and Ocean Circulation Explorer data corrected for crustal density heterogeneities and based on combining Runcorn's and Hager and O'Connell's theories for the mantle flow model: (A) the northward stress component, (B) the eastward stress component and (C) the stress-vector orientation and intensity. *(From Eshagh M, Steinberger B. Tenzer R, Tassara A. Comparison of gravimetric and mantle flow solutions for lithospheric stress modelling and their combination. Geophys. J. Int. 2018;213(2):1013–1028.)*

this theory and satellite gravity gradiometry data, or the gravitational tensor measured by a satellite.

Mathematical modelling of the relations between the elements of the gravitational and stress tensors is simple because both tensors are defined in a common frame, the LNOF, and they have similar spherical harmonic expressions. In this case, obtaining simple integral equations with well-behaving kernels is expected. In the following, we divide our derivations into two parts. First, the relations between the diagonal elements of the stress and gravitational tensors are obtained; in the second part, the focus will be on finding the relations between the shear stresses and the off-diagonal elements of the gravitational tensor.

9.7.1 Diagonal elements of stress and gravitation tensors

The diagonal elements of the gravitational tensor are considered here as satellite gradiometry data. The goal is to obtain integral equations connecting these elements to the corresponding ones of the stress tensor. The Laplace coefficients of the second-order radial derivatives of the disturbing potential $T_{zz,n}$ has the following relation to those of T for a point at the geocentric distance r above the Earth's surface:

$$t_n = \frac{R^2}{(n+1)(n+2)} \left(\frac{r}{R}\right)^{n+3} T_{zz,n} \tag{9.31}$$

where $t_n = t_n\,(R,\theta,\lambda)$ and $T_{zz,n} = T_{zz,n}\,(r,\theta,\lambda)$. Substitution of Eq. (9.31) into Eq. (8.109) and simplification of the result yield:

$$\tau_{zz,n} = \frac{R^2}{r'}\left(\frac{r}{R}\right)^{n+3}\frac{\tilde{\lambda}K_n^1 + 2\tilde{\mu}K_n^2}{(n+1)(n+2)}T_{zz,n} \tag{9.32}$$

where $\tau_{zz,n} = \tau_{zz,n}(r',\theta,\lambda)$ is the Laplace coefficient of the normal stress τ_{zz}, at the surface of sphere with the radius r' which is the geocentric distance of a point inside the lithosphere, and $\tilde{\lambda}$ and $\tilde{\mu}$ are elasticity parameters.

To obtain an integral equation for determining τ_{zz}, Eq. (9.32) should be solved for $T_{zz,n}$:

$$\frac{r'}{R^2}\frac{(n+1)(n+2)}{\tilde{\lambda}K_n^1 + 2\tilde{\mu}K_n^2}\left(\frac{R}{r}\right)^{n+3}\tau_{zz,n}(r',\theta,\lambda) = T_{zz,n}(r,\theta,\lambda) \tag{9.33}$$

By the integral formula for $\tau_{zz,n}$ (see Eq. 1.11), and taking summation over n from both sides of Eq. (9.33), we obtain:

$$\frac{r'}{4\pi R^2}\iint_\sigma \sum_{n=2}^{\infty}\frac{(2n+1)(n+1)(n+2)}{\tilde{\lambda}K_n^1 + 2\tilde{\mu}K_n^2}\left(\frac{R}{r}\right)^{n+3} P_n(\tilde{x})\tau_{zz}(r',\theta',\lambda')d\sigma = T_{zz}(r,\theta,\lambda) \tag{9.34}$$

Eq. (9.34) is an integral equation for inverting T_{zz} at the satellite level to τ_{zz} inside the lithosphere at a constant geocentric distance of r'. No closed-form formula for the kernel of this integral equation can be derived.

Deriving direct relations between $(T_{xx,n}, T_{yy,n})$ and $(\tau_{xx,n}, \tau_{yy,n})$ is impossible without involving another gradient, but for the differences between these two pairs, two similar equations for $T_{xx,n} - T_{yy,n}$, and $\tau_{xx,n} - \tau_{yy,n}$ can be written:

$$T_{xx,n} - T_{yy,n} = \frac{1}{R^2}\left(\frac{R}{r}\right)^{n+3}\left(\frac{\partial^2 t_n}{\partial\theta^2} - \cot\theta\frac{\partial t_n}{\partial\theta} - \frac{1}{\sin^2\theta}\frac{\partial^2 t_n}{\partial\lambda^2}\right) \tag{9.35}$$

$$\tau_{xx,n} - \tau_{yy,n} = \left(\frac{2\tilde{\mu}}{r'}\right)K_n^5\left(\frac{\partial^2 t_n}{\partial\theta^2} - \cot\theta\frac{\partial t_n}{\partial\theta} - \frac{1}{\sin^2\theta}\frac{\partial^2 t_n}{\partial\lambda^2}\right) \tag{9.36}$$

By comparing Eqs. (9.35) and (9.36), the relation between $T_{xx,n} - T_{yy,n}$ and $\tau_{xx,n} - \tau_{yy,n}$ will be obtained:

$$\tau_{xx,n} - \tau_{yy,n} = \frac{2R^2}{r'}\left(\frac{r}{R}\right)^{n+3}\tilde{\mu}K_n^5(T_{xx,n} - T_{yy,n}) \tag{9.37}$$

Solving Eq. (9.37) for $T_{xx,n} - T_{yy,n}$ yields

$$\left(\tau_{xx,n} - \tau_{yy,n}\right)\frac{r'}{2R^2}\left(\frac{R}{r}\right)^{n+3}\frac{1}{\widetilde{\mu}K_n^5} = T_{xx,n} - T_{yy,n} \tag{9.38}$$

By writing $\tau_{xx,n} - \tau_{yy,n}$ in integral form, based on Eq. (1.11), the following integral equation can be derived:

$$\begin{aligned}&\frac{r'}{4\pi R^2}\iint_\sigma \sum_{n=2}^{\infty}\frac{2n+1}{2\widetilde{\mu}K_n^5}\left(\frac{R}{r}\right)^{n+3} P_n(\widetilde{x})\left(\tau'_{xx} - \tau'_{yy}\right)\mathrm{d}\sigma \\ &= T_{xx}(r,\theta,\lambda) - T_{yy}(r,\theta,\lambda)\end{aligned} \tag{9.39}$$

where $\tau'_{xx} - \tau'_{yy} = \tau_{xx}(r',\theta',\lambda') - \tau_{yy}(r',\theta',\lambda')$.

Eq. (9.39) means that the difference between the normal horizontal stresses is obtainable from the difference between the corresponding elements from the tensor of gravitation.

To derive a relation between $\tau_{xx,n}$ and $T_{xx,n}$, and $\tau_{yy,n}$ and $T_{yy,n}$ we rewrite Eqs. (8.110) and (8.111) in the form:

$$\left.\begin{matrix}\tau_{xx,n}\\ \tau_{yy,n}\end{matrix}\right\} = \frac{\widetilde{\lambda}K_n^1 + 2\widetilde{\mu}K_n^3}{r'}t_n + \left(\frac{2\widetilde{\mu}}{r'}\right)K_n^5\begin{cases}\dfrac{\partial^2 t_n}{\partial\theta^2}\\[2ex] \cot\theta\dfrac{\partial t_n}{\partial\theta} + \dfrac{1}{\sin^2\theta}\dfrac{\partial^2 t_n}{\partial\lambda^2}\end{cases} \tag{9.40}$$

If we write Eqs. (2.130) and (2.131) in the spectral form, we obtain:

$$\left.\begin{matrix}\dfrac{\partial^2 t_n}{\partial\theta^2}\\[2ex] \cot\theta\dfrac{\partial t_n}{\partial\theta} + \dfrac{1}{\sin^2\theta}\dfrac{\partial^2 t_n}{\partial\lambda^2}\end{matrix}\right\} = (n+1)t_n + r^2\left(\frac{r}{R}\right)^{n+1}\begin{cases}T_{xx,n}\\ T_{yy,n}\end{cases} \tag{9.41}$$

Now, we substitute Eq. (9.41) into Eq. (9.40) and simplify the result to:

$$\begin{aligned}\left.\begin{matrix}\tau_{xx,n}\\ \tau_{yy,n}\end{matrix}\right\} &= \frac{\widetilde{\lambda}K_n^1 + 2\widetilde{\mu}\left(K_n^3 + (n+1)K_n^5\right)}{r'}t_n \\ &+ \left(\frac{2\widetilde{\mu}}{r'}\right)K_n^5 R^2\left(\frac{r}{R}\right)^{n+3}\begin{cases}T_{xx,n}\\ T_{yy,n}\end{cases}.\end{aligned} \tag{9.42}$$

Inserting Eq. (9.31) into Eq. (9.42) and simplifying the result yields:

$$\left.\begin{matrix}\tau_{xx,n}\\ \tau_{yy,n}\end{matrix}\right\}=\frac{R^2}{r'}\left(\frac{r}{R}\right)^{n+3}\left\{\frac{\widetilde{\lambda}K_n^1+2\widetilde{\mu}\left(K_n^3+K_n^5(n+1)\right)}{(n+1)(n+2)}T_{zz,n}+2\widetilde{\mu}K_n^5\begin{pmatrix}T_{xx,n}\\ T_{yy,n}\end{pmatrix}\right\}. \tag{9.43}$$

Let us divide Eq. (9.43) into two parts:

$$\left.\begin{matrix}\tau_{xx,n}\\ \tau_{yy,n}\end{matrix}\right\}=W_n^1+\left\{\begin{matrix}W_{xx,n}^2\\ W_{yy,n}^2\end{matrix}\right.. \tag{9.44}$$

Comparison of Eqs. (9.43) and (9.44) reveals:

$$W_n^1=\frac{R^2}{r'}\left(\frac{r}{R}\right)^{n+3}\frac{\widetilde{\lambda}K_n^1+2\widetilde{\mu}\left(K_n^3+K_n^5(n+1)\right)}{(n+1)(n+2)}T_{zz,n} \tag{9.45}$$

and

$$\left.\begin{matrix}W_{xx,n}^2\\ W_{yy,n}^2\end{matrix}\right\}=\frac{R^2}{r'}\left(\frac{r}{R}\right)^{n+3}2\widetilde{\mu}K_n^5\left\{\begin{matrix}T_{xx,n}\\ T_{yy,n}\end{matrix}\right.. \tag{9.46}$$

Now, Eq. (9.45) is solved for $T_{zz,n}$ and Eq. (9.46) for $(T_{xx,n}, T_{yy,n})$:

$$\frac{r'}{R^2}\frac{(n+1)(n+2)}{\widetilde{\lambda}K_n^1+2\widetilde{\mu}\left(K_n^3+K_n^5(n+1)\right)}\left(\frac{R}{r}\right)^{n+3}W_n^1=T_{zz,n} \tag{9.47}$$

and

$$\frac{r'}{R^2}\frac{1}{2\widetilde{\mu}K_n^5}\left(\frac{R}{r}\right)^{n+3}\left\{\begin{matrix}W_{xx,n}^2\\ W_{yy,n}^2\end{matrix}\right.=\left\{\begin{matrix}T_{xx,n}\\ T_{yy,n}\end{matrix}\right.. \tag{9.48}$$

According to Eq. (1.11), W_n^1 can be written in an integral form by inserting it into Eq. (9.47); after taking summation over n from degree 2 to infinity, we obtain:

$$\frac{r'}{4\pi R^2}\iint\limits_{\sigma}\sum_{n=2}^{\infty}\frac{(2n+1)(n+1)(n+2)}{\widetilde{\lambda}K_n^1+2\widetilde{\mu}\left(K_n^3+K_n^5(n+1)\right)}\left(\frac{R}{r}\right)^{n+3}P_n(\widetilde{x})W^1(r',\theta',\lambda')\mathrm{d}\sigma$$
$$=T_{zz}(r,\theta,\lambda). \tag{9.49}$$

Eq. (9.49) is an integral equation for inverting on-orbit $T_{zz}(r,\theta,\lambda)$ to $W^1(r',\theta',\lambda')$ on a regular grid at the constant geocentric distance r'. No closed-form formula can be found for the kernel function of this integral equation; its spectral form should be directly used.

Similarly from Eq. (9.48), two integral equations for recovering W^2_{xx} and W^2_{yy} can be obtained from T_{xx} and T_{yy}:

$$\frac{r'}{4\pi R^2}\iint_\sigma \sum_{n=2}^{\infty}\frac{2n+1}{2\widetilde{\mu}K_n^5}\left(\frac{R}{r}\right)^{n+3}P_n(\widetilde{x})\begin{Bmatrix} W^2_{xx}(r',\theta',\lambda') \\ W^2_{yy}(r',\theta',\lambda') \end{Bmatrix}d\sigma = \begin{Bmatrix} T_{xx}(r,\theta,\lambda) \\ T_{yy}(r,\theta,\lambda) \end{Bmatrix}. \tag{9.50}$$

Therefore, by inverting T_{zz}, W^1 is obtained and by T_{xx} and T_{yy}, respectively, W^2_{xx} and W^2_{yy}; this means that the normal stresses will be:

$$\left.\begin{matrix} \tau_{xx} \\ \tau_{yy} \end{matrix}\right\} = W^1_n + \begin{Bmatrix} W^2_{xx} \\ W^2_{yy} \end{Bmatrix}. \tag{9.51}$$

9.7.2 Off-diagonal elements of stress and gravitation tensors

Here, the goal is to derive integral equations for inverting T_{xz} and T_{yz} to shear stresses τ_{xz} and τ_{yz}, respectively. By comparing Eq. (9.23) and Eqs. (8.112) and (8.113), we obtain:

$$\left.\begin{matrix} \tau_{xz,n} \\ \tau_{yz,n} \end{matrix}\right\} = \frac{R^2}{r'}\left(\frac{r}{R}\right)^{n+3}\widetilde{\mu}\frac{K_n^4}{n+2}\begin{Bmatrix} T_{xz,n} \\ T_{yz,n} \end{Bmatrix}. \tag{9.52}$$

Solving Eq. (9.52) for $T_{xz,n}$ and $T_{yz,n}$ yields

$$\frac{r'}{R^2\widetilde{\mu}}\frac{n+2}{K_n^4}\left(\frac{R}{r}\right)^{n+3}\begin{Bmatrix} \tau_{xz,n} \\ \tau_{yz,n} \end{Bmatrix} = \begin{Bmatrix} T_{xz,n} \\ T_{yz,n} \end{Bmatrix}. \tag{9.53}$$

According to Eq. (1.11), and taking summation over n from degree 2 to infinity, the following integral equations are obtained:

$$\frac{r'}{4\pi R^2}\iint_\sigma \sum_{n=2}^{\infty}\frac{(2n+1)(n+2)}{\widetilde{\mu}K_n^4}\left(\frac{R}{r}\right)^{n+1}P_n(\widetilde{x})\begin{Bmatrix} \tau_{xz}(r',\theta',\lambda') \\ \tau_{yz}(r',\theta',\lambda') \end{Bmatrix}d\sigma = \begin{Bmatrix} T_{xz}(r,\theta,\lambda) \\ T_{yz}(r,\theta,\lambda) \end{Bmatrix}. \tag{9.54}$$

Based on the similarity of the mathematical formulae of T_{xy} and τ_{xy}, we can write

$$T_{xy,n} = \frac{1}{R^2}\left(\frac{R}{r}\right)^{n+3}\left(\frac{1}{\sin\theta}\frac{\partial^2 t_n}{\partial\lambda^2} - \cot\theta\frac{\partial t_n}{\partial\theta}\right) \quad (9.55)$$

$$\tau_{xy,n} = \frac{\widetilde{\mu}}{r'}K_n^5\left(\frac{1}{\sin\theta}\frac{\partial^2 t_n}{\partial\lambda^2} - \cot\theta\frac{\partial t_n}{\partial\theta}\right). \quad (9.56)$$

Following a procedure similar to what was done to obtain the previous integral equations, the integral is derived for inverting T_{xy} to τ_{xy}:

$$\frac{r'}{4\pi R^2}\iint_{\sigma}\sum_{n=2}^{\infty}\frac{2n+1}{\widetilde{\mu}K_n^5}\left(\frac{R}{r}\right)^{n+3}P_n(\widetilde{x})\tau_{xy}(r',\theta',\lambda')\mathrm{d}\sigma = T_{xy}(r,\theta,\lambda). \quad (9.57)$$

9.7.3 Simple application of integral equations for inverting real on-orbit Gravity Field and Ocean Circulation Explorer data to lithospheric stress

Here, some parts of Eshagh's (2017) study on inverting satellite gradiometry data to the stress tensor insider the lithosphere are presented. The real satellite gravitational tensor measured by GOCE in Nov. 2009 over South America, limited between latitudes 65°S and 20°N and longitudes 90°W and 25°W is selected. To implement our theory, elements T_{zz}, T_{xz}, T_{yz}, T_{xy} and $T_{xx} - T_{yy}$ of the gravitational tensor, measured by GOCE, are inverted to τ_{zz}, τ_{xz}, τ_{yz}, τ_{xy} and $\tau_{xx} - \tau_{yy}$, respectively at a depth of 35 km below the Earth's surface where a huge earthquake happened on Feb. 27, 2010, assuming that the lithospheric depth is 100 km.

9.7.3.1 *Gravity Field and Ocean Circulation Explorer data and coverage*

The real elements of the gravitational tensor in the LNOF measured by GOCE should be converted to the corresponding derivatives of the disturbing potential (T) by reducing for the influence of the normal gravity field GRS80 (Moritz, 2000). In addition, a rate of 10 s along the orbit is considered, In Nov. 2009, the GOCE data covered the area well with 22,284 values, with the 10-s rate, for each element of gravitational tensor. The idea is to estimate each stress element with a resolution of 1° × 1° over the area corresponding to 5525 values.

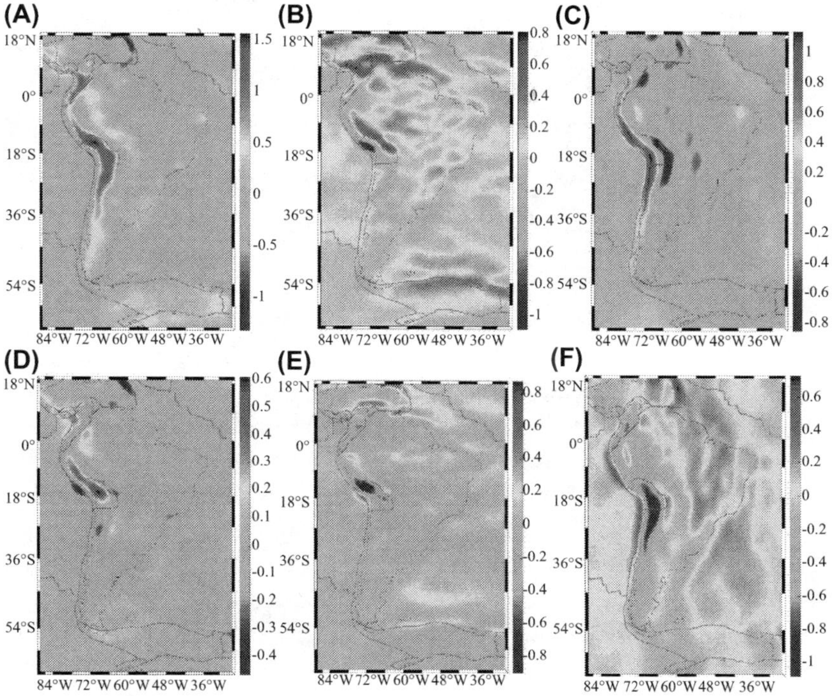

Figure 9.10 Gravity Field and Ocean Circulation Explorer gravity gradients: (A) T_{zz}, (B) T_{xz}, (C) T_{yz}, (D) T_{xy}, (E) T_{xx} and (F) T_{yy} [E].

Fig. 9.10 shows maps of the elements of the gravitational tensor over the study area. For better visualisation, we have interpolated the GOCE on-orbit data over a regular grid with a resolution of $5' \times 5'$. Fig. 9.10A is the map of T_{zz} and Fig. 9.10B shows T_{xz} representing the variations of T along the north–south direction. Regions with large gravity changes are elongated in the east–west direction. The map of T_{yz} is presented in Fig. 9.10C, showing gravity variations along the northeast direction. The tectonic boundary and coastal lines of the active continental margin, or the Andes subduction zone, are clear there. This means that the eastward changes of gravity field are considerable along this line. Therefore, gravitational and tectonic features that are elongated northwards or southwards are more pronounced in the map of T_{yz}. Also, the continental shelf in the eastern part of the continents is recognisable in this map. T_{xy}, presented in Fig. 9.10D, shows the slopes of the gravity features in both north–south and east–west directions; this detects the corners of the anomaly features of the gravity field. T_{xx}, in Fig. 9.10E, represents the inflation points of the gravity field in the north—south directions and shows the border of the anomaly feature, which is elongated east—west. T_{yy} shows the inflation points in the east—west direction. Those gravitational features elongated north—south are more pronounced.

9.7.3.2 Recovery of lithospheric stress from on-orbit Gravity Field and Ocean Circulation Explorer data

In this section, integral equations developed for stress recovery are applied and solved to invert the elements of the gravitational tensor measured by GOCE to the corresponding elements of the stress tensor at a depth of 35 km. These integrals should be discretised according to the resolution of the stress recovery, which is $1° \times 1°$ and organise a systems of equations whose observations are the elements of the gravitational tensor and its unknown stress values. Such systems are ill-conditioned. To solve them, regularisation methods should be applied (see Chapter 4). In this study, the Tikhonov Regularisation (Tikhonov 1963) combined with the L-curve (Hansen, 1998) and/or quasioptimal (Hansen, 2007) methods is applied (Hansen, 1998). To reduce the effect of the spatial truncation error (Eshagh, 2011), the inversion is performed over the whole area, but the results in the central part, which is smaller by 5 degrees from each side, are selected as the solution.

Fig. 9.11A is a map of recovered τ_{zz} from T_{zz}. The values of this stress are small compared with the rest of the stresses. This is normal because the value of this stress at the sub-lithosphere and the lithosphere's surface is assumed to be zero. Therefore, large values cannot be expected for this stress inside the lithosphere.

Fig. 9.11B and C are maps of the recovered τ_{xz} and τ_{yz} from on-orbit GOCE T_{xz} and T_{yz}, respectively. τ_{xy} is shown in Fig. 9.11D. Finally, Fig. 9.11E illustrates the map of $\tau_{xx} - \tau_{yy}$. For interpretations and further details about the inversion process, see Eshagh (2017).

9.8 Inter-satellite tracking data and stress

Inter-satellite tracking data can also be applied to determine the gravity field and its temporal changes. Compared with the satellite gradiometry data, they are more complicated from a mathematical point of view, because of the involvement of two satellites, whereas the gravitational tensor is measured by only one satellite. In this section, two types of inter-satellite tracking data are considered: the low-low range rates and

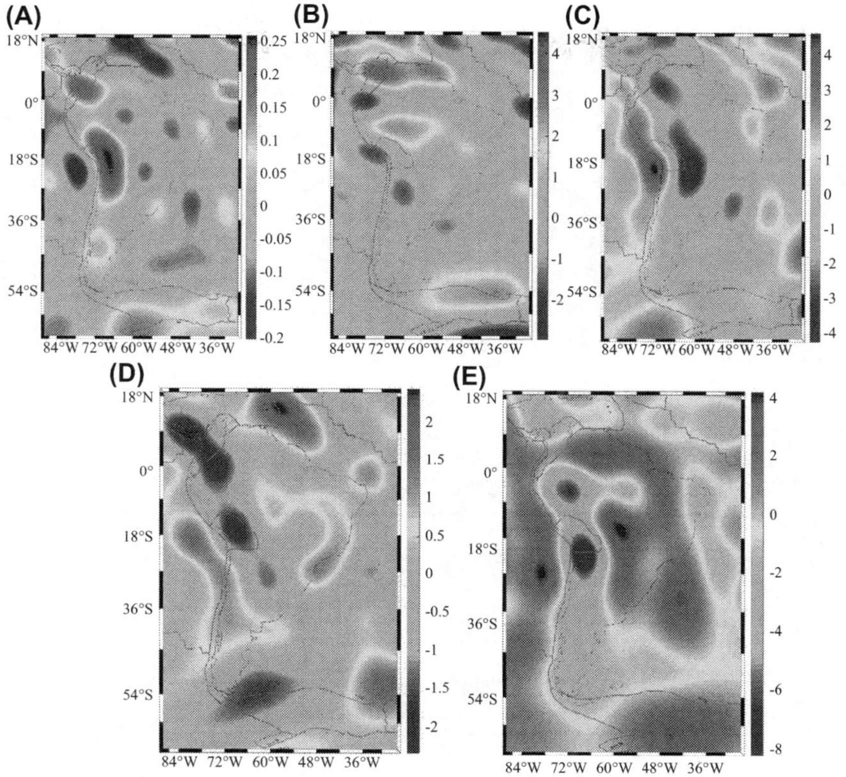

Figure 9.11 Recovered (A) τ_{zz}, (B) τ_{xz}, (C) τ_{yz}, (D) τ_{xy} and (E) $\tau_{xx} - \tau_{yy}$. (MPa).

line-of-sight (LOS) measurements between two satellites in the same orbit. In this section, the procedure is discussed for determining sub-lithospheric shear stresses resulting from the mantle convection and lithospheric stress from inter-satellite range rates and LOS measurements.

9.8.1 Inter-satellite low-low range rates and sub-lithospheric shear stresses

The mathematical relation between inter-satellite low-low range rates and anomalous quantities was presented in Chapter 2. Consider Eq. (2.113 and 3.75) and its spectral form for T measured at satellites 1 and 2 with positions $(r_1, \theta_1, \lambda_1)$ and $(r_2, \theta_2, \lambda_2)$, respectively:

$$\nu_0\dot{\tilde{\rho}} - \Upsilon = \sum_{n=2}^{\infty} \sum_{m=-n}^{n} t_{nm} \sum_{i=1}^{2} (-1)^i \left(\frac{R}{r_i}\right)^{n+1} Y_{nm}(\theta_i, \lambda_i) \tag{9.58}$$

where v_0 is the average velocity between the twin satellites, $\dot{\tilde{\rho}}$ is the range rates measured between satellites, Υ is the contribution of the normal gravity field to $v_0\dot{\tilde{\rho}}$, and i is the satellite number.

Let us solve Eq. (9.3) for t_{nm}:

$$t_{nm} = \frac{1}{k}\frac{n+1}{2n+1}\left(\frac{R - D_{\mathrm{Lith}}}{R}\right)^{n+1} \tau_{z,nm} \tag{9.59}$$

By inserting Eq. (9.59) into Eq. (9.58), we obtain

$$v_0\dot{\tilde{\rho}} - \Upsilon = \frac{1}{k}\sum_{n=2}^{\infty}\frac{n+1}{2n+1}\sum_{m=-n}^{n}\tau_{z,nm}\sum_{i=1}^{2}(-1)^i\left(\frac{R - D_{\mathrm{Lith}}}{r_i}\right)^{n+1} Y_{nm}(\theta_i, \lambda_i). \tag{9.60}$$

To obtain an integral equation for recovering τ_z, the integral form of the global spherical harmonic analysis, Eq. (1.8), should be used:

$$\tau_{z,nm} = \frac{1}{4\pi}\iint_{\sigma}\tau_z(r', \theta', \lambda')Y_{nm}(\theta', \lambda')\mathrm{d}\sigma. \tag{9.61}$$

By inserting Eq. (9.61) into Eq. (9.60), taking summation over m, applying the addition theorem of spherical harmonics (1.9) and summation over n from the results, we reach

$$\frac{1}{4\pi k}\iint_{\sigma}\tau_z(r', \theta', \lambda')\Psi(r_1, r_2, \psi_1, \psi_2)\mathrm{d}\sigma = v_0\dot{\tilde{\rho}} - \Upsilon \tag{9.62}$$

with the kernel

$$\Psi(r_1, r_2, \psi_1, \psi_2) = \sum_{n=2}^{\infty}(n+1)\sum_{i=1}^{2}(-1)^i\left(\frac{R - D_{\mathrm{Lith}}}{r_i}\right)^{n+1} P_n(\tilde{x}_i) \tag{9.63}$$

where $\tilde{x}_i = \cos\psi_i$. The closed-form formula of this kernel function is

$$\Psi(r_1, r_2, \psi_1, \psi_2) = \sum_{i=1}^{2}(-1)^i\left[s_i\left(\frac{1}{D_i} - 1 - s_i\tilde{x}_i\right) + s_i^2\left(\frac{\tilde{x}_i - s_i}{D_i^3} - \tilde{x}_i\right) - s_i - s_i^2\tilde{x}_i\right]. \tag{9.64}$$

where $s_i = (R - D_{\mathrm{Lith}})/r_i$ and $D_i = \sqrt{1 + s_i^2 - 2s_i\widetilde{x}_i}$.

By solving the integral equation Eq. (9.62), τ_z is estimated at a sub-lithosphere with a geocentric radius r'. The recovered τ_z should be differentiated numerically to obtain the northward and eastward shear stresses at the sub-lithosphere.

9.8.2 Inter-satellite line-of-sight measurements and sub-lithospheric shear stresses

As shown in Chapter 2, the LOS measurements are range, range rate, and range acceleration measured between the twin satellites moving in the same orbit in addition to the velocity vector differences between them. To find the integral relation between this type of inter-satellite data, Eq. (2.126) is written in the following form:

$$\ddot{\widetilde{\rho}} - \frac{(\delta\dot{\mathbf{r}}_{12})^2 - \dot{\widetilde{\rho}}^2}{\widetilde{\rho}} - \Sigma = \sum_{n=2}^{\infty}\sum_{m=-n}^{n} t_{nm} \sum_{i=1}^{2} \frac{(-1)^i}{r_i}\left(\frac{R}{r_i}\right)^{n+1} \left(-(n+1)Y_{nm}(\theta_i,\lambda_i)\mathbf{e}_z + \left(\mathbf{X}_{nm}^{(2)}(\theta_i,\lambda_i)\right)\right)\bullet\mathbf{e}_{\mathrm{LOS}} \tag{9.65}$$

where $\mathbf{e}_{\mathrm{LOS}}$ is the unit vector point from satellite 1 to satellite 2. $\mathbf{e}_z$ is the unit vector along the z-axis of the LNOF. ' • ' means the inner product operator, and $\mathbf{X}_{nm}^{(2)}(\theta_i, \lambda_i)$ stands for the vector spherical harmonics of degree n and order m at a satellite with the colatitude θ_i and longitude λ_i. Note that the left-hand side of Eq. (9.65) is a function of the positions of both satellites. Eq. (9.65) is suitable for recovering t_{nm} or the SHCs of the disturbing potential. If the goal is to determine the SHC of T, which are needed to compute the sub-lithospheric shear stresses resulting from mantle convection based on the Runcorn theory, Eq. (9.65) should be reduced for the normal gravity field by computing the contribution of the normal gravity field to this type of satellite gravimetry data and subtracting it from the data (Eq. 3.78). If Eq. (9.59) is inserted into Eq. (9.65), we obtain:

$$\ddot{\widetilde{\rho}} - \frac{(\delta\dot{\mathbf{r}}_{12})^2 - \dot{\widetilde{\rho}}^2}{\widetilde{\rho}} - \Sigma = \frac{1}{k}\sum_{n=2}^{\infty}\frac{n+1}{2n+1}\sum_{m=-n}^{n} \tau_{z,nm} \sum_{i=1}^{2} \frac{(-1)^i}{r_i}\left(\frac{R - D_{\mathrm{Lith}}}{r_i}\right)^{n+1} \left(-(n+1)\mathbf{X}_{nm}^{(1)}(\theta_i,\lambda_i) + \left(\mathbf{X}_{nm}^{(2)}(\theta_i,\lambda_i)\right)\right).\mathbf{e}_{\mathrm{LOS}} \tag{9.66}$$

After rearranging Eq. (9.66) we derive:

$$\frac{1}{4\pi k}\iint_{\sigma}\tau'_z\sum_{n=2}^{\infty}\frac{n+1}{2n+1}\sum_{i=1}^{2}\frac{(-1)^i}{r_i}\left(\frac{R-D_{\mathrm{Lith}}}{r_i}\right)^{n+1}$$
$$\left(-(n+1)\sum_{m=-n}^{n}Y_{nm}(\theta',\lambda')\mathbf{X}_{nm}^{(1)}(\theta_i,\lambda_i)+\right.$$
$$\left.+\sum_{m=-n}^{n}Y_{nm}(\theta',\lambda')\mathbf{X}_{nm}^{(2)}(\theta_i,\lambda_i)\right).\mathbf{e}_{\mathrm{LOS}}\mathrm{d}\sigma=\ddot{\tilde{\rho}}-\frac{(\delta\dot{\mathbf{r}}_{12})^2-\dot{\tilde{\rho}}^2}{\tilde{\rho}}-\Sigma \tag{9.67}$$

from addition theorem of spherical harmonics (1.9) we can write:

$$\frac{1}{4\pi k}\iint_{\sigma}\tau'_z\sum_{n=2}^{\infty}\frac{n+1}{2n+1}\sum_{i=1}^{2}\frac{(-1)^i}{r_i}\left(\frac{R-D_{\mathrm{Lith}}}{r_i}\right)^{n+1}\left(-(n+1)P_n(\tilde{x}_i)\mathbf{e}_{z_i}+\right.$$
$$\left.+\frac{\partial P_n(\tilde{x}_i)}{\partial\theta}\mathbf{e}_{x_i}+\frac{\partial P_n(\tilde{x}_i)}{\sin\theta\partial\lambda}\mathbf{e}_{y_i}\right).\mathbf{e}_{\mathrm{LOS}}\mathrm{d}\sigma=\ddot{\tilde{\rho}}-\frac{(\delta\dot{\mathbf{r}}_{12})^2-\dot{\tilde{\rho}}^2}{\tilde{\rho}}-\Sigma \tag{9.68}$$

where $\tau'_z=\tau_z(R-D_{\mathrm{Lith}},\theta',\lambda')$.

According to the addition theorem of the scalar and vector spherical harmonics (1.9), (1.61) and (1.62), the following integral equation will be obtained to recover τ_z from this type of inter-satellite tracking data:

$$\frac{1}{4\pi k}\iint_{\sigma}\Psi^{\mathrm{LOS}}(r_1,\alpha_1,\psi_1,r_2,\alpha_2,\psi_2)\tau_z(R-D_{\mathrm{Lith}},\theta',\lambda')\mathrm{d}\sigma$$
$$=\ddot{\tilde{\rho}}-\frac{(\delta\dot{\mathbf{r}}_{12})^2-\dot{\tilde{\rho}}^2}{\tilde{\rho}}-\Sigma \tag{9.69}$$

with

$$\Psi^{\mathrm{LOS}}(r_1,\alpha_1,\psi_1,r_2,\alpha_2,\psi_2)=\sum_{n=2}^{\infty}(n+1)\sum_{i=1}^{2}\frac{(-1)^i}{r_i}\left(\frac{R-D_{\mathrm{Lith}}}{r_i}\right)^{n+1}$$
$$\left(-(n+1)P_n(\tilde{x}_i)\mathbf{e}_{z_i}+\frac{\mathrm{d}P_n(\tilde{x}_i)}{\mathrm{d}\psi}(\cos\alpha_i\mathbf{e}_{x_i}-\sin\alpha_i\mathbf{e}_{y_i})\right)\cdot\mathbf{e}_{\mathrm{LOS}} \tag{9.70}$$

This kernel function has a complicated mathematical form and is of the bipolar type.

Once τ_z is computed, its northward and eastward derivatives can be computed numerically from it.

9.8.3 Example: application of Gravity Recovery and Climate Experiment–type data for recovering sub-lithospheric shear stresses

Here, the study performed by Sprlak and Eshagh (2016) is presented as an application of the theory for sub-lithospheric shear stress determination. They used the true orbits of the twin-satellites of the Gravity Recovery and Climate Experiment (GRACE) and employed EGM08 to simulate the range rates and LOS measurements. Both methods were applied under the same conditions (e.g., the same orbits, resolution, mean depth of lithosphere and gravity models). The difference between these solutions depends on the type of observables and the mathematical models connecting the observables to τ_z. The LOS measurements are more sensitive to higher frequencies of the stress than the range rates, owing to involvement of the partial derivatives of T difference. Simulated inter-satellite data are free of error, but in reality, the measured velocities and orientation parameters are mixed so that the error of the LOS measurements are complicated.

Fig. 9.12A and B are maps of the simulated range rates and the LOS measurements over the Himalayas in the GRACE orbits, respectively, from EGM08 to degree and order 360. The range rates presented in Fig. 9.12A have positive values in the southern part of the area; over India and the

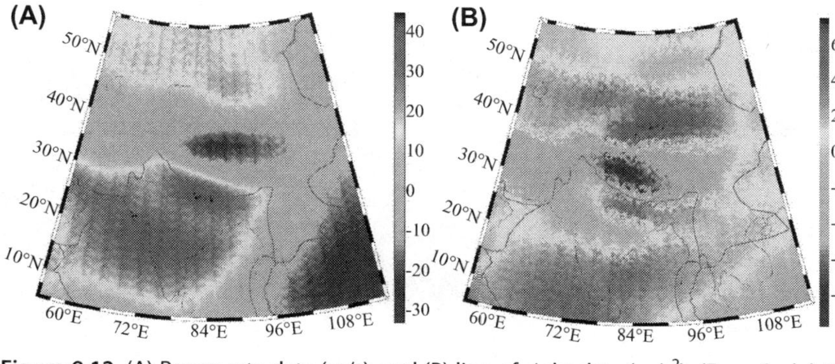

Figure 9.12 (A) Range rate data (m/s), and (B) line-of-sight data (m/s^2). *(From Sprlak M. and Eshagh M., Local recovery of sub-crustal stress due to mantle convection from satellite-to-satellite tracking data, Acta Geophysica 64 (4), 2016, 904–929.)*

Indian Ocean and over the topographic features of Tibet, they have small negative values. The map of the LOS measurements presented in Fig. 9.12B contains more details and is not as smooth as the map of Fig. 9.12A. Again, over the Himalayas, this type of satellite gravimetry data has negative values; positive ones are seen in the northern and southern parts of this area.

The simulated inter-satellite tracking data have been inverted after discretising the integral equations and solving them for τ_z using Tikhonov regularisation. The resolution of recovery was 1×1 degree and the central part of the area, which is smaller than the whole area by 5 degrees, is selected as the recovery area to minimise the effect of the STE (see Section 4.2). The recovered τ_z from the inter-satellite data, presented in Fig. 9.12A and B, are shown in Fig. 9.13A and B, respectively. As observed, τ_z recovered from the range-rate data (Fig. 9.13A) is smoother than that estimated from the LOS measurements (Fig. 2.13B).

Maps of the magnitudes of sub-lithospheric shear stresses determined by numerical differentiation of the recovered τ_z are presented in Fig. 9.14A and B. Fig. 9.14A is the sub-lithospheric shear stress magnitude determined after performing the numerical differentiation of τ_z presented in Fig. 9.13A, whilst Fig. 9.14B from Fig. 9.13B. As Fig. 9.14 shows, the magnitude of shear stress is smaller when they are estimated from the inter-satellite range rates.

Sprlak and Eshagh (2016) considered a depth of 30 km for recovering τ_z, which is not realistic. In addition, the effects of the mass and structural heterogeneities inside the lithosphere have not been removed from the inter-satellite tracking data. The low frequencies of the gravity field that are mainly from the core and deep interior structure of the Earth should have been

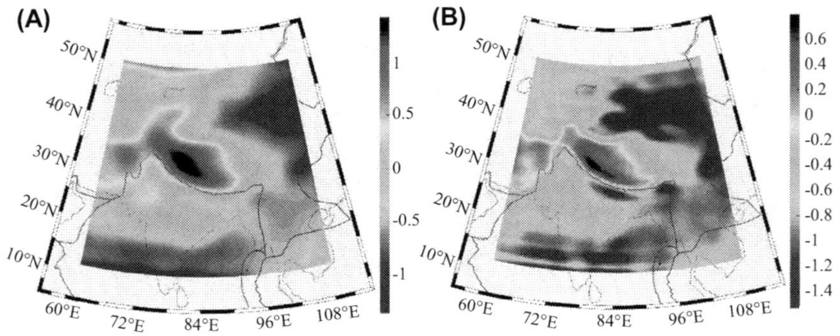

Figure 9.13 τ_z recovered from (A) range-rate data, and (B) line-of-sight data (MPa).

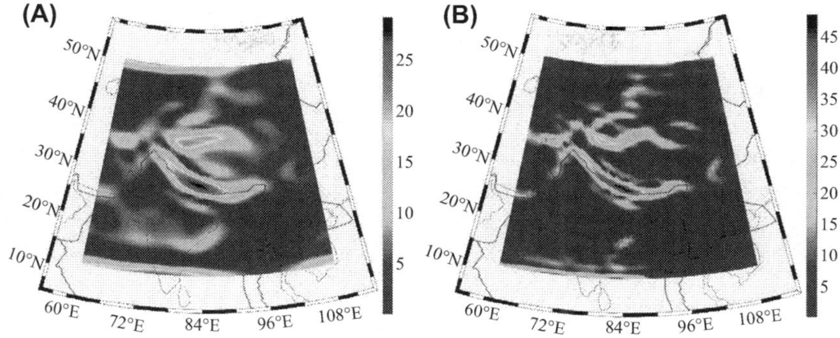

Figure 9.14 Magnitude of sub-lithospheric shear stress determined from (A) range rates, and (B) line-of-sight data (MPa).

removed from the measurements. Recovering high frequencies of sub-lithospheric stress is not meaningful because stress caused by mantle convection is seen at low frequencies. Therefore, range rates should be more suited to τ_z recovering at the base of the lithosphere. However, recovering such stresses from other types of satellite gravimetry data is possible if the contribution of the mass and structural heterogeneities in the Earth's interior and lithosphere are properly reduced.

9.9 Determination of lithospheric stress tensor from inter-satellite tracking data

Determination of the elements of the stress tensor from inter-satellite data is not as straightforward as it is from satellite gradiometry data. The main reason is that inter-satellite measurements are function of the positions of twin satellites following each other along their orbit. Developing integral equations to recover the gravity field parameter geoid height and gravity anomaly/disturbance from such data is not complicated, but when the desired parameters are the elements of the tensor of stress, the mathematical derivation will be long and complicated because of the involvement of first- and second-order partial derivatives of the spherical harmonics. In such a case, finding integral equations with simple kernels is not simple. In this section, our goal is to present such complicated derivations and formulae for the inter-satellite range rates and LOS measurements. Applying the mathematical models and performing the inversion of real data are left to interested readers.

9.9.1 Determination of elements of lithospheric stress tensor from low-low inter-satellite range-rates

These measurements are functions of the coordinates of two satellites. The potential difference between the satellites is related to the inter-satellite low-low range rates. However, the potential difference is a scalar quantity between satellites, which eases the mathematical derivations. Let us start the discussion with the simplest case of determining τ_{zz} from this type of inter-satellite measurement. The spectral form of Eq. (8.109) is

$$\tau_{zz,nm} = \frac{\tilde{\lambda} K_n^1 + 2\tilde{\mu} K_n^2}{r'} t_{nm}. \tag{9.71}$$

It is enough to derive t_{nm} from the inter-satellite low-low range rates and insert it into Eq. (9.58):

$$r' \sum_{n=2}^{\infty} \frac{1}{\tilde{\lambda} K_n^1 + 2\tilde{\mu} K_n^2} \sum_{m=-n}^{n} \sum_{i=1}^{2} (-1)^i \left(\frac{R}{r_i}\right)^{n+1} Y_{nm}(\theta_i, \lambda_i) \tau_{zz,nm} = v_0 \dot{\tilde{\rho}} - \Upsilon. \tag{9.72}$$

In the case of inserting the integral formula of $\tau_{zz,nm}$, according to Eq. (1.8), into Eq. (9.72), taking summation over m, applying the addition theorem of the spherical harmonics (1.9), and finally take summation over n, the integral equation will be obtained:

$$\frac{r'}{4\pi} \iint_{\sigma} \sum_{n=2}^{\infty} \frac{1}{\tilde{\lambda} K_n^1 + 2\tilde{\mu} K_n^2} \sum_{i=1}^{2} (-1)^i \left(\frac{R}{r_i}\right)^{n+1} P_n(\tilde{x}_i) \tau_{zz}(\theta', r', \lambda') d\sigma = v_0 \dot{\tilde{\rho}} - \Upsilon. \tag{9.73}$$

Now, let us develop the integral equation to recover τ_{xz} and the range rate data. However, writing a equation similar to Eq. (9.71) is impossible owing to the presence of the derivative of the spherical harmonics with respect to the latitude. In this case, the equation will be (see Eq. 8.112):

$$\tau_{xz,nm} Y_{nm}(\theta, \lambda) = \frac{\tilde{\mu} K_n^4}{r'} t_{nm} \frac{\partial Y_{nm}(\theta, \lambda)}{\partial \theta}. \tag{9.74}$$

After solving Eq. (9.74) for t_{nm}, we obtain:

$$t_{nm} = \frac{Y_{nm}(\theta, \lambda)}{\frac{\partial Y_{nm}(\theta, \lambda)}{\partial \theta}} \frac{r'}{\widetilde{\mu} K_n^4} \tau_{xz,nm}. \tag{9.75}$$

Inserting the integral expression of $\tau_{xz,nm}$ (Eq. 1.8) into Eq. (9.75), and after further simplification, an integral equation can be developed to determine τ_{xz} from $\nu_0 \widetilde{\dot{\rho}} - \Upsilon$. The same procedure can be applied to develop similar integral equations to invert $\nu_0 \widetilde{\dot{\rho}} - \Upsilon$ to the rest of the elements of stress tensor. Let us summarise them in the integral equation:

$$\frac{r'}{4\pi\widetilde{\mu}} \iint_{\sigma} \Psi_j^* \tau_j(\theta', \lambda') \mathrm{d}\sigma = \nu_0 \widetilde{\dot{\rho}} - \Upsilon, \ \ j = xz, yz, xy, xx - yy \tag{9.76}$$

with

$$\Psi_j^* = \sum_{n=2}^{\infty} \left(\frac{1}{K_n^4}\right) \sum_{m=-n}^{n} \sum_{i=1}^{2} (-1)^i \left(\frac{R}{r_i}\right)^{n+1} Y_{nm}(\theta_i, \lambda_i) \frac{C_{nm}}{C_{j,nm}}, \quad j = xz, yz, xy, xx - yy \tag{9.77}$$

where

$$C_{nm} = Y_{nm}(\theta, \lambda) Y_{nm}(\theta', \lambda') \tag{9.78}$$

$$C_{xz,nm} = \frac{\partial Y_{nm}(\theta, \lambda)}{\partial \theta} \tag{9.79}$$

$$C_{yz,nm} = \frac{\partial Y_{nm}(\theta, \lambda)}{\sin \theta \partial \lambda} \tag{9.80}$$

$$C_{xy,nm} = \frac{\partial^2 Y_{nm}(\theta, \lambda)}{\sin \theta \partial \theta \partial \lambda} - \cot \theta \frac{\partial Y_{nm}(\theta, \lambda)}{\sin \theta \partial \lambda} \tag{9.81}$$

$$C_{xx-yy,nm} = \frac{\partial^2 Y_{nm}(\theta, \lambda)}{\partial \theta^2} - \frac{1}{\sin^2 \theta} \frac{\partial^2 Y_{nm}(\theta, \lambda)}{\partial \lambda^2} - \cot \theta \frac{\partial Y_{nm}(\theta, \lambda)}{\partial \theta}. \tag{9.82}$$

The integral Eq. (9.76) is simple to solve when its complicated kernel function is computed numerically. Generating such kernels is time-consuming because they cannot be written in terms of the Legendre polynomials, and the summations over n and m should be performed. In addition, they involve the derivatives of the spherical harmonics. Finally, each kernel is the function of coordinates of four points and computation and integration points as well as the twin satellites.

9.9.2 Determination of elements of stress tensor from inter-satellite line-of-sight measurements

Mathematical development of the relation between the lithospheric stress tensor and the inter-satellite LOS measurements is more complicated than the case of using only the intersatellite range rates. Determination of τ_{zz} from this type of data is the simplest case. According to Eqs. (9.65) and (9.71), we obtain:

$$\ddot{\tilde{\rho}} - \frac{(\delta\dot{\mathbf{r}}_{12})^2 - \dot{\tilde{\rho}}^2}{\tilde{\rho}} - \Sigma = \sum_{n=2}^{\infty} \frac{r'}{\tilde{\lambda}K_n^1 + 2\tilde{\mu}K_n^2} \sum_{m=-n}^{n} \tau_{zz,nm} A_{nmi} \tag{9.83}$$

where

$$A_{nmi} = \sum_{i=1}^{2} \frac{(-1)^i}{r_i} \left(\frac{R}{r_i}\right)^{n+1} \left(-(n+1) Y_{nm}(\theta_i, \lambda_i) \mathbf{e}_z + \mathbf{X}_{nm}^{(2)}(\theta_i, \lambda_i) \right) \cdot \mathbf{e}_{\text{LOS}} \tag{9.84}$$

Note that A_{nmi} is a function of both the satellites' positions and the unit vector pointing from satellite 1 to satellite 2. According to Eq. (1.8) we can write:

$$\ddot{\tilde{\rho}} - \frac{(\delta\dot{\mathbf{r}}_{12})^2 - \dot{\tilde{\rho}}^2}{\tilde{\rho}} - \Sigma = \frac{r'}{4\pi} \iint_{\sigma} \tau_{zz}(r', \theta', \lambda') \sum_{n=2}^{\infty} \frac{1}{\tilde{\lambda}K_n^1 + 2\tilde{\mu}K_n^2} \sum_{m=-n}^{n} A_{nmi} Y_{nm}(\theta, \lambda) \mathrm{d}\sigma \tag{9.85}$$

By taking the advantage of the addition theorem of spherical harmonics we have:

$$\frac{r'}{4\pi} \iint_{\sigma} \tau_{zz}(r', \theta', \lambda') \sum_{n=2}^{\infty} \frac{1}{\tilde{\lambda}K_n^1 + 2\tilde{\mu}K_n^2} \sum_{i=1}^{2} \frac{(-1)^i}{r_i} \left(\frac{R - D_{\text{Lith}}}{r_i}\right)^{n+1} \left(-(n+1) P_n(\tilde{x}_i) \mathbf{e}_{z_i} + \frac{\partial P_n(\tilde{x}_i)}{\partial \theta} \mathbf{e}_{x_i} + \frac{\partial P_n(\tilde{x}_i)}{\sin\theta \partial \lambda} \mathbf{e}_{y_i} \right) . \mathbf{e}_{\text{LOS}} \mathrm{d}\sigma = \ddot{\tilde{\rho}} - \frac{(\delta\dot{\mathbf{r}}_{12})^2 - \dot{\tilde{\rho}}^2}{\tilde{\rho}} - \Sigma \tag{9.86}$$

Finally, the following integral equation is derived:

$$\frac{1}{4\pi}\iint_{\sigma} \tau_{zz}(\theta', \lambda')\Psi^{**} d\sigma \tag{9.87}$$

where

$$\begin{aligned}\Psi^{**} = \sum_{n=2}^{\infty} \frac{r'(2n+1)}{\widetilde{\lambda} K_n^1 + 2\widetilde{\mu} K_n^2} \sum_{i=1}^{2} \frac{(-1)^i}{r_i}\left(\frac{R}{r_i}\right)^{n+1} \Bigg(-(n+1)P_n(\widetilde{x}_i)\mathbf{e}_{z_i} \\ + \frac{\partial P_n(\widetilde{x}_i)}{\partial \psi}(\cos\alpha_i \mathbf{e}_{x_i} - \sin\alpha_i \mathbf{e}_{y_i})\Bigg)\cdot \mathbf{e}_{\mathrm{LOS}}\end{aligned} \tag{9.88}$$

and by applying the integral equations to recover the rest of the elements of the stress tensor, we have:

$$\frac{r'}{4\pi\widetilde{\mu}}\iint_{\sigma} \Psi_j^{**} \tau_j(\theta', \lambda') d\sigma = \ddot{\widetilde{\rho}} - \frac{(\Delta\dot{\mathbf{r}}_{12})^2 - \dot{\widetilde{\rho}}^2}{\widetilde{\rho}} - \Sigma, \quad j = xz, yz, xy, xx-yy \tag{9.89}$$

where

$$\left.\begin{matrix}\Psi_{xz}^{**} \\ \Psi_{yz}^{**}\end{matrix}\right\} = \sum_{n=2}^{\infty} \frac{1}{K_n^4} \sum_{i=1}^{2} \left(\frac{(-1)^i}{r_i}\left(\frac{R}{r_i}\right)^{n+1} \left(\left\{ \begin{matrix} \dfrac{1}{C_{xz,nm}} \\ \dfrac{1}{C_{yz,nm}} \end{matrix} \right\} \sum_{m=-n}^{n} \left(-(n+1) C_{nmi}\mathbf{e}_{z_i} + \mathbf{C}_{nmi}\right) \cdot \mathbf{e}_{\mathrm{LOS}} \right)\right) \tag{9.90}$$

$$\left.\begin{matrix}\Psi_{xy}^{**} \\ \Psi_{xy-yy}^{**}\end{matrix}\right\} = \sum_{n=2}^{\infty} \frac{1}{K_n^5} \sum_{i=1}^{2} \left(\frac{(-1)^i}{r_i}\left(\frac{R}{r_i}\right)^{n+1} \left\{ \begin{matrix} \dfrac{1}{C_{xy,nm}} \\ \dfrac{1}{2C_{xx-yy,nm}} \end{matrix} \right\} \sum_{m=-n}^{n} \left(-(n+1) C_{nmi}\mathbf{e}_{z_i} + \mathbf{C}_{nmi}\right) \cdot \mathbf{e}_{\mathrm{LOS}} \right) \tag{9.91}$$

with

$$C_{nmi} = Y_{nm}(\theta', \lambda') Y_{nm}(\theta_i, \lambda_i) \tag{9.92}$$

$$\mathbf{C}_{nmi} = Y_{nm}(\theta', \lambda') \mathbf{X}_{nm}^{(2)}(\theta_i, \lambda_i) \tag{9.93}$$

Acknowledgements

I am thankful to Dr Martin Pitonak for his help with processing the real GOCE data over South America.

References

ESA, 1999. Gravity Field and Steady-State Ocean Circulation Mission. Report for Mission Selection of the Four Candidate Earth Explorer Missions, ESA SP-1233(1). ESA Publications Division, ESTEC, Noordwijk, The Netherlands.

Eshagh, M., 2011. The effect of spatial truncation error on integral inversion of satellite gravity gradiometry data. Adv. Space Res. 47, 1238–1247.

Eshagh, M., 2014. From satellite gradiometry data to sub-crustal stress due to mantle convection. Pure Appl. Geophys. 171, 2391–2406.

Eshagh, M., 2015. On the relation between Moho and sub-crustal stress due to mantle convection. J. Geophys. Eng. 12, 1–11.

Eshagh, M., 2017. Local recovery of lithospheric stress tensor from GOCE gravitational tensor. Geophys. J. Int. 209 (1), 317–333.

Eshagh, M., Romeshkani, M., 2015. Determination of sub-lithospheric stress due to mantle convection using GOCE gradiometric data over Iran. J. Appl. Geophys. 122, 11–17.

Eshagh, M., Tenzer, R., 2015. Sub-crustal stress determined using gravity and crust structure models. Comput. Geosci. 19, 115–125.

Eshagh, M., Hussain, M., Tiampo, K.F., 2016. Towards sub-lithospheric stress determination from seismic Moho, topographic heights and GOCE data. J. Asian Earth Sci. 129, 1–12.

Eshagh, M., Steinberger, B., Tenzer, R., Tassara, A., 2018. Comparison of gravimetric and mantle flow solutions for lithospheric stress modelling and their combination. Geophys. J. Int. 213 (2), 1013–1028.

Farr, T.G., et al., 2007. The shuttle radar topography mission. Rev. Geophys. 45 (2), RG2004. https://doi.org/10.1029/2005RG000183.

Hansen, P.C., 1998. Rank-deficient and Discrete Ill-Posed Problems: Numerical Aspects of Linear Inversion. SIAM, Philadelphia.

Hansen, P.C., 2007. Regularization tools version 4.0 for Matlab 7.3. Numer. Algorithm. 46, 189–194.

Heidbach, O., Tingay, M., Barth, A., Reinecker, J., Kurfess, D., Mueller, B., 2010. Global crustal stress pattern based on the World Stress Map database release 2008. Tectonophysics 482, 3–15.

Hinze, W.J., 2003. Bouguer reduction density, why 2.67? Geophysics 68 (5), 1559–1560.

Laske, G., Masters, G., Ma, Z., Pasyanos, M.E., 2013. Update on CRUST1.0—a 1-degree global model of Earth's crust. Geophys. Res. Abstr. 15. Abstract EGU2013-2658.

Moritz, H., 2000. Geodetic reference system 1980. J. Geod. 74, 128–133.

Pavlis, N.K., Holmes, S.A., Kenyon, S.C., Factor, J.K., 2012. The development and evaluation of the Earth gravitational model EGM2008 (EGM2008). J. Geophys. Res. Solid Earth 117, B04406.

Phillips, R.J., Ivins, E.R., 1979. Geophysical observations pertaining to solid-state convection in the terrestrial planets. Phys. Earth Planet. Inter. 19 (2), 107–148.

Runcorn, S.K., 1964. Satellite gravity measurements and laminar viscous flow model of the Earth mantle. J. Geophys. Res. 69, 4389–4394.

Runcorn, S.K., 1967. Flow in the mantle inferred from the low degree harmonics of the geopotential. Geophys. J. R. Astron. Soc. 14, 375–384.

Sprlak, M., Eshagh, M., 2016. Local recovery of sub-crustal stress due to mantle convection from satellite-to-satellite tracking data. Acta Geophysica 64 (4), 904–929.

Tassara, A., Swain, C., Hackney, R., Kirby, J., 2007. Elastic thickness structure of South America estimated using wavelets and satellite-derived gravity data. Earth Planet. Sci. Lett. 253 (1–2), 17–36.

Tenzer, R., Eshagh, M., 2015. Subduction generated sub-crustal stress in Taiwan. Terr. Atmos. Ocean. Sci. 26 (3), 261–268.

Tikhonov, A.N., 1963. Solution of incorrectly formulated problems and regularization method. Soviet Math. Dokl. 4, 1035–1038. English translation of Dokl. Akad. Nauk. SSSR, 151:501–504.

CHAPTER 10

Satellite gravimetry and applications of temporal changes of gravity field

10.1 Time-variable gravity field

The gravity field of the Earth is not constant in time because of the dynamic nature of the Earth. Large earthquakes and hydrological signals are two main sources of changes that affect the long-wavelength structure of the gravity field. Satellite gravimetry is a suitable approach to these purposes because of the high altitude of satellites from the Earth's surface, which diminishes the high frequencies of the gravity field. In addition to the satellite altitude, the type of satellite sensor has an important role in recovering the gravity field. For example, satellite gradiometry is more suitable for recovering high frequencies of the gravity field rather than its temporal changes. However, the situation is different for intersatellite low-low tracking data measured between twin satellites moving in the same orbit around the Earth. Such type of data can provide precise information about the low-frequency content of the gravity field and its temporal changes. Theoretically, any satellite gravimetry data should be able to be used to determine these temporal variations, but for practical reasons, instruments or sensors sensitive to all frequency bands of the gravitation are not yet available.

Generally, the temporal changes of the gravity field can be presented by the spherical harmonic series:

$$\delta V(r,\theta,\lambda)=\sum_{n=0}^{\infty}\left(\frac{R}{r}\right)^{n+1}\sum_{m=-n}^{n}\delta v_{nm}Y_{nm}(\theta,\lambda) \tag{10.1}$$

where δv_{nm} is the spherical harmonic coefficients (SHCs) of the changes of the gravitational potential, r stands for the geocentric radius of any point outside the Earth's surface, and $Y_{nm}(\theta,\lambda)$ is the spherical harmonics with

Satellite Gravimetry and the Solid Earth
ISBN 978-0-12-816936-0
https://doi.org/10.1016/B978-0-12-816936-0.00010-4

degree n and order m, with arguments of the co-latitude θ and the longitude λ. R stands for the radius of the spherical Earth.

In Chapters 2–4, methods for recovering the gravity field from satellite gravimetry data were extensively discussed. From theoretical point of view, the time-dependent gravity field can be recovered from these data collected during specific time intervals. By performing a similar process for another time interval, a new gravity field is obtained. The difference between these two fields will be nothing else that the gravity field change. Such change can be determined in terms of SHCs if the data have with global coverages. Disturbing potential, gravity anomaly/disturbance can also be inverted locally using integral equations for such purposes.

10.2 Hydrological effects and equivalent water height from time-variable gravity field

As explained, one of the main sources of temporal variations of the gravity field of the Earth is the hydrological mass transport over continents. Generally, inland water comes from soil moisture (SM), snow water equivalent (SWE), canopy (CAN), and groundwater storage (GWS). A main purpose of satellite gravimetry missions of the Gravity Recovery and Climate Experiment (GRACE) (Tapley et al., 2004) and GRACE Follow-On (GRACE-FO) (Flechtner et al., 2014) is to monitor continental water. Therefore, it is important to see how the time-variable gravity models are determined from these missions are related to hydrological mass transport over continents.

10.2.1 Gravitational potential of a surface mass

Wahr et al. (1998) modelled the SHCs of the gravitational potential of surface mass from a thin spherical shell with a certain density and thickness. In this section, we present the mathematical foundations to derive their equation. Assume that the solid Earth is covered by a thin shell with a density of $\delta\rho$ and thickness of δh, according to the Newton integral, Eq. (5.1), the gravitational potential of this thin shell, or the surface mass, is:

$$\delta V^{\text{surface mass}}(r,\theta,\lambda) = G\iint\limits_{\sigma}\int\limits_{R}^{R+\delta h} \frac{\delta\rho(r',\theta',\lambda')}{l} r'^2 \mathrm{d}r' \mathrm{d}\sigma \tag{10.2}$$

where $\mathrm{d}r'$ and $\mathrm{d}\sigma$ are the radial and surface integration elements, R is the Earth's radius, and G is the Newtonian gravitational constant. l is the distance between any point inside the thin shell and the computation point outside it. The reciprocal distance l^{-1} can be expanded in terms of Legendre polynomials of degree n (Eq. 5.3). Substitution of this expansion into Eq. (10.2) yields:

$$\delta V^{\text{surface mass}}(r,\theta,\lambda)=G\iint_{\sigma}\sum_{n=0}^{\infty}r\int_{R}^{R+\delta h}\delta\rho(r',\theta',\lambda')\left(\frac{r'}{r}\right)^{n+2}\mathrm{d}r'P_n(\tilde{x})\mathrm{d}\sigma \tag{10.3}$$

Now, consider the radial integral. When $r=R$, and the integration point is at the upper bound of the thin shell,

$$\left(\frac{r'}{r}\right)^{n+2}=\left(\frac{R+\delta h}{R}\right)^{n+2}\approx\left(1+\frac{\delta h}{R}\right)^{n+2}\approx 1. \tag{10.4}$$

By defining the surface density by:

$$\delta\rho^{\text{surface}}(\theta',\lambda')=\int_{R}^{R+\delta h}\delta\rho(r',\theta',\lambda')\mathrm{d}r', \tag{10.5}$$

Eq. (10.2) will change to:

$$\delta V^{\text{surface mass}}(R,\theta,\lambda)=GR\sum_{n=0}^{\infty}\iint_{\sigma}\delta\rho^{\text{surface}}(\theta',\lambda')P_n(\tilde{x})\mathrm{d}\sigma \tag{10.6}$$

Based on the Legendre integral (1.11), or addition theorem of spherical harmonics (1.9), Eq. (10.6) will be:

$$\begin{aligned}\delta V^{\text{surface mass}}(\theta,\lambda)&=4\pi GR\sum_{n=0}^{\infty}\frac{\delta\rho_n^{\text{surface}}(\theta,\lambda)}{2n+1}\\&=4\pi GR\sum_{n=0}^{\infty}\frac{1}{2n+1}\sum_{m=-n}^{n}\delta\rho_{nm}^{\text{surface}}Y_{nm}(\theta,\lambda)\end{aligned} \tag{10.7}$$

where (Eq. 1.8):

$$\delta\rho_{nm}^{\text{surface}}=\frac{1}{4\pi}\iint_{\sigma}\delta\rho^{\text{surface}}(\theta',\lambda')Y_{nm}(\theta',\lambda')\mathrm{d}\sigma \tag{10.8}$$

The gravitational potential of the surface mass directly influences the gravity field. Its indirect effects will result from loads on the solid Earth and thus its gravitational attraction. This means any change in the gravitational potential of the Earth can be presented as the integral of these two potential changes. If the changes in gravity field is shown by δv_{nm}, it will be

$$\delta v_{nm} = \delta v_{nm}^{\text{surface mass}} + \delta v_{nm}^{\text{solid Earth}} \tag{10.9}$$

where $\delta v_{nm}^{\text{surface mass}}$ is the SHC of the potential of the surface masses and changes in the gravitational potential of the solid Earth owing to the surface load $\delta v_{nm}^{\text{solid Earth}}$, which is

$$\delta v_{nm}^{\text{solid Earth}} = k_n \delta v_{nm}^{\text{surface mass}} \tag{10.10}$$

where k_n stands for Love numbers for a viscoelastic Earth (Han and Wahr, 1995). By inserting Eq. (10.10) into Eq. (10.9), we obtain:

$$\delta v_{nm}^{\text{solid Earth}} = (1 + k_n)\delta v_{nm}^{\text{surface mass}}. \tag{10.11}$$

Eq. (10.11) is in fact the potential resulting from surface mass; therefore, by inserting it into Eq. (10.7), one reads:

$$\begin{aligned} \delta V^{\text{surface mass}}(R, \theta, \lambda) &= 4\pi GR \sum_{n=0}^{\infty} \frac{1 + k_n}{2n + 1} \delta\rho_n^{\text{surface}}(\theta, \lambda) \\ &= 4\pi GR \sum_{n=0}^{\infty} \frac{1 + k_n}{2n + 1} \sum_{m=-n}^{n} \delta\rho_{nm}^{\text{surface}} Y_{nm}(\theta, \lambda). \end{aligned} \tag{10.12}$$

Eq. (10.12) is the spherical harmonic expression of the gravitational potential of the surface density. This means that if $\delta\rho_{nm}^{\text{surface}}$ is available, their gravitational effects can be determined by this equation.

10.2.2 Effect of hydrological masses

Different hydrological models have been provided to monitor the global hydrological regimes. Such models provide the surface density of the SM, SWE, CAN, and GWS. Summation of all these surface densities provides the surface density of the total water storage (TWS):

$$\delta\rho^{\text{TWS}} = \delta\rho^{\text{SM}} + \delta\rho^{\text{SWE}} + \delta\rho^{\text{CAN}} + \delta\rho^{\text{GWS}}. \tag{10.13}$$

By dividing these surface densities by the density of water, the equivalent water height (EWH) will be obtained. The surface densities presented in Eq. (10.13) can be written in terms of their corresponding SHCs, to have them comparable to gravity models:

$$\delta\rho_{nm}^{\text{TWS}} = \delta\rho_{nm}^{\text{SM}} + \delta\rho_{nm}^{\text{SWE}} + \delta\rho_{nm}^{\text{CAN}} + \delta\rho_{nm}^{\text{GWS}} \tag{10.14}$$

where (Eq. 1.8):

$$\left(\delta\rho^{i}\right)_{nm} = \frac{1}{4\pi}\iint_{\sigma} \delta\rho^{i}(\theta', \lambda') Y_{nm}(\theta', \lambda') d\sigma, \tag{10.15}$$

$$\text{where, } i = \text{SM, SWE, CAN, or GWS}$$

The spherical harmonic synthesis of the surface densities yields the EWH once it is divided by the density of water ρ_w:

$$\delta h_{\mathrm{Hyd}}^{\mathrm{TWS}}(\theta, \lambda) = \frac{1}{\rho_w}\sum_{n=0}^{\infty}\sum_{m=-n}^{n}\left(\delta\rho_{nm}^{\mathrm{SM}} + \delta\rho_{nm}^{\mathrm{SWE}} + \delta\rho_{nm}^{\mathrm{CAN}} + \delta\rho_{nm}^{\mathrm{GWS}}\right) Y_{nm}(\theta, \lambda) \tag{10.16}$$

This EWH can also be determined from the spherical harmonic of the gravity field changes:

$$\delta h_{\mathrm{Grav}}^{\mathrm{TWS}}(\theta, \lambda) = \frac{1}{4\pi G R \gamma}\sum_{n=0}^{\infty}\frac{2n+1}{1+k_n}\sum_{m=-n}^{n}\delta v_{nm} Y_{nm}(\theta, \lambda) \tag{10.17}$$

where γ is the normal gravity.

The linear trend of the TWS extracted monthly by the GRACE time-variable gravity models from Apr. 2002 to Jun. 2017 is presented in Fig. 10.1.

The hydrological models provide water cycle in the land. The one with high consistency with the GRACE time-variable gravity models is the Global Land Data Assimilation System (GLDAS) (Rodell et al., 2004), which contains several land surface models including Neural Optimisation Applied Hydrology (NOAH) (Coppola et al., 2003), the Community Land Model (Dai et al., 2003), MOSAIC (Koster and Suarez, 1996), and Variable Infiltration Capacity (VIC) (Liang et al., 1994, 1996; Cherkauer et al., 2003). This model has several temporal and spatial resolutions. The newest version of the GLDAS model, version 2.1 provided by NOAH, is often used because of its good agreement with the GRACE gravity models. Fig. 10.2A,B presents the summation of SM, SWE, and CAN, considered the TWS for Jan to Feb. 2000 and 2019, respectively, over the land area. The changes in the TWS during these 19 years are significant when comparing these maps.

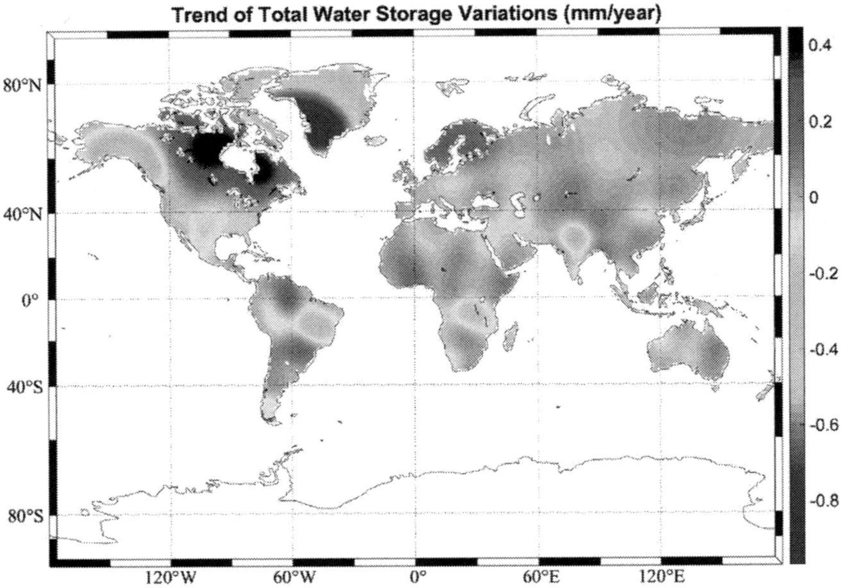

Figure 10.1 Trend of variations of total water storage over continents (mm/year). *(Computed by Fatolazadeh.)*

10.2.3 Determination of hydrological parameters from satellite gravimetry

Here, the SHCs presented in the time-variable gravity field models, determined from satellite gravimetry data, are used for modelling the hydrological parameters. Our main assumption is that no additional changes occur in the gravitational potential of the solid Earth and the changes of the gravity field are only the result of the hydrological mass transport. By considering Eqs. (10.16) and (10.17) and assuming that $\delta h_{\mathrm{Hyd}}^{\mathrm{TWS}} = \delta h_{\mathrm{GRAV}}^{\mathrm{TWS}}$, which is not far from reality, the following relations between the SHCs of the hydrological parameter of surface density and changes in the gravity field of the Earth are obtained:

$$\delta\rho_{nm}^{\mathrm{SM}} + \delta\rho_{nm}^{\mathrm{SWE}} + \delta\rho_{nm}^{\mathrm{CAN}} + \delta\rho_{nm}^{\mathrm{GWS}} = \frac{\rho_w}{4\pi G R \gamma}\frac{2n+1}{1+k_n}\delta v_{nm}. \tag{10.18}$$

Eq. (10.18) means that the summation of surface densities of SM, SWE, CAN and GWS are derived from the time-variable gravity models and not

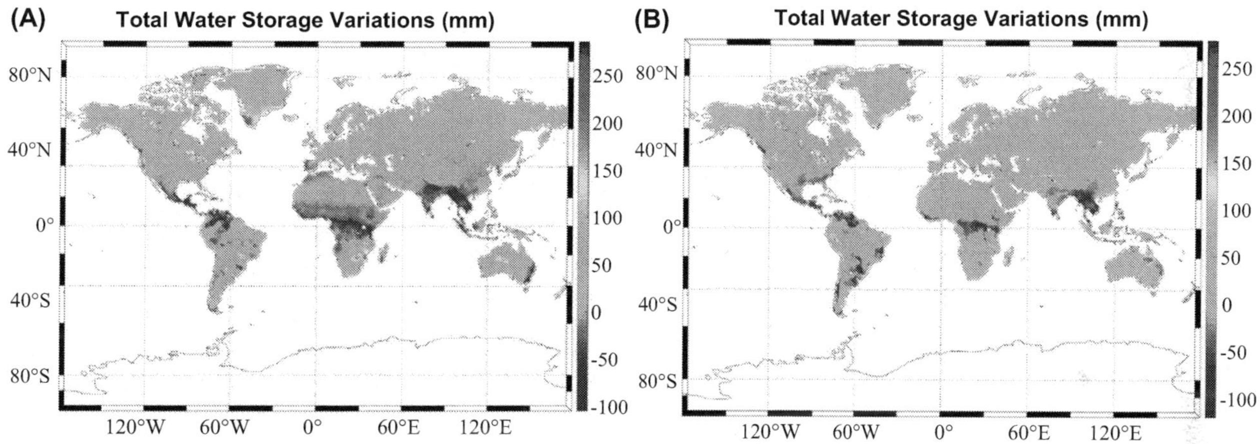

Figure 10.2 (A) Total water storage variation from (A) Jan. to Feb. 2000 and (B) Jan. to Feb. 2019. *(Ccomputed by Fatolazadeh.)*

each hydrological parameter. Therefore, if the goal is to determine one of these surface densities, the rest should be available from external sources or hydrological models. For example, the GLDAS model provides all hydrological parameters except the GWS. Therefore, by subtracting the effects of these parameters from the time-variable gravity models, determined, for example, by GRACE or GRACE-FO, the GWS signals are determined. In Fig. 10.3, the trends of the GWS variation in 15 years from Apr. 2002 to Jun. 2017 were determined for Iran based on that mentioned principle using the GLDAS and GRACE models. The positive values of the GWS trend are seen in the northeast of the study area and outside Iran. The trend is negative mainly in Iran and decreases southwest of the area, meaning a linear loss of the GWS during these 15 years. The Caspian Sea, Persian Gulf and Oman Sea have been masked out from the map because the hydrological models do not contain information over oceans and seas. In addition, the procedure of determining mass variations over such areas is totally different.

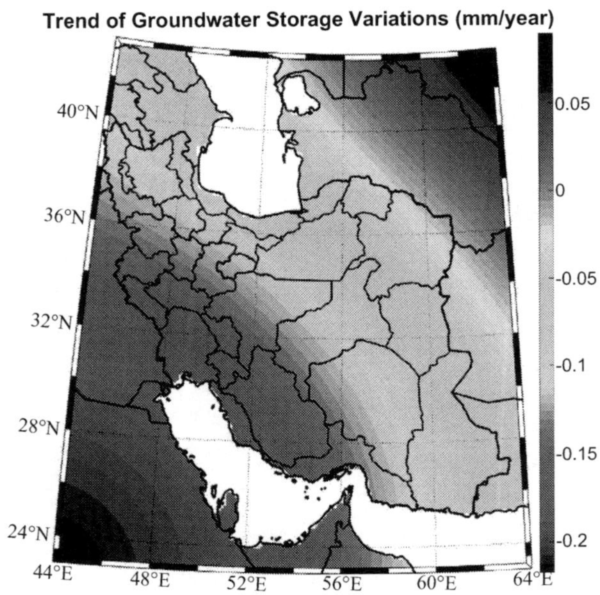

Figure 10.3 Trend of groundwater storage in Iran, determined from Gravity Recovery and Climate Experiment time-variable gravity models and GLDAS models from Apr. 2002 to Jun. 2017. *(Computed by Fatolazadeh.)*

10.2.4 Earthquake monitoring and time-variable gravity field

This section presents a part of the study conducted by Hussain et al. (2016) involving the earthquake that occurred in 2005 in Kashmir. The time-variable gravity models of GRACE contain mass variations owing to different phenomena. By removing the gravitational effects of the hydrological signals from these models, the changes caused by earthquakes are more pronounced. To study the Kashmir earthquake, the GLDAS model is used to compute the hydrological effects on the gravity field models (Fig. 10.4) in the month in which the earthquake occurred. Fig. 10.4A,B

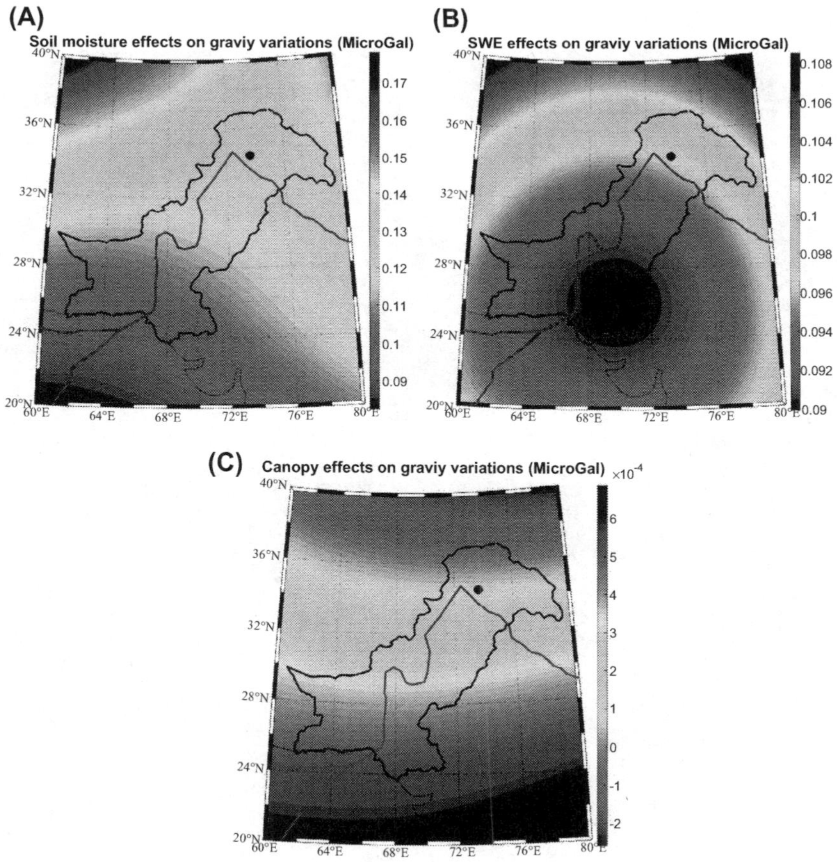

Figure 10.4 Effect of (A) soil moisture, (B) snow water equivalent (SWE) and (C) canopy on gravity over Pakistan. The *red line* represents the tectonic boundaries. *(Computed by Fatolazadeh.)*

show maps of the SM and SWE, respectively. The largest value of the SM is seen in the northwest of the area, and the smallest one is in the southern part, which has a warm climate. The SWE is seen over the northern part, close to the mountains. The effect of CAN is seen in Fig. 10.4C in the southeast part, in India. However, this effect is considerably smaller than the effects of SM and SWE. The place where the Kashmir earthquake happened is marked with a small circle.

After removing the effect of hydrological signals from the time-variable gravity models of GRACE, the geoid changes between each 2 months are computed for 2 months before and after the earthquake. The results are presented in Fig. 10.5. The black points over the map are where the

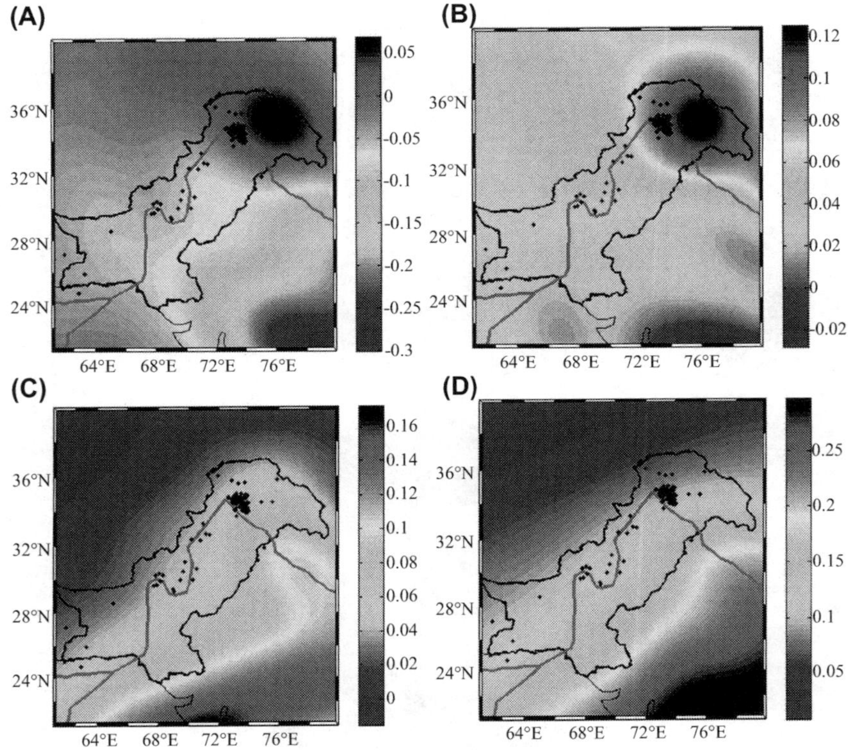

Figure 10.5 Changes in geoid height before and after the Kashmir earthquake. The *red line* represents the tectonic boundaries during (A) August, (B) September, (C) October and (D) November (metres). *(From Hussain M., Eshagh M., Zulfiqar A., Sadiq M. and Fatolazadeh F., 2016. Changes in gravitational parameters inferred from time variable GRACE data—a case study for October 2005 Kashmir earthquake, J. Appl. Geophys. 132, 174–183.)*

earthquake occurred in 2005. As observed, Kashmir experienced frequent earthquakes during that period. Fig. 10.5A is the map of the geoid change from August to November; it shows positive values in the northern part, but maximum changes of the geoid seem close to the Kashmir earthquake epicentre. The positive value in this area means that the geoid height increased before the earthquake because of changes in the interior mass inside the Earth. Fig. 10.5B shows the period in which the earthquake occurred; a significant change in geoid height is clear. Figs. (10.5C) and (10.5D) are maps of geoid changes after the earthquake stress release containing no remarkable change in geoid height.

Fig. 10.5A shows that during August, large changes in geoid heights were seen around the earthquake epicentre point. Such a change was even more pronounced in September, 2 months before the earthquake. After the earthquake occurred in October geoid variations in October and November were smooth and without special concentration. Changes in the geoid heights resulted from the mass and structure variations owing to the earthquake.

10.3 Surface mass changes over ocean and satellite gravimetry

Time-variable gravity models and satellite altimetry data can be used to study permanent sea level changes. These changes are the result of several factors. Changes in storage or the volume of water (e.g., by rain or the presence of the steric sea level (SSL)) are due to the changes the structure of atoms of ocean; temperature and salinity are two other most contributing factors in change of structure of atoms in molecule water. Satellite gravimetry data do not have the ability to determine these parameters, but the total volume changes (non-steric and steric components) can be measured by satellite altimetry data. To compute the surface mass changes over oceans, the time-variable gravity models should be reduced for the SSL signals. This means

$$\Delta\eta_{\text{Ocean}}(\theta,\lambda,t)=\Delta\eta_{\text{SLA}}(\theta,\lambda,t)-\Delta\eta_{\text{SSL}}(\theta,\lambda,t) \tag{10.19}$$

where $\Delta\eta_{\text{Ocean}}$ is the total equivalent water thickness variations in oceans, $\Delta\eta_{\text{SLA}}$ is the sea level anomaly measured by satellite altimetry and $\Delta\eta_{\text{SSL}}$ is SSL variations. To compute SSL, the World Ocean Atlas (WOA), containing several oceanic variables including temperature, salinity, silicate, oxygen, phosphate and nitrate, is used.

As the two of most highly contributing factors to oceanic mass changes, temperature and salinity are inserted into the International Equation of State of Sea Water (Eq. 10.20), and the surface density is normally computed based on data from one of the three maximum depths of 5500, 1500 and 800 m, provided by WOA, to the surface of the ocean. Temperature and salinity at different depths are available as functions of co-latitude and longitude. Therefore, surface density of ocean can be determined at each depth as a function of changes in co-latitude and longitude, time, depth, temperature and salinity. On the other hand, if the mean of the determined surface is subtracted from the surface densities and the results are radially integrated from the selected maximum depth to the oceanic surface and then divided by density of water (1027 kg / m^3), the SSL will be obtained:

$$\Delta\eta_{\mathrm{SSL}}(\theta,\lambda,t)=\frac{1}{\rho_w}\int_{-h}^{\mathrm{surface}}\left[\delta\rho(\theta,\lambda,t,z,T,S)-\overline{\delta\rho}\left(\theta,\lambda,t,z,\overline{T},\overline{S}\right)\right]\mathrm{d}z \tag{10.20}$$

In Eq. (10.20), $\Delta\eta_{\mathrm{SSL}}$ is the equivalent water thickness changes for SSL, ρ_w is the standard density of water, h is the selected maximum depth (e.g., 5500 m), $\delta\rho$ is the surface density as a function of latitude, longitude, time (t), depth (z), temperature (T) and salinity (S), and $\overline{\delta\rho}$ its mean value finally dz is the radial integration element. By subtracting the estimated $\Delta\eta_{\mathrm{SSL}}$ from the $\Delta\eta_{\mathrm{SLA}}$ measured by satellite altimetry (Eq. 10.19), the effect of oceanic surface mass changes $\Delta\eta_{\mathrm{Ocean}}$ is obtained. The SHCs of the determined oceanic effect $(\Delta\eta_{\mathrm{Ocean}})_{nm}$ should be computed and converted to the SHCs of the gravity field and removed from them. In this case, the reduced coefficients will be suitable for studying the Earth's interior mass changes owing to tectonic activities. Fig. 10.6 shows the oceanic effects extracted by the Jason-1 satellite altimeter in Jan. 2005 and the WOA 2005 model. The radial integral (Eq. 10.20) has been solved numerically by the Simpson integration method with a radial resolution of 1 m.

10.4 Determination of land uplift caused by postglacial rebound

During the ice age, there were huge ice caps over some parts of the Earth's surface (e.g., Hudson Bay and Fennoscandia). After the ice melted, the Earth's lithosphere was constantly under isostatic rebound to come back to

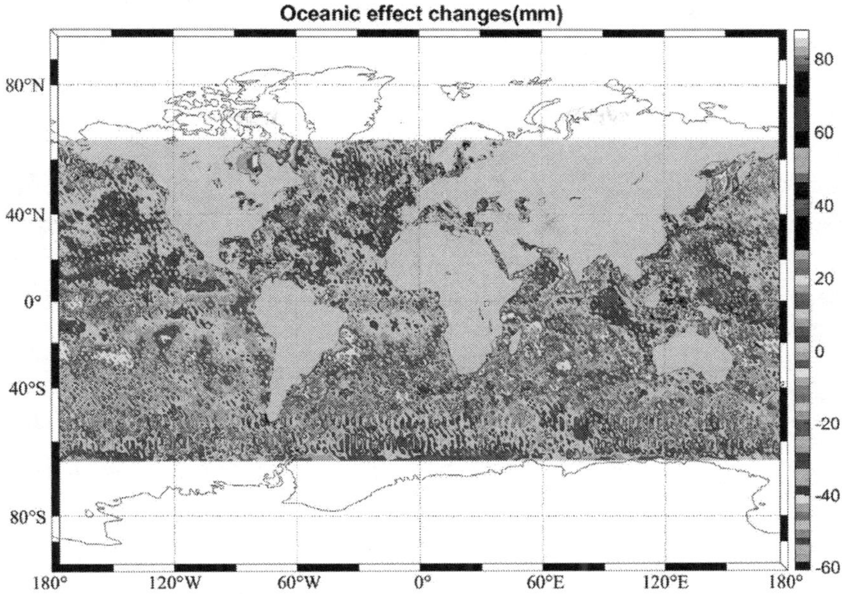

Figure 10.6 Ocean surface changes. *(Computed by Fatolazadeh.)*

equilibrium, the status before the ice cap existed. This phenomenon is known as postglacial rebound. In this section, our goal is to develop mathematical foundations to determine the land uplift from time-variable gravity field models. Here, the isostatic compensation potential of the ice cap after its complete melting is (Sjöberg, 1983, Sjöberg and Bagherbandi, 2013):

$$\begin{aligned}\delta V^{\mathrm{CMP}}(r,\theta,\lambda) = &- G\iint\limits_{\sigma}\int\limits_{R-d(\theta',\lambda')}^{R} \rho^{\mathrm{C}}(\theta',\lambda')\frac{r'^2 dr'}{l}\,d\sigma \\ &- G\iint\limits_{\sigma}\int\limits_{R-\widetilde{D}_0-d(\theta',\lambda')}^{R-\widetilde{D}_0} \Delta\rho(\theta',\lambda')\frac{r'^2 dr'}{l}\,d\sigma = I_1 + I_2\end{aligned} \tag{10.21}$$

where $\widetilde{D}_0$ is the mean compensation depth, $\Delta\rho$ the density contrast between the crust and upper mantle, and ρ^{C} the crustal density. d stands for the lithospheric depression caused by the ice cap load. Finally, l is the distance between the computation and integration point.

Using the Legendre expansion of l^{-1} (see Eq. 5.3) and inserting it into the first integral yield, performing the radial integration and considering its limits:

$$I_1 = -G \iint_{\sigma} \rho^{C}(\theta', \lambda') \frac{1}{r^{n+1}} \sum_{n=0}^{\infty} \frac{R^{n+3}}{n+3} \left[1 - \left(1 - \frac{d(\theta', \lambda')}{R} \right)^{n+3} \right] P_n(\tilde{x}) d\sigma. \tag{10.22}$$

Because $d(\theta', \lambda')$ is considerably smaller than R, the binomial term in the parentheses can be expanded by the Taylor series up to linear term with acceptable accuracy. By inserting this series into Eq. (10.22) and using Eq. (1.9), we obtain:

$$\begin{aligned} I_1 &= -4\pi G R \sum_{n=0}^{\infty} \left(\frac{R}{r} \right)^{n+1} \frac{(\rho^{C} d)_n}{2n+1} \\ &= -4\pi G R \sum_{n=0}^{\infty} \left(\frac{R}{r} \right)^{n+1} \frac{1}{2n+1} \sum_{m=-n}^{n} (\rho^{C} d)_{nm} Y_{nm}(\theta, \lambda). \end{aligned} \tag{10.23}$$

Eq. (10.23) presents the potential of depression of the upper bound of the crust. The second integral of Eq. (10.21), I_2, is in fact the depression at the lower bound. Performing similar mathematical derivations as was done for the upper boundary, the lower depression will be:

$$\begin{aligned} I_2 &= -4\pi G R \sum_{n=0}^{\infty} \left(\frac{R - \tilde{D}_0}{r} \right)^{n+2} \frac{(\Delta\rho d)_n}{2n+1} \\ &= -4\pi G R \sum_{n=0}^{\infty} \left(\frac{R - \tilde{D}_0}{r} \right)^{n+2} \frac{1}{2n+1} \sum_{m=-n}^{n} (\Delta\rho d)_{nm} Y_{nm}(\theta, \lambda). \end{aligned} \tag{10.24}$$

For $r = R$, and inserting Eqs. (10.23) and (10.24) into Eq. (10.21), we have:

$$\delta V^{\mathrm{CMP}}(R, \theta, \lambda) = -4\pi G R \sum_{n=0}^{\infty} \frac{1}{2n+1} \left((\rho^{C} d)_n + \left(\frac{R - \tilde{D}_0}{R} \right)^{n+2} (\Delta\rho d)_n \right). \tag{10.25}$$

Solving Eq. (10.25) for d is not straightforward unless we assumed that the density of the crust and the density contrast between the crust and upper mantle is constant. In this case, Eq. (10.25) changes to:

$$\begin{aligned} \delta V^{\mathrm{CMP}}(R, \theta, \lambda) &= -4\pi G R \sum_{n=0}^{\infty} \frac{\kappa''_n}{2n+1} d_n \\ &= -4\pi G R \sum_{n=0}^{\infty} \frac{\kappa''_n}{2n+1} \sum_{m=-n}^{n} d_{nm} Y_{nm}(\theta, \lambda) \end{aligned} \tag{10.26}$$

where

$$\kappa_n'' = \rho^C + \left(\frac{R - \widetilde{D}_0}{R}\right)^{n+2} \Delta\rho \tag{10.27}$$

By dividing Eq. (10.26) by the normal gravity γ, we change it to equivalent geoid height

$$\begin{aligned}\Delta N(\theta,\lambda) &= -\frac{4\pi GR}{\gamma}\sum_{n=0}^{\infty}\frac{\kappa_n''}{2n+1}d_n \\ &= -\frac{4\pi GR}{\gamma}\sum_{n=0}^{\infty}\frac{\kappa_n''}{2n+1}\sum_{m=-n}^{n} d_{nm}Y_{nm}(\theta,\lambda)\end{aligned} \tag{10.28}$$

Let assume that the geoid change has a spherical harmonic expansion. If this series is inserted into Eq. (10.28), the following spectral relation will be obtained between the geoid change and crustal depression:

$$\Delta N_{nm} = \frac{4\pi GR}{\gamma}\frac{\kappa_n''}{2n+1}h_{nm} \quad \text{where } h_{nm} = -d_{nm}. \tag{10.29}$$

By taking the derivative of Eq. (10.29) with respect to time, we obtain a relation between the rate of change of the geoid and the land uplift:

$$\Delta\dot{N}_{nm} = \frac{4\pi GR}{\gamma}\cdot\frac{\kappa_n''}{2n+1}\dot{h}_{nm}. \tag{10.30}$$

By solving Eq. (10.30) for $\dot{h}_{nm}$, the SHCs of the land uplift, and rate, and after performing the spherical harmonic synthesis from the results, we obtain

$$\dot{h}(\theta,\lambda) = \frac{\gamma}{4\pi GR}\sum_{n=0}^{\infty}\frac{2n+1}{\kappa_n''}\sum_{m=-n}^{n}\Delta\dot{N}_{nm}Y_{nm}(\theta,\lambda). \tag{10.31}$$

Fig. 10.7 shows the geoid rate determined by a linear fitting to the monthly GRACE gravity models computed by the Centre of Space Research (CSR) for 15 years from Apr. 2002 to Jun. 2017. The positive rates are seen over the Hudson Bay in Canada and Fennoscandia. This means that the geoid heights are increasing owing to postglacial rebound. Large negative rates are visible over Greenland, meaning that the geoid heights are decreasing over the area owing to melting ice and global warming. This means that the volume of the water that enters the ocean from the continental ice can also be monitored by the GRACE time-variable gravity models.

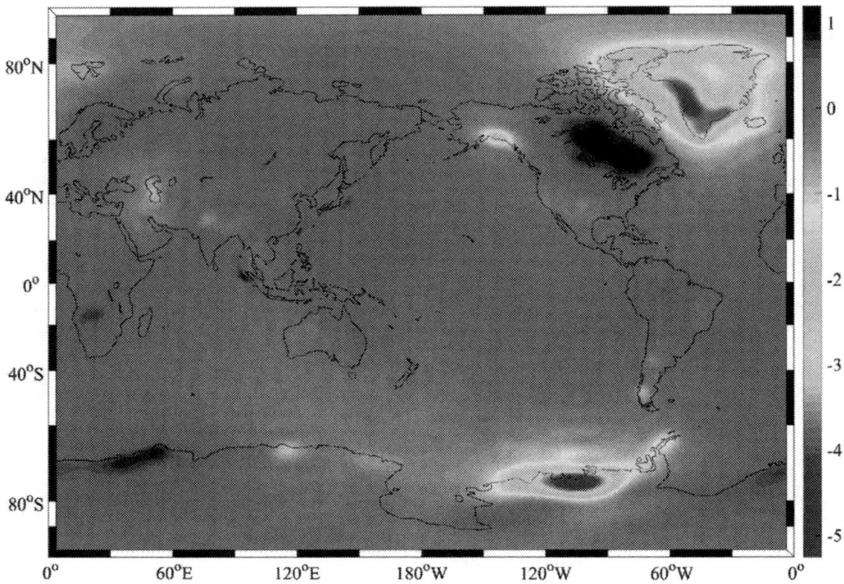

Figure 10.7 Geoid height rates determined by linear fitting to geoid heights determined by Centre of Space Research Gravity Recovery and Climate Experiment time-variable gravity models from Apr. 2002 to Jun. 2017 (mm/year).

The map of the upper mantle density of the CRUST1.0 model, which will be used to compute the land uplift in Fennoscandia is presented in Fig. 10.8A. In Fig. 10.8B, the geoid height rates are presented from 2002 to 2016 determined by linear fitting to the geoid heights determined by the GRACE time-variable gravity model of CSR. The highest rates, reaching 0.4 mm/year, are seen around the Gulf of Bohemia, where the centre of the ice may have been. This geoid rates and the density of upper mantle of CRUST1.0 are used in Eq. (10.31) to determine the land uplift rates. A constant density of 2670 kg/m^3 and a mean compensation depth of 30 km were used in this computation. The results are presented in Fig. 10.8C. The uplift rates reach about 12 mm/year, but after performing a Gaussian filter, they are reduced to about 9 mm/year, in close agreement with the rates determined by the Global Navigation Satellite System (GNSS) data shown rates reaching 11 mm/year.

10.5 Determination of upper mantle viscosity

The method presented by Sjöberg and Bagherbandi (2013) is presented and discussed in this section. They assumed that the Earth is spherical and the

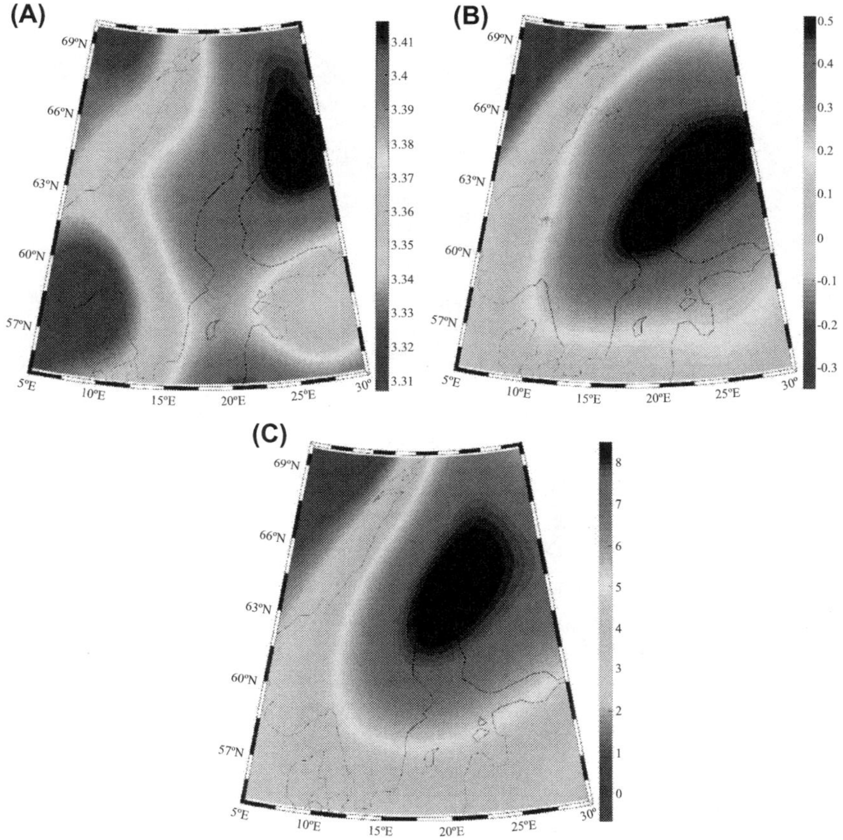

Figure 10.8 (A) Upper mantle density of CRUST1.0 (g/cm^3), (B) Geoid height rates determined by Centre of Space Research time-variable gravity models of Gravity Recovery and Climate Experiment from 2002 to 2016 (mm/year) and (C) Land uplift rates determined from Eq. (10.31) for a mean compensation depth of 30 km, constant crustal density of 2670 kg/m^3 and CRUST1.0 density of upper mantle (mm/year).

lithosphere is a solid shell on a viscous upper mantle. They used the following equation presenting the relation between the depression and time (Walcott, 1980):

$$h_n = (h_0)_n e^{\frac{-t}{\tau_n}} \tag{10.32}$$

where h_n is the Laplace coefficient of depression owing to load with degree n of the depression at time t, $(h_0)_n$ is that of the total depression at time 0, and τ_n is the spectral delay relation time.

By taking the natural logarithm of both sides of Eq. (10.32), we obtain:

$$\ln(h_n) = \ln\left((h_0)_n\right) + \frac{-t}{\tau_n}. \tag{10.33}$$

Taking the derivative of Eq. (10.33) with respect to time leads to:

$$\frac{\dot{h}_n}{h_n} = \frac{-1}{\tau_n}. \tag{10.34}$$

Solving Eq. (10.34) for τ_n yields:

$$\tau_n = -\frac{h_n}{\dot{h}_n} \tag{10.35}$$

The relation between viscosity $\widetilde{\eta}$ and delay time τ_n is (Cathles, 1971, 1975; Peltier, 1974; Sjöberg and Bagherbandi, 2013):

$$\tau_n = \frac{\widetilde{\eta}}{\gamma \rho_m R}\left(2n + 4 + \frac{3}{n}\right) \tag{10.36}$$

where γ is the normal gravity and ρ_m is the density of the upper mantle.

According to Eqs. (10.35), (10.30) and (10.29) we can write:

$$\widetilde{\eta}\dot{h}_{nm} = -\frac{\gamma^2 \rho_m}{4\pi G} \frac{(2n+1)\Delta N_{nm}}{\kappa_n''(2n+4+3/n)}. \tag{10.37}$$

After performing the spherical harmonic synthesis of both sides and solving the results for the viscosity, we have:

$$\widetilde{\eta} = -\frac{\gamma^2 \rho_m}{4\pi G \dot{h}(\theta, \lambda)} \sum_{n=1}^{\infty} \frac{2n+1}{\kappa_n''(2n+4+3/n)} \sum_{m=-n}^{n} \Delta N_{nm} Y_{nm}(\theta, \lambda). \tag{10.38}$$

Bjerhammar et al. (1980) found that the geoid depression in Fennoscandia is related to the degrees of the geoid, from 10 to 70, but Shafiei Joud et al. (2017) concluded that the contribution of land uplift caused by the postglacial rebound to the geoid height is seen between 10 and 23 degrees. This means that Eq. (10.38) can be written as:

$$\widetilde{\eta} = -\frac{\gamma^2 \rho_m}{4\pi G \dot{h}(\theta, \lambda)} \sum_{n=10}^{23} \frac{2n+1}{\kappa_n''(2n+4+3/n)} \sum_{m=-n}^{n} N_{nm} Y_{nm}(\theta, \lambda). \tag{10.39}$$

The land uplift $\dot{h}(\theta, \lambda)$ can be derived from re-levelling data, or GNSS measurements, or it can be derived from the time-variable gravity fields derived from satellite gravimetry.

Bjerhammar et al. (1980) and Sjöberg and Bagherbandi (2013) presented a simple formula for determining viscosity. To show how Eq. (10.39) can be simplified to their solution, assume $\left(R - \widetilde{D}_0\right)/R \approx 1$, leading to $\kappa_n'' = \rho_m$. In addition, the term $(2n+1)/(2n+4+3/n) \approx 1$, by these assumptions and approximations, Eq. (10.39) will change to:

$$\widetilde{\eta} = -\frac{\gamma^2}{4\pi G}\frac{N_{10-23}(\theta, \lambda)}{\dot{h}(\theta, \lambda)}, \tag{10.40}$$

where N_{10-23} is the geoid computed between 10 and 23 degrees.

This is the equation presented by Bjerhammar et al. (1980) and Sjöberg and Bagherbandi (2013), with the difference of considering the maximum degree to be 23 instead of 70. In fact, 23 is obtained by performing a correlation analysis between the geoid derived to different maximum degrees and the land uplift model determined by the GNSS measurements. Fig. 10.9A, shows the geoid model over Fennoscandia computed from the GOCO03S model (Mayer-Guer et al., 2012), from 10 to 23 degree. By comparing this map with that of the land uplift model presented in Fig.10.8C, their high correlation is clear. In addition, in the case of using the summations of the series of Eq. (10.39) and multiplying the result by the

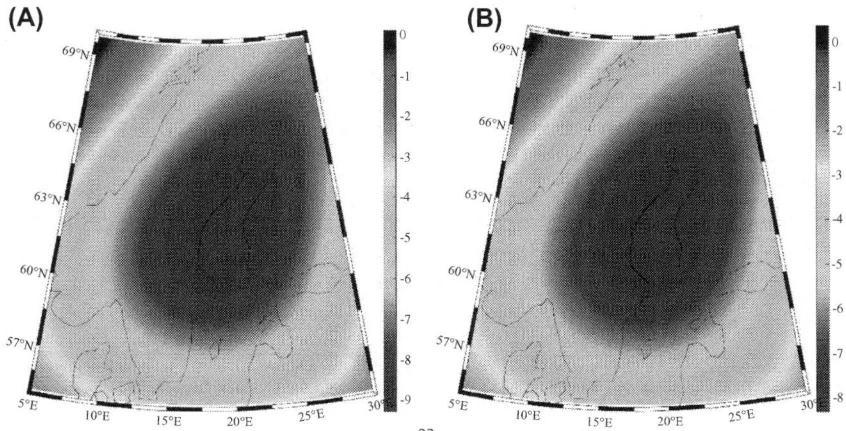

Figure 10.9 (A) N_{10-23} and (B) $\rho_m \sum_{n=10}^{23} \frac{2n+1}{\kappa_n''(2n+4+3/n)} \sum_{m=-n}^{n} N_{nm} Y_{nm}(\theta, \lambda)$ (m).

mantle density, the map presented in Fig. 10.9B is obtained, which is similar to the map of the geoid presented in Fig. 10.9A, but the minimum value reaches −8 m, whereas in Fig. 10.9A it reaches −9 m. In the case of applying Eq. (10.40), the mean viscosity of the upper mantle will be $(5.0 \pm 0.2) \times 10^{21}$ Pa, and in the case of using Eq. (10.39), it will be $(6.0 \pm 0.3) \times 10^{21}$ Pa.

10.6 Gravity strain tensor and epicentre points of shallow earthquakes

The strain tensor is determined from displacement fields. If we consider that an equipotential surface (geoid) changes or displaced, another type of strain tensor can be defined which is named gravity strain tensor. In fact, the displacement vector is the gradient of the gravitational potential. Here, the theory presented by Dermanis and Livireratos (1983) to determine this tensor is explained and discussed.

Consider two surfaces, 1 and 2, and an infinitesimal distance, ds_1, at the first one and the corresponding distance, ds_2, at the second one. The squares of these distances at both surfaces can be written in the following quadratic forms:

$$ds_1^2 = \mathbf{ds}_1^T \mathbf{G}_1 \mathbf{ds}_1 \tag{10.41}$$

$$ds_2^2 = \mathbf{ds}_2^T \mathbf{G}_2 \mathbf{ds}_2 \tag{10.42}$$

where $\mathbf{G}_1$ and $\mathbf{G}_2$ are metric tensors of surface 1 and 2, respectively. ds_1 and ds_2 are the infinitesimal coordinates. If the second surface is considered to be the deformed form of the first one, the difference between the squared displacements will be obtained by subtracting Eq. (10.41) from Eq. (10.42):

$$ds_2^2 - ds_1^2 = \mathbf{ds}_2^T \mathbf{G}_2 \mathbf{ds}_2 - \mathbf{ds}_1^T \mathbf{G}_1 \mathbf{ds}_1. \tag{10.43}$$

Consider the coordinate the vector s_2 at the second surface to be a function of the coordinate the vector s_1 at the first one or non-deformed surface:

$$\mathbf{s}_2 = \mathbf{s}_2(\mathbf{s}_1) \tag{10.44}$$

The total differential of Eq. (10.44) is:

$$\mathbf{ds}_2 = \frac{\partial \mathbf{s}_2}{\partial \mathbf{s}_1} \mathbf{ds}_1 \tag{10.45}$$

Therefore, a mathematical relation between the infinitesimal coordinate vector before and after deformation is obtained. Substituting Eq. (10.45) into Eq. (10.42) yields

$$\mathrm{d}\mathbf{s}_2^{\mathrm{T}}\mathbf{G}_2\mathrm{d}\mathbf{s}_2 = \mathrm{d}\mathbf{s}_1^{\mathrm{T}}\left(\frac{\partial \mathbf{s}_2}{\partial \mathbf{s}_1}\right)^{\mathrm{T}}\mathbf{G}_2\frac{\partial \mathbf{s}_2}{\partial \mathbf{s}_1}\mathrm{d}\mathbf{s}_1. \tag{10.46}$$

If Eq. (10.46) is inserted into Eq. (10.43), we obtain:

$$\begin{aligned}\mathrm{d}s_2^2 - \mathrm{d}s_1^2 &= \mathrm{d}\mathbf{s}_1^{\mathrm{T}}\left(\frac{\partial \mathbf{s}_2}{\partial \mathbf{s}_1}\right)^{\mathrm{T}}\mathbf{G}_2\frac{\partial \mathbf{s}_2}{\partial \mathbf{s}_1}\mathrm{d}\mathbf{s}_1 - \mathrm{d}\mathbf{s}_1^{\mathrm{T}}\mathbf{G}_1\mathrm{d}\mathbf{s}_1 \\ &= \mathrm{d}\mathbf{s}_1^{\mathrm{T}}\left(\left(\frac{\partial \mathbf{s}_2}{\partial \mathbf{s}_1}\right)^{\mathrm{T}}\mathbf{G}_2\frac{\partial \mathbf{s}_2}{\partial \mathbf{s}_1} - \mathbf{G}_1\right)\mathrm{d}\mathbf{s}_1\end{aligned} \tag{10.47}$$

When the space is Euclidian, the metric tensor will be identity matrix $\mathbf{G}_1 = \mathbf{G}_2 = \mathbf{I}$; this changes Eq. (10.47) to

$$\mathrm{d}s_2^2 - \mathrm{d}s_1^2 = 2\mathrm{d}\mathbf{s}_1^{\mathrm{T}}\mathbf{S}\mathrm{d}\mathbf{s}_1 \tag{10.48}$$

where

$$\mathbf{S} = \frac{1}{2}\left(\left(\frac{\partial \mathbf{s}_2}{\partial \mathbf{s}_1}\right)^{\mathrm{T}}\frac{\partial \mathbf{s}_2}{\partial \mathbf{s}_1} - \mathbf{I}\right) \tag{10.49}$$

is the general definition of the strain tensor, which can be derived from the displacements.

To obtain the gravity strain, consider $\partial\mathbf{s}_2/\partial\mathbf{s}_1$ to be equivalent to $\partial\nabla V_2/\partial\nabla V_1$, where V_1 and V_2 are the gravitational potentials before and after deformation, and ∇ stands for the gradient operator; see Eq. (1.45). By inserting them into Eq. (10.49), the gravity strain tensor will be:

$$\mathbf{S} = \frac{1}{2}\left(\left(\frac{\partial \nabla V_2}{\partial \nabla V_1}\right)^{\mathrm{T}}\frac{\partial \nabla V_2}{\partial \nabla V_1} - \mathbf{I}\right). \tag{10.50}$$

Let us write the derivatives of the gradients as

$$\frac{\partial \nabla V_2}{\partial \nabla V_1} = \frac{\partial \nabla V_2}{\partial \mathbf{s}_1}\frac{\partial \mathbf{s}_1}{\partial \nabla V_1} \tag{10.51}$$

By defining $\dfrac{\partial \nabla V_1}{\partial \mathbf{s}_1} = \mathbf{B}$ and $\dfrac{\partial \nabla V_2}{\partial \mathbf{s}_1} = \mathbf{b}$, and inserting them into Eq. (10.51) and the result into Eq. (10.50), we obtain:

$$\mathbf{S} = \frac{1}{2}\left(\mathbf{B}^{-1}\mathbf{b}\mathbf{b}\mathbf{B}^{-1} - \mathbf{I}\right) \tag{10.52}$$

where

$$\mathbf{B}=\begin{bmatrix} V_{xx}(t_1) & V_{xy}(t_1) & V_{xz}(t_1) \\ V_{xy}(t_1) & V_{yy}(t_1) & V_{yz}(t_1) \\ V_{xz}(t_1) & V_{yz}(t_1) & V_{zz}(t_1) \end{bmatrix} \text{ and } \mathbf{b}=\begin{bmatrix} V_{xx}(t_2) & V_{xy}(t_2) & V_{xz}(t_2) \\ V_{xy}(t_2) & V_{yy}(t_2) & V_{yz}(t_2) \\ V_{xz}(t_2) & V_{yz}(t_2) & V_{zz}(t_2) \end{bmatrix}. \tag{10.53}$$

where t_1 and t_2 are the time before and after deformation, respectively. In fact, **B** and **b** matrices are the gravitational tensor before and after deformation; see Eqs. (1.95)−(1−100) for the mathematical expressions for their elements in terms of SHCs.

Dilatation and maximum shear strain of the gravity strain tensor are, respectively,

$$\Delta=\lambda_{\max}^{\text{eig}}+\lambda_{\min}^{\text{eig}} \tag{10.54}$$

$$\gamma=\lambda_{\max}^{\text{eig}}-\lambda_{\min}^{\text{eig}} \tag{10.55}$$

where $\lambda_{\max}^{\text{eig}}$ and $\lambda_{\min}^{\text{eig}}$ are the largest and smaller eigenvalues of the gravity strain tensor.

Fatolazadeh et al. (2019) showed that the maximum shear strain of the gravity strain tensor can be used to determine shallow earthquake epicentre points. Here, the theory presented by Dermanis and Livieratos (1983) is applied to the Bam earthquake in Iran, which occurred at a depth of 8 km on Dec. 26, 2003. One-year monthly solutions of gravity models of the GRACE mission before and after the earthquake are used for this purpose. The effect of the hydrological signals should be removed from the gravity models before computing the gravity strain tensor to make sure the resulted gravity strain tensor and changes do not result from the hydrological signals. In addition, considering two long periods before and after the earthquake is a key factor for determining the earthquake epicentre point (Fatolazadeh et al., 2019).

Fig. 10.10 is the map of the maximum shear strain obtained by considering the gradients of the potential for 6 months before and after the earthquake. The location of the earthquake epicentre point reported by the US Geological Survey is shown by the cyan circle on the map of Iran. The values of the maximum shear strain are given in the background of this map, which show that the large value of the maximum shear strain gives a good approximation of the location of the earthquake epicentre.

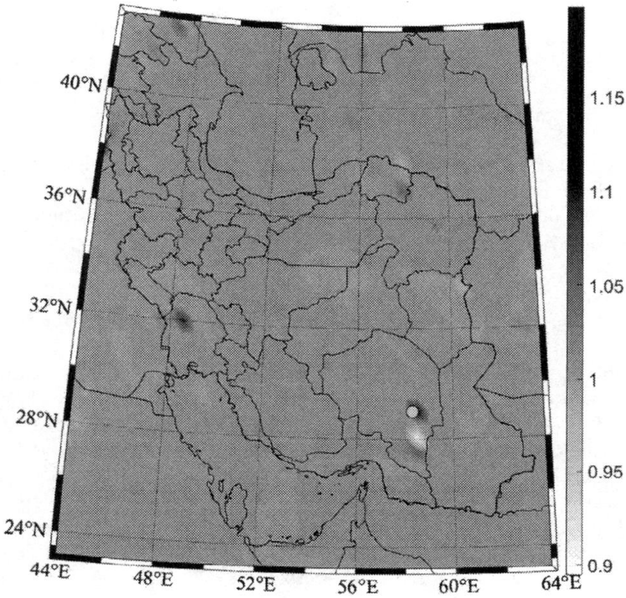

Figure 10.10 Maximum shear strain derived from gravity strain tensor for Bam earthquake. *(Computations done by Fatolazadeh.)*

Acknowledgements

The idea of adding a chapter about temporal variation of gravity field and its application came from Farzam Fatolazadeh, University of Sherbrooke, Canada, when he was with me for a research sabbatical. He kindly performed the computations and generated the maps and improved some parts of the text related to hydrology. I would like to express my great gratitude and special thanks to Farzam for motivating me to write this chapter and his help in completing it. Professor Mohammad Bagherbandi is appreciated for his advice regarding the land uplift and viscosity determination sections. I also thank Amy Shapiro, acquisition editor of this book at Elsevier, for encouraging me to add this chapter.

References

Bjerhammar, A., Stocki, S., Svensson, L., 1980. A Geodetic Determination of Viscosity. The Royal Institute of Technology, Stockholm.

Cathles, L.M., 1971. The Viscosity of the Earth's Mantle. PhD thesis. Princeton University.

Cathles, L.M., 1975. The Viscosity of the Earth's Mantle. Princeton University Press, Princeton.

Cherkauer, K.A., Bowling, L.C., Lettenmaier, D.P., 2003. Variable Infiltration Capacity (VIC) cold land process model updates. Glob. Planet. Chang. 38 (1–2), 151–159.

Coppola, E., Szidarovszky, F., Poulton, M., Charles, E., 2003. Artificial neural network approach for predicting transient water levels in a multilayered groundwater system under variable state, pumping, and climate conditions. J. Hydrol. Eng. 8 (6), 348–359.

Dai, Y., Zeng, X., Gordon, R.E., Bonan, B., Bosilovich, M.G., Denning, A.S., Dirmeyer, P.A., Houser, P.R., Niu, G., Oleson, K.W., Schlosser, C.A., Zong-Liang, Y., 2003. The common land model. Bull. Am. Meteorol. Soc. 84, 1013–1023.

Dermanis, A., Livireratos, E., 1983. Applications of deformation analysis in geodesy and geodynamics. Rev. Geophys. Space Phys. 21, 41–50.

Fatolazadeh, F., Kalifa, G., Javadi Azar, R., 2019. Determination of Earthquake Epicenter Based on Invariant Quantities of GRACE Strain Gravity Tensor. Scientific Reports. Submitted.

Flechtner, F., Morton, P., Watkins, M., Webb, F., 2014. Status of the GRACE follow-on mission. In: Marti, U. (Ed.), Gravity, Geoid and Height Systems. International Association of Geodesy Symposia, 141. Springer, Cham.

Han, D., Wahr, J., 1995. The viscoelastic relaxation of a realistically stratified earth, and a further analysis of postglacial rebound. Geophys. J. Int. 120 (2), 287–311.

Hussain, M., Eshagh, M., Zulfiqar, A., Sadiq, M., Fatolazadeh, F., 2016. Changes in gravitational parameters inferred from time variable GRACE data—a case study for October 2005 Kashmir earthquake. J. Appl. Geophys. 132, 174–183.

Koster, R., Suarez, M., 1996. Energy and water balance calculations in the Mosaic LSM. NASA Tech. Memo. 104606 (9), 76 (1996).

Liang, X., Lettenmaier, D.P., Wood, E.F., Burges, S.J., 1994. A simple hydrologically based model of land surface water and energy fluxes for GCMs. J. Geophys. Res. 99 (D7), 14415–14428.

Liang, X., Wood, E.F., Lettenmaier, D.P., 1996. Surface soil moisture parameterization of the VIC-2L model: evaluation and modifications. Glob. Planet. Chang. 13, 195–206.

Mayer-Gürr T., Rieser D., Höck E., Brockmann J.M., Schuh W.-D. Krasbutter I., Kusche J., Maier A., Krauss S., Hausleitner W., Baur O., Jäggi A., Meyer U., Prange L., Pail R., Fecher T. and Gruber T. (2012) The new combined satellite only model GOCO03s, International Symposium on Gravity, Geoid and Height Systems, GGHS 2012, Venice, Italy, 9-12 December 2012.

Peltier, W.R., 1974. The impulse response of a Maxwell Earth. Rev. Geophys. Space Phys. 12, 649–669.

Rodell, M., Houser, P.R., Jambor, U., Gottschalck, J., Mitchell, K., Meng, C.J., Arsenault, K., Cosgrove, B., Radakovich, J., Bosilovich, M., Entin, J.K., Walker, J.P., Lohmann, D., Toll, D., 2004. The global land data assimilation system. Bull. Am. Meteorol. Soc. 85, 381–394.

Shafiei Joud, M.S., Sjöberg, L.E., Bagherbandi, M., 2017. Use of GRACE data to detect the present land uplift rate in. Fennoscandia 209 (2), 909–922.

Sjöberg, L.E., 1983. Land uplift and its implications on the geoid in Fennoscandia. Tectonophysics 97, 97–101.

Sjöberg, L.E., Bagherbandi, M., 2013. A study on the Fennoscandian post-glacial rebound as observed by present-day uplift rates and gravity field model GOCO02S. Acta Geod. Geophys. Hung. 48 (3), 317–331.

Tapley, B.D., Bettadpur, S., Ries, J., Thompson, P.F., Watkins, M.M., 2004. GRACE measurements of mass variability in the earth system. Science 305, 503–505. https://doi.org/10.1126/science.1099192.

Walcott, R.I., 1980. Rheological models and observational data of glacio-isostatic rebound. In: Mörner, N.-A. (Ed.), Earth Rheology, Isostasy and Eustasy. Wiley, Chichester.

Wahr, J., Molenaar, M., Bryan, F., 1998. Time variability of the Earth's gravity field: hydrological and oceanic effects and their possible detection using GRACE. J. Geophys. Res. 103, 30229–32205.

Index

Note: 'Page numbers followed by "f" indicate figures and "t" indicate tables.'

D

E

K

L

Made in the USA
Monee, IL
13 September 2026

63138395R00275